MW00901970

SCIENCE BUT NOT SCIENTISTS

How Everything Began: CHANCE or CREATION?

VERNON L. GROSE

A true story of a private citizen caught in the crossfire between world renowned scientists, publishers, news media, educators and religious leaders over how science textbooks describe the origins of the universe, life, and man.

Bloomington, IN

Milton Keynes, UK

AuthorHouse™
1663 Liberty Drive, Suite 200
Bloomington, IN 47403
www.authorhouse.com
Phone: 1-800-839-8640

AuthorHouse™ UK Ltd.
500 Avebury Boulevard
Central Milton Keynes, MK9 2BE
www.authorhouse.co.uk
Phone: 08001974150

All artwork and exhibits are used by letters of permission.

First published by AuthorHouse 10/11/06

Library of Congress Control Number:

ISBN: 1-4259-6992-5 (dj)
ISBN: 1-4259-6991-7 (s)

Printed in the United States of America
Bloomington, Indiana

This book is printed on acid-free paper.

Dedication

To Phyllis, my lifemate for over 55 years who brought into this world our six children – Rhonda, Brenda, Lynnda, Wesley, Bradley, and Nanette who, in turn, conceived our 28 grandchildren and 3 great-grandchildren. Her selfless support and encouragement has been a priceless treasure.

Acknowledgements

Resurrection of this book from an obscure and dusty shelf where it had lain untouched for about 30 years defies reason.

Several played a role in it. Royal Australian Navy Captain Michael Anthony Houghton started it all by asking if I would mind if he located a publisher for it. (He had overheard my mentioning once that I had written such a book.) Next, long-time friend John B. Mumford — recognizing the uselessness of a 700-page *typewritten manuscript* in today's publishing world — provided not only his high-speed scanner to convert it into electronic format but his constant encouragement to publish. Finally and most remarkably, Keith A. Godwin edited and translated the scanned version into book configuration required for publication. This latter effort – involving hundreds of hours of Keith's time and inspiration – wove together his technical expertise, creative insight, and intellectual ferment to produce this resurrected story.

The conceptual artistry of my son Bradley of *Bradley Grose Design* in visualizing a descriptive and vivid image for the book's cover is deeply appreciated.

Of course, all this remarkable restoration was built on the untiring dedication, commitment to error-free typing, and faith in its ultimate message contributed by my secretary Shirley Y. Markus.

OVERVIEW

PART I
"Conflicting Context"

Prologue, Chapter 1 and Chapter 2

The stage is set for *battle*... war about origins has been brewing for years... *mythology* about the origin of the universe, life, and man pervades *science* textbooks... yet *science* can never determine these *origins*... scientists insist that evolutionary dogma rather than genuine science be taught... children are seduced by scientists in the name of science!

PART II
"The Incredible Struggle"

Chapter 3 through Chapter 9

A father responds to a newspaper editorial about how science is taught in public schools... combat is triggered and an unbelievable saga unfolds... the California Board of Education invites the father to help them restore objectivity... intrigue, deception, ridicule, and threats erupt... notable scientists attempt to retain *mythology* as the only explanation for *origins*... dogmatism and bad science exposed and refuted... down deep, the struggle is about man's best intellectual effort – in the name of science — to deny the existence of God.

PART III
"Is the War FOREVER?"

Chapter 10 and Epilogue

Science won this time... Big Bang, life from a soup of dead amino acids, and apes becoming human are *myths*... declared to be no more than *private beliefs* of certain scientists... conflict between *naturalism* (everything began by chance) will apparently always be pitted against *theism* (everything began with a Creator).

Contents

FOREWORD

SCIENCE BUT NOT SCIENTISTS

Wernher von Braun

Space exploration, with which much of my life has been spent, is simply the latest chapter in a continuing scientific revolution that dates back to Copernicus, Kepler and Galileo. It was their observations, and their ability to draw inferences from what they saw, that moved the Earth from its exalted position as the center of the universe. That was a profound blow to the human ego, and it called for some very painful reassessments of man's place in the universe. But it was a very necessary step in the continuing search for truth and the advancement of knowledge.

Thanks to their resolute spirit and refusal to be intimidated by those holding great power, every school child today is aware that our small planet, its neighbors in the solar system, and even our Sun itself, are in what someone has described as the suburb of a minor galaxy in a dynamic universe populated by galaxies and super-galaxies in numbers that probably surpass our comprehension.

The average citizen today, of course, has far more scientific information at his disposal than did those greatest of intellects of earlier times. Yet paradoxically, I think that there has never been a greater need for increased understanding and appreciation of science. It has been said that, although the choice of direction for our civilization will be determined through the democratic process, it is there that the problem begins. To make rational choices, the average citizen must understand the nature and role of science at a time when its breadth and complexity are increasing almost exponentially.

Conversely, the scientist, at a time when he can barely keep up to date in his specialty, must not isolate himself in his parochial interest. Instead, he should see his profession as a part of the larger world, to evaluate himself and his work in relation to all forces, especially the humanities, which shape and advance society. The need, then, is for an educational process resulting in more scientific literacy for the layman, and more literacy in the humanities for the scientists. It is also important that the layman not attach too much importance to the scientist's opinions on issues outside their special disciplines. Scientists are not experts in everything just because they are scientists.

Science But Not Scientists, by means of example rather than hypothetical supposition, focuses on this problem of scientific literacy. Man in this scientific age is free only to the extent he has a grasp on himself and his surroundings. Freedom — the ability to speak, think, act and vote intelligently — is based largely on our ability to make choices growing out of our understanding of the issues involved. With each advance of science, there is an invitation to more understanding. This is the essence of the burden borne by all peoples since the dawn of humanity. There must be widespread understanding of the role of science in modern society, both as to its limits and our dependence on its basic function as a tool for our survival. This is the imperative for scientific literacy.

How do we encourage scientific literacy? I think the problem is how to instill in students a permanent desire to learn. All youth is endowed with curiosity from the very beginning. What can the education process do, not only to keep this natural curiosity alive, but to make it a permanent part of the individual drive?

Students should be encouraged, beyond learning facts, to be intrigued by objects and events in their environment, as well as to become aware of and responsive in a positive manner to beauty and orderliness in their environment. They should be taught to willingly subject their data and ideas to criticism of their peers while acquiring a critical, questioning attitude toward inferences, hypotheses and theories. Early in education, they should be led to recognize the limitations of scientific modes of inquiry and the need for additional, quite different approaches to the quest for reality, including the search for answers to questions like the origin of the universe, life and man. Ultimately, they should be instilled with an appreciation for the interrelatedness of science, technology and society.

This is essentially the scientific method. By learning the scientific method, students will understand its role in society and at the same time learn to think for themselves. Learning to think for oneself, in turn, imparts a deep sense of freedom. Once tested, an appetite for it is formed which may well endure throughout life.

But if our young people are going to gain this appetite, our schools, our colleges, our universities, must bear an ever greater responsibility. All too many times in the past, education — particularly in the scientific disciplines — has placed extremely heavy emphasis on transmitting the established knowledge of the past. There has been a tendency for teachers to assign reading, and to encourage rote learning, instead of taking the admittedly more difficult path of encouraging students to think for themselves.

The mainspring of science is curiosity. Since time immemorial, there have always been men and women who felt a burning desire to know what was under the rock, beyond the hills, across the oceans. This restless breed now want to know what makes an atom work, through what process life reproduces itself, or what is the geological history of the moon.

Because scientists, regardless of their scientific specialties, have been universally schooled to seek out basic, underlying causes which integrate their observations of nature, they are likely at times to extend their cause — seeking beyond the limits of true science. This, of course, is the case whenever scientists attempt to define origins. These events, whether they be the origin of the universe, life or man, are historic, unique happenings which can never occur more than once. They all lack on-the-spot observation which scientists value so highly.

In recent years, there has been a disturbing trend toward scientific dogmatism in some areas of science. Pronouncements by notable scientists and scientific organizations about "only one scientifically acceptable explanation" for events which are clearly outside the domain of science — like all origins are — can only destroy the curiosity of those who must carry on the future work of science. Humility, a seemingly natural product of studying nature, appears to have largely disappeared — at least its visibility is clouded from the public's viewpoint.

Extrapolation backward in time until there are no physical artifacts of certainty that can be examined, requires sophisticated guessing which scientists prefer to refer to as "inference." Since hypotheses, a product of scientific inference, are virtually the stuff that comprises the cutting edge of scientific progress, inference must constantly be nurtured. However, the enthusiasm that encourages inference must be matched in degree with caution that clearly differentiates inference from what the public so readily accepts as "scientific fact." Failure to keep these two factors in balance can lead either to a *sterile* or a *seduced* science.

Vernon Grose, in tracing out in ***Science But Not Scientists*** his personal involvement in the vortex of these two forces, illustrates one more time the humanity of scientists — their likelihood of being just as prejudiced and bigoted as anyone untrained in science. He properly calls for *objectivity* rather than *scientific consensus.* He rightly urges that *message* rather than *messenger* should be scrutinized and tested for validity. Science will be the richer and humanity the ultimate beneficiary by heeding this clarion call.

APPROVED FOR PUBLICATION:

Wernher von Braun

Wernher von Braun

20 Oct 1975

Date

Figure 1. Wernher von Braun, father of America's space program, discusses NASA's Space Shuttle with Vernon L. Grose — 30 January 1976

PROLOGUE

Suppose someone approached you with a public relations scheme that guaranteed, for the investment of only four hours of your time, the following results:

1. You would be quoted (mostly in error) in periodicals like *Time*, *US News & World Report*, *Newsweek*, and *National Observer*.
2. You would be threatened personally by the Dean of the Graduate Division of one of the largest universities in the world.
3. You would be appointed a State Commissioner for textbook selection in the largest state in the nation, determining textbook criteria for over four million public school children.
4. Your personal action would produce *Associated Press* and *United Press International* news releases nationwide for more than three years.
5. A single action you took would be viewed as a "throwback to the Middle Ages" by a large and prestigious group of people — while simultaneously being cheered by hundreds of thousands as a "return to sanity."
6. Johnny Carson would discuss with William F. Buckley, Jr. on the NBC *Tonight Show* a proposal you suggested.
7. Two paragraphs — consisting of 127 words you wrote — would lead to over 200 individual revisions by 8 major publishers in their science textbooks.
8. Nineteen Nobel laureates in science would be mobilized and sign a joint resolution to combat a position that they were erroneously led to believe you held.
9. You would be invited to address students in universities and colleges across America on a subject you had never formally studied.
10. A proposal you made would be featured on the *ABC Evening News* by Howard K. Smith.
11. Your words would be reprinted worldwide in respected science journals like *Science*, *Nature*, and *Scientific American*.

12. The National Academy of Sciences (the most prestigious scientific body in America and perhaps the world), for the first time in its 110-year history, would pass a resolution to block what they erroneously thought you were proposing.

Would you believe that proposition? Well, I wouldn't either — except that it actually happened to me. The story, truly stranger than fiction, of how all these events came to pass will be unraveled in this book.

Are you wondering how those 4 hours were spent? I simply read a newspaper editorial, wrote a letter to the editor, located the names and addresses of the ten members of a State Board of Education, reproduced that letter for each Board Member (attaching an explanatory cover letter), and mailed the eleven letters. Elapsed time: *about four hours.*

Obviously, this four-hour investment turned out to be only a token of what I would ultimately spend. It was like a small down payment. In fact, I became "hooked" into an unbelievable scenario that demanded thousands of additional hours of my time. It could have, perhaps, even become a lifelong career if I had chosen to make it so.

As the title suggests, this book is about *science.* It is not a science book, per se. Rather it is a vignette or portrait of science, as well as an insight on those who produce science — the *scientists.*

Had this book been written in the 16th century, it would have been viewed as an outcry in defense of Galileo, Copernicus and others who were being persecuted by the church for their heretical views.

Likewise, if it had been written in 1925 at the time of the Scopes Trial in Tennessee, its readers would have considered it one-sided in favor of Clarence Darrow's position over that of William Jennings Bryan.

It is amazing, ironic and regrettable that it should be necessary at this late date to once again sound a clear call for freedom of thought.

Even worse, the call must be sounded in the field of *science* — that field of study so committed to and dependent upon toleration of all viewpoints for its advance. However, that necessity is upon us!

By recounting my personal experiences in a whirlwind of ideological conflict and deciding to share them with the world at large, I hope to focus attention on a wide-ranging, seven-fold objective:

1. To restore freedom of expression in science.
2. To distinguish *scientists* from *science* in public understanding.
3. To expose a little-recognized yet serious prejudice of some scientists.
4. To reveal the vital relationship between science and philosophy.

5. To foster understanding of the *limitations* of science.

6. To cleanse science education of the *religion of scientism.*

7. To encourage dialogue between scientists and the public.

Such an ambitious objective could appear as unnecessary, impossible, or incredible — depending on the initial viewpoint of the observer. Hopefully, my objective will be achieved in the minds of all readers regardless of their original perspective.

By no means is it intended that this book be a heavy research text. Because I hope that it will be widely read by high school and undergraduate college science students, as well as many others, it is deliberately written in a light, personal style. Rather than being either a theoretical or philosophical treatise, the book is pragmatic and provides a chronological record of a current scientific controversy.

To be more accurate, this book is an account of the current *form* of an unresolved conflict among scientists that has alternately smoldered and blazed ever since Charles Darwin published his *The Origin of Species* in 1859. The conflict last erupted in a brilliant flash of fire in the 1925 Scopes Trial and then subsided into a quiescent yet unquenched state, leading many scientists to believe that it had been eternally quelled.

One of the interesting and remarkable aspects of this conflict is that the deflagration changed form while smoldering between 1925 and 1969. When it again erupted into flame in 1969, it had moved geographically from the fundamentalist Bible Belt of middle America to the largest and most progressive state in the nation — home of 19 Nobel laureates in science, as well as the base for the world's greatest technological achievement — Project Apollo.

More significantly, it had moved away from an arena where science and religion had been pitted against each other — reappearing as a confrontation based entirely on *scientific evidence.*

This change in form, as well as the unexpected re-ignition of a fire long thought extinguished, caught much of the scientific community off guard — particularly those committed to denial of any conflict. So great was the surprise that, instead of examining the form of the controversy, those threatened by the reappearance of fire immediately attacked using the same arguments that had successfully (or so they thought) put out the fire in 1925. Obviously, since the fire had changed character, their arguments were unsuccessful — and even ridiculous. *So the fire continues to burn unabated.*

From the very beginning of my personal involvement, the issue on which this book focuses has been frequently compared with the Scopes

Trial of 1925. Newspapers, radio, television and scientific journals have all suggested that it was a re-run of that famous trial.

As the background for that test case against evolution held a half-century earlier in Dayton, Tennessee is recounted, perhaps some readers will recognize similarities between John Scopes' involvement and my own — both of us embroiled in an issue much larger than any single individual who participated in it.

It was in Frank Robinson's drugstore in Dayton, Tennessee one day in the spring of 1925 that some of John Scopes' friends gathered to sip nickel lemonades and plot his arrest. They had in mind a test case in the courts that would bring down a law that made it a crime in Tennessee to teach evolution — "that man has descended from a lower order of animals." Of course, they needed someone to serve as the guinea pig in this test case. And they further needed to have that person's permission.

John Scopes, age 24, was a likely candidate to act in this capacity. He taught science at the local high school, doubled as football coach and was a generally popular figure about town, although a few disapproved of the fact that he smoked cigarettes and danced.

When Scopes was contacted at a nearby tennis court and urged to join his friends at the drug store, he was really uncertain as to whether he had even *taught* evolution. Some of his friends reminded him that he had served as a substitute teacher in a biology class for the last two weeks of the school year. When they asked whether or not he had discussed evolution, he answered, "Well, we reviewed for final exams, as best as I can remember."

The textbook that he had used, Hunter's *Civic Biology*, contained the theory of evolution. So John Scopes assumed that he must have taught it, and — on that basis — agreed to stand trial. Once they had his concurrence, the meeting in the drug store quickly broke up so that a warrant for Scopes' arrest could be sworn out.

In the meantime, Scopes returned to his tennis match.

That low-key, rather innocent beginning for an event that would ultimately go down in the annals of science education was not unlike the introduction I experienced in the struggle described in this book.

Certainly writing one's first letter to an editor cannot be considered to be a momentous event — unless it results in a conflagration like that to be revealed. Just as with Scopes, I was an unqualified, illogical, and unlikely actor to be placed on center stage. This is meant neither to overstate nor understate our roles. Instead, it serves to register a significant point that might be otherwise overlooked — that a global issue was at stake in both cases.

While I gradually came to accept the news media's perception that I was involved in a modern version of the Scopes Trial, I agreed for exactly the opposite reason for which they were making the comparison. If you recall the issue in the 1925 Scopes Trial, it was that evolution could not be taught in science classrooms because Biblical creation had *exclusive rights* that were threatened by evolution.

Now there is no possible way to truthfully state that the issue that arose in 1969 in California (hereafter noted as the "Science Textbook Struggle") was a replay of that offered earlier in Tennessee.

In fact, the issue was precisely *reversed!* In place of the Biblical account being universally taught in the public schools of California — as it was in 1925 in Tennessee, the general theory of evolution was the exclusive version being taught in 1969. Likewise, Clarence Darrow, representing Scopes in that trial, was arguing that the then current version in Tennessee (Biblical creation) was not a totally adequate explanation for origins and needed additional alternative explanations to fully explain the findings of science.

Exactly the inverse of that argument was proposed in California in 1969. It was on the basis of this *reversed analogy* that my first public statement on 13 November 1969 contained the observation that I would have been on Clarence Darrow's side rather than that of his opponent, William Jennings Bryan, at the Scopes Trial.

Another parallel between the Scopes Trial and the Science Textbook Struggle is that two of the leading legal giants of the land were locked in mortal combat in a *courtroom* over the issue raised by John Scopes.

The Science Textbook Struggle, on the other hand, was not resolved in the courtroom but in the *educational arena.* But it did engage the attention and involvement of the biggest giants in science — even 19 Nobel science laureates.

A further similarity between these two conflagrations is that the issue of *academic freedom* was at stake. Indoctrination appeared to be inevitable in both cases, although the two theories had reversed themselves in the driver's seat in the intervening 44 years. By 1969, evolution was being dogmatically and emphatically indoctrinated in the minds of all public school children — *to the total exclusion of any other alternative.*

Although many other similarities might be drawn between these two cases, an additional obvious one is that John Scopes technically lost his case and yet in reality won it in the ultimate sense. Likewise, the news media reported that the Science Textbook Struggle was finally put to rest in favor of the prevailing philosophy (evolution).

But the truth is that those who were protesting dogmatism and indoctrination in the science classroom won a great victory. *Textbooks will never again be the same.*

More than forty years after the end of his trial, Scopes published in his memoirs a remarkable statement that I believe still applies in the Science Textbook Struggle:[1]

> The basic freedoms defended at Dayton are not so distantly removed; each generation, each person must defend these freedoms or risk losing them forever.

Although I confess that my initial reason for getting involved in the struggle was to preserve objectivity rather than the loftier concept of freedom, I must now agree most wholeheartedly with Scopes in his concern.

In addition to reminding me of John Scopes, my involvement in this struggle caused me to empathize with a person much earlier in history — Galileo. Of course, neither Scopes nor I would claim any of Galileo's greatness. It was his *humanity*, rather, with which I identified.

Some time before the Science Textbook Struggle arose, my wife, six children and I visited Pisa, Italy where Galileo was born. He and I happen to share (again in essence rather than degree) common interests as church organists — in addition to the study of mathematics and science.

On a windy February morning, my family wound its way to the top of the Leaning Tower of Pisa to gain a spectacular view of the Mediterranean to the west beyond the Cathedral of Pisa that stands adjacent to the tower. It was while observing a giant lamp swinging back and forth in that cathedral that Galileo had made his first important scientific contribution at the age of 20 — the law of the pendulum.

Turning toward the east, we could see in the distance the city of Florence where he lies buried. It was both sobering and inspiring to contemplate the genius of Galileo in this setting that he had known so well.

As we descended the tower, we observed the plaque at its base describing his experiments that established his law on falling bodies. In particular, I considered once more the mental tortures to which he and his colleagues at the dawn of scientific inquiry were subjected by the religious authorities of that day.

As the Science Textbook Struggle developed, my thoughts again returned to Galileo. I was earlier convinced that Galileo had lived in a far less tolerant age than the one in which we live today. In his day, most leaders of European thought were suspicious of anyone who dared speak against popular belief based on the teachings of those considered to be wise men.

But then, questions began to enter my mind. Was Galileo's age *really* any less tolerant than mine? Was the religious totalitarianism of the Middle Ages different in character, magnitude and position from that which science represents in today's technological society?

What is the difference in treatment between that accorded Galileo and Velikovsky? Or Copernicus and Shockley? Are the forces to protect status quo within science today any different from the forces that existed in the Middle Ages within the church?

Of course, there are some differences between the Middle Ages and today. In the first place, we do not burn people at the stake any longer. Secondly, we rarely excommunicate people from an ecclesiastical body, thereby subjecting them to eternal damnation. Third, we have no formal method for demanding recantation of belief. *Or do we?*

Is the denial of the right to speak on a college or university campus in any way analogous to burning a person at the stake? Is blackmailing of publishers and university faculties — if they allow a controversial speaker the right to be heard — the modern version of excommunication? Is the consistent refusal of the popular press and the news media to tell the truth about an issue or their frequent policy of publishing only one side of an argument in any sense a modern equivalent to forcing recantation?

Perhaps some of these questions may become more meaningful as the issue around which this book is written unfolds.

There is a disturbing irony in an obsession that has been exhibited by many scientists with whom I have had confrontation in this struggle. They seem to be dedicated to an exclusive, snobbish "elitism" — stating emphatically and even derisively that only a very small portion of the population that works first-hand on a daily basis with scientific investigation should be allowed to make any judgments concerning science.

This thinking is very reminiscent of the attitude that prevailed in the Middle Ages with respect to theology, whereby only the *clergy* were allowed first-hand access to the Bible because the Church had decreed officially that the lay public could not possibly understand the Bible. That attitude has prevailed in some factions of the Church to this very day.

While there is undoubtedly truth in the idea that it does take specific training to understand a particular field of study, the Church, on the other hand, went a step further in the Middle Ages. It demanded that the source material; i.e., the Bible, *not* be read under *penalty* for fear that people would get the wrong idea or the wrong interpretation of the Bible.

It is that same paranoia or fear that became obvious among scientists with increasing frequency during the Science Textbook Struggle. "Only the scientific elite should be allowed to speak about science," they cried. Specific illustrations of this perversion are frequent in this book.

In contrast to the popular image of science as a source of *answers*, this book is filled with *questions* — not only about the "scientifically-valid facts" of science, but also about scientists as *people*. When a person becomes concerned about how origins are discussed in science textbooks, as I have been recently, a number of unresolved but basic philosophic issues arise:

- The proper role of science in society and education
- The accountability demanded of elected or appointed officials versus some specific field of professional study like science
- The viability of a theological concept of an Almighty God in a scientifically-dominated society
- The part that textbooks should play in the educational process (when compared to the teacher, for example)
- Whether or not laws (scientific or otherwise) can change beliefs
- The types of societal problems for which the scientific inquiry process can be expected to produce solutions

And, of course, the age-old one of *science versus religion.*

Obviously, none of the questions arising from these philosophic issues were settled in this controversy. It would be ideal if they could have been, but the fact that they were not seems to emphasize that they may never be subject to universal resolution.

As all good people in sales know, two errors must be avoided —overselling and underselling. Both errors have their dangers. Particularly since World War II, scientists have enjoyed a "wonder boy" complex. Anthony Standen, nearly three decades ago, said it well:[2]

> "The world is divided into Scientists, who practice the art of infallibility, and non-scientists, sometimes contemptuously called 'laymen,' who are taken in by it. The laymen see the prodigious things that science has done, and they are impressed and overawed."

Whether deliberately or inadvertently, scientists have been guilty of the error of *overselling*. This is evident in television advertising where the white-smocked scientist is the final word on everything from headaches to landing on the Moon.

Instead of appearing as a divergent field of increasingly more questions instead of answers, scientists project the convincing message to the lay public that science is virtually home-free on answers to anything that the human race desires to have or know. If anyone rises to suggest that there are limits to the scientific approach, there is almost a Pavlovian response of hysterical defense among scientists.

Perhaps the current anti-science ground swell among youth all over the world can be somewhat correlated with the God-like complex often projected by the scientific community of solely possessing ultimate truth. Yes, even young people are capable of discerning the error of overselling.

So when scientists proclaim themselves as prophets of final truth, youths can see that scientists are far from answering the pressing problems of humanity related to such phenomena as racial prejudice, hatred, murder, inability to cope with mental stress, and all other frailties of the human race.

This high priesthood of science has perhaps tricked or at least enticed other fields of endeavor such as religion, music, art and education into thinking that tacking the adjective "scientific" in front of their name enhances their credibility. Thereby, the physical sciences have been either mimicked or emulated by the so-called *social* sciences.

I would be disappointed if this book were to be viewed as contributing to the anti-science movement that now is fashionable. Neither is it intended to be an endorsement of irrationalism and all that that term connotes. Yet the distortion of the orthodox role and capability of science does call for corrective action – *not to kill it but to set it on a true course.* This corrective effort must be undertaken, in large part, by the articulating mechanisms in society — particularly the news media.

My first mass involvement with newspapers, magazines, radio and television occurred in the Science Textbook Struggle. True, my name had appeared in print in the past — for winning a photo contest, for various promotions received as an executive in industry, for presentations made at technical meetings, as well as involvement in events like an auto accident, athletics or a public speech. However, the amount of that exposure had been relatively limited. Without hesitation, I could say that none of those sparse reports had ever been *totally* accurate. Those small errors, on the other hand, were different in character than those I experienced once I became involved in the issue of teaching science.

Hopefully free of the paranoia attributed to former Vice President Spiro Agnew regarding the news media, I nonetheless have a strong impression based on the Science Textbook Struggle that news media are far more motivated by speed of reporting and power to shape reader opinion than by accuracy or truthfulness. I can appreciate this marketing emphasis because of my business background where profit is primary.

So I do not care to open up a broad front of warfare with news reporters. But I came to realize that *how* news is presented on radio, TV, or in written format tends to carry a ring of factual certainty — possessing great authority well beyond its rationally proper role in influencing opinion.

This supra-authority of news media, of course, is not new. Undoubtedly, it has been around since the Gutenberg press. It needs to be recognized again, however, in the context of this struggle because it contributed to serious error in public opinion. In turn, that error wreaked all types of havoc in the minds of many folks and many organizations like the National Academy of Sciences, American Association for the Advancement of Science, and other bodies of responsible citizens who should not have been drawn into battle had they either taken the time to examine the truth in the issues or had reacted on any other information than reported accounts.

Public response to news — regardless of medium — would likely not atrophy if truth and accuracy were favored over sensationalism. Freedom of the press based in the First Amendment is frequently hailed as the cornerstone of liberty, but seldom is there any countervailing *responsibility* cited or championed. Is there *any* responsibility for the press — or are ratings and sales its only goals?

Beyond inaccuracy and untruthfulness of news reporting, another factor compounded the news of the Science Textbook Struggle — deliberate and distorted *polarization.* Everyone resorts to polarization from time to time. Children early in life learn to classify "good guys" and "bad guys." Politicians cannot avoid labels of liberal or conservative. Polarizing people into groups can at times be very helpful for rapid communication concerning complex issues. But it can also *distort* so that truth suffers.

Beyond the *convenience* provided by polarization, it also is used to jog people off dead center. Creative folks often use it by emphasizing an aspect or feature that they believe to be generally overlooked by the public while not acknowledging what is already recognized — hoping to get an audience to see both sides. Because they present only one of two poles, creative people can be branded as lop-sided.

In the Science Textbook Struggle, polarization has been used extensively — but hardly for the productive purposes of emphasis or education. News coverage has given every appearance of deliberate distortion via polarization by pitting *scientists* on one hand against *creationists*, as though these were two exclusive groups.

This clever but inaccurate polarization was apparently intended to negate the need to address a very real issue — *scientific interpretation of observed data.* While such polarization could have been somewhat valid during the Scopes Trial in 1925, it simply does not reflect the thinking of today. Admittedly, there may be a larger number of articulate non-scientific spokesmen arguing in favor of recognizing the possibility of creation than there are laymen articulating the cause of general evolution. But it is simply a distortion of fact to suggest that scientists are on one side of the question and creationists on the other.

This book will recount that there have been theologians, housewives, educators and others who publicly argued the position of teaching evolution exclusively in public schools to explain the origin of the universe, life and man. In the popular press, these folks were lumped together and called *scientists.*

In contrast, the vast majority of those speaking in favor of opening up options for origins beyond the general theory of evolution have been scientists who hold advanced degrees — most of them doctorates — in various scientific disciplines. They are conducting experiments, making original first-hand observations and publishing the results of those observations in scholarly scientific journals, as well as addressing scientific meetings where they discuss their findings with colleagues in their fields. Yet they are called *creationists*!

I have come to abhor the terms *creationist* and *evolutionist.* Whatever usefulness they may have had in the past, they are counterproductive today. The *popular* news media are primarily responsible for the misinformation they connote. But the *scientific* press also employs them to generate useless and distracting rhetoric.

Were I to be forced to align myself with one or the other of these two terms, I would probably label myself an *evolutionist* — since a preponderance of scientific evidence to date supports a theory where changes have occurred over time *within species.* With that general theory, I am in total agreement.

My only hesitancy in being dubbed an evolutionist would be if I also were obligated to concur in the idea that Darwinian evolution provides *any* explanation for origins of the universe, life, and man. I contend that the *general* theory of evolution breaks down at that point — not because of mechanistic inadequacies but because *no* scientific theory (while remaining scientific) can describe non-repetitive historical events such as origins.

Stepping momentarily out of the context of science to discuss the data that scientists have gathered that might relate to primordial events, there is certainly equal or greater reason to postulate that origins of many living groups may have been discrete, abrupt, discontinuous, full-fledged *beginnings* rather than products of gradualism that the general theory of evolution demands of its adherents.

More expansive discussion will occur in the book about what renowned paleontologist George Gaylord Simpson has defined as *micro-*evolution, *macro-*evolution, and *mega-*evolution. In short, I am a micro-evolutionist – but neither a macro nor a mega-evolutionist.

Summarizing the role that the news media have played in the Science Textbook Struggle, they have, in the main, been both *inaccurate* and *biased* — while deliberately inciting the public to rally about one of two fallacious poles. Apparently, this distortion produced the desired results — sales, listeners, and viewers. The price paid, on the other hand, by the public — whether as school children, publishers, public servants, taxpayers, or scientists — was and continues to be unwarranted.

The authority and credibility of *all* our institutions — government, businesses, industries, churches, schools and homes — are fast disappearing. Therefore, to appeal for some type of moral responsibility to exist within the written news media, one could anticipate a rapidly eroding conviction amongst the general public that such responsibility is even needed.

Nonetheless, I feel strongly that if our civilization (and particularly scientific activity) is to survive, it will be necessary to restore a sense of *responsibility* within the management of the written media to balance its *unabated liberty* to print whatever it chooses to print. Basis for this severe accusation of news media irresponsibility is evident throughout the book.

This is an *autobiographical* account. There are many hazards in making that choice. Even recognizing some of those hazards, I still may not be able to avoid all of them. Hopefully, the book evades self-justification or defense of my admitted personal bias and prejudice.

One of the means I have employed to hopefully overcome autobiographical pitfalls is to be candid about my personal involvement. My shortcomings, errors, reversals of thought, and defeats have greater validity in this issue than those events in which I perceived personal triumph.

Honesty forces me to confess that I have enjoyed this struggle virtually from its outset. Of course, there have been moments of anguish sporadically interspersed with this enjoyment. Fortunately, I have the chemistry of a crusader. And I have paid the price that all crusaders pay — misunderstanding by others, personal despair, loneliness, and tremendous unseen investment of psychic and physical resources.

On the plus side however, the experiences in this struggle have been most rewarding. Entering the fray rather naively as simply a concerned parent and taxpayer, my vast ignorance of so many aspects of the struggle forced me to convert it into a learning experience.

Readers will share this progressive learning experience with me by noting a gradual transition in my public statements, the honing of my arguments, and the ultimate focus reached in my address in November 1972 before the State Board of Education. I not only *listened* seriously to my opponents — I *changed* in response to their arguments.

There is no evidence, on the other hand, that those opposed to what they *thought* I was proposing ever listened to me. I am not bitter about this — only disappointed.

In addition to the learning directly germane to the Science Textbook Struggle, I also learned much about American politics, public reaction to news media, the educational process, and the complexities of preparing textbooks. Most of all, my understanding of human behavior on the societal scale was greatly enlarged. Finally and happily, my entire family shared the learning adventure with me.

Today's complex and demanding society is constantly changing. And it is impossible for any person to know enough about anything. If ignorance is a sin, it is a universal sin. There is no way that anyone could have been prepared to play the role forced upon me in the Science Textbook Struggle. That is not to excuse my many errors. Instead, I propose that the role was both unique and extremely complex. No two persons would have played it in an identical way.

This recounting of my involvement in the struggle retains the same perspective I held while the sequel was taking place. For example, I claim no scientific expertise — and I never did. Likewise, I represent no pressure group or coalition — and never did. I seek no censorship or overthrow of scientific theory — and never did. I desire no religious teaching in the science classroom — and I never did during the struggle.

Probably the most prevalent reaction shown by my scientific opponents was to attack my credentials in science. There were two reasons why I consistently set my scientific credentials at naught.

First, I do not possess that type of education or experience in science that would be acceptable to those committed to defending and retaining *status quo* in science.

Second (and I believe more importantly), I wanted the scientific community to address and resolve *issues* rather than *personalities*. This latter reason was almost entirely ignored. Scientists who imagined themselves to be threatened by my actions simply refused to examine *issues* – attacking instead my already-admitted inadequate *scientific credentials*.

It was essential during the Science Textbook Struggle that my personal education and experience in science be openly exposed so that my involvement not be granted greater scientific credibility than warranted — by parties on either side of the controversy.

My undergraduate degree was in physics. Extensive graduate studies in physics, chemistry, physiology, and mathematics followed. Several years as an applied physicist were spent in fracture mechanics of complex electronic equipment in high-intensity vibratory environments. My numerous professional publications have been nearly all in technology rather than science, although one in *medical* science was widely recognized.[3]

Beyond this involvement in science and technology, I have taught physics, chemistry and mathematics at the graduate level in three major universities — in Europe as well as the United States. My role as a consultant has involved work with researchers at the National Academy of Sciences and National Bureau of Standards on scientific issues like combustion products of polymeric materials and associated human toxicology. I have addressed National Academy of Sciences conferences and been a member of its National Research Council panels for several years.

That concludes the only reference to my science-related credentials. From this point forward, the focus will hopefully remain on *issues*.

In any field of endeavor, its best criticism comes from within the field itself. This is true for at least two reasons. First, there is credibility between members of that field because they are engaged in similar work. So there is higher probability of gaining acceptance of an idea (particularly a controversial one) within that field than to permit someone from outside to make observations and have them accepted. Second, those within that field best know its strengths and weaknesses.

Despite this reasoning, there is danger that criticism will not likely be forthcoming in an elitist field like science. This danger increases if that field gains — as science has — a public status of being powerful and authoritative. When members of any profession fail to expose and cleanse its internal error, it becomes necessary for outsiders to expose it — particularly when it persists.

This book is intended to do that. It is based on the ancient principle that it is difficult to see oneself. Its title separates a *field* from its *participants* — pointing out that, while *science* is committed to objective truth, *scientists* are just as subject to humanity's weaknesses as anyone else.

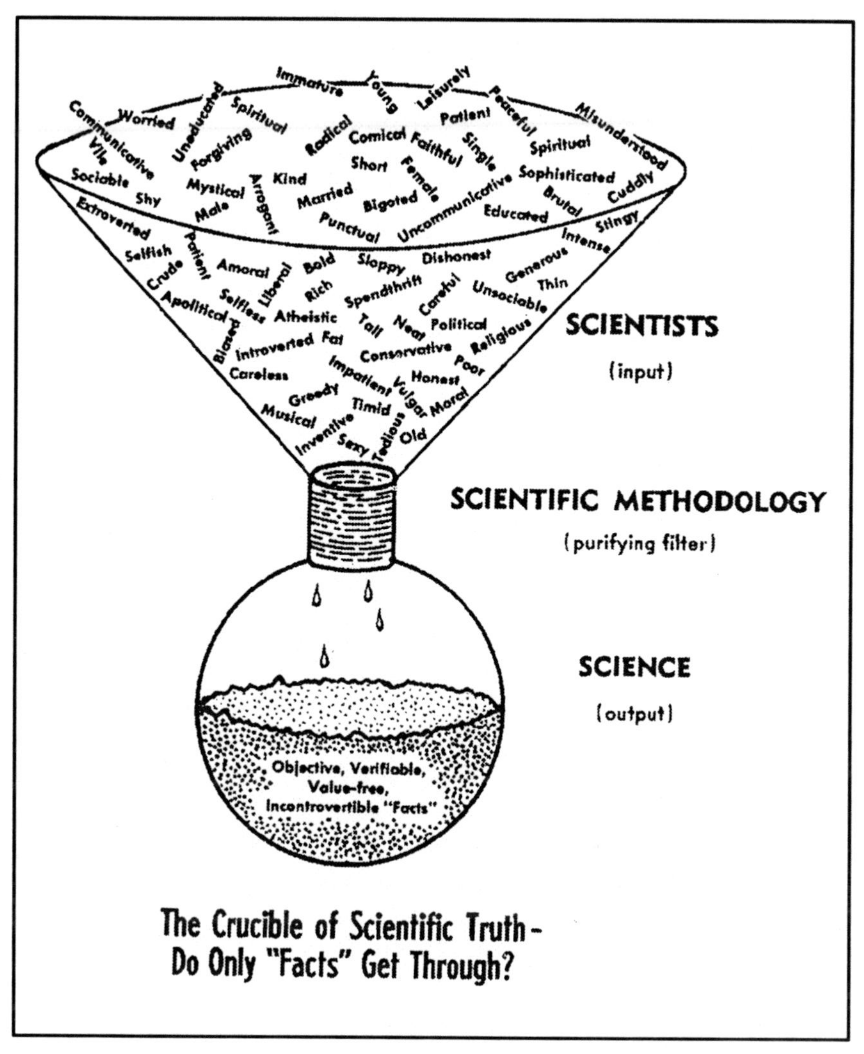

Figure 2.

The Crucible of Scientific Truth in Figure 2 illustrates the diversity of humanity that contributes to the body of knowledge known as science. Several myths are also depicted there.

Myth 1: The *human characteristics* of scientists are magically stripped away in their daily work by working in science.

Myth 2: Scientists are people who all end up thinking alike about science.

Myth 3: There is a *singular* "scientific method" that automatically purifies the observe-measure-test-analyze-discard-discuss-challenge-publish process that scientists employ so that only objective, verifiable, value-free, incontrovertible "facts" remain.

Regarding the so-called scientific method, entire books have been written to attempt to define it. However, there is a "filtration" of scientific work — consisting of several stages — that produces the recognized scientific body of knowledge.

First, there is a stage of observing natural phenomena, noting and publishing their interrelationships. Next, the community of scientists responds to the published work. Then there is the ultimate endorsement and confirmation by the scientific community of published work. Thereby, over a process of iteration and time, scientific "truth" is established.

One of the key premises of this filtration is that all subjectivity be removed. The heterogeneity representative of all scientists is tacitly assumed to expose and eliminate error. However, like all human endeavors, there is residual error that persists over time.

Even the entire scientific community, at times, can be duped or subjected to collective peer pressure that forces it to accept error. Hans Christian Andersen's classic, "The Emperor's New Clothes," applies equally to science as to other fields.

Quite often, scientists are poor students of *history*. In the Science Textbook Struggle, many scientists fail to see the very close parallel between the suppression of scientific truth by the medieval Church and the current refusal to consider other alternatives to what scientific elitists have already decided is "fact."

The sorry episode of the Piltdown Man illustrates how easily the body politic of science can be swayed to believe a hoax — if it fits a particularly attractive and popular pet theory that is widespread.

Likewise, the prejudice in the selection/rejection of data by those who believe that the earth was created in 4004 BC is equally indicative of this phenomenon.

If any field of study demanded objectivity, it would appear to be science. Unfortunately, science cannot claim purity in its search for objectivity. So one of the goals of this book is to expose some of the forces that militate against it. The battle against prejudice is as old as mankind. It is not likely that this battle will be won in my lifetime.

Nonetheless, I believe that I must light one more small candle toward that end.

Chapter 1

WONDER ABOUT ORIGINS

"It is impossible to contemplate the spectacle of the starry universe without wondering how it was formed." — *Poincare*

I am a transcontinental commuter — a California resident who works in Washington, D.C. Jet travel makes this ridiculous lifestyle possible.

For the past few years, I have spent the equivalent of several months out of each year in Washington. In addition to teaching numerous short courses at The George Washington University School of Engineering and Applied Science, I have been a consultant to Federal agencies like the National Transportation Safety Board on aircraft accident investigation, the National Aeronautics and Space Administration regarding risk management, and the Department of Agriculture on systems management.

I have also been a member of the 5-man Washington METRO rapid transit Board of Consultants that meets frequently. Likewise, many trips to Washington have been necessary to meet with a National Academy of Sciences panel on which I have served for over four years. At last count, I made my 148th flight between Washington and the West Coast.

Washington, despite my many visits, continues to inspire me. It is "history-in-the-making." Much of its charm for me, however, lies not in its being daily at the hub of world-shaking events, but in the marvelous preservation of monuments and buildings that enable one to trace out the origin of our country. With very little imagination, anyone can be transported right back to the beginning events of the Nation.

PAST IS PROLOGUE

The primary repository of information about America's origin is in the National Archives in Washington. It is one of my favorite buildings — not only for the important documents it contains but also for four handsome statues that surround it. Often, I have walked by this stately building and admired them. Each of the four figures sits on a pedestal on which is inscribed a quotation.

James Earle Fraser carved the two figures — one male and the other female — that face Constitution Avenue. The female figure represents Heritage, while the male statue symbolizes Guardianship. Heritage holds a child and a sheaf of wheat, representing the primary purpose of government in preserving the home. My admiration of this beautiful statue includes the quotation on its pedestal. It reads: "The Heritage of the Past Is the Seed That Brings Forth the Harvest of the Future."

The other side of the National Archives faces Pennsylvania Avenue where two figures — again one male and the other female — represent the Future and the Past. The female figure, symbolizing the Future, sits on a pedestal on which are inscribed the words quoted from Act 2, Scene 1, of Shakespeare's *The Tempest,* "What is Past is Prologue." The pedestal of the male figure, representing the Past, bears the words of Confucius, "Study the Past." Both of these figures are the work of the noted sculptor, Robert Aitken.

These three quotations engraved on statuary pedestals seem to indicate that the future is influenced, if not determined, by the past. Yet our society seems to run away from the past as fast as possible. Obsolescence is our password.

Young people — whose language, dress codes, and musical tastes are so rapidly changeable — might think of the National Archives as containing dry, useless, and musty old information that some history teacher has tried to force them to memorize. Even the word "Archives" suggests something ancient and unrelated to today.

In an age when national and international economics demand that everyone frequently exchange money with one another so that an index like a "gross national product" may increase, emphasis is obviously placed on *new* things. Madison Avenue continues to dream up new gimmicks and sales techniques to whet the buyer's appetite for devices, services, foods, clothing, and recreations that are *new.* The "in" thing, the "latest" look, the "with-it" generation all bespeak *newness* — something apparently not done in the past.

How then could all that has transpired in the past be "The Seed That Brings Forth the Harvest of the Future?" Why should we "Study the Past?" What makes "What is Past is Prologue" a true statement?

Is the *future* really related to the *past?* Aren't modern trends independent of the past? Why should progressive, developing people be interested in the past? Don't you have to "get rid of the past" in order to progress? These sayings *must* be in error.

So forget the past, and go forward.

Civilization is Old-Fashioned

But we *can't* forget the past. Whether or not you *want* to forget it, the past keeps popping back into view. How could Madison Avenue sell us something without comparing it to what it does *differently,* does *better,* or does *faster* than what we have done in the past? Progress must be compared with something *previous*. Development means going from some past stage to a future one.

Civilization is an ongoing process. This process has the properties assigned to the mathematical term "vector" — that is, it has both *direction* and *magnitude*. Civilization can be thought to be going in a certain direction. Likewise, it can be more or less civilized.

However, in both dimensions of direction and magnitude, the past is inexorably wound and twisted around the concept of civilization. Therefore, we can say that civilization is "old fashioned." It is fashioned from the old.

Consider for a moment the subject of *history*. It is certainly a field of study that looks backward to the past. Some who have been critical of history have said, "We learn nothing from history — except that we learn nothing from history."

On the other hand, this criticism does not deter historians from documenting what has gone on in the past and attempting to synthesize a rationale for all that they see happening in the world. Did you ever notice that serious historians often reject the interpretation of current events because they believe that such analysis must await the fermentation of time in order to get a proper perspective of what has been happening?

In other words, they value the *distant* past over the *recent* past!

But history isn't the only field that looks at the past. Consider the study of *law*. Very little if any of the curriculum in law school is devoted to projecting where the law should be going.

Future laws receive very little attention. In contrast, the law student is forced to study case after case after case from the *past*. Court decisions, even those made in the Supreme Court of the United States, are inevitably based on some *previous* ruling or incident (or a combination thereof).

Why all this interest in what judges of the past — most of them now dead — have thought? Why don't judges simply "call 'em as they see 'em?" Because of a vital but unspoken assumption.

It is tacitly and inherently assumed that, in the past, judges have judged correctly. To deviate from these obviously-correct precedents would be hazardous for justice. So the law is also fashioned from the past and is thereby old-fashioned.

Architecture, at first glance, could appear to be a field free and independent from the past. Modern high-rise buildings are springing up not only in America, but also in some of the most unlikely cities internationally. When my wife and I made a trip around the world recently, some of the most modern hotels in which we stayed were in so-called under-developed nations.

The architecture of the Dusit Thani Hotel in Bangkok hardly has an equal in the United States. Taipei's Grand Hotel is unmatched anywhere for its opulent, daring design. Having traveled in about 30 countries, I have never seen more striking architecture than the Oberoi Sheraton Hotel on Marine Drive in Bombay.

The contrast of these architectural wonders with the setting of squalor from which they have risen is unbelievable. Any link between the destitute neighborhoods and these monumental works of architecture seems to be totally missing.

Yet each one of these buildings has a continuous lineage traceable to ancient buildings in Greece or the Pyramids of Egypt. Architecture is most sensitive to the past. No architect would dare say, "I don't care what has been done in the past. I'm going to start fresh with no inhibitions and by ignoring all previous designs."

Why? Because the past actually determines and directs all future architecture. Even Frank Lloyd Wright paid strict attention to the past.

What about *medicine*? Is *it* not free from the past? A modern hospital seems to be totally unrelated to the world of Hippocrates or even to Florence Nightingale. Radical surgical breakthroughs, like heart transplants, appear entirely free from any precedents. But are they? Not really. They can be traced backward into the past by an unbroken chain of medical practices. Step by step, the field of medicine is linked to the past.

Perhaps *technology* is "free and clear" from any influence of the past. Surely Abraham Lincoln would stand overwhelmed, after a three-hour buggy ride from the White House to a meadow now called Dulles International Airport, as he beheld a monstrous 747 jet airliner descend from the sky and disgorge nearly 500 persons who had been looking down on Pike's Peak from high above the Rocky Mountains at the instant his buggy was rolling through the White House gates.

By being absent from the human scene for little more than 100 years, he would be unable to link this technological wonder with anything he recognized in 1865. He might even insist that there could be no relationship between this marvel and the past. Yet, most school children today can trace, step by step, the technology that links the horse and buggy with interplanetary spacecraft!

Concepts like design safety margins for construction of technological products are firmly based on earlier work. While there are those who appear to deviate from previous practice and ignore the past, their innovative contribution represents a very small modification of long-standing custom.

Backward Science?

American culture is overwhelmed by the contribution that science has made in the past century to our way of life. Some of the awe with which science is viewed by laymen is justified. Much of it is not. The next chapter discusses this imbalance in depth.

One premise that might be difficult for many to accept is that *science* — just like history, law, architecture, medicine and technology — is "old-fashioned." It is fashioned from and based upon the past. Scientists look backward far more than they look forward.

The average person has come to accept, and even *expect*, scientific developments as the inevitable harbingers of the future. "If you want to know what life will be like in fifty years, read today's science fiction," is often heard. "The key to the future lies in scientific research," others proclaim.

If there is *any* problem facing humanity — even a cure for cancer — just secure government funding for scientific research, sit back and wait. The solution is inevitable because science is the current "keeper of the keys" to the future. This futuristic image of science could easily persuade many to believe that the past does not affect this marvelous prodigy of human knowledge. Permit me the privilege of bursting this blinding bubble of balderdash.

The backbone of science — that upon which all scientific progress is dependent — is a compendium of scientific laws. These laws have no substance *per se*. They are simply documented agreements about observations of nature that have been made in the *past*.

Of course, there are some ground rules that govern the agreements that have been reached. Among these ground rules is the requirement that the agreed observations must be free from exceptions. In other words, they must be *universal*.

They must also be able to be challenged for their *truthfulness*. Further, they should be useful in predicting the interrelationship of future observations. The *validity* of scientific laws is enhanced as they continue to explain the relationship of current observations to ones that have been made in the past.

Therefore, science is just as involved in the past as any other field of study. Its distinction from other human endeavors lies not in its separation from the past, but with its approach and classification of knowledge. Since history is well recognized as dealing with the past, it might be well to compare *history* and *science*.

Both fields *record* the past, but science tends to go further than history. It attempts to explain *how* natural phenomena and happenings may have occurred. It seeks the interrelationship of past events in order to predict future events.

Science also seems to project further into the past than history does. For example, recorded history is thought to be only about 3-4000 years old. Science, on the other hand, describes events in terms of *billions* of years ago.

How can science do something that history cannot? It uses *extrapolation.* That means that known observations or information are extended beyond what is *known* to the *unknown* by means of assumption or inference. It should be obvious that this activity has lots of hazards.

The *biggest* hazard of extrapolating backward into the distant past beyond actual data is that the data themselves get increasingly scarce. The further back in time we go, the less data there are (to extrapolate between). Pretty soon, *it's all extrapolation and no data.*

This problem of "disappearing data" as we go backward through time is generally compounded by a universal characteristic we all have — *curiosity.* The more remote or inaccessible actual facts about an event of the past become, the more human curiosity will drive us to let our imagination run loose. Then the door is wide open to genuine mythology. (Incidentally, mythology is neither harmful nor anti-science — it is just *outside* of science.)

If we could stop with mythology — even mythology mistakenly accepted as science, we would be safe from the final difficulty that generally results — *dogmatism.* Would you believe that scientists become dogmatic at a rate *inverse* to actual knowledge they possess about an event in the past? It's true!

This phenomenon causes me to propose the **Law of Scientific Dogmatism**: *"To the degree that undisputed facts are unavailable or unknowable, scientific statements can be expected to become dogmatic."*

Though perhaps amusing, this Law describes a frightening situation as scientists approach the *ultimate* limit — where it all started!

The Ultimate Past

Even the smallest child wonders, "Where did it all begin?" There is hardly a parent who has not pondered at length before responding to such a profound question. Likewise, there is hardly a parent who feels that he or she can adequately answer that question.

There are probably as many answers to the child's question as there are people answering it. For centuries, philosophers and theologians were most often those offering answers.

However, in the past two centuries, many people have come to look to *scientists* for the answer. And scientists have not been bashful in proposing answers for origins — even when they had no data whatever upon which to base their answers!

Curiosity about origins has resulted in great diversity in science education concerning this subject around the world.[4] In Israel, science students are taught a variety of views — those that exclude the concept of a Creator of the universe, life, and man as well as those that include one. This is also true in Great Britain, Italy, France, Germany and Ethiopia.

Is there anything really different about speculation concerning origins — even when *scientists* do the speculating? Isn't such speculation the same as speculation about the Ming dynasty, Nero, or Greek antiquity? No — for many reasons.

ONLY ONCE!

What makes speculation about *origins* unique from other speculations concerning the past? Several things.

First, origins can occur only *once*. They are not repeatable. Only one event can qualify as the "original" event. An origin stands *alone* between nothing and something.

Secondly, it follows that origins cannot be verified as to *how* they occurred. This is particularly true for the origin of the universe, life, and man. Verification necessarily requires a *verifier*. For these three origins, there could not possibly have been a verifier.

Third, origins involve not only "how" (which science *attempts* to explain) but also "why" (which science can *never* explain). In other words, speculation regarding origins inevitably involves a "belief system" like those offered by philosophy, metaphysics, or religion.

Allow me to explain by using a crude illustration. As we go backward through time, physical evidence becomes more scarce. All antique dealers know that the *price* of antiques goes up with their *age*, every other factor being constant. Why? Because they are more scarce.

Now, if we want to speculate about life at the time of the Revolutionary War in America, we can find literally thousands of artifacts — furniture, personal letters, buildings, dishes, paintings and music — that existed at that time. There is a chronological sequence of events concerning these artifacts that can verify that they are authentic.

If we continue backward another thousand years in time, we do not have nearly as many authentic artifacts. If we speculate about life in central Europe in 700 A.D., we do so at a greater risk of error than we did with the Revolutionary War period. In other words, we have to fill in between the artifacts with much more imagination than we did earlier.

Therefore, the relationship between *artifacts* (observable data) and *speculation* (a scientific hypothesis) becomes less certain as we go backward in time. Figure 3 is a simplistic attempt to illustrate this relationship. Note that the "data" (21 separate observations or measurements) are identical for Hypothesis A, Hypothesis B, and Hypothesis C.

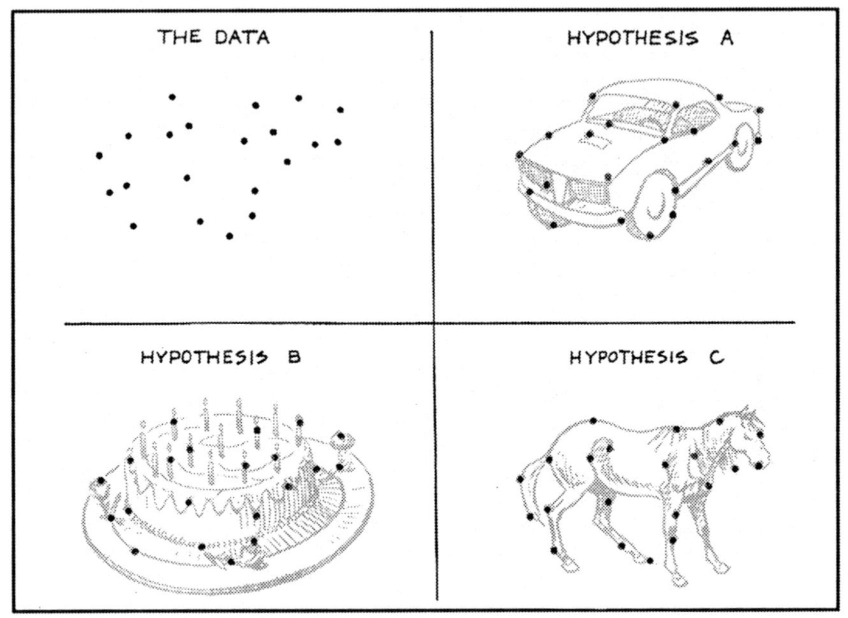

Figure 3. Which Hypothesis is the *Right* One?

Looking at those 21 data points, a *physicist* might see a physical system like an automobile. A *chemist* might interpret the same 21 data points as a chemical system comparable to a baked cake, and a *biologist* might view the 21 identical observations as a biological system such as a horse.

It should be no mystery then that the *interpreter* of data has considerable influence on the *interpretation* of data. Just as "Beauty is in the eye of the beholder," it can be said, "Hypotheses are in the eye of the interpreter." This critical problem is discussed at length later.

So continuing our walk backward through time, we actually come to a stage where there are *no* human artifacts. This is true, incidentally, long before we get to the *very first person* who ever lived (whoever, wherever, and whenever he or she might have been).

You can rest assured that, whenever a scientist says *anything* about the origin of man, he is doing so with the same amount of information or data possessed by a baker, barber, or banker — exactly *none!*

The origin of *life*, of course, probably preceded the origin of *man* by some extended time. The same reasoning used in discussing the origin of *man* applies to the origin of *life.* And since the origin of the *universe* is a likely predecessor of life by another lengthy period of time, the *"all belief — no facts"* argument becomes even more binding.

Therefore, the only truly objective statement that scientists, speaking entirely within the context of science, can make about the origin of the universe, life, and man is, "*We don't know and never will know how they occurred.*"

But guess what? Many scientists, either deliberately or inadvertently, make dogmatic statements about these three origins. Ironically, they are most descriptive and dogmatic about the origin for which there is the least possibility of having any data to extrapolate — the origin of the *universe!* Recall the Law of Scientific Dogmatism?

What a Blast!

Pick up any modern grammar school textbook about science. Look for discussion of the origin of the universe. What will you find? Not just *words* that describe it, but a *picture*! And not a picture like those of Napoleon, Caesar or Plato — in black and white. It will be in the unsurpassed glory of *color!*

What does this marvelous masterwork depict? A giant fireball in the process of disintegrating. Every possible, imaginable material thing in the universe was supposedly contained in a single, primordial atom billions of years ago. Sometime, between 10 and 20 billion years ago, depending on the textbook publisher, this atom is presumed to have convulsively exploded.

Pieces of material of various sizes are believed to have been instantaneously flung in all directions. Like a giant fireworks display, the bigger chunks continued to explode and fling off chunks of themselves in all directions.

Some had a delayed reaction and waited for several billion years before firing. Some chunks were hotter than others. This convulsive activity has gradually slowed down over the billions of years, and now all the particles large and small are supposedly going away from one another at a slowing-down pace.

This is called the "Big Bang" theory. Some blast, no?

The Big Bang is ideal for stimulating an artist to use every color in the rainbow when painting it. The pictures generally start at the center as white-hot and end in the blackness of space. Of course, there could not have been "space," as we commonly think of it, wherein the fireball was suspended. But that does not bother the artist.

Between the white center and the black surrounding, every imaginable color can be and generally is shown. With modern art techniques, the picture even closely resembles an actual photograph.

This technicolor tableau undoubtedly sparks the imagination of the young child. The stimulating simulation, however, may initiate a perversion in the mind of that child that can never be erased. He will be hooked on a lie.

The falsehood is not that the origin of the universe might well have resembled the bombastic blast being shown in four-color press, although there is rapidly-growing doubt as to the validity of this description, the far-fetched fiction is that anyone *knows* that it happened — especially a *scientist!*

I want to emphasize at this point that, although this account of the origin of the universe (as well as the accounts of the origins of life and man appearing later) may strike the reader as humorous, it is neither my intent nor desire to make the account appear ridiculous. In fact, I personally deplore the frequent attempts by well-meaning but ill-informed people to belittle serious work by being cute, facetious or boorish about it — particularly when they are obviously ignorant of that work.

There are many things that appear foolish only to fools — the body movements of a symphony conductor during the finale of a great musical masterpiece, the mating activities of birds and animals, or the dress and appearance of a surgeon during surgery.

Therefore, because something *seems* to be incomprehensible or foolish, it does not mean that it *is*. Brilliant people have seriously studied all three origins being described in this chapter for many years. Though the descriptions could appear to be frivolous, the work on which they are based is extensive and erudite.

On the other hand, when attempting to translate highly specialized knowledge into a language that can be universally comprehended, any author must compromise. Generally, the compromise involves *simplifying* something quite complex.

It is this very problem that science textbook writers also face. Since they have chosen to introduce concepts like the Big Bang theory to children in the first two or three years of schooling (a questionable practice in my opinion), they are forced to simplify both the concept and the language.

There are two reasons why I am using these oversimplified accounts of origins. First, most of the children will be *indoctrinated* – not *educated* – by this single description, with no alternative views presented. Secondly, only a small percentage of all children will ever study the accounts in greater depth.

This means that these descriptions ultimately become the non-scientific public's understanding of such momentous and consequential events. Such understanding is both infantile and unquestioning. The end result, in terms of public support of science as well as nurturing communication between scientists and the non-scientific world, cannot help but be counterproductive.

If the accounts concerning origins presented in this chapter are offensive because of their simplicity (and they are certain to be for some), please recognize that they represent what is being taught to young minds in science classrooms.

The stories are not my personal inventions. Hopefully, the reader will recognize what I consider to be an insidious danger — that such accounts may do more damage to science by their dogmatic nature than if textbooks had remained totally silent about origins.

This forgery of fact — that science produces knowledge of origins — is both subtle and deceitful. Starting with the origin of the universe, it follows in the youngster's mind that science can easily account for everything that occurs subsequently! If scientists can know in such accurate detail about this very first event that occurred 10-20 billion years ago, what could be hidden from them?

Many questions about this picture race through one's mind. *How* does anyone know that this happened? Since no one could have been there to view it, who told the artist what it looked like? Is this the unanimous view of *all* scientists? What *alternative* ideas about the origin of the universe were rejected prior to accepting this one? On what *basis* were they rejected? Upon what *data* (provable, repeatable observations) is the picture based? How *long* has this picture been accepted as the accurate one? Are there any *difficulties* or *uncertainties* about this picture being the ultimate truth?

The travesty perpetrated by this beautiful but preposterous artist's dream is two-fold. First, it ignores the questions just asked. Secondly, it masks the imponderable nature of the origin of the universe in a cloak of smug, scientific certainty.

The school child is thereby conditioned to become dogmatic. The mystery, intrigue and enigma that have traditionally goaded scientists to pursue knowledge are stultified at a tender age. It would not be surprising — should this practice continue — if science itself would be transformed into a rigid, intransigent dogma.

In fairness, it must be admitted that this "certainty without facts" which a colorful picture of the "Big Bang" depicts must be charged primarily against textbook publishers and authors and only secondarily against scientists *per se.*

However, why are scientists so unconcerned about this dogmatism? Could it be that they agree with the certainty portrayed? If scientists won't discipline those who distort science, *who should?*

Without being technical about it and risking the charge of oversimplification, the whole Big Bang idea is based really on the work of one man — Edwin P. Hubble — an American astronomer who, about forty years ago, postulated the idea that the universe was expanding.

When studying galaxies, Hubble noticed that the color of each one was always shifting consistently. The color always changed to one closer to the red end of the spectrum. This consistency he assumed was a Doppler shift in color arising from a difference in velocity between our galaxy and the distant one under observation. Every galaxy seemed to be flying away from every other galaxy since all the differences in velocity were positive.

Therefore, Hubble proposed that the universe was expanding. He also derived a relationship between distance and the redshift that was linear. Hubble's new law did something else. It halfway suggested that, since we are obviously not at the beginning of time, anyone who so desired could, in a sense, *reverse* time and find out exactly when all these galaxies must have existed in one common point.

Despite the simplicity of it all, it is ironic that calculation by means of backtracking has produced a wild group of estimates for this common point running anywhere from 3 to 20 billion years.

Of course, once one got all these galaxies gathered back to a common point, there had to be proposed a reason or method for them to start to separate. This is where the Big Bang fits. For some as yet unexplained reason, all of the galaxies were incompatible and one day flew apart in a terrific explosion.

(It has always bewildered me that no textbook or science journal poses the question of the origin of the primordial atom. After all, if one is dreaming up fairy tales, why not start at the *beginning?*)

If the Big Bang theory were only as universally accepted among scientists as it appears to be in full-color pictures in the children's textbooks, there would be no need for the rash of scientific symposia being held in recent months and years on the subject of the origin of the universe.

As an example, the Smithsonian Astrophysical Observatory recently sponsored a symposium in Cambridge, Massachusetts on the meaning of Hubble's Law and its predictive validity. Here's the report in the scientific press:[5]

No overwhelming consensus emerged, and none was really expected. In the opinion of George Field, the organizer of the meeting, the question may never be completely resolved. The reason is that there are *so many uncertainties and loose ends in the data, and so many assumptions to be made* in drawing conclusions from them, that two equally competent observers can come up with virtually opposite conclusions from essentially the same data.

This statement should immediately cause a flashback to Figure 2. Identical *data* — with diametrically opposed *hypotheses.* Why? "So many uncertainties and loose ends in the data, and so many assumptions to be made."

Although this quote was directed at the *future* of this universe instead of its *origin,* objective scientists would admit that the comment applies equally to both. After all, Hubble's Law represents at best only 40 years of measurement out of billions in either direction — future or past. In fact, much criticism has been leveled at the tiny number of measurements that have been made of the redshift of galaxies:[6]

Of the uncounted galaxies in the sky, only 3,000 have had their redshifts measured, and most of these have been special-interest items. A systematic field of redshifts, those in a given volume around our own galaxy, which would make a regular sample, extends only to 200. One of the great future needs is a much more exhaustive red shift catalogue. Getting it with ground-based observations is difficult, because each measurement is time consuming and must compete for telescope use with more glamourous observation programs — Mt. Palomar does about 20 a year.

The two major factors that cause so much disagreement among astronomers and astrophysicists are (1) the actual density of matter throughout the universe and (2) the role of mutual gravitational attraction of the matter in the universe. If the matter is not particularly dense and the gravitational attraction is not particularly consequential, the universe will expand forever.

On the other hand, if matter is fairly dense in the universe and the gravitational attraction between this density is sufficient, the particles, stars or galaxies in the universe could slow down to the point where they would be attracted to one another on a faster and faster basis. Eventually all these fragments would come back together, colliding in a reverse of the Big Bang idea.

There is no prospect in the foreseeable future that either of these two ideas will be able to be resolved on any basis other than belief.[7] While the age of the computer has enabled scientists to speed up much of the calculation for modeling of the universe, the assumptions required to put such models into the computer so influence the resultant answers that the scientists end up frustrated by the conflicting conclusions.

Returning momentarily to Hubble's law about the linear relationship between distance to a galaxy and its color, a mathematician at the Massachusetts Institute of Technology, I. E. Segal, proposed at the same meeting that this expanding universe hypothesis is "all wet."[8] Again, the scientific press had an interesting report of the response scientists gave Segal's statistical analysis:[9]

> The assertion was greeted by the assembled astrophysicists with a chill as cold as intergalactic space. After the formal close of the session, a heated argument ensued between Segal and several prominent astrophysicists over a number of points, including whether the galaxies whose redshifts are known are a fair sample for statistical analysis.

Another difficulty with the Big Bang theory that is often debated among scientists is how to reconcile the existence of heavier elements as a product of a single explosion of the primordial atom.

For example, supernova explosions are the explanation of how heavy elements originate – in contrast to the uniformitarianism implied by the Big Bang idea. Computer models of supernova explosions have recently been developed that describe how such phenomena might have occurred in conflict with the Big Bang explosion and expansion concept.[10]

One of the most interesting developments regarding the Big Bang theory occurred during the opening session of the 1975 Annual Meeting of the American Physical Society. Allen D. Allen, President of Algorithms, Incorporated, a California physics research company, presented a paper entitled, "The Big Bang is Not Needed."[11] His paper also received widespread coverage in the popular press.[12]

The primary and most disturbing aspect of Allen's paper for those who are committed to the Big Bang theory is his mathematical proof that Hubble's Law, that describes the expansion of the universe, is virtually independent of *initial* cosmic states. In particular, he shows that Hubble's Law does not imply the singular initial state generally credited to it — the universe crammed together into one single atom.

Therefore, there is no need to postulate a Big Bang in order to counteract the overwhelming gravitational field that would accompany such a super-dense initial state; i.e., a single primordial atom. The Big Bang is not necessarily wrong. It is simply one explanation of what may have occurred in the past, and there are possibly *many* equally valid explanations!

I have spent several hours in discussion with Allen concerning his theory. He offers convincing proof that there are infinitely many simple and non-contrived initial states for the universe, all of which could lead to expansion under Hubble's Law.

"The most respected authority on galaxies in the world, whose book is the primary one used in universities today, does not believe in the Big Bang theory," Allen said to me recently. "However, he can never be expected to deny the Big Bang theory *publicly* for several reasons."

"First, he presently has nothing with which to replace the old theory. Secondly, it could prove professionally embarrassing for him to be required to deny something that he has taught so convincingly for so long. Third, many scientists of eminence like this fellow must change their position so slowly as to not be noticed. Although he has agreed in private conversation with me that there is no validity to the Big Bang theory, we can expect only a gradual shift in position."

Allen then outlined a probable pattern of retreat in the scientific world from the Big Bang. The reader may wish to track his prophecy!

1. In the near future, the following statement, "The universe originated in a Big Bang..." will be changed to, "Scientists believe that..." The full-color pictures of the Big Bang will still be prominent.

2. Within a few more years, "Scientists believe that..." will be softened to, "Most scientists believe..." The full-color pictures will remain.

3. A few years later, a much softer statement will prevail, "Many scientists believe..." The full-color pictures will become less frequent.

4. The next stage will be reached when the textbooks read, "It is widely believed by scientists..." By that time, perhaps only black and white pictures will be shown.

5. "A likely probability for the origin of the universe is thought to be..." will come next. By this time, few, if any, pictures will be included.

6. An even later version will read, "One way the universe may have originated has been described as..."

7. The final position will be, "A view formerly held by some scientists was..." Obviously there will be no pictures by this time.

So far, it has been shown that scientists are in wide disagreement as to whether or not the universe is open (expanding) or closed (oscillating).[13]

There is just as much difference of opinion about the amount of mass that is in the total universe and the effect of gravitational attraction between these particles of mass, however large or small. Also, there are diverse views about where the heavy elements in the universe have come from.

Finally, the Big Bang theory actually may be totally unnecessary to explain what little we have been able to observe. Therefore, it would appear justified to say that the science textbooks are in error by presenting *mythology* rather than *science* in discussion of the origin of the universe.

Not only do scientists hold widely different views about the origin of the *universe* but even of *elements* of that universe like our solar system. This fact is best confirmed by consulting Exhibit 63 (*) that contains seven troublesome and unanswered questions about the origin of the solar system that were discussed recently at an international meeting in France.

If anyone ever thought that scientists are in agreement on how even the sun and its planets were formed, Exhibit 63 would probably influence them to think otherwise. It speaks of "valid criticisms of several of the leading models for the evolution of the solar nebulae" as well as the "still weakly-developed state of the Genesis art."

Not only is there a wide variety of opinion regarding the origin of the *universe* and the *solar system* within the scientific community — that is not comprehended by the general public, there are also widely variant scientific theories for the origin of the *earth!*

One theory that counters the gradual aspect of uniformitarianism (whereby billions of years are required to form planets like the earth) has been proposed by two seismologists from California Institute of Technology.[14] They contend that the earth was created in a 10,000-year "instant."

Even the *moon's* origin sparks considerable controversy and disagreement within scientific circles. Would you believe that differences of opinion exist between prominent Nobel Prize winners on this subject? Hannes Alfven, who won the 1970 Nobel Prize in physics, has proposed a theory — along with oceanographer Gustaf Arrhenius — that the moon is much too large to have evolved "normally." They believe that the moon is an amalgamated collection of five to ten moons that formerly revolved around the earth![15]

Another Nobel laureate, chemist Harold Urey, believes that the moon broke away in one big chunk from the earth about 4.5 billion years ago.[16] This is an older theory, having been earlier proposed by Sir George Darwin about the turn of the century.

The important thing for every person to remember is that there is no *scientific* proof of how the universe, solar system, earth or moon originated — and there never will be! The beautiful and colorful pictures in books illustrate *myths*, rather than depicting what actually happened.

The dogmatism conveyed by a full-color portrait of an event for

* By way of explanation, the reader is advised that the exhibits mentioned from time to time in the text appear in the Exhibits at the end of the book. They are included for those who wish to either verify statements being made or study individual points of interest in greater depth.

which there are absolutely no actual data may well destroy the very soul of science. By projecting — particularly to young virgin minds — an image consisting of pure speculation and no facts in the name of science, textbook writers appear to return to the witchcraft, voodoo and sorcery from which science claims to have delivered us!

Lucky Strike

Imagine a large pot of chicken broth simmering over an open hearth on a threatening stormy evening. Suddenly, the house is struck by lightning. The bolt of lightning zips down the chimney, passes through the broth, and presto — the chicken in the broth becomes alive again!

Ridiculous, you say. Maybe so. But don't be hasty in rejecting this possibility. If you are, you will *really* have trouble with understanding the origin of life as taught in science classes today.

That imaginary story is actually much more believable than the one we teach children in school about how life began. You see, the chicken in the broth was once alive and is only being *resurrected.* Resurrect means to "form back again" or return to a former position.

Guess what science textbooks say about the origin of *life*? They say that about three billion years ago, some *dead* amino acid-like molecules (the exact types are greatly debated) got together accidentally in some pools and formed a soup. While sitting there together in the pool, those dead molecules "interacted" with oxygen and some other "elemental constituents of the earth." However, none of these particles that were getting together had ever been alive. Suddenly, a lightning strike hit this soup and presto — "the first organization of matter which possessed the properties of life" emerged! *Pretty lucky strike.*

"That is a disgusting, insulting oversimplification of a marvelously complex hypothesis representing the brilliant synthesis of diverse information from paleontology, genetics, biogeography and biochemistry," I can hear some eminent scientist cry (see Exhibit 27).

"Perhaps — but I was only quoting from a masterful work (see Exhibit 5) that 15 distinguished scientists and science educators prepared over a four-year span to guide the preparation of science textbooks for the school children in the largest state of our country," I would reply.

The origin of *life* obviously came some time after the origin of the *universe.* Therefore, it should be easier to obtain facts or data concerning how it happened. Yet, the origin of life is taught, in accord with the Law of Scientific Dogmatism, with *less* dogmatism than the origin of the universe!

There may be several reasons for this.

First, it may not be as easy for an artist to depict this momentous event because the first *living* cell (?) probably looked much like its *dead* predecessors. Beside, what did any of these cells (?) — dead *or* alive — look like?

Second, there is not universal agreement among scientists that the introduction of energy among the dead progenitors was necessarily in the form of a lightning strike.

One gets the feeling in reading scientific literature about the origin of life that some scientists would prefer to have this stupendous "emergence" be a more gradual, inevitable process — in order to avoid the "lucky strike" syndrome.

"Shucks, folks, don't get *that* excited — it was *inevitable*, given enough time," they claim. Here's how the American Chemical Society's prestigious journal describes it:[17]

> In the current scientific view of evolution the transition from lifeless chemicals to living, self-replicating things was an "accident" only in the mathematical-probability sense that a lot of time and energy were required. In fact, a great many, if not all, scientists involved in the study of evolution would agree that, given proper conditions and enough time, the transition from nonliving to living was inevitable."

Nonetheless, that precise moment that dead matter gave birth to life is a unique, binary and spectacular miracle. The "luck" requirement is not reduced by stretching out time... or by having all scientists *agree* on something they know nothing about!

Third, children and the non-scientific public have seen photographs of real explosions, so the "Big Bang" has some credibility with them. But only in fairy tales have they ever heard of something equivalent to the story given in science for the origin of life.

Objectivity, of which science is the chief advocate, appears to be badly twisted and bent to force belief in this "frog turning into a prince." Dogmatism generally requires something a bit more believable.

So much for the popular belief of life's origin in the textbooks. What do scientists say to each other about this mysterious happening? Plenty. And it's not all in agreement, either.

The famed British physicist, William Thomson, who later became Lord Kelvin, reportedly could not conceive of "living organisms arising from dead matter." He is not alone, even among scientists today. Frequently, it is implied that if Lord Kelvin had been aware that vast time intervals between changes were available, he would have changed his mind. I doubt it.

Stretching the time scale does absolutely nothing in terms of the amount of change that matter must necessarily undergo to convert from inanimate to animate. For example, a football play executed in real time is no different, in substance, than the same play seen in slow motion on instant replay. The same amount of energy is expended, the same ground is covered, the same players are involved, and the same officiating occurs.

The divergence in opinion among scientists about the "lucky strike" story appearing universally in science textbooks centers about several key unresolved issues. Probably the fundamental issue is whether (1) life originated here on earth (in some version of the "lucky strike" story) or (2) was transported to earth from elsewhere in the universe.

The latter theory has been popular, off and on, among scientists since the turn of the century. Its primary appeal is that it allows scientists to avoid the preposterous alternative — that dead matter produced life all by itself. The microbes that might have drifted in from outer space are called "panspermia." They supposedly could travel from one planetary system to another, propelled by the pressure of starlight. Carl Sagan of the University of California has calculated that these panspermia could be propelled between Mars and Earth if they varied from 0.2 to 0.6 millionths of a meter in diameter.[18]

The biggest drawback to the panspermia theory is radiation from the sun that would likely kill any unprotected organism leaving or approaching Earth, according to Sagan. It is on this basis that he has personally ruled out this means of originating life on Earth.

On the other hand, partly because many astronomers believe that there is a good probability that life exists elsewhere in the universe and, as a corollary, that it is probable that some of these presumed beings are ahead of us in intelligence and civilization, Leslie E. Orgel of the Salk Institute finds the panspermia idea more valid than the "lucky strike" one.

Along with Francis H. C. Crick, a Nobel laureate, Orgel recently proposed a "directed panspermia" theory.[19] This theory proposes that intelligent beings from elsewhere deliberately "infected" the earth. This could explain why we have only one genetic code instead of many.

Since I have been personally involved in manned space flight from Project Mercury through Apollo-Soyuz, I have been intrigued with the fact that NASA's *top goal* in space is "to search for life beyond earth." Probably not many taxpayers were aware of this. (If they had known, their apparent disenchantment with the space program may have come much sooner than it did.)

NASA did not propose this goal on its own — it was urged on them by the National Academy of Sciences in a report prepared in 1965. This report was most optimistic about life elsewhere:[20]

Martian exploration is designed to test the hypothesis that living things probably will exist on planets which have environments and histories like that of earth. So far this theory is based on a single case — the earth.

Its foundations lie in the naturalist view of evolution first stated a century ago, the report said.

Beyond that, in the 1950's and thereafter, experiments were conducted which show that all biological compounds necessary for life can be synthesized by the application of energy to the chemicals presumed to have existed in the primitive environment of earth.

"The general tenet that life involves no qualitative novelty — no elan vital (no vital force such as divine action) — goes hand in hand with the more explicit proposition that it is the molecular organization, as such, of living things that alone distinguishes them from the non-living," the report stated.

This kind of organization is improbable, since disorder is more the rule than the exception. Yet life exists on earth, and the same logically must be expected elsewhere, given a similar environment...

"We believe it entirely reasonable that Mars is inhabited with living organisms and that life independently originated there," it concluded.

I doubt whether the non-scientific public realizes that the reason most scientists choose an *Earth* origin for life over the panspermia concept is not based on data, but on the process of elimination of alternatives! In other words, it is not that the "lucky strike" story or some other version of it is a *good* explanation — it is just *less bad* than the other alternatives.

Listen to what Sagan says:[21]

Today, it is far easier to believe that organisms arose spontaneously on the Earth than to try to account for them in any other way. Nevertheless, this still is a statement of faith rather than of demonstrable scientific knowledge. Scientists have only sketchy notions of how this evolution might have occurred.

Do these statements convince *you* that how life originated has been finally and completely determined? Would their logic be persuasive enough for *you* to authorize a simple story to be universally taught to fertile young minds in the name of scientific truth?

Why can't science textbooks tell the real truth — that there is no way for scientists to *ever* know how, when, where or why life originated? Is there anything wrong with textbooks expressing the same uncertainty that Sagan expresses? Maybe it would be dangerous...

Please understand my plea. I am for continued scientific research on the intricacies of mechanisms associated with life.

The books should still teach about biochemical experiments wherein life processes are reduced to apparently fundamental building blocks of life involving, for example, ammonia, methane, proteins and nucleic acids. By no means am I proposing censorship or book-burning campaigns.

I only want *honesty*! Is there an unwritten law that demands that, "Scientists must never admit to the public that there are unknowables in science?" The Nobel Prize-winning chemist, Harold C. Urey, who is mentioned throughout this book, attempted to explain why many scientists have chosen to believe the "lucky strike" idea without any data:[22]

> All of us who study the origin of life find that the more we look into it, the more we feel that it is too complex to have evolved anywhere. We all *believe as an article of faith* that life evolved from dead matter on this planet. It is just that its complexity is so great, it is hard for us to imagine that it did. (Emphasis added)

Few scientists, particularly those of Urey's eminence, have ever been so candid. It is refreshing to hear such open admission. Evidently, the news media were also intrigued by his frankness because it was reported:[23]

> Pressed to explain what he meant by having "faith" in an event for which he had no substantial evidence, Dr. Urey said his faith was not in the event itself so much as in the physical laws and reasoning that pointed to its likelihood.

Why can't science textbooks be this honest? It would not bring discredit on science if children's minds were left open on the origin of life instead of being slammed shut with "fantasy based on faith." Having the humility to be this honest would even help to offset the cold arrogance that people often see in scientists.

So far, we have discussed only one aspect of the origin of life over which scientists disagree — whether life arose here or was transferred from elsewhere. But there are many other unresolved aspects, including:

1. The location of life's origin — whether it occurred at one singular location or was simultaneously widespread over the earth.

2. The exact chemical constituents of the first living matter (cell?).

3. The energizing element or activity that converted dead matter into living matter.

4. The date (even giving or taking a few hundred million years!) on which life began.

Unfortunately, the style of this book permits only brief highlighting of these aspects, rather than an in-depth treatment. Rest assured that scientific literature is replete with substantive debate around these critical points of uncertainty.

Regarding the *location* of the very first life on earth, most science textbooks are not too precise. However, lack of precision does not mean lack of dogmatic statements about *where* life began. In fact, the Law of Scientific Dogmatism is once more confirmed in accounts of the origin of life.

Let me illustrate this dogmatism with an actual incident. In 1972, a proposal was made to change a statement about life's beginning in Harcourt Brace Jovanovich's 3rd grade science textbook from "It is known that life began in the sea" to "Many scientists believe that life may have begun in the sea." (See proposed change number 2 in Exhibit 49.) You should have heard the howl! Prominent scientists, individually as well as collectively, acted as though science would be irreversibly rolled back to the Dark Ages if that change were to be made. Many thought the call for such a change was part of an insidious plot to discredit science. (Later chapters will elaborate their hysterical reaction.)

Though the location is never stipulated, what were the exact *chemical constituents* of that first bit of living matter? Scientists certainly are not in agreement on this important fact either.

One currently popular scientific view is that the dead progenitors of the first living bit were formed by condensation of simple gases in the earth's atmosphere. Thereby, speculation on what the earth's atmosphere contained becomes critical to life. However, some scientists like D.E. Nicodem of Rensselaer Polytechnic Institute believe that the atmosphere was much more likely composed of carbon dioxide and nitrogen than the ammonia and methane that many other scientists propose.[24]

Another theory about the early earth atmosphere is proposed by a team of chemists at the University of Maryland. They suggest that "great earth-colliding comets and thunder over turbulent primordial seas helped greatly to set off the chemical precursors to life on earth."[25] These events could have possibly introduced additional but unknown interactions among the dead elements in the "primordial soup" to aid the miracle of life's origin.

A slightly different example of unresolved questions about the actual make-up of the "parents" of first life is the debate on whether proteins preceded nucleic acids like DNA and RNA, or vice versa.

Both of these macromolecules are needed by living creatures. The nucleic acids carry necessary information to form proteins, and proteins are required in production of nucleic acids.

But which came first? The Director of the Max Planck Institute for Physical Chemistry in Goettingen, Germany, Manfred Eigen, believes that *both* proteins and nucleic acids had to have developed simultaneously and randomly. However, he further believes that a special kind of behavior for these macromolecules during their subsequent generation caused them to develop in a *preordained manner* (with a precise pattern) rather than *randomly* (without a known pattern) after the first ones "happened" into existence.[26]

But surely scientists know what *energy source* produced the magic transformation of dead material into something alive! Not really. Oh, there are many who feel that it was a lightning strike. But others prefer ordinary sunshine. One thing you can bank on — there's no *unanimous* verdict.

There is probably an interrelationship between how strongly the individual scientist feels about the *inevitability* of life's spontaneously arising and the type of energy source he would vote for. If he *really* believes that life was an inevitable occurrence, then he would probably vote for ordinary sunshine (in order to avoid the "lucky strike" or accidental connotation). On the other hand, that choice would tend to force a conclusion that, rather than a "first particle" of living matter, there must have been a widespread simultaneous eruption of life throughout the world when exactly the right environmental conditions were reached.

Of course, there is a large segment of the scientific community that has, as yet, been unable to believe the idea that life was an inevitable occurrence. The "lucky strike" story, though incredible, requires less faith than the belief in "automatic" life due only to certain unnamed environmental conditions. Since a lightning strike suggests less dependence on the inevitability idea, the initial appearance of life on earth would tend to be limited to a single, unique location.

This first place in the world where lightning struck after the soup had been brought to the "right" condition by environmental forces would be an important place indeed! Wherever that place is, it would not be surprising if a bright young science student proposed erecting a monument marking the very spot — like there is in Kitty Hawk, North Carolina commemorating man's first powered flight.

If "it is known that life began in the sea..." as we dogmatically teach our schoolchildren, then maybe Congress could be persuaded to authorize the erection on that spot in the sea of a commemorative memorial on the 3,000,000,000th anniversary of life.

But then the child might begin to wonder which *sea*... Adriatic, Mediterranean, Red... maybe even the *Dead* Sea?

Okay, so scientists aren't agreed on: (1) whether life arose here on earth or elsewhere, (2) the location of the first life on earth, (3) what the dead "parents" of life were made of, or (4) what the mysterious energizer that turned death into life was. But at least they must know *when* this event took place!

It's a shame to let you down, but even *that* is up for grabs. In fact, the date just got revised again the other day. Soviet paleontologist Boris Timofeyev, by discovering a little old (real little and *real* old) monocellular weed in a deep hole (4 miles deep) in the Ukraine, was able to push the date back a bit.[27]

Would you believe that by one little discovery he revised the date by one *billion* years? Anytime revisions that large can be made in a simple factor like a *date*, it is likely that the science of life's origins will not become an *exact* science in the immediate future.

In summarizing the discussion on life's origin, I am certain that many readers are curious why I have not yet discussed the remarkable progress that scientists have made, in biochemistry and other life sciences, toward *laboratory* creation of life.

"Surely if scientists can create in a test tube some chemical compound which possesses properties normally associated with life, we will know for certain exactly how life began on earth," some will exclaim. It hardly seems necessary to point out the fallacy in such a statement. But maybe an analogy would help us.

Suppose an engineering team could be assembled and given a government contract to reproduce Stonehenge, the ancient monument in England, in exactly the condition it is in today. This team obviously would include surveyors, stonecutters, draftsmen, masons, civil engineers, report writers, crane operators, laborers, budget analysts, truck drivers, managers, accountants, and union representatives.

They would utilize power equipment like cranes, pneumatic hammers and chisels, trucks, railroads and forklifts.

They would undoubtedly perform soil surveys and lay subterranean foundations. There would be materials specialists who would sample the original Stonehenge blocks and then conduct a search for identical stone.

Logistical planners would make cost effectiveness tradeoff studies to determine the most efficient manner for bringing the selected stone from the quarry to the site of Stonehenge II.

Progress reports to the government contracting officer would be made on a monthly basis. Contract renegotiation would undoubtedly be necessary as the project wore on.

Ultimately, there would be a ribbon-cutting ceremony for the opening of Stonehenge II.

Now would you be willing to believe that having reproduced the original Stonehenge that these people could explain how it was built the first time? Would you think that the methodology utilized for Stonehenge II resembled in any manner the methodology utilized for Stonehenge I? Of course, there is a very small probability that Stonehenge I was constructed in an *absolutely identical* fashion to that of Stonehenge II. No one knows for sure...

Scientists will justly deserve praise and honor for their brilliance should they produce a substance possessing life-like characteristics — if and when that occurs. However, it will not reveal *how* life originated.

It will suggest only one of many possible combinations of materials and energy that can *today* produce life-like behavior. For those seeking more scientific background on this general subject, an organic chemist at the University of Illinois, A.E. Wilder Smith, has published some thought-provoking material worthy of further study.[28]

Early But Never First

There was a *very first* human being. Think about it for a while.

Was it male or female? Bisexual, homosexual or heterosexual? Caucasoid, Mongoloid, Negroid, Australoid, or Capoid? Black, yellow, red or white? Tall, short or medium? Fat, skinny or average? Hairy or hairless?

The list of interesting questions goes on and on...

How long was this person *alone* — with no other humans?

What *language* did it speak?

What was the date on the *very first day* for Human Number One? May 16th, 35,025,836 BC?

Where did this remarkable person first set foot on the earth?

Did it start life as a newborn *baby* (from non-human parents) or did it become converted as an *adult* from some apelike creature?

Was there some unique environmental *"happening"* (similar to the "lucky strike") that introduced some magic ingredient into an advanced ape to produce this remarkable person?

Earlier we established that an origin implies that unique *moment in time* when something not previously existing started to exist. The origin of the *universe* (that moment when nothingness was suddenly replaced with existence) and the origin of *life* (that moment when life began among dead matter) are two fundamental and important origins. From the viewpoint of the non-scientific public, the third origin of great consequence is the origin of *man.*

What do science books say about this "very first human"? Absolutely *nothing*! Why not? Because they *know* nothing about it. But, according to the Law of Scientific Dogmatism, they are dogmatic about the *origins* (note the plural rather than singular) of man.

Logically, it would seem that observable data for these three origins — universe, life, and man — would vary in quantity. The most data should be available for the most *recent* origin — that of man.

Less data should be available for life's origin and the least amount of data should occur for the beginning of the universe. Logic would also seem to stipulate that the more we know about something, the more dogmatic we could be about it. However, as the Law of Scientific Dogmatism has postulated, science textbooks fly into the face of this logic.

Two violations of this logic occur. First, science textbooks are *most* dogmatic about the origin of the universe, the event for which there should be the *least* data. Second, there is absolutely *no* data for the origin of man, the event for which there should be the *most* data. Since we have already discussed the first violation of logic, let's examine the second one.

As of today, the only piece of data that exists for the very first human is an *unfilled gap*. "Now, I hope you're not going to start beating that dead horse of 'missing links' again," I can hear some anthropologist moan.

"No, I'm not," I'd reply.

The term "missing links" generally forms a picture in the public's eye of a long chain of creatures, running from an animal closely resembling a gibbon in the zoo to modern man dressed in a business suit. This chain is very nearly complete except for a few links that are missing.

However, the description that usually accompanies this chain of full-color pictures declares — with unabashed optimism — that the rarely missing links should be found within the next few years. There is no question that the links *exist* — they simply haven't been *found* yet!

Of course, when one is seeking data like "links," four possibilities exist:

1. The data exist and ultimately will be found.
2. The data exist but will never be found.
3. Although existing at one time, the data no longer exist because they were destroyed.
4. The data never did exist.

Only one of these four possibilities — No. 1 — will ultimately produce data. Yet that single possibility causes scientists to continue to search. If the possibility were only one in a *million*, I believe that scientists would and should continue searching. After all, origins are extremely rare events (they can occur only once!). The probability of finding a skull or any other fossil evidence of Human One (the very first human) therefore must be *infinitesimally small!*

By pointing out other possibilities beside "existing but still missing links" I hope to accomplish two things:

a. Open up some objectivity about an event that can never be reproduced with surety.

b. Show that it is "belief in the existence of something not yet seen" that urges scientists to continue searching.

If there are absolutely no hard facts available for Human One, and the probability of finding even one smidgeon of fact is extremely small, it is no wonder that science textbooks say nothing about this unique creature.

What do the books say about man's *beginning*? Generally, there is a long introductory discussion filled with words like "proto-man," "prehuman," or "pre-men" that are the equivalents of the inanimate or dead matter from which life arose in the origin of life.

These creatures apparently had some physical characteristics similar to those possessed by human beings. They walked upright on their hind legs and used primitive tools. But they lacked at least one dimension considered critical for being classified as human — adequate *cranial capacity*.

The stage is thereby set for that fateful day when the *very first human* will step on the earth. The long-awaited debut is about to occur. The crowd is hushed with anticipation. The spotlight goes on the curtain, the timpani give a roll, the trumpets blast a fanfare, the curtain goes up — and guess what?

It's a "no-show!" Instead of introducing this unique creature, the drama begins centuries and many generations *after* the event we were promised to see!

As we have already said, Human One is skipped over. Instead of the *first* man, discussion is filled with words like *dawn* man. *Ancient*, *early*, *primitive*, *ancestral*, *prehistoric* and even *earliest* man are described — but never *first* man.

Just as was true with the origins of the universe and life, scientists simply know *nothing* about the origin of man. They don't know the *date*. They have no idea where Human Being No. 1 first set foot on the earth. They do not know whether it had non-human *parents* or was a creature that had a unique start without parents. They do not know its *sex*, *skin color*, or *size*. They do not know *why* it suddenly appeared on the earth. The list goes on and on...

Would you buy a ticket to a baseball game that arbitrarily started with two outs in the bottom half of the fourth inning with a score of 6-1 in favor of the visitors? For those not enthusiastic about baseball, that could be a blessing.

But most baseball fans prefer to know how the game began. The scientific story of man's origin starts with a "zero information period" or a blank space on time's tape. This odd beginning for man must puzzle a thinking child — especially when the textbooks are so dogmatic about man's ancestors. Any story that starts in the third chapter gets off to a bad start.

Let's examine some of the assumptions scientists have made about man's origin, so we can better understand *why*, without any data, they can be so *dogmatic* about how humans began.

Probably, the most basic assumption is that every living thing on earth — trees, plants, birds, alligators, coral, dinosaurs, and man — has *one common ancestor*. What is it? That tiny "whatever" that zapped up alive out of the dead soup at the origin of life!

It must have been a miraculous microscopic marvel because it apparently contained all the secret latency for butterfly wings, dinosaur skeletons, human eyes, and the bat's sonar sensing system.

What potential it possessed! Try to remember that this little speck of life was so small that about 100 of them could walk abreast on a human hair! Talk about microminiaturization! And, of course, there was only *one*. (One of the great mysteries that science books never discuss is how "one became many.")

To make a very long (3,000,000,000-year) story short, this super-small speck managed to split without dying (a feat never described in textbooks). The two parts that split grew enough so that they too could split. Remarkably, they apparently split about 50%-50% so that both could eventually grow and become "whole and equal specks."

(The school child cannot help but wonder what would have happened if this first split had been 80%-20%.)

After while, these specks began to look different from one another instead of being identical (again no reason is given for this capability of the speck). The specks continued to look more and more different from one another. Some got large while others remained small. Some got colorful while others remained colorless. And on and on for what is currently estimated to be 3,000,000,000 years.

Size, shape, color, longevity, length, height, temperature, mobility, strength, speed, reproductive mechanism, sensory capability and weight are simply differences that were inevitable because they were programmed within that tiny speck which started it all.

In this marvelous lineage of proliferating change, man is a definite late-comer. Some enterprising science student, with the help of a pocket calculator, can compute just how late humanity arrived on the scene.

Given that 3,000,000,000 years have passed since life originated and that "modern man" (a term often used but never defined) got here about 30,000 years ago, the history of life was 99.999% complete before man arrived.

To put that figure in terms of a person's life span of 70 years, if your moment of birth was equivalent to the origin of life on earth, modern man would not arrive until 9.77 *seconds* before your 70th birthday! With all that history preceding humanity, a child cannot help but wonder why the discussion skips from ape-like creatures right over Human One to its distant descendants.

Are the science textbooks *deliberately excluding* known scientific evidence about Human One? Is there specific scientific knowledge about Human One that is *available elsewhere*? Not really.

In fact, there is very little, if any, indication in scientific literature that scientists are even *concerned* about Human One. Instead, there is a "blurring" or diffusion on the subject. In place of a *single* origin for man, the literature discusses "origins" of man. While listing myriads of unknowns and unsolved mysteries about mankind's beginning, scientists will simultaneously state with unqualified certainty that they know man's ancestors.

The primary problem for the scientist who is committed to the "unbroken chain" idea — that is, that all living things can be shown to be *reproductively* related — is *taxonomy*, the classification of plants and animals into various groups like phyla or species.

On first consideration, the "amoeba-to-man chain" should be *one long continuous story*. The first chapter should read "amoeba," the last read "man," and the intermediate chapters should have titles like "fish," "reptiles," or "birds. But it isn't that simple.

For one thing, we still have amoeba with us today — they didn't just pass life on to the next most complex creature and then cease to exist. Therefore, no one can write the end of the chapter called "amoeba." The same difficulty exists for *all* the groups of living things, with a few exceptions.

So instead of a smooth and logical book, a better illustration of the history of life might be a *tree*. This allows us to account for both those living families that have continued from their beginnings until the present time, as well as those that have ceased to exist (like dinosaurs).

Of course, the taproot for this tree is that miraculous first bit of living matter discussed earlier. The tough decisions in drawing a tree occur in deciding what living things form a legitimate branch, where branches leave the main trunk, which boughs form a true branch, and whether a particular bunch of living things is a limb or a twig (dead end).

Scientists have much greater difficulty with the concept of the origin of man than they do with either the origin of the universe or the origin of life. These latter two origins obviously mark a distinct "stop-start" or "then-now" event. Because man cannot be given any such individual attention if he is simply the latest version of living thing (and these living things all reproduce one another), scientists tend to duck the issue of a *single* origin for man.

Instead, they speak of *multiple* origins. Once in a great while, some scientist will address the singularity of man's origin.

The 1972 Faculty Research Lecture at the University of California in Berkeley was delivered by Sherwood Washburn, a renowned anthropologist and Chairman of the Department of Anthropology. The title of this lecture, that subsequently was taped for the Voice of America, was *"MAN: On The Origin of the Species."*[29] Although there is ample literature dealing with the origins of man, I wish to quote extensively from Washburn's lecture because of his stature in the field and the title he selected.

First of all, Washburn may surprise the non-scientific public by admitting that fossils and similarity of anatomy between apes and man (the only reasons most people have heard) have not been *scientifically* convincing that these two groups belong on the same branch of life's tree.[30]

> Many scientists, particularly in the last 50 years, have been unwilling to accept the conclusion that man and ape are so closely related. A wide variety of other theories have been supported by different kinds of evidence. None of the traditional methods (comparative anatomy, paleontology) could force agreement between the various scientists.

However, Washburn continues to believe that man and ape are closely related. So, he offers *biochemical* and *behavioral* reasons why ape-like creatures and man might be considered to be close relatives. In particular, he feels that chimpanzee behavior studies should revise the focus on when man might have originated.

Specifically note how he defines "the origin of man:"[31]

> In my opinion, one of the major results of the (chimpanzee) field studies is to fundamentally change the whole setting of the origin of man. The origin of man — that is the separation of the lineage leading to man and ape — was considered to be his coming to the ground from the trees. It has been suggested that coming to the ground might have been due to desiccation and contraction of the forests, that this event separated man from the arboreal apes. Yet our nearest living relatives are still on the ground; the gorilla moves, feeds, and even often sleeps on the ground; the chimpanzee feeds largely on fruit in the trees, but moves from one

> feeding place to another on the ground. The problem of the origin of man is centered not on how he came to the ground, but on how a forest-living, knuckle-walking ape became a biped.

His definition for the origin of man, then, is nowhere near the concept of discovering and describing *Human One*. Instead, he is trying to discover and describe a *junction* between two branches in a tree. How impersonal!

Also note his overwhelming conviction that "apes begat men." This, of course, is the same faith that other scientists exercise when believing that plants became animals — a subject due for serious questioning later.

Interestingly, Washburn then proceeds to outline the problem of proving that there was an ancestral ape that lived an intermediary life between contemporary apes and ancient man. Even though the title of his lecture would lead us to believe that he would address the problem of the *very first man*, it becomes obvious that he is instead searching for an *animal group* that would apparently form the crotch where the "ape branch" and the "man branch" fork in the tree of life.

He had already established that there were four possible sources of evidence being offered to prove that man is the direct descendent of apes: (1) paleontology (fossils), (2) comparative anatomy (similar physical structure), (3) biochemistry (genetic codes and protein compositions), and (4) behaviorism (field studies of chimpanzees). He then lists the difficulties posed by each source.

Of *paleontology* he says:[32]

> One can imagine an ancestral ape of 5 to 10 million years ago living...a life intermediate between the contemporary apes and ancient man. Most unfortunately, so far no relevant fossils have been found to help settle the matter. The direct evidence must come from evidence in the ground...The fossils are too few in number and too fragmentary to settle the question of just when the common ancestors of African apes and men lived...The controversies stem from the history of the discoveries, the fragmentary nature of many of the fossils, and the uncertainty of the dating.

Despite these fossil uncertainties, Washburn later says, "The main outline of the events (comprising the uniquely human way of life) seems reasonably clear from the fossil record."[33] Which way is it? Do fossils aid or hinder definite conclusions?

Washburn then comments on the *comparative anatomy* evidence:[34]

> For anatomical reasons, the transformation of a knuckle-walking ape into a bipedal man is far less of a transition than it appears to be at first sight. Human arms and trunk are still very similar to those of the apes.

> Professional football linemen adopt a knuckle-walking position, the only creatures other than chimpanzee and gorilla to do so...The controversies over just how they (various pre-human creatures) should be interpreted and classified, or just how many kinds there were, should not be allowed to obscure the fact that a great deal is now known about the kinds of creatures which are ancestral to man.

Again an apology for the uncertainty and obscurity of data but capped by an assurance that firm conclusions can be drawn. What permits this merging of incompatibles? A *faith* or *belief.*

Concerning difficulties with the third source of evidence, *biochemistry*, Washburn comments:[35]

> As important as the chemical information is for solving questions of relationship, it tells nothing about the nature of our ancestors, their way of life, or their ecology.

Since Washburn is an eminent anthropologist, we could expect that he might favor behaviorism as the best of the four sources of evidence that man is a direct descendent of apes. It is true that he is enthusiastic about the contributions of recent chimpanzee studies. But even *behaviorism* poses some difficulties:[36]

> Little of the possible behavioral reality leaves even indirect traces in the ground. There is wide latitude for reasonable disagreement on the course of even such major...events as the origin of the human ability to speak.

While we certainly laud Washburn's candid admission of the vast shortage and limitation of scientific data that might lead to a conclusion of apes being ancestral parents of human beings, his confident optimism in that conclusion — particularly in the face of such missing or insecure information, defies logical explanation.

One of Washburn's students, F. Clark Howell, has written some of the most widely-read material on the "ape-to-man-sans Human One" concept. Howell, Professor of Anthropology at University of Chicago, is the author of *Early Man*, an amazing compilation of mythology disguised as science.[37]

It is not likely that there has ever been as much speculation supported by so few facts as this *Time-Life* series contains. It is a virtual potpourri of moral preachments, wild speculations and unwarranted assumptions mixed with a few facts, attractive photographs and mythological artistry in full color.

Consider these moralistic statements made in the name of science:[38]

> Here — and not in moon shots — lies man's greatest challenge. For the first time in the two-billion-year history of life will come an opportunity

to attempt to combine the good of the species with the good of the individual, a dilemma that has not been resolved very well in the past and is certainly not being resolved *very* well by the human species today. (page 176)

Our society, from the point of view of a coldly rational intellect observing us from outer space, is unforgivably stupid in its aggressiveness, its prejudices, its superstitions and in its misuse of the powers and opportunities that our technology has given us. (page 170)

Or if poetic speculation is desired, *Early Man* certainly furnishes it wrapped in a scientific cloak:[39]

To tell the story (of how apes became man) properly, we must look behind apes to monkeys and, behind them, to the earlier animals from which monkeys sprang, because traits that would later begin to emerge as distinctly human are believed to have had their origins in the shapes and behavior of these shadowy creatures. (Page 31)

Most (pithecuses or apelike creatures) exist in lamentably small fragments; and which definitely belongs with which is still being unraveled. What they do make clear, from the abundant and varied fossil beds of the Fayum in particular, is that the primate tree had nothing like a central trunk but was a luxuriant vine with many shoots and tendrils growing side by side, sometimes withering and dying, sometimes branching. And in those branches we see extremely ancient prosimianlike types, more advanced monkeylike and apelike types, and finally a group that belongs definitely in the ape line alone and from which it is possible to begin to pick out the directions from which the apes themselves would go. (Page 38)

In case full-color pictures are more to your liking, *Early Man* furnishes beautiful detailed artwork like a 9"x14" phantasm entitled, "The Hunt Is Re-Enacted." This picture, consisting *entirely of conjecture*, carries a caption that typifies the entire book:[40]

The great glut is over. The two bands of *Homo erectus*, enjoying the rare double sensation of a full stomach and the stimulus of strange company, have gathered for warmth around the dwindling cooking fires. Within a circle of coals symbolizing the fire drive, a hunter prances about, draped in the gory skin of an infant elephant. Under his arms he clutches its mother's tusks. The sight of these trophies has inspired others to re-enact the hunt — leaping, shouting and thrusting with wooden spears. For the children squatting in the foreground this is a kind of schooling. Here they will absorb the tradition of the hunt. For the adolescents it may also be a way of courting, with the males showing off their bravery to impress prospective mates. When the bands part to struggle through the winter ahead, certain females of one band will follow the males of another. Though *there is no concrete evidence for this scene* from the sites, it is enough to

> remember that these were human beings after all, and the kinds of activities that their descendents would — and still do — engage in, must have begun somewhere, (Emphasis added)

It is this type of mythology that objective scientists will recognize as being identical to the Greek, Hindu, and medieval myths from which science supposedly rescued modern man.

There is no question that, for young impressionable minds, a *picture* of a myth is especially misleading. Perhaps the most unscientific and incredible feature in *Early Man* is a foldout set of 15 drawings of monkey-like animals supposedly becoming modern man. It is introduced by the amazing admission on page 41:[41]

> (These drawings are) pieced together from the fragmentary fossil evidence. It is a revealing story, not only for the creatures it shows, but also because it *graphically illustrates how much can be learned from how little*: the seemingly chaotic collection of bones... can give a quite complete picture...Many of the figures shown here have been built up from far fewer fragments – a jaw, some teeth perhaps...and thus are *products of educated guessing* (emphasis added).

These drawings obviously can be seen and interpreted by a very young child as being authentic representations of scientific fact long before that child can read the captions accompanying them.

By christening a few chips of bone with an incomprehensible and formidable Latin name like *Australopithecus* or *Dryopithecus*, an overwhelming aura of authority can be impressed on a child. If one reads the captions beneath each of these fanciful fictions, the amount of guessing used is distressing. In fact, the pictures are almost *entirely* guesses! Consider these examples:

> *Proconsol* — "Known from numerous fragments adding up to almost complete skeletons..."
>
> *Dryopithecus* — "Though its skeleton is tantalizingly incomplete it can be fairly described from a few jaws and teeth..."
>
> *Ramapithecus* — "This hominid status is predicated upon a few teeth, some fragments of jaw and a palate unmistakably human in shape..."
>
> *Solo Man* — "Solo man is recognized so far only from two shin bones and some fragmentary skulls..."

Not only are the figures in this lineup *imaginary*, but figures are included in this group whom the author admits are unrelated to each other! *Pliopithecus*, *Proconsul*, *Dryopithecus*, *Oreopithecus*, *Paranthropus*, *Solo*

Man, and *Rhodesian Man* are admitted "dead ends."

They do not form a *sequence* of life, yet they are called "the stages of man's long march from apelike ancestors to *sapiens*!" Why are they included? Only the author knows. This blatant deception could end quite easily if the *facts* were separated from the *speculation*. Imagine this same series of figures if the fossil fragments were shown solidly and everything else were only dotted!

We would then better understand the introduction of *Early Man* where Harvard anthropologist William Howells says of the author:[42]

> Howell...has not been afraid to apply *imagination* where it counts. He offers his *educated suggestions* on day-to-day problems and general conditions of life in the ancient past, and on the actual uses man made of his tools. *Without such a view, the fossils of man are nothing more than bones* (emphasis added).

Howell and Washburn have subsequently collaborated in another *Time-Life* series, *The Emergence of Man*, which perpetrates the same type of mythology in the name of science. It even includes "photo-paintings" showing how late *Australopithecus* (the "missing link" between ape and man) looked two million years ago.

The *link* is missing (in fact, it is not known that such a creature ever existed) — yet a full-color photo-painting of it is shown! Where are those scientists who are supposed to keep science free from superstition and mythology? Something really slipped through the filter here (remember *"The Crucible of Scientific Truth"* in the Prologue?).

Scientific literature on the origin of man can be summarized as follows:

1. There is no discussion of Human One, the very first human being.

2. There is open admission of a lack of evidence that would conclusively connect any specific ancient or current ape-like creature with humans. Therefore, no definite point in the yet incomplete chain of creatures where ape supposedly became man can be established.

3. There is dogmatic assertion that man had apelike parents.

The layman naturally wonders, "How can scientists proceed from the second summary point to the third one on a *scientific* basis?" He cannot. For if he pursues the origin of man objectively beyond the second point, he cannot logically reach the third point without selecting one of two contrasting possibilities:

1. That there is an unbroken reproductive chain or continuum running between ape and man.
2. That man is actually a creature entirely separated in lineage from apelike creatures.

Unfortunately, there is no *scientific* way to choose between these two with the presently available data. The key to resolving these two possibilities lies with Human One. And science knows *nothing* about Human One!

But since Human One is so unique, it is also a highly *improbable* creature. Not only is its sudden appearance an improbable event (that we have not been able to prove), but finding any conclusive evidence of this individual is even more improbable.

These improbabilities have naturally favored the pursuit of the first possibility (of a continuum or "unbroken chain" between ape and man) instead of the second — (that man is a unique creature). Even if the continuum were never totally complete, its advocates argue, logic would increasingly favor this possibility, as pieces or links were found, over having to find the criteria and ingredients for a brand-new creature.

But is that really true? Would a totally complete continuum be an *asset* in concluding that man's ancestors were apes? It might raise even more difficult questions like:

1. What arbitrary criteria would be used to *separate* Human One from its parents (the most advanced apes)?
2. *Was* there a Human One? Which *sex* was it?
3. How long did Human One exist alone prior to being *joined* by other humans?
4. If Human One was not alone but part of a larger group of beings (Human Firsts) who all were distinctly different from their parents, how many were in *that* group? (Remember, they would all have to be born at the same instant. Otherwise, the *eldest* would be Human One!)
5. What environmental or other *external source* was introduced between the last generation of apes and the first generation of man to make this remarkable conversion (similar to the energy that was apparently introduced into dead matter to convert it into a living speck)?
6. Was this ape-to-man conversion a freak *accident* or chemically *inevitable?*

7. At the moment when Human One — or that unique group that was distinctly human — first appeared, was it (or were they) the only offspring of their parents (apes), or did these first humans have brothers and sisters that remained apes like their parents?

8. If Human One was the only human for a long time, could it mate with one of its ape siblings (non-human "brother" or "sister")? If so, why can't humans mate today with gibbons in the zoo?

9. Who was the more remarkable — Human One or its ape parents? *Siring* a marvelous breakthrough may be a greater feat than simply *being* that breakthrough — especially if the breakthrough was inevitable, given enough time and energy. (It would be a shame to honor the wrong party.)

These questions are certainly going to irritate some scientists who will say that they are inept, illegitimate, illogical or inane.

Nonetheless, they are questions a young, inquiring science student would postulate — particularly after reading the dogmatic mythology of man's origin in current science textbooks. Why not answer these questions in the science textbooks, and thereby complete the mythology instead of leaving the reader hanging?

Sooner or later scientists will have to quit begging the question on the uniqueness of man's origin by stretching it over millions of years. The old wornout chronology of *Ramapithecus* (95% ape/5% man), *Australopithecus* (80% ape/20% man), *Advanced Australopithecus* (75% ape/ 25% man), *Homo Erectus* (65% ape/35% man), *Early Homo Sapien* (50% ape/ 50% man), *Neanderthal* (30% ape /70% man) *Cro-Magnon* (15% ape /85% man) *et cetera* has to go.

This concept of gradualism — that avoids the sticky question of man's uniqueness — is not only intellectually dissatisfying but outside scientific validation. Just as it is difficult to be "a little bit pregnant," it is difficult to be "a little bit human." However primitive some people may be today, there is never any difficulty distinguishing them from apes.

EVENTS EVOKING ENIGMA

Origins are not only *unique* events. They are *critical* events. They form the baseline for our understanding of everything. The age-old questions that have occupied philosophers for centuries, "Who am I? Why am I here?" are inextricably tied to *origins* — particularly the three origins we have examined.

Hopefully, we have shown that science cannot describe or explain these three origins. Though scientists have unraveled many mysteries about nature, their hands are tied when it comes to providing explanations of these events so pungent with enigma. Because of their success in many areas and their common human ego, they often suffer the temptation of over-extending their knowledge, thereby unwittingly succumbing to the same delusion that ignorance brings to anyone — "they know not that they know not."

Science must be protected from well-meaning but misguided scientists who fail to realize the limitations of scientific inquiry. This protection is essential not only for the future of scientific progress but also for the proper relationship between scientists and the non-scientific public.

Meaning Versus Mechanism

Picture the most beautiful sunset you have ever seen. Myriads of shades of pink, gold, red and purple with narrow shafts of the setting sun breaking through clouds of indescribable shapes and forms to paint them with gilded edges. This inspiring and breathtaking sight causes you to be overwhelmed until tears begin to blur the enchanting scene.

What is the *scientific* explanation of your reaction? There is probably *no* scientific answer that can explain it. Certainly there are some valid contributions that science could make toward understanding your reaction — like the reason for the colors being predominantly at the red or longer wave-length end of the solar spectrum, the types, composition and altitude of the clouds, the angle in minutes and degrees of arc that the cloud formations subtend on your eye, and the influence of wind patterns on the cloud formations.

The scientist would be at a loss however to explain the tears in your eyes. There could be *many* reasons — your eyes might be sensitive to bright light, your contact lenses could be irritating, you might have a head cold, the scene possibly could bring back a nostalgic memory, or you are just "emotionally moved by the beauty." While scientists could describe *mechanisms*, they have no inkling as to *meaning* of the sunset to you.

This simple analogy illustrates the profound truth that mechanism and meaning are two aspects of one and the same *reality*. There are continuing discussions among scientists as to what constitutes reality.

However, we can assume that the origin of the universe, life, and man were *real* events. They *did* occur. There *was* a first moment in the history of the universe (unless one believes that matter is *eternal* — a thought difficult to think). There *was* a very first moment for life. Likewise, there *was* a very first human being.

As discussed earlier, it should be evident that these moments cannot be described by scientists — not just because they happened so long ago, had no observers, left no specific data, and are not repeatable. All those factors are *mechanistic*. The true reality of origins also involves *meaning* beyond the *mechanisms*. Determining such meaning is a scientific impossibility.

Even if scientists were lucky enough to stumble upon a vault that contained bound, laboratory notebooks filled with identical first-hand observations by two qualified witnesses to all three origins — complete with dates, locations, chemical compositions, noise levels, temperatures, humidities, times and every other possible measurement — scientists could not fully explain the reality of origins. They would have none of the critical clues as to the *meaning* of the events. There is a thoughtful discussion of this point in *Beyond Science* by a British biochemist, Denis Alexander.[43]

Reconsider the ancient questions that are so dependent upon origins, "Who am I?" and "Why am I here?" Is it not obvious that the answers to these questions would be quite different, for example, if life and humanity were to be considered simply chemical inevitabilities as opposed to being the result of planned design by a benevolent Intelligence who desires interaction with humanity?

Incidentally, this is by no means to propose a religious concept within a scientific discussion. Quite to the contrary, it illustrates the point that scientists dare not address either of these two options because science provides neither methodology nor access to knowledge that could resolve them.

Recall Stonehenge. Even if we believed that Stonehenge I and Stonehenge II were constructed in an identical manner, *why* was Stonehenge I created in the first place? What did it *mean* to its creator? Don't ask a scientist.

Destiny from Descent

What *difference* does it make where we came from? We're here, aren't we? Why all the fuss about origins, anyhow? We can move forward from the "here-and-now" regardless of how we got here!

That kind of reasoning makes just as much sense as saying, "It doesn't matter what Aristotle, Copernicus, Newton, Darwin or Einstein thought. Science can move forward on what we *now* know!"

Origins are as important to civilization as early scientists are to science. Civilization and science are quite similar. Consider this description of science:[44]

> Science is not a mere register of facts; and indeed our minds do not (like a cash register) tabulate a series of facts in a natural sequence one after

> the other. Our minds connect one fact with another — they seek for *order* and *relationship* — and in this way they arrange the facts so that they are linked by some *inner law* into a *coherent network*.
>
> Essentially, it is the very process of establishing *order* and *relationship* among facts that has been systemized and further developed into a set of procedures we have chosen to call science a systematic way of answering questions. As a result of this process, an *organization* of knowledge is being built.
>
> The facts are observable, but their *organization* is not. Man invents an *organization* which seems to fit the facts he observes (emphasis added).

If we were to substitute "civilization" for "science" and "historical events" for "facts" in this description, we would have *a very* fine explanation of civilization. The *order*, *relationship*, *organization*, or *coherent network* that interrelates what can be observed is very important for both science and civilization.

We *believe* something about both scientific facts and historical events that ties them all together. We have *faith* about the future of both that is based on our belief. The next time you visit the Jefferson Memorial in Washington, read Thomas Jefferson's words that are engraved on the inner walls. They reveal a very strong faith, belief and conviction that the destiny of civilization is dependent on how we perceive the origins of life and humanity.

Just as Shakespeare's saying, "What is past is prologue" is a valid observation, so the saying, "Destiny is dependent on descent" speaks truth.

Is this not the basis for the Afro-American, Asian-American, Native American and Chicano studies which are currently so popular in educational curricula? Advocates of these courses in school are certainly not saying, "It does not matter what our heritage *was*. We are now Americans." Quite the opposite!

They are insisting that self-respect and self-worth are very dependent on the correct (and perhaps even a *proud*) understanding of ancestry and historical development.

Consider once more two alternatives mentioned earlier:

> *Alternative A*: Both the origin of life and the origin of man were chemical accidents or inevitabilities (the choice is yours).
>
> *Alternative B*: Both the origin of life and the origin of man were deliberate actions planned and executed by a benevolent Intelligence who wishes to have fellowship with people.

There are obviously many variations of these two basic alternatives.

Further, other basic alternatives undoubtedly exist. However, with respect to *destiny* (as contrasted with *meaning* — the context in which these two were earlier discussed), the following points are worthy of consideration:

1. Science cannot determine the *truth* of either alternative (as already discussed).

2. Both alternatives are *equally* unscientific.

3. Both alternatives equally influence the *order*, *relationship organization*, or *coherent network* that historical events comprise and which we perceive as civilization — in the past (history) as well as in the future (destiny).

4. The influence of these alternatives is not likely to lead to *identical* conclusions as to where our destiny lies.

If these points are valid, we could conclude that:

1. There are contrasting, if not *conflicting*, belief alternatives available to us regarding the origin of life and the origin of man.

2. *Destiny* is dependent, in some degree, on which belief alternative we select.

3. The *truth* or *validity* of any selected belief alternative concerning origins cannot be established, confirmed or denied by science.

Our civilization, blessed as it has been by science in the past century, now stands *seduced* — in the name of the same force that liberated it from the bondage of superstition only a few years ago!

Certain scientists have transformed science into a dogma by asserting, with a haughty air of finality, that Alternative A is now a proven *scientific* fact. "All options on destiny are now closed," they claim.

Don't you believe it!

Wonderment or Certainty?

We live in a crazy, mixed-up world. Anxieties, compulsions, obsessions, phobias, and frustrations seem to be on the increase everywhere. Many experts who study these disturbing phenomena often conclude that an "unresolved state of ambivalence" lies at the root of such disorders. In other words, we can't "make up our minds."

We are constantly undecided. We waver back and forth but can never choose, "I like *this*, but on the other hand..." "I shouldn't eat that, but..." "I'll try it just this one time..."

Often, the best advice seems to be, "Make a choice — it doesn't matter which one — and stick with it!" Replace *wonderment* with *certainty*.

This is perhaps sound advice in many instances. Particularly, when the alternatives between which we are wavering are both inconsequential — like whether to wear brown or blue clothing to work on Tuesday. Make an arbitrary decision and "get it over with." Be decisive! Don't dilly-dally around.

There may be other occasions, on the other hand, where remaining undecided has a few virtues. "Be slow to anger," is ancient wisdom. Shakespeare said, "Too swift arrives as tardy as too slow." The greater the consequences, the more that hesitation makes sense.

We have already established that the origins of life and man are loaded with consequence. They are important in influencing our *perception* of who we are and why we find ourselves here. They are also vital in establishing the *meaning* of life.

Likewise, these origins determine the *destiny* of our civilization. Therefore, it seems that we should not be hasty in jumping to conclusions about such origins.

When you stop to think about them, origins are really puzzling. They are like mysterious, baffling riddles. They intrigue us. In one way, they appear simple and yet because there is no satisfactory way to explain them, we know that they must be *very* complex events. Just about the time that you feel that you might have a very good explanation for them, a volley of new questions bursts forth in your mind.

The key question at this point is this: would it be wise to replace all the *wonderment* about origins with the *certainty* offered by Alternative A discussed earlier?

Before you answer that question, consider the following:

1. Alternative A is no more valid (in terms of scientific provability) than any other alternative, including Alternative B.

2. Alternative A has been proposed by a group of people (i.e., scientists) who have no way of knowing or proving whether it is valid. It is simply a matter of faith or belief with them.

3. Endorsing Alternative A as being certain closes off all debate — a cornerstone of objectivity — on alternative explanations for the origins of life and man.

Many questions have been asked so far. Not many answers have been offered to those questions. That is how science really is — *many more questions than answers*. Every time an answer is found, it produces lots more new questions. That is why an inquiring mind is so vital in science. Those who are uncomfortable with questions should never tangle with science.

Questions raised by reading science textbooks, including some asked in this chapter, can be classified or grouped as follows:

1. Unanswerable questions — (Example: "What was the chemical composition of the first living matter?")
2. Unanswered questions that are possibly answerable in the future — (Example "Does life exist anywhere else beside on earth?")
3. Unanswered questions that possibly have current answers — (Example: "How did Human One differ from its parents?")
4. Questions answered without scientific basis — (Example: "Where did life begin?")

Questions in groups 1 and 4 above imply that science might be limited. "I thought that science has answers for everything," some of the non-scientific public might reflect.

If you are among them, read on…

Chapter 2

SCIENCE IS LIMITED?

"Nothing is so firmly believed as that we least know." — *Montaigne*

"Incredible," I mused as I gazed out of a small window in an old Franciscan monastery standing on a prominent seacoast hill in La Rabida, Spain.

Looking out at the Rio Tinto flowing placidly into the Gulf of Cadiz and on into the Atlantic, I tried to picture three small wooden ships, the largest only 90 feet long, sailing out over the horizon on a bright August day with a total of 87 men aboard.

None of these ships had an engine, and the cooking was done with wood in a firebox on the wooden deck. On board, there was no way to measure distance traveled, and the only navigational instrument available was a crude quadrant that was not accurate when the ship rolled, which it did constantly. The man in charge of this little fleet, Christopher Columbus, knew less about celestial navigation than a young sailor in boot camp does today.

Columbus was fired with one passion — to reach India by going west. It is only accidental that he had set sail from Palos de la Frontera, a tiny village just three or four miles up the Rio Tinto from the old monastery. His dreams of a voyage already had been rejected by John II, King of Portugal. However, to cover all bets, Columbus had his brother Bartholmew making proposals to King Henry VII of England and Charles VIII of France while he journeyed with his only son, Diego, to Spain to sell the idea to Isabella and Ferdinand.

Less than 25 miles from the Portugese border, Diego and his father came upon the monastery at La Rabida. The Franciscan friars conducted a school for small boys there. Columbus decided to leave Diego in La Rabida while he went to make his appeal to the King and Queen. However, he later returned and spent several months at the monastery awaiting the royal decision.

When the royal approval of his plans reached Columbus, he secured his ships and began recruiting the crews. As we walked through the courtyard of the ancient church, Fray Juan Perez in Palos de la Frontera where he did his recruiting, wonderment flashed through our minds as to how he convinced these sailors to join him. What did he tell them? Did they laugh at him? Likewise, at an old well dating to Roman times in that village, we had fun imagining how the crews filled large pots with the last fresh water to be taken aboard the flotilla.

Once in the winter and again in the spring, my family and I made trips to La Rabida from Seville, where I was teaching a graduate course for the University of Southern California. La Rabida is not easy to locate. It is not on a major road. In fact, not many maps of Spain even show it. Profuse with historical significance, both La Rabida and Palos de al Frontera are virtually unknown.

I often remarked when I was there that if the monastery were located in the United States, it would exist in a carnival atmosphere. The capitalist instinct could not leave it alone. All the sideshow accouterments of the typical "tourist trap" — hot dog stands, guides dressed in 1492 costumes, cotton candy, cheap souvenir shops, blaring loudspeakers, noisy roller coasters, huge parking lots, and of course, the inevitable entrance fee would be part of the scene.

As difficult as it may seem to believe, our family was the only visitor both times that we visited the old monastery. The last portion of the road leading to it consists of unpaved dusty reddish soil. We drove into the grounds and parked only feet away from the door where Columbus and his son ate a piece of bread and drank a jug of water prior to Columbus' departure to make his proposal to Queen Isabella and King Ferdinand. It was easy to picture those days just before he set out to make history.

But could he have dreamed how far he would *miss* his objective? The ignorance of Columbus, as he departed the security of the known for the hope of the unknown, was not only incredible but overwhelming. The basic assumptions he made were not even close to being correct. He believed, in ignorance, that La Rabida was only 3,000 miles from Japan. (It is nearly four times that distance.) His belief that India was closer to the west of Spain than it was to the east was also in error by a large margin. (The west coast of India is about 4900 miles east and about 16,500 miles west of La Rabida.) And, of course, he knew nothing of the American continents lying between Spain and India.

The incredibility of this scene, to me personally, lay in the subject of the graduate course — "Space Technology" — I was teaching only 50 miles east of Columbus' sailing point.

This course covered topics like astronautical objectives in the solar system, space physics, lunar and interplanetary trajectories, interaction of space phenomena and planetary atmospheres, astrogeology, experimental aerodynamics, kinematics, dynamics, inertial navigation, rocket propulsion, and space medicine. Not only were these subjects obviously unknown to Columbus, but he had never heard of Copernicus, Galileo,

Kepler, or Newton — men upon whose work these subjects were based.

Just prior to our first visit to La Rabida, Apollo 8 had made man's first journey behind the Moon. Our family met privately with Frank Borman, Apollo 8 Commander, in Madrid shortly after this momentous achievement as he was making a worldwide goodwill tour. This firsthand discussion with someone who had been 240,000 miles from Earth increased the contrast with Columbus.

The elation over Borman's journey reached much higher heights in Spain than any I had observed, before or since, in the United States regarding space flight. Although we have visited Columbus' birthplace and early home in Genoa, Italy and his Italian ancestry is well publicized, most Spaniards consider him as one of *theirs*.

Similarly, as James Michener points out in his *Iberia*, they also nationalize El Greco, the Greek. In the typical Spanish spirit of joy and celebration, they christened Borman "Christopher Columbus II" and showered many honors on him as they compared him with Columbus. I was making my own private comparison — and becoming more bewildered!

The distance spanned in less than 500 years between Columbus' fleet — the Santa Maria, Pinta, and Nina — and Apollo 8 was very difficult to comprehend as I contemplated the vista out that monastery window. The distance obviously was measurable in more than *miles*. It was in terms of knowledge, understanding, and comprehension of natural forces.

How was this knowledge, understanding and comprehension obtained? Was it gained in an orderly progression from previous learning? Not really.

At the time of Columbus, the world was steeped in medieval scholasticism very much aligned with monastic thought. Based on Aristotelian logic and Judeo-Christian teachings that had accumulated over centuries, this scholasticism demanded adherence to traditional doctrines and methods. It could be considered an *accrual* system of learning.

All that had gone on before, once established, was no longer subject to revision or repeal. New information, particularly as it related to natural phenomena, had to conform to what had already been adopted. If it didn't, it was branded heretical and the authors were either executed or persecuted.

Shortly after Columbus, there began to arise a type of thinking that proposed a different basis for determining the truth of an observation. Instead of checking an "observed" fact against the accrued deliberations and traditions of the past, it was checked strictly on whether the fact was consistently repeatable under identical conditions.

Observations were freed thereby from a "value system" based on revelation. Admittedly oversimplified, this description of a shift in thinking about natural phenomena has been called the beginning of the rise of scientific inquiry.

There were glimpses of scientific thought during Columbus' day. For example, Prince Henry of Portugal (who is better known as Henry the Navigator) had established an observatory and school of navigation in Sagres, Portugal about the time of Columbus' birth. Many of the buildings in this school are still standing, high on a rugged bluff on the southwestern-most point in Europe. During my visit to this historic site, I learned that most of the famous Portuguese explorers, like Vasco Da Gama and Ferdinand Magellan had studied there. It is also possible that Columbus attended this school.

However, despite the existence of such a school, we can assume that Columbus lived during pre-scientific days. So the distance between Columbus and Apollo 8 could be considered to be bridged by what we now have come to know as *science.*

CAN'T ARGUE WITH SUCCESS!

Modern science has been a blessing, not only in earth and space exploration, but in almost every avenue of human life. Start to recount the impact of science on your own life. It is virtually impossible to even *know* what science has accomplished. Technology, that child of science spawned to apply the basic findings of science to the needs of mankind, has heaped benefits on all of us so rapidly yet so subtly that we should stop occasionally and try to recount them.

Rather than simply enumerate the *things* of science and technology it would be better to perhaps review their *influence* on the quality of life. Physical labor has been reduced, our health continues to improve, longevity is increasing, we enjoy geographical mobility, and a much broader awareness of the world about us now exists. Science and technology have been a central force in all these developments. Without science, we could well be living in conditions much like Columbus had.

The Middle Ages (dating from the fall of the Roman Empire in 476 A.D.) are often and rightly called the Dark Ages. For about 1,000 years, Western Europe was characterized by widespread ignorance, lack of progress and superstition. It is certainly outside the scope of this book to discuss why these conditions developed and prevailed for so long. It was this setting, however, from which scientific inquiry began to rise.

The price paid for shaking off the shackles of a society steeped in superstition was high indeed. Often it was paid in blood. This ugly era is one that all of us would like to erase from history. However, we must recollect it just long enough to properly realize what science has accomplished.

Shook Off Superstition

We are *all* superstitious — even perhaps a Nobel laureate in science, when a black cat crosses his path on Friday the 13th as he steps aboard the next commercial airliner to fly the same route on which a disaster has just occurred — especially if he spills the salt during his meal as he sits in Row 13.

You might chuckle a bit about this allegory, but you also might be as surprised as my wife and I were when we first visited Tokyo and learned that there are no 4th, 9th or 13th floors in Japanese hotels and hospitals. We were told that 4 in Japanese means death, 9 means agony, and the combination of death and agony occurs at 13.

While we still have the residuals of superstition persisting in our society, the world is far less superstitious than it was at the rise of scientific inquiry. At that time, superstition was not just a cultural carryover — it included the threat of incurring the wrath of supernatural beings! It was *necessary* in those days to respond with a "Gesundheit" or "God bless you" immediately following a sneeze, or the sneezer's spirit that had departed as a result of the sneeze would not return into the body.

Beyond the individual reaction to superstition, however, there was the much more serious superstition regarding cosmology which depended on a *Weltanschauung* or conception of the universe and of life itself. Almost every school child is aware of the torment and abuse Copernicus suffered when he suggested that Ptolemy had been wrong 1300 years earlier about the Earth being the center of the universe. Obviously, Copernicus' suggestion upset the applecart of tradition. The experts had already decided that Ptolemy was right, had built all subsequent thought on that "fact," and there could be no other way to view the earth.

In the Prologue, I express my profound admiration for Galileo and Copernicus as typical representatives of that small minority who paid (and continue to pay) such a horrible price for simply urging prevailing thinkers to consider observable, repeatable facts instead of traditional beliefs. All of us today inherit the immeasurable benefits of those who had courage to challenge conventional caveats. Whether science *shook off* or simply *replaced* superstition could be debated. Regardless, we live in a far less superstitious age than Columbus did.

With such a revolution in cosmology as Copernicus and his

contemporaries were able to set off, many personal benefits came to humanity. One resulting benefit affects the state of health we all enjoy. This blessing from science is most easily recognized when traveling in some foreign culture (with ever-present *Lomatil*).

Legislation like the Pure Food and Drug Act is firmly based on scientific research, rather than wive's tales and "home brew" cures concocted from superstition. Even Western Europeans (whose foods generally disturb American stomachs the least), when visiting in our home, often express amazement at the relative ease with which one can eat in restaurants from coast to coast in the United States and never fear contaminated food.

Or consider the conquering of diseases that science has made possible. Not too many remember that Alexander the Great, the world's first conqueror, died at 33 of malaria. The Roman Emperor Titus, who sacked Jerusalem in 70 A.D., died at only 41 years of age. Even Columbus' contemporary explorer, Hernando DeSoto, was a victim of a raging fever that took his life at 42. Musicians lament the loss of Franz Schubert at only 31 years from typhus and Wolfgang Mozart at 35 from recurrent rheumatic fever — two brilliant careers prematurely cut short by diseases that science has now brought under control.

One of my many quirks is a long-standing avocation of visiting cemeteries. I think that they are an overlooked gold mine of history. Combining my historical interests with photography, I have sought out small but significant burial grounds such as Sleepy Hollow Cemetery in Concord, Massachusetts, where famous American authors like Nathaniel Hawthorne, Ralph Waldo Emerson, and Henry David Thoreau lie together in the sod. Wandering through an old Jewish cemetery in Worms, Germany, dating to 900 A.D., I was able to photograph, on a cold and miserable winter day, barely-visible engravings on some of the headstones which now tilt all directions.

Some burial places are not nearly so obscure, of course — the Taj Mahal for one. It must claim the honor of being the most pretentious "tomb for two" in the world — the Shah Jahan and his favorite wife who died at an early age while giving birth to her 14th child in 19 years of marriage. I nearly destroyed my photographic budget once I laid my eyes on its beauty. Even graves of Old Testament patriarchs that lie across the Valley of Kidron from the east wall of Jerusalem have been captured by my history-seeking lens.

What could be historically significant about such a morbid avocation? Several things — but the one that most reflects the impact of science is the increase in life span or longevity in modern times. Science has lengthened life for all those who exchanged superstition for scientific knowledge.

Shrank The World

Three years before my father Wesley was born in 1896, man set a new speed record of an unthinkable 100 miles per hour in Locomotive 999. Only 75 years later, and well within my father's lifetime, man had traveled 25,000 miles per hour on his way to the Moon. This increase in velocity of *250 times* the speed record at Dad's birth, illustrates another impact that science has brought to the human race.

It might be easier to picture how much this change in velocity has shrunk the world by a crude illustration. Imagine that the earth was the size of a basketball when my father was born. The escape velocity of Apollo 8 as it left earth orbit to go to the Moon shrank that basketball, in terms of velocity, down to a ball less than 0.04 inches in diameter — about the diameter of the ball in a ballpoint pen! That's some shrink!

An even more astounding reduction that science and technology have achieved is in communications. My mother Pearl recounts the day in 1928 I was born in Spokane, Washington. Dad, a lifelong Democrat, and Dr. D.F. Sells, the physician who was waiting to deliver me in the single bedroom of our modest home, had their ears glued to a small primitive radio most of that very hot June day, as mother advanced in labor.

They were listening to the Democratic National Convention being held in Houston, Texas. Governor Al Smith of New York, the first Roman Catholic candidate for President, was being nominated. It was the first American political convention to be broadcast by radio. By 1948, only twenty years later, such conventions were viewed by millions on television. Another 20 years later, over a billion people were seeing television pictures from the Moon in real time!

Even *nations* are classified in terms of the impact that science and technology have made on their culture and economy. Words like *undeveloped* and *underdeveloped* are employed when referring to the degree that scientific advancements have been adopted or incorporated by such nations. Having lived in Spain for an extended period of time, I find myself cringing when I hear that beautiful country referred to as a "backward" or "underdeveloped" land, as is frequently done in the United States. Many so-called uncultivated nations could teach Americans (the "developed" ones) a great many lessons about human relations, individual worth, and life fulfillment. Nonetheless, science has assumed a dominant role in ranking the maturity of nations.

Another inheritance from science is the broadened awareness of the rest of the world through technological wonders like high-fidelity sound systems or mass printing capabilities for books and other written media. Cultural exposure and cross-cultural sensitivity has increased on such a broad scale that it overwhelms us.

Much of the music of composers like Bach, Handel or Beethoven was ahead of its time, not only in terms of musicianship, but also instrumentally. Those masters never did hear their own music played as well as we do today when we purchase a tape or record! Whereas only royalty or the extremely wealthy ever had an opportunity to hear such wonderful works during the lifetime of the composer, we now can have superlative performances of these master-works at our fingertips by simply flipping a switch.

By facsimile, portions from the rarest of books can be transmitted in seconds across thousands of miles — with no wear or tear on the original. Books that in Columbus' time were laboriously hand-copied and available to only a very few to read, can now be purchased for the price of a meal. And the list could go on and on...

Deduct the Debits?

Unfortunately, the coin of science, like all other coins, has two sides. Not everything that science has contributed to humanity can be viewed as an asset. There have been *debits* as well. Before we can sum the *contribution* of science, we must deduct its *debits*.

The same distinction should be made between *science*, as a process or way of thinking, and the *products* of science that was made earlier as we recounted the success of science.

Science, per se, has no personality or substance. For that reason, it is probably incorrect to speak of either its assets or liabilities. It is more correct to discuss the impact or results that have come to humanity through *technology* — that activity that translates the conclusions that scientists reach into tangible products — whether these products be machines, laws, edibles or environments. Obviously some of these products are commonly seen and touched (like pharmaceuticals) while others are only abstractions of the mind (like the threat of nuclear war).

There is an irrational anti-science mood in some parts of society today. Because it is reasonless, we cannot find its motivation. It has been attributed to many factors — breakdown in hierarchical authority, unfulfilled personal ambition, frustration over self-perceived insignificance — to name a few.

I prefer not to include this mood in the list of debits to be subtracted from the gains of science. On the other hand, there may be contributors to this negative attitude from some of the debits to be discussed.

Let's list a few problems in society which are frequently attributed to science:

1. Anxieties and resultant stress diseases.
2. Thermal, noise, air, soil and water pollution.
3. Malutilization of natural resources.
4. Erosion of moral values (or denial of the spiritual dimension of man).
5. Urban blight.
6. Threat of nuclear war.
7. Massive drug addiction.
8. Depersonalization of the individual.
9. Widespread anarchy, terrorism and violence.
10. Overpopulation.

This list of troubles obviously cannot be charged entirely to science and technology. Many of these perplexing issues are discussed in greater depth from a perspective of human behavior in a paper I was invited to present to the First Western Space Congress.[45]

It would be easy and even understandable for scientists to wash their hands of any responsibility by blaming *non-scientists* for the dilemma this list represents. Nonetheless, a major portion of all these problems (a) did not exist prior to the rise of scientific inquiry, (b) could not exist without the current technological tools provided by scientific research, and (c) appear to be increasing in severity and magnitude with every technological advance.

That portion of each of the 10 problems listed that cannot legitimately be charged against science, per se, could be described as the *moral* or *judgmental* aspect.

For example, scientists cannot be blamed for air pollution. Blame instead the politicians, the automobile manufacturers, petroleum industries, automobile owners — *anyone* but the scientist. He did not force all these various parties to make the decisions they made. He only provided a chemical energy source.

Or how about that hallmark of social progress and sound sanitation, the flush toilet?

> "To a visitor from another planet," Harold H. Leich is quoted in the *Bulletin of the Atomic Scientists*, "it would seem incredible that human beings who are intelligent enough for space travel solve their problems of personal hygiene by putting their body wastes into the public drinking water and then spend billions in futile efforts to restore the water to its original condition."[46]

The typical user of a flush toilet uses about 80 gallons of drinkable water to carry away 1 gallon of body wastes! Certainly scientists can propose ways to re-purify this contaminated water. But the problem started with the help of technology emanating from science. Is the scientist to be *blessed* or *cursed* for this wasteful process? Probably both.

Anything a scientist discovers about nature can be rightly or wrongly applied to humanity. The *application* is a matter of value judgment. And science does not address values — moral or otherwise. Therefore, many scientists resent being charged with the problems arising from application of their findings.

It may be that this dilemma concerning values should be one of the debits — if not the major one — to be deducted from the assets of science. Rene Descartes, the French philosopher-mathematician, who is one of the major figures contributing to the rise of scientific inquiry in the early 17th century, probably was guilty of overestimating the capability of science. His thinking, often called Cartesian rationalization, drew the following criticism from the renowned Swiss psychiatrist and author, Paul Tourniere:[47]

> Cartesian rationalization believed it had attained absolute objectivity by resolutely rejecting all value judgments, recognizing only facts, causes, and effects, and excluding *a priori* all moral judgments. But of course every a priori assumption involves a lack of objectivity. Man's moral drama so dominates the problem of man that if science is forbidden to have anything to do with it, science has no contact with life. It constructs systems which are satisfying to reason, but have nothing to say to the real anguish of man. They leave him to fight his inner battle alone, and he is always defeated.
>
> Cartesianism has brought about a fundamental divorce between the spiritual and the material, and this is the disease from which our modern world is suffering.

Tournier seems to be suggesting that science is limited — that "it's not all that it's cracked up to be." Whether or not science has been oversold, it is definitely limited in its ability to determine truth — one of its avowed goals. Despite all the accomplishments it rightly claims — and even overlooking its contributions to the problems of society, the benefits that science can potentially offer mankind are of a limited nature. In fact, the *limitations* of science are twofold.

The first limitation of science is entirely a constitutional one. Science does not — and never will — constitute *all* truths. Scientific statements involve only one *type* of truth. The truths approachable through scientific methods have been described by H.A. Nielsen as having five characteristics:[48]

1. They are truths supposedly available to the entire interested public. Theoretically, anyone can discover them, only one or a few people are required to discover and publish them for everyone, and they do not need continual rediscovery.

2. They are truths addressed to "whom it may concern" rather than to individuals on the personal level. No one is forced to take note of them as though he had received a personal message.

3. They are truths that, once established, are passive rather than imperative. They simply rest on library shelves, ready for those who may find them useful. Even if they remain unread forever, they do not force themselves on anyone.

4. They are truths that are independent of their discoverer. Since anyone could have discovered them, they carry no mark or imprint of their finder. They could have carried anyone else's name, if they are identified by their revealer's name, just as readily without changing the content or importance of the discovery.

5. They are truths which cannot properly address, scold, or accuse anyone by name, or do anything to force thought or behavior into conformity. Their authority deals exclusively with their content.

These constitutional ground rules for science obviously exclude another large class of truths — truths involving *the individual person.* The latter are sometimes called "Socratic" truths. The Greek philosopher Socrates felt impelled to remind Athenians of both the importance and elusiveness of these truths that reveal a person to himself. Nielsen divides Socratic truth into two parts or types:[49]

1. "Truths that a person carries within himself and which determine how he treats himself as well as others." These truths about what we really think, like the truths of science, are true regardless of anyone else's opinion. They differ, however, from scientific truths because they cannot be discovered and passed on to a person by any other person.

2. "Truths that do not originate within ourselves but supposedly come from outside sources." These truths teach us about our ultimate origin, values, and future. Obviously, outside sources are often conflicting and are backed by various claims of authority — each

> offering to answer ultimate questions. It is this type of Socratic truth that addresses "meaning" which was declared to be outside of science in Chapter 1.

So science offers no solution to Socratic truths. At least, it should not. Obviously, there are scientists who will sometimes forget this constitutional limitation and make dogmatic pronouncements about Socratic truth in the name and authority of science. The issue of origins discussed in Chapter 1 involves Socratic — not scientific — truth. *Science is and must remain scrupulously neutral concerning the origins of the universe, life, and man.*

So much for science's constitutional limitation. A *twofold* limiting of science was mentioned earlier. The second limitation is the *fallibility of scientists.* This limit cannot be as clearly defined as the constitutional one.

However, the *Crucible of Scientific Truth* in the Prologue illustrates it. We need not overemphasize the *humanity* of scientists. That they have bias, prejudice, bad judgment, ego and intellectual barriers that limit their ability to be good scientists is understood. Scientists themselves, unfortunately, also provide ample proof of their fallibility.

Further, just because some scientist may have made a remarkable contribution to scientific truth and have been honored by a Nobel Prize or election to the National Academy of Sciences, he is not exempt from overstepping science's constitutional limits.

In fact, such scientific heroes may be more susceptible than their lesser-known colleagues due to their heady honors. Of them it is most likely to be said, "They don't know that they *can't* know!"

When totaling the score for science, remember that *all* its contributions to humanity do not carry plus signs — deduct the ones with a *minus* sign.

SUCCESS BREEDS POWER

Anyone or anything that has influence, force or authority is considered powerful. *Science* is powerful. On the other hand, it has already been said that reality is only partially accessible by the scientific method. Science has no personality, no tangible substance, no leader, no financial assets, no home office, no nationality, no official voice.

How then can science be *powerful*? What makes it powerful? How did it become powerful?

Ask Hertz Rent-A-Car, IBM, Boeing, Xerox, AFL-CIO, or Volkswagen why they have power in the marketplace. Their answer? Success. They *outperform* their competitors. They *accomplish* their goals. They *attain* supremacy.

Science has done the same.

When you stop to think about it, the success of science is most remarkable. It has quite a circumscribed existence. In studying the nature of the universe, scientists have no capability to change any events of the past. They observe and attempt to interrelate their observations. They seek to find a cohesive explanation that is universal. Yet they cannot revise that which they are observing. If they have a perfect explanation for all their observations save one, they cannot discard, overlook, or change that one.

What happened, happened.

Therefore, nature is not at the mercy of the scientist. Quite the opposite. "Well, Hertz Rent-A-Car is at the mercy of customers," you may remonstrate. True. However, if Hertz management fails to get the customer response they desire, they can change products and become Hertz Rent-A-House or Hertz Rent-A-Suit.

Science is stuck with the universe.

So, in spite of its tougher ground rules, science has been successful and thereby become powerful. The tremendous surge of interest and involvement in science since 1940 is truly astounding. It is said that more than half the scientists who have ever lived are alive today. Much of the concern for science has been attributed to World War II, where technological aspects of warfare overshadowed the role of the combatant. Also in the United States, widespread response to the GI Bill for military veteran higher education resulted in an upsurge of college enrollment in courses related to science and technology. This exposure, in turn, significantly increased scientific literacy and understanding.

Subtle Shadows

There are several pitfalls that accompany success. One is *disorientation* — losing touch with values, ethics and mores of the real world. Another is *pride* — an exaggerated estimate of your own worth. Frequently, *insensitivity to others* also accompanies success. Perhaps the ultimate pitfall — one that has snared military geniuses, great politicians, business tycoons and even famous athletes — is *induced infallibility*.

This pitfall is insidious — it gives no warning, comes on imperceptibly and subtly, and is deadly. By the time the victim might sense it, it is already too late.

Infallibility is so seductive that it often appears to be totally justified in the mind of the seduced. Rather than making a frontal assault whereby it could be easily identified, infallibility slips in on the blind side. The dazzling, blinding light of success in which science has basked in recent years has begun to produce subtle shadows of infallibility.

One of the early spotters of science's subtle shadows was a British chemist, Anthony Standen. He published in 1950 a brilliantly amusing, highly informative debunking of science entitled, *Science Is A Sacred Cow.*[50] It was received by the scientific community with about the same gratitude that General George S. Patton showed one of his combat-fatigued soldiers in a Sicilian hospital tent in July 1943.

From time to time since Standen's alert, there have been other authors — in and out of science — who have attempted to expose the onset of infallibility in science. Biology, probably because of its preoccupation with and almost total dependence upon a belief system that can be neither proven nor falsified — "That all living things form an unbroken chain from a unique source," seems to be the most frequent target for accusations of infallibility.

One book that disclosed fallibility of biologists was published in 1960 by a British professor in the Department of Physiology and Biochemistry at the University of Southhampton, G. A. Kerkut.[51] This treatise proposed that there are seven basic and critical assumptions that biologists make which are not capable of experimental verification; i.e., they cannot be tested.

Therefore biologists *believe* these seven pre-suppositions as a matter of *faith.* The obvious similarity between these beliefs and the beliefs into which Copernicus ran headlong during the Middle Ages is discomforting. Even worse, I have not been able to get any biologist to even *read* Kerkut's book! It seems to be on their "banned books" list.

A graduate of the Harvard Law School, Norman Macbeth, published a critique of the biologist's belief system from a quite different perspective in 1971. He described his approach as follows:[52]

> Courtroom experience during my career at the bar taught me to attach great weight to something that may seem trivial to persons not skilled in argumentation — the burden of proof. The proponents of a theory, in science or elsewhere, are obligated to support every link in the chain of reasoning, whereas a critic or skeptic may peck at any aspect of the theory, testing it for flaws. He is not obligated to set up any theory of his own or to offer any alternative explanations. He can be purely negative if he so desires.

While he was studying biology as an outsider, Macbeth was quick to discern that biologists reacted not unlike any other party who claims infallibility:[53]

> I have been rather surprised to discover that many biologists dispute the propriety of a purely skeptical position. They assert that the skeptic is obligated to provide a better theory than the one he attacks. Thus Professor Ernst Mayr of Harvard rules out admittedly valid objections on

> the ground that the objectors have not advanced a better suggestion. I thought at first that this was a personal foible of Mayr's, but it has recurred in so many other places that it must be a widespread opinion.
>
> I cannot take this view seriously. If a theory conflicts with the facts or with reason, it is entitled to no respect...Whether a better theory is offered is irrelevant.

When I have attempted to test the open-mindedness of biologists by suggesting that they read Macbeth's book, they ask, "Is he a *biologist*?"

I am obligated to say, "No, he is an attorney."

Their immediate response is, "How could he *possibly* know anything consequential about biology?"

I wonder how the monks responded to Copernicus . . .

In Chapter 1, considerable emphasis was placed on the inability of scientists to address the "meaning" of anything they observed. This point was reaffirmed earlier in this chapter when Nielsen separated Socratic truth from scientific truth.

This exclusion is a tough blow for many scientists. Some who can't endure this limitation ignore it. At that point, infallibility begins to emerge. Therefore, in its scientific version, infallibility can be detected by the implication that "science will soon have answers for everything." I have had recent discussion with some biologists and chemists who, with straight face and apparent sincerity, insist that morals and ethics will soon be *scientifically* determined!

Omnipotence of Objectivity

I often ask my students in the university classroom, "What do we mean by the term *objective*?" Without much thought, they immediately respond with, "True, certain, factual, *scientific*..." They are wrong. But why do they almost always answer the question that way?

Actually, "objective" means anything external to the person who is observing it — or distinguishing between something known or perceived by a person and something existing only in the mind of that person.

Therefore, "objective" could be considered to be detached, impersonal, and free from bias or prejudice. From there, it is only one more small step, then, to "scientific."

However, it is a giant step from "scientific" to *true*. Remember that science consists of only one *type* of truth. But to those who would oversimplify, the Middle Ages were "subjective" times as contrasted with modern times which, thanks to science, are "objective." *Ipso facto*, "subjective" is uncertain, prejudiced or judgmental while "objective" is true, certain, factual and scientific.

Untrue as this generalization may be, it is nevertheless a popular view.

Science can be defined as "a body of important truths that have been *objectively* determined." Nielsen's five characteristics of scientific truth listed earlier describe what is meant by "objectively determined." Knowledge so obtained has particular value to all of us because it is equally *applicable* to all of us.

Contrasted with political, social, or religious truths (that obviously do not apply to everyone), scientific truths should not suffer internally from sectarian tensions or opposing parties. Part of the power of science emanates from its distinctive singularity, unity, and universality.

The subtle shadows of infallibility among scientists portend an end to this solidarity of science. Science is beginning to have its Democrats and Republicans, too. One issue that threatens to split scientists is related to *objectivity*.

Scientists are in great disagreement over how science "*objectively* determines truth." Some feel that the concept of objective determination should be *expanded* and become more complex while others would like to see it *reduced* and simplified.

The retiring President of the American Association for the Advancement of Science in 1965, Don K. Price, noted that, at that time, there was an intellectual rebellion over how scientists determine truth. He said that many scientists were resisting "the change from systems of thought that were concrete but complex and disorderly, and that often confused what is with what ought to be, to a system of more simple and general and provable concepts."[54]

In other words, scientists rankle at being hemmed in by the ground rules Nielsen specified earlier. They would like to stretch or expand beyond "what *is*" over into "what *ought to be.*" Of course, this involves a clear invasion into the territory of Socratic truth.

Gerald Holton, a Harvard physics professor who also bridges between history and science in his writings, has given some fancy names to the two groups that are involved in this rebellion. That group of scientists which is urging an *expansion* of the traditional or constitutional view of science Holton calls "New Dionysians." Holton describes this group as follows:[55]

> With all the differences among them, they are agreed in their suspicion or contempt of conventional rationality, and in their conviction that the consequences flowing from science and technology are preponderantly evil. Methodology is not their first concern; they think of themselves primarily as social and cultural critics. But they would "widen the spectrum" of what is considered useful knowledge as a precondition of

other changes they desire. They tend to celebrate the private, personal, and, in some cases, even the mystical.

The opposition to New Dionysians Holton names the "New Apollonians." They are depicted thus:[56]

> They advise us to take precisely the opposite path — to confine ourselves to the logical and mathematical side of science, to concentrate on the final fruits of memorable successes instead of on the turmoil by which they are achieved, to restrict the meaning of rationality so that it deals chiefly with statements whose objectivity seems guaranteed by the consensus in public science. They would "shrink the window" emphatically, discarding precisely the elements which the other group takes most seriously.

On one hand, it appears that the New Dionysians want science to be *widened* in scope so that other types of truth may be gathered under the umbrella or label of *science.* They do not want to accept, perhaps, the idea of total truth (or reality) being divided among several sources. The New Dionysians also would champion the cause of social sciences being given equal status with the physical sciences. This integrative, all-embracing effort would increase the power of science and is typical of holistic thinking. It is branded, however, by the New Apollonians as *irrationality.*

On the other hand, the New Apollonians seem to want to *narrow* the scientific definition of truth sufficiently that all traces of mysticism, metaphysics and other fantasies would be excluded from science. They are purists, somewhat analogous to political conservatives. Their thinking is branded, in derision by the New Dionysians, as *rationality* – a cold, dehumanizing activity.

Since this book is predominantly directed at how science textbooks present science, I would quote Holton's observation of what these two opposing parties contribute to textbook preparation:[57]

> In truth, those scientists who write textbooks all too often make matters worse by providing distortions of their own, namely by presenting severely rationalized versions of the scientific process...Usually, they provide indeed only "candles of information"; at worst, their products appear to have the character of unchallengeable, closed (and therefore essentially totalitarian) tracts. It may well be that the anarchic reconstructions of science on the part of the new Dionysians are largely reactions to the view of science that emanates from the rationalistic reconstructions of the new Apollonians and of text writers — for that is perhaps all they (like most students) ever get to see of "science."

So, although science is based on objectivity, how *objective* it should be cannot be objectively resolved. In one direction (that of the New

Dionysians), scientists appear to invade the territory of Socratic truths by incorporating value-determination within science. In the other (that of the New Apollonians), scientists seem to urge that values be excluded from consequence within science.

My personal reaction to this controversy over objectivity is mixed. I find myself aligned with the New Dionysians in their observation that there is much of reality in life (including the whole value hierarchy of society) that scientists are ignoring. However, I join the New Apollonians in their fight to keep scientists from meddling in Socratic truth — because there are no objective grounds for such meddling!

It would be an ironic turn, indeed, for scientists to incorporate within the body of science the worlds of philosophy, metaphysics, and religion after their turbulent struggle to extricate and then disassociate science from these schools of thought. Only if there were to be demonstrated some capability (as yet unknown to me) for scientists to objectively determine "good," "bad," "evil," and other values would I support those who are urging that the boundaries of objectivity be expanded.

As earlier mentioned, I completed my undergraduate work in physics — one of the so-called "hard" or physical sciences. The chauvinism of physicists and chemists, with regard to the "soft" or social sciences, has always amused me.

The pride of the physical sciences lies in their higher degree of rationality or objectivity. Yet in the more advanced specialties within the field of physics — for example, the physics of high-energy particles — the experimental work is not conducted in a manner where any of the five senses can observe results. Instead, the sophistication of testing hypotheses demands abstract inference gained only through modeling.

To illustrate, even though a young child today can draw a picture of an atom (a nucleus with electrons zipping around it in orbit), such a picture is not a representation of anything that any scientist has ever seen. It is an *imaginary* model. It may *appear* objective because it has no human values of behavior associated with it (like the social sciences do), but it is simply fantasy — *not* true, certain or factual.

Allen D. Allen, a theoretical physicist quoted earlier, published a controversial paper, "Does Matter Exist?" in 1973.[58] In it, he proposed that the traditional "basic unit of matter" idea — that there are elemental particles like molecules, atoms, electrons, neutrons, *et cetera* on down to the still-sought "quark" proposed by physicist Murray Gell-Mann — should be abandoned. There seems to be no end to the search for "the most basic unit of matter." Allen has an alternative:[59]

> Rather than using *objects* — such as particles — for raw material, nature uses the fundamental *laws* of physics — such as the law of the conservation of momentum, which states that the faster you drive, the harder it is to stop. So long as we obey these laws, we can produce any matter we have the equipment to create. This concept, that ultimately the world is constructed from principles rather than from units of matter, is almost theological in character. Yet it is now an established (if competing) theory in the mainstream of theoretical physics.

While *objectivity* has been a source of power for science, it has not made science omnipotent — if for no other reason than because it cannot be totally attained. Rationalistic thought must start with assumption, and such assumptions always involve the absence of objectivity.

Therefore, it is entirely proper as well as necessary for scientists to say, "I don't know." An astronomer at the University of Minnesota, William J. Luyten, expressed this point well in *Science*:[60]

> During the past few years impressive evidence has been obtained about the existence of bodies with masses not much larger than that of Jupiter circling around other stars, but we do not yet know whether these are planets, or star-like objects, or different from either — they cannot as yet be seen. Arguing from general principles one might say that life could well exist outside Earth, but it seems to me that the only definite statement that is now scientifically tenable is that we do not know: we can neither prove nor disprove it.

Summarizing the power of objectivity, it would be well to note the following excerpt from an editorial in another technical journal:[61]

> Misleading are some of the terms associated with scientific endeavor, such as "prove," "demonstrate," "establish," etc.; terms which have come to suggest elimination of all doubt or question; final, unambiguous "fact" or "truth"...For example, "prove" really means to "test," to "try," to examine — *not* to eliminate all doubt. Science is a continuing *process* of observation, speculation, testing, checking, and questioning. Science does *not* create or "establish" truth; it endeavors only to discover it.

Stunted by Power??

Comparative anatomy is that branch of science that studies and classifies the anatomical or structural similarities of various animals. By noting these *similarities* between animals that are otherwise quite different, scientists attempt to determine the function of structures that might not be evident by independent observation. Common functions that may be inferred by similar *anatomy* are sometimes taken a step further to infer common *ancestry*.

Embryology, another branch of science, studies the unborn fetus or embryo of animals as it develops from fertilization to birth. Because the embryo changes as it develops, embryologists attempt to classify or divide this pre-birth development into distinct stages. These stages, like the similarities in comparative anatomy, are then used by some scientists to infer many possibilities of interrelationship between otherwise diverse groups of animals.

Although both of these branches of science are havens for those who generate much of the mythology regarding origins that has been denounced in Chapter 1, my intent in mentioning them at this point is to introduce a similar type of analogy regarding the influence of science on the *history* of man.

This analogy is credited to Blaise Pascal, a 17th century French philosopher with whose name are associated a number of principles in the mechanics of fluids. Tournier, the Swiss psychiatrist cited earlier, quotes Pascal as saying:[62]

> All the generations of men, following each other in the course of so many centuries must be considered as one man who continues always to subsist and is constantly learning.

Tournier adapts Pascal's statement into the context of a psychiatrist's analysis of the history of a single life. Just as the comparative anatomist and the embryologist have divided and classified observations in their fields, Tournier divides the history of man into stages or ages that a single individual undergoes as he progresses from birth to death.

In a chapter entitled, "The Inner Conflict of Modern Man," he conducts a hypothetical analysis of a person (who represents the history of man) requiring healing.[63]

When a sick person consults a psychiatrist, he is normally asked to reveal what he knows of his childhood and his adolescence. In other words, it is necessary to understand this person's development in the same manner that the scientists in comparative anatomy and embryology analyze developmental stages.

Here is Tournier's conclusion of the first stage in mankind's development:[64]

> The childhood of mankind is Antiquity. Our sick person was a child prodigy. Antiquity has all of the characteristics of a child prodigy, who seems to be able effortlessly and spontaneously to discover the purest, the truest, and greatest treasures, especially in the realm of art and poetry and dreams, just as if complete masterpieces came gushing forth from his childlike, innocent soul...Childhood, Antiquity, is the age of poetry.

Tournier then goes on to compare the Middle Ages with the *school age* of man. As we all know, a child is taught during his school days under a system of authoritarianism and obedience. He accepts without argument what his parents and his teachers tell him. Also, it is generally during the school age that whatever religion he is going to have, he accepts.

So it was with the Middle Ages. Mankind grew up with a thought system that the church imposed on it. With a childlike lack of criticism that is prevalent during the school age, it accepted all teachings of the church without criticism and assumed that its teacher was without fault. We all recall those days for ourselves when we believed that those instructing us were perfect and knew everything. While we may have disobeyed these people, we still did not question their authority.

At some point in this school age, there emerges an independence that we associate with the next stage — adolescence. Here is Tournier's description of that age:[65]

> A flood of new knowledge, the intoxication of learning, and the yearning for personal experience confront the adolescent with a thousand problems, which, so it seems to him, his parents escaped. He rises up against them; he revolts. He demands the right to think for himself and not in accord with the system of traditional thought, the right to follow his own opinions rather than the authority of others. He sits in judgment upon his parents and finds that they themselves do not practice the morals they inculcate in him. He argues about everything and exults when his parents confess that they have no answers to the insatiable questions he asks.
>
> Can we not compare this crisis of adolescence to that which was set off by the Renaissance?...Since the Renaissance, mankind has taken the opposite view of the world from that which had been taught by Antiquity and the Middle Ages. For a spiritual, religious, and poetic view of the world it substituted a scientific, realistic, economic view.

Adolescence, for most of us, is remembered as a turbulent time of our life. For many, it was a time of passionate study as well as entertainment of extreme and contradictory doctrines. It often resulted in violent reaction to parental claims — particularly those rules that parents handed down that appeared rigid, logical, or derived from faith.

In addition, the young adolescent generally discourages the *values* in which he was raised. There is scoffing of parents. Moral and social conformity is seen as hypocrisy. Everything represented by the people he admired in his childhood is now denounced with bitter resentment.

Tournier compares the centuries man has lived since the Renaissance to these critical years of adolescence. Adolescence is not unhealthy. Neither is it evil. Such a crisis time in life is both necessary and normal.

However, the day comes when the adolescent begins to discover many of the treasures of his childhood. He will often return to the faith in which he grew up and the principles that he learned in childhood. In a sense, he *rediscovers* these truths. In so doing, he adopts them as his own personal views, rather than those of his parents or teachers. Psychologists describe this phenomenon as *integration*.

If integration is delayed, however, there can develop an illness. Psychiatrists define the illness as the "neurosis of defiance." Tournier proposes that the development of human history has reached this point. It is not normal adolescence in which the world finds itself today. Mankind, in many respects, is sick.

Whether or not you agree with Tournier in his diagnosis, it might be well to consider the reasons why he has decided that the *power of science* has stunted human development and induced the "neurosis of defiance."

Tournier bases his diagnosis on four major points:

1. *Neurosis is primarily distinguished by anxiety.* The neurotic will often parade behind a facade of bravado while he hides within himself a deep anxiety. He may or may not be aware that he is entirely dissatisfied with himself. One only needs to listen to the six o' clock news every night on television to recognize both the bravado and the deep anxiety of modern man. Jean-Paul Sartre seems to accurately describe our day and age with the saying, "Man is anxiety."[66]

2. *A further characterization of neurosis is its sterility.* Neurotics may have big dreams, but these are merely compensations and escapes. They do not produce anything useful, and they do not relieve the person from his anxiety. Tournier observes that the modern world has some true values — elite, literary, artistic, spiritual — which have been set aside (if not *abandoned*) and not allowed to play an effective role in society's destiny.

3. The most tragic aspect of neurosis is that the very effort expended by the neurotic to save himself actually ends up destroying him. If neurotics try to open themselves up to anyone close to them in order to end misunderstanding, they go about it in such a manner that the misunderstandings are multiplied and their loneliness is aggravated. The modern world suffers this same paradox in that the efforts it makes to save itself bring it to ruin. The very actions the world takes to avert war cause it to plunge into war. When attempting to guarantee material security, these very endeavors disrupt the economy and increase its misery. Social activists, attempting to free man from slavery, plunge the world into struggles which are more frequently worse than the original condition. Neurotics desire to be loved so much that they do things that stir up hatred of themselves.

4. *The source of neurosis is an unconscious inner conflict.* Carl Jung once said, "The neurotic is sick because he is not conscious of his problems."[67]

> The psychiatrist attempts to help the neurotic advance beyond his outward *apparent* problems to his inmost *real* problem. Mankind today thinks that it has eliminated the world of values, the world of poetry, the world of moral consciousness. But they are only *repressed* — temporarily out of view, only to pop back at most inopportune moments.

One can easily disagree with Tournier. Perhaps he is being too simplistic. Undoubtedly, he cannot legitimately claim to see the world in its entirety more accurately than any other person. Yet, much of what he proposes makes sense.

Consider one more time the Renaissance. That period of time is almost universally characterized by historians as an abrupt rejection by humanity of most of what had been allowed to guide it previously. There is not only *observation* but also *endorsement* of the idea that value judgments had been rejected. Likewise, metaphysical intuition was no longer to be trusted after the Renaissance. Poetic inspiration, supernatural revelations, and all other subjective evaluations were to be discarded so that civilization could build solely on the material realities and objective knowledge produced by science.

Today's world appears to be solely determined by economics, science, technology, and politics. Philosophic, artistic, moral, and religious problems are debated by specialists in obscure corners, but there is no feeling by the public that these debates influence the destiny of society.

On the other hand, such problems have not been able to be wholly suppressed in the minds of individuals. Tournier suggests that they have merely been repressed into the unconscious. A person is a neurotic when he has repressed something without having really eliminated it.

Could the modern world actually be stunted into a "neurosis of defiance" by the success-bred power of science?

POWER BREEDS DANGER

America's Founding Fathers had a rather funny idea. "Let's break up the power of government into three parts instead of letting one leader have it all," they agreed. For them to think that way, their history books must have had some missing chapters.

Hadn't the Founding Fathers ever heard of all the great nations and alliances that had strong leaders in whom all power was centralized? What about the empire of Alexander the Great or the Roman Empire under the Caesars? Did the Founding Fathers deliberately want America to be *weak* instead of strong like these empires? Were they ignorant of the saying, "Divide and conquer"? Maybe they had no aspirations for the future of America! Were they the first ones to believe Volkswagen's motto, "Think small"?

It's not likely that these men were advocating weakness or attempting to guarantee impotence. In fact, their writings reveal a strong belief that their idea would produce a long and prosperous posterity. Well, did they see a virtue then in dispersed, unfocused leadership? With such diffused leadership, to whom would the citizens look for inspiration? And whom would they be able to blame for failures?

Rather than *virtue* in scattered leadership, it was the *vice* of centralized leadership that inspired the Founding Fathers to make their unusual agreement. Yes, they *had* read of Alexander the Great and of the Caesars. In fact, it was their *first-hand acquaintance* with powerful tyrants that persuaded them to advocate an approach with the earmarks of being slow, unresponsive, uninspiring, clumsy and unmanageable. And it was worth risking all those attributes just to avoid the *dangers of power.*

But power, per se, is not *evil.* Powerful men have done so many *good* things with their power. While living in Germany, I learned to appreciate many admirable things that Adolf Hitler did with the power he had. The network of superhighways known as Autobahns that he established was an idea that most other nations have since copied. He breathed a spirit of pride into the German people through pageantry, art and music. His work programs provided employment for the unemployed. He heavily supported science and technology.

But, of course, Hitler also used his power in other ways...

Due to the potential perversity of those who hold power — rather than the properties of power itself, our Founding Fathers showed great wisdom in diffusing the "terror of tyranny" before it could get established in America. That's the time to pull the stinger — before the power structure gets built. After power has been granted, it's like "taking candy from the kids" to get it back to the diffused state.

It's already too late for science. Science, as we said earlier, is already powerful. Its success bred power. But the breeding process continues. Power also breeds. It breeds *danger!*

Tempting Temples

Thailand should really be called Temple Land. Temples are everywhere! One day, my wife and I climbed the ornate and colorful main prang (spire) of Wat Arun, Thailand's Temple of Dawn, to look across the Chao Phraya River at the almost indescribable sight of the prangs of over 300 Buddhist temples in Bangkok. But that forest of prangs revealed less than 1% of Thailand's temples. There are over 30,000 such temples in that country!

Not all temples have prangs that make it easy to see and identify them. In fact, temples vary greatly in architecture, depending on the god or gods they honor or who is believed to reside in them.

There are even temples that have no walls or other visible structure! They are just as real as though they were visible, however. Within these shrines, worship is just as devout, the veneration is just as sincere, and the piety is just as fervent. They, too, have their fetishes, graven images, and icons. Sacrifice is offered on unseen — but very real altars.

Have you ever seen *Temple Science*? Maybe not. But it exists! It is difficult to see with the naked eye because it is one of the invisible ones just described. It is not nearly as old as Wat Arun in Bangkok. Only in modern times did some scientists begin erecting it. How worship is conducted at Temple Science will be described in later chapters. Sacrifice on its altar will be illustrated by specific incidents. Not only will the reverence of scientists be depicted but also their demands that all men everywhere pay homage with them.

"How *ridiculous* can it get? It sounds like an allegory gone ape," I can hear some disturbed scientist say. It could be. But I wasn't the first person to spot Temple Science. Over a decade ago this statement appeared in *Science*:[68]

> Science may have to enlarge its house, to accept that it is not a temple but a kind of rambling, unfinished, temporary shelter.

Perhaps I may be misreading Arthur Kornberg, the Stanford biochemist who shared the 1959 Nobel Prize in medicine for his work in synthesizing DNA, but it seems that this statement also has a worshipful tone:[69]

> What is remarkable is that science enables ordinary people to express their creative talents in a global and purposeful way. Their humble probings, so picayune individually, combine to exert irresistible forces in exposing the grand designs of nature.

Am I alone in detecting a parallelism between Kornberg's science and theological hierarchies? Did not the medieval church enable "ordinary" Martin Luther to express his creative talent in a global and purposeful way? Or did not Thomas Aquinas, John and Charles Wesley or perhaps Billy Graham — all very ordinary men – combine their probings with others "to exert irresistible forces in exposing grand designs"?

Even the news media occasionally detect or connote a theological twist to science:[70]

> Laymen — and most of us in this highly technical world are laymen — are at the mercy of researchers and technicians who set themselves up to interpret their particular part of nature to us. Over the years, they — the *high priests* of science and technology — have convinced us that we should leave the fact-finding to them. And to a large part we have. In return, they have promised to be "objective," to tell the truth no matter how or whom it hurts. With the truth, freshly gathered from the mouths of the experts, politicians can make laws, and laymen can form opinions and make decisions with confidence. All is fine, until one of the experts decides that he knows more than the decision-makers and that he no longer need be objective. That he can be *preacher* as well as researcher. (Emphasis added)

The more I observe scientists, the more I become aware that they too share the common traits that they have assigned to everyone else but themselves — defense of tradition alone, close-mindedness, commitment to a belief system, hostility to challenge, and blind faith.

Some scientists are so quick to assert, "When a scientist *believes* something, he does so on a completely different basis than a *religious* person." Or, "If we scientists have faith that such-and-such will happen during a laboratory experiment, *our* faith has no relationship to *religious* faith."

Let's try to examine any difference by defining faith as follows:[71]

> Faith means putting our full confidence in the things we hope for; it means being certain of things we cannot see.

What, may I ask, is the distinction between the faith of (1) a nuclear physicist's experiment in a linear proton accelerator in California where he is seeking to find a short-lived nuclear particle reportedly observed previously on one occasion in an experiment in New York and (2) a child kneeling in prayer beside her bed, asking God to answer, one more time, a prayer answered on previous occasions? Are they not both exercising the same kind of faith?

We must be careful, obviously, to avoid carrying this analogy too far. For example, their faith is based in two different premises — the physicist in a report and the girl in a Person. But with respect to *faith* itself, there is no difference.

Both the physicist and the child possess a mental commitment and take action based on prior evidence that they accept as valid. Why, then, the emphatic insistence on some undefined difference? Personally, I would be pleased to see a reduction in the hostility and contempt that has traditionally separated science and religion. Recognizing this common reference point of faith could help.

It should be obvious that, by discussing their commonalities, I am not proposing a merger of science and religion — only an understanding that could lead to a reduced suspicion between the two. To that end, it would also be helpful if both parties would quit judging the other as though they were in the 19th century.

There is about as much difference between 19th century *religion* and that of today as between 19th century *science* and today's version. Scientists often quote the 19th century version of religion as though it represented current thought, and churchmen are guilty of the same mistake about science. Alexander treats this subject well.[72]

The role of tradition, common to both science and religion, needs to be reassessed. The scientist may be quick to recall how Copernicus suffered almost exclusively because of religious tradition. On the 500th anniversary of his birth, the National Academy of Sciences commissioned a volume of essays on the meaning of Copernicus' work. The editor of this volume, mathematician Jerzy Neyman, had an interesting observation about tradition:[73]

> Along with other domains of human behavior and activity, (scientific) research is also affected by routines of thought and, particularly, of premises. Here again many routines are very useful by providing important economy of mental effort. But again there are exceptions. It does happen that some premises of our thought, acquired through traditional learning in schools and universities, have no other backing than tradition. And the longer the tradition of a commonly accepted premise, the more difficult it is, psychologically, to notice its routineness and to question its validity...Cases when a generally accepted premise of thought is suddenly identified as a mere dogma with no backing but tradition occur even now in many domains of study. Ordinarily, after an unavoidable period of resistance on the part of some scholarly establishment, a new fruitful field of research is opened.

To continue to trace this parallel between science and religion is not essential. In fact, it is my intent to simply point out the resemblance of the awe afforded science to that associated with theological trappings.

If there had been no religious temples to emulate, I propose that many scientists would be inclined to create one for science. Even in those situations where scientists would not have erected Temple Science, non-scientists have tended to do so for them. Due to their isolation from scientists, the public creates myths that would attribute super-human power to scientists:[74]

> The public must realize that scientists are human and that "experts disagree" in this supposedly objective realm just as they do in art, politics, and ethics. More scientist-layman contact is the only way to destroy the myth that scientists are, depending upon the latest horrors of CBW or miracle cures from the laboratory, either gods or devils.

Thoughtful scientists must constantly maintain a two-pronged attack against erection of Temple Science.

First, they must discourage the public from constructing the edifice in gratitude to them. The humility required for this resistance does not come easily.

Secondly and perhaps more importantly, they must be vigilant that their over-zealous colleagues in science do not build such a temple on their own. The unfounded zeal that needs to be moderated by responsible scientists to preclude temple construction is illustrated by two examples.

The first is by Captain Jacques-Yves Cousteau, the famed French oceanographer:[75]

> It is reasonable to predict that man will tame the forces of nature, the forces even of the universe. He will master gravity and send the apple back up into the tree. He will exchange matter and energy between normal and anti-matter universes. He will protect the moon by changing its orbit, and ultimately push stars around.
>
> Perhaps the greatest achievement will be to change life itself and thrust himself into immortality. Immortal man will take on the task of repairing planet Earth from the ravages of the earlier mortals — us. He will start the process of evolution again when he has cleaned up the oceans and the atmosphere...There will be no violence in this world of the future. Even today, some scientists claim they could suppress violence in humans at birth if they were permitted to do so...Science has all the keys to save the future for mankind.

Isaac Asimov, a biochemist who has written over 100 books on science and science fiction, provides the second example. In an article discussing *cloning* (the production out of a body cell of an individual of a whole, new individual with genetic equipment identical to that of the original), Asimov proposes that the human body, like an automobile, could have its natural lifetime extended by the availability of spare parts:[76]

> And think of the psychological effect! Each person will know that his or her heart, kidneys, lungs, glands are not all there is; that there are replacements awaiting the need. What a vast load of fear and anxiety will be removed.
>
> Remember, too, that the new organs can supply new cells for new cloning. Assuming that all this works (and, of course, we can't be sure that there may not be insuperable difficulties that develop), would it mean that human beings will live forever?

There may be no way to forever preclude the building of Temple Science. The power that success bestowed on science may demand such recognition. Man's need to worship something — even something of his own creation — seems to be incurable.

How ironic that the temple be named *Science*!

Hang Heretics?

Not only do scientists, as well as non-scientists, try to build temples for science, but some scientists are considered *heretics* by their colleagues. As you may recall, a heretic is a person who holds beliefs opposed to orthodox or official doctrine.

Scientists are not bashful in recalling how the medieval church persecuted heretics like Nicholas Copernicus and Galileo Galilei. They often point how important, though tragic, it was for those heroes of science to stand firm in their conviction that the orthodox or official position of the church was in error.

They decry the stupidity of church leaders who refused to examine facts, and who, instead, used their positions of authority and power to stifle the truth.

They further claim that without an open forum, where all ideas are examined freely and without prejudgment, mankind is doomed to superstition and darkness. And they are absolutely correct. We *all* know that...

Of course, heresy is *nasty*. It's like hard, dried egg yolk stuck on a supposedly clean dinner fork. It must be dealt with. There is no gracious way to wipe it off. You can't slip the fork under the table and quietly clean between the tines while no one is looking. Someone is bound to spot you.

Heresy demands *action*. And the options are never nice. On one hand, you can persecute or murder the source. On the other, you can agree to examine the heresy. Not much choice in between. Neither choice flatters orthodoxy or the established truth. Oh certainly, one can *ignore* the heresy, but like the egg yolk on the fork, it doesn't just go away. The longer the heresy persists, the more embarrassed the official "truth-sayers" are likely to become.

Unfortunately for heretics, those in power find it much easier to *attack heretics* than to *consider the merits of the heresy.*

In many respects, heresy is like *crime*. It is committed by a small minority against a vast majority. It frightens the majority. The majority reacts and defends itself against it. Yet, the majority still desires to be fair with criminals.

So maybe it would be helpful to propose a "heretic justice system" analogous to our "criminal justice system." Without doubt, it would have been much easier for Copernicus and Galileo if such a system to insure heretic justice had existed in their day. To set up a heretic justice system, certain basic questions would have to be resolved:

1. What constitutes heresy — is it just *any* disagreement with official consensus?
2. What constitutes *official* consensus (that from which heresy supposedly deviates)?
3. Who *establishes* the official consensus position?
4. On what *basis* is the official consensus position established?
5. Whose *responsibility* is it to level charges of heresy?
6. Is the heretic *innocent* until proven guilty or vice versa?
7. What courtroom or *process* exists for resolving heresy?
8. What is the *goal* of the heretic justice system – punishment or rehabilitation?
9. If punishment is the goal, what *types* are sanctioned for heretics?
10. If rehabilitation is the goal, *how and by whom* is the heretic rehabilitated?

The first difficulty in establishing a heretic justice system — especially for scientists — may be in proving my earlier statement, "Some scientists are considered heretics by their colleagues."

Not *scientists* crying "heretic!" Surely not in this *enlightened* age! Besides, scientists are well known for their *open-mindedness* and willingness to entertain even the wildest speculations. They are *objective. They* don't have a belief system held together only by tradition, superstition, threat, and fear of excommunication.

Perhaps that's right. But let's check the record, just to be sure.

For some readers, it may seem odd that the cases of Immanuel Velikovsky or William B. Shockley have not been discussed until now. Their names automatically flash "heretic" in many minds. Even by discussing them at this point, I do so only in the limited sense of illustrating the *possibility* of current scientific heresy — not by way of endorsing their respective positions. Lack of endorsement, however, is not based on my fear of being allied with dissidents, but on my unfamiliarity with details of their work.

Consider first Immanuel Velikovsky, a Russian physician and psychoanalyst born in 1895. He published a book, *Worlds In Collision*, in 1950 that rewrote ancient history, questioned the concept of an immutable solar system, challenged the amoeba-to-man idea, and made statements that "the scientific community" found outrageous.

A popularized account of Velikovsky's impact through his book and subsequent ones he wrote like *Ages in Chaos* (1952) and *Earth in Upheaval* (1955) appeared in a recent *Reader's Digest.*[77] For those interested in studying in depth the reaction of prominent scientists to *Worlds in Collision*, there are some thoughtful documentary articles in *Pensee*, a quarterly published by the Student Academic Freedom Forum.[78]

The magnitude of antagonism and outright warfare that erupted when Velikovsky's "worlds" collided with the Congregation of the Holy Office of Science — that tribunal of self-appointed defenders of scientific truth — cannot be properly sketched in this book. A synopsis of one aspect is included for illustrative purposes.

More than a dozen publishers had refused to publish *Worlds In Collision* at the time the Macmillan Company published it in 1950. Furious controversy developed over the book even before its appearance. Academicians and scientists — those who write and buy textbooks — threatened and intimidated Macmillan to the point that they transferred publication of the book to the Doubleday Company. It was already the nation's number one bestseller at the time of transfer!

One of the leaders of the scientific inquisition of Velikovsky was the astronomer Harlow Shapley, Director of the Harvard Observatory. He probably deserves the credit for forcing the then President of Macmillan, George Brett, to transfer *Worlds In Collision* to Doubleday. In addition, the Macmillan editor for the book, James Putnam, who had been with Macmillan 25 years, was fired over his decision to publish Velikovsky's book.[79]

Here's how scientific persecution works. Shapley wrote Putnam on 25 January 1950 this "subtle" message:[80]

> It will be interesting a year from now to hear from you as to whether or not the reputation of the Macmillan Company is damaged by the publication of *Worlds In Collision*...Naturally you see that I am interested in your experiment. And frankly, unless you can assure me that you have done things like this frequently in the past without damage, the publication must cut me off from the Macmillan Company.

Yet, only nine months later, Shapley denied in the *Harvard Crimson* any part in suppressing publication of the book:[81]

> The claim that Dr. Velikovsky's book is being suppressed is nothing but a publicity promotion stunt...Several attempts have been made to link such a move to stop the book's publication to some organization or to the Harvard Observatory. This idea is absolutely false.

Parenthetically at this point, I wish to express my gratitude to Macmillan for not only choosing to publish *Science But Not Scientists*, that is also controversial but differently so than Velikovsky's — but also for including this account of intimidation over a quarter century ago.

Of course, the management responsible for that decision has not been with the company for many years, but it is my hope that this book will restore to Macmillan whatever reputation for courage and freethinking that it may have had damaged by that earlier decision.

Returning to the sordid sequel of persecution, the scientific illuminati were not through with Velikovsky after they forced the transfer to Doubleday. Another prominent astronomer, Fred Whipple, who succeeded Shapley as Director of the Harvard Observatory, threatened the Doubleday Company on 30 June 1950:[82]

> ...Oddly enough, in its anti-scientific account of the book, *Newsweek* has unwittingly done the Doubleday Company a considerable amount of harm. They have made public the high success of the spontaneous boycott of the Macmillan Company by scientifically-minded people...In any case, since I believe that the Blakiston Company is owned by the Doubleday Company, which controls its policies as well as the distribution of its books, I am now then a fellow author of the Doubleday Company along with Velikovsky. My natural inclination, were it possible, is to take *Earth, Moon and Planets* (editorial note: written by Whipple) off the market and find a publisher who is not associated with one who has such a lacuna in its publication ethics. This is not possible however, so the next best that I can do is to turn over future royalty checks to the Boston Community Fund and to let *Earth, Moon and Planets* die of senescence. In other words, there will be no revision of *Earth, Moon and Planets* forthcoming so long as Doubleday owns Blakiston, controls its policies and publishes *Worlds in Collision.*

To reaffirm both his intransigence and his high regard for the truth, Whipple wrote 20 years later on 2 July 1970:[83]

> With regard to Mr. Velikovsky's *Worlds In Collision* there is no change in my attitude or in the situation since the book was first released nearly a decade (sic) ago. There is no truth to allegations that I sought to dissuade the Doubleday Company from publishing this book or any other book...

Even Harold C. Urey, the Nobel laureate in chemistry whose openness we lauded in Chapter 1, felt impelled to jump on Velikovsky. He wrote in a letter on 7 March 1969:[84]

> Velikovsky is a tragedy. He has misguided people like you in great numbers, and my advice is to shut the book and never look at it again in your lifetime.

Yet, by his own admission, *Urey has not read Velikovsky's books!*

So much for the inquisition of Velikovsky. After 25 years, it is only beginning to abate somewhat. In studying scientific heresies, it becomes evident that there are at least two types.

Type A heresy is thinking that violates or runs counter to an established or widely-recognized position among scientists.

Type B heresy is thinking that, in addition to containing *Type A heresy*, is viewed by some scientific authorities as having potentially damaging social or moral consequences if it is pursued to a feared conclusion.

Immanuel Velikovsky was guilty of Type A heresy.

A contemporary illustration of Type B heresy is furnished by William B. Shockley, who won (along with John Bardeen and Walter H. Brattain) the 1956 Nobel Prize in Physics for development of the transistor.

In 1969, the *Harvard Educational Review* carried an article, written by educational psychologist Arthur R. Jensen of the University of California at Berkeley, that contended that the American Negro is intellectually inferior to the American white in some measurable ways. Shockley subsequently derived some genetic theories that purport to account for Jensen's observations. He theorizes that intelligence is *hereditary* rather than determined by *environment.*

His theories — known as "dysgenics" — describe "retrogressive evolution through the disproportionate reproduction of the genetically disadvantaged." Because the social dogma of racial equality is enormously important to every aspect of our government today — legislative, judicial and executive, Shockley's theories are not only obscene and repugnant to politicians. They are *immoral!*

The Associate Editor of the *New York Times*, Tom Wicker, asked two provoking questions in 1973 about Shockley's "repugnant" theories:[85] (a) To what extent is a free society obligated to create opportunities for expression of such ideas, and (b) Do universities and publications have an obligation to extend him (Shockley) a respectable forum for his theories?

These two questions induced Albert R. Hibbs, a scientist at the California Institute of Technology, to discuss Shockley's theories with some of his colleagues. Here was their response:[86]

> In a number of discussions with my friends and colleagues, I have found that the majority find the genetic theories of Dr. Shockley as personally abhorrent as does the editorial writer whom I have quoted. One of my friends, an eminent medical researcher and a man of liberal view, responded, "Oh I know about his stuff. He's just a racist!" Well, maybe he is. And of course "racist" is a highly pejorative word these days. But

> even if we *question his motives*, does that *disprove his concepts*? (Emphasis added)

Hibbs, in his stimulating article, "Inquisition, Repression and Ridicule," proposes that there are three ways used today to silence heresy:[87]

> I believe it (the story of Galileo) has some special meaning for us. I believe we must recognize that times and bureaucracies have really not changed so much, whether the bureaucracies are governmental, religious, educational, or scientific. Still today, when an individual disturbs the establishment, or deviates too loudly and too effectively from the accepted wisdom of a large and bureaucratically organized group, we find ways to silence him — sometimes by *ridicule*, (although this has often proved a very weak weapon, and one which frequently turns against its user), sometimes by *repression*, and, if all else seems to fail, by *legal action*. (emphasis added)

Applying these three methods of suppression to Shockley, Hibbs evaluates his status as a scientific heretic:[88]

> Right now Dr. Shockley seems to be going through the ridicule phase. He is being shouted down at public lectures, insulted by newspaper writers, and occasionally a university cancels his lectures. But as we have seen in past cases, this seldom works. Will the next step be repression? What form will it take? Will the scientific and educational establishments that you and I represent take part in it? And if that fails, will the law be used next? The philosophy behind our current laws on this matter is clear: there are no racial differences in mental capability. Differences in capability appearing between the races are due to environmental factors only. This is the official position of the federal government — Administration, Congress, and the Courts. It is just as official as the position of the Holy Office in 1633 that the earth stands still and the sun moves around it.
>
> In fact, the situation now may be even more rigid. The dogma of the church regarding the solar system in the 17th century was really not a central issue in the structure of the bureaucracy.

So to summarize, Type B heresy in science carries double jeopardy. It not only ignores or rejects the majority scientific viewpoint — it threatens to embarrass science in front of the whole world, crossing the scientific frontier over into moral, social and political arenas in the exalted name of science.

William F. Buckley, Jr. puts his finger on the embarrassment that Shockley has caused academicians:[89]

> A contact with Dr. William Shockley serves to remind one how ill poised the academic community is to cope with its own. First they tell you that

> the colleges are havens for all ideas, that no one has anything to fear in the open society because we shall seek the truth and endure the consequences. Then along comes someone like Dr. Shockley, whose ideas grate on fashionable sensibilities, and the academic community makes a fool of itself by its own standards. Meanwhile, the same community that deplores Dr. Shockley, who is merely an inventive dilettante playing chess with genetics, listens with great respect to Dr. B. F. Skinner who, if his notions about the nature of man were ever believed, would make Dr. Shockley the Mr. Nice-Guy of the academic season.
>
> I do not suggest that Dr. Shockley is harmless. He is in two senses harmful. First, he is a live carrier of scientific hubris. Second, his palaver encourages an Archie Bunkerite racial invidiousness.

As we will discover in later chapters, the scientific elite demand that anyone proposing an idea (particularly one counter to *their* position) must possess the proper academic qualifications.

Note that Velikovsky is a physician and psychoanalyst — far afield when discussing astronomy, geology and paleontology. Likewise, Shockley may be a Nobel laureate — but in *physics* not genetics.

"It can reasonably be argued that, unless a scientist speaks on the subject of his expertise, he is not professionally entitled to serious attention or academic credit," is often proclaimed by scientists.

Hibbs observes:[90]

> It is interesting that one of the charges raised against Galileo in the long process of bringing him before the Inquisition was that he was discussing matters of theology and natural philosophy, whereas he should stick to his own field; namely, mathematics.

"Okay, so Velikovsky and Shockley may be modern-day scientific heretics. But you have declared heresy is a *general* threat to the scientific power structure. What proof do you have?"

Good point.

The British science journal *Nature* recently reported an "eloquent plea from a distinguished group of scientists" including the renowned French biochemist, Jacques Monod, winner of the 1965 Nobel Prize for Medicine and Physiology for elucidating the replication mechanism of genetic material.

What was this eloquent plea? It was a "Resolution on Scientific Freedom Regarding Human Behavior and Heredity." (*American Psychologist* carried this resolution in its July 1972 issue.)

Why were these distinguished scientists signing a resolution calling for "scientific freedom?" Because freedom was *missing!*

"Do you mean that scientists were being denied the right to speak?" you might exclaim.

"Apparently so."

"On what subject?"

"On whether people of some races are genetically less intelligent than others."

This is the way the Editor of *Nature* explained it:[91]

> There are many professional scientists who hold that the subject is so full of danger that, for the time being at least, investigations of the inheritance of various attributes by the people of different races should be postponed...
>
> Few will quarrel with the principle that science must be open. To do so is to deny the lessons of recent centuries that the consequences of suppression are invariably confusion and malevolence. Equally, however, there is no doubt that in fields in which the results of research are liable to misinterpretation, the scientific community has a duty not merely to gain further understanding of what are often complicated problems but also to insure that the results of their research are accurately and moderately interpreted to the world at large...Dr. William Shockley...has for example done a great deal of harm with his recurring proposals that the National Academy in the United States should investigate the matter and then produce an authoritative statement of the position. For the truth is that only the most dogmatic adherents of the hereditary view would think that the time is ripe for a final assessment of the relative importance of nature and nurture — the old-fashioned terms — in, say, the determination of intelligence.

Note the subtle way that heresy is described in this editorial. "The scientific community has a *duty*" to stop (or maybe just postpone) research because "the subject is so full of danger."

Very noble indeed. But from whence springs this sense of duty? How do we know — ahead of the research results — that they are going to be *dangerous?* Dangerous to *whom?* What is the *danger?*

What if Shockley had proposed "that the National Academy of Sciences investigate defoliation in Viet Nam (shortly after the Defense Department initiated it) and then produce an authoritative statement of the position."

Would "the time be ripe for a final assessment" of *that* subject? Why? How does one know — ahead of proposing — that "the time is ripe?" Who decides *when* this appropriate time has arrived?

Evidently not even Nobel laureates like Jacques Monod understand this sudden shift in ground rules — that there are now sacrosanct subjects that science may not investigate.

"All right, whether we understand *why* or not, there is some never-never territory that we must avoid," we can hear these scientists say.

"How will we know what has been approved by the all-knowing barons of science for further research?"

Whether or not it is the "official" method of notification, do you know how the scientists *were* notified?

Listen to their complaints:[92]

1. Those who now emphasize the role of heredity in human behavior are being subjected to "suppression, censure, punishment and defamation." This practice is similar to, and in principle the same as, the suppression of Galileo in Italy, the criticism of Darwin in England a century ago, the persecution of Einstein and his works in Nazi Germany, and the suppression of Mendelian genetics in Stalin's Russia two decades ago.

2. These scientists' published positions are misrepresented.

3. Scientists who are committed to the *environmentalist* view of how observed differences arise in intelligence mount attacks to prevent research intended to throw light on the *inherited* basis of racial differences. Even political militants, who are not always scientists, are able to successfully attack this research.

4. Many scientists who are of the heredity persuasion in human behavior feel obligated to remain silent in public for fear of giving offense. The result is a "kind of orthodox environmentalism dominating the liberal academy. "

5. Scientists are inhibited from turning to explanations based on inheritance which, in their view, are already valuable and which call for further and more energetic research.

Did not Lysenko (the villain who subordinated science to Soviet politics) accuse geneticists of maintaining a theory radically at odds with dialectical materialism, which made it thereby *necessarily false?* How, in principle, does "dialectical materialism" differ from "genetic equality of all races"?

Of course, this comparison should not be construed as endorsing any conclusion, one way or the other, regarding genetic determination of intelligence. But in what respects are the racial environmentalists any different from Lysenko?

Lest you think that scientific inquisition is directed exclusively at the sticky issue of the role that heredity plays in determining intelligence, the account of an unmerciful persecution of an Austrian experimental biologist, Paul Kammerer, should indicate that the scope of heresy is not limited to genetic issues that have social implications.

Arthur Koestler, the renowned novelist, gives intriguing insight into how Kammerer was driven to suicide by scientific colleagues who would not examine Kammerer's experimental evidence, and, instead, declared his work to be faked.[93] The key to Kammerer's heresy? It threatened the established amoeba-to-man concept. George Steiner of *The Sunday Times* describes Koestler's account as, "A study of the cruelties, of the fanatical detestations rife in the academic scientific establishment."

While heresy is a threat in many areas of science, it appears to follow a pattern. It is most fierce where the possibility of *values for society* could result from scientific research or conclusions. Anything that could conceivably contribute to the *meaning* of life also is subject to suppression. Could it be that the suppressors hold *private* values or meanings that they view as threatened by scientific findings?

One recent scientific discovery *stunned* physicists but inquisition is *not* likely to be mounted (because it does not threaten social or political values). It is the recently-found "monopole."[94] Reputedly one of the most unusual and long sought-after objects in the universe, this small particle (only *evidence* of the particle was found rather than the speck itself) is supposedly the fundamental unit of magnetic charge.

It is analogous to the electron in electricity and has never been found until very recently. As yet, no one has suggested that the monopole "is so full of danger that further investigation should be halted." They have not even proposed that "the time is not yet ripe" for assessing its value to the study of magnetism. Why not? Maybe because no scientist sees it threatening some preconceived social or political values...

Did it ever occur to you that heresy could not exist without some *doctrine* against which it is measured? Going one step further, it is the *power* that the official doctrine represents that makes heresy consequential. The greater the power of the Establishment, the greater threat heresy represents (in terms of toppling precedent).

Science is powerful. Scientists, erroneously thinking that they speak for science, preach a doctrine that in their minds is just as exempt from question as any medieval papal decree.

A clue to scientific devotion was once expressed by T. H. Huxley: "The tragedy of science is a beautiful hypothesis slain by an ugly fact."

Risk of Rigor Mortis

Compared to other fields of study like history, philosophy, theology, or economics, the *birth* of science is quite readily identified. This is particularly true if we mark the start of what is frequently called "modern science." The conditions — location, time, cultural setting and intellectual climate — surrounding the birth of modern science are quite well documented.

If science were to be likened to a living organism with a known birth, would it be inconceivable to propose that it might ultimately *die*? After all, feudalism is considered quite dead today after thriving in the Middle Ages. Even colonialism, much younger than feudalism, also appears to have all the earmarks of death. No organism is guaranteed eternal life, is it?

Assuming that science will die some day, what signs or indications should precede that fateful day? If such signs were to occur a sufficient time prior to the last gasp, we might be able to extend the life of science! Geriatrics, that child of science that specializes in diseases of old age, could offer us hope.

Let's see if any of the signs that accompany human aging might be present in science today. Some indicators of old age (apart from the physical evidences) are:

1. Being set in one's ways (obstinacy)
2. Knowing it all (bigotry)
3. Refusing to entertain new ideas (dogmatism)
4. Dwelling on the "good old days" (reminiscence)
5. Demanding agreement with one's ideas (insecurity)

Checking the first sign, is science becoming "set in its own ways?" Are there signs of its resisting change?

A recent editorial in *Science* discussed a meeting sponsored by the National Academy of Sciences known as "Academy Forum on Experiments and Research with Humans: Values in Conflict." Naturally this forum could be expected to involve criticism of science. But here's the reaction to criticism:[95]

> Wanting (on the part of scientists) were a sense of shared humility and a willingness to confront the facts honestly. Solutions to problems concerning the essential elements of the human conditions are never perfect but always involve compromise, resulting in frustration and heartbreak for some.
>
> Scientists are naturally defensive before critics who seem indifferent to the grandeur of the scientific achievement. But critics of science can be expected to be hostile toward those who seem to believe that science is a self-vindicating enterprise, not accountable to the public. Science survives at the pleasure of the public, which supports it, and if it is coming under even closer public scrutiny, this is as much a result of its

> success as of its failings...To question what science should do and how science should do it is not to be against science. Such questioning is at the heart of scientific methodology.

William R. Thompson, a renowned entomologist and Director of the Commonwealth Institute of Biological Control in Ottowa, was asked to write an introduction to the sixth edition of Charles Darwin's *The Origin of Species*. Thompson included in his introduction a persuasive discussion of the great divergence of opinion among biologists concerning Darwin's work, pointing out that the divergence is due to unsatisfactory evidence to support Darwin's position. He urged that the non-scientific public should be apprised of the disagreements among scientists on this subject.

In so doing, Thompson ran headlong into opposition to such exposure. He responds:[96]

> This situation, where scientific men rally to the defence of a doctrine they are unable to define scientifically, much less demonstrate with scientific rigour, attempting to maintain its credit with the public by the suppression of criticism and the elimination of difficulties, is abnormal and undesirable in science.

Does *bigotry*, the second sign of old age, exist in science, too? Do scientists claim to "know it all?" Philip H. Abelson, Editor of *Science*, wrote an editorial, "Bigotry in Science," in which he expressed shock at this trait among scientists:[97]

> One of the most astonishing characteristics of scientists is that some of them are plain old-fashioned bigots. Their zeal has a fanatical, egocentric quality characterized by disdain and intolerance for anyone or any value not associated with a special area of intellectual activity...To achieve success one must concentrate on performing a series of specific tasks with complete rigor. Putting the blinders on is a great help toward this accomplishment. The trick is to know how and when to take them off. One must be able to specialize but one must be able to escape the web of his own rationalizations. Many have not the will or wit to do this.

After this editorial appeared, there was an interesting letter to the editor, written by a scientist at the Brookhaven National Laboratories that appeared in *Science*:[98]

> Since scientists are people, it seems much more likely that their capacity for bigotry is fixed long before they attain even undergraduate status. It therefore cannot really be astonishing that some scientists, like some butchers, bakers, or candlestick makers, are bigots...By the time a bigot has grown up to be an unhumble scientist, it is probably too late for salvage...The only really effective method is prevention, and this is a sad thing: that we cannot lead our children through the undergrowth of life experience into their places in an orderly, free, and responsible society

without somehow passing on to them our own prejudices, fears, and bigotry.

Although scientific dogmatism was discussed in Chapter 1, an observation by Ernst Mayr, Alexander Agassiz Professor of Zoology at the Museum of Comparative Zoology of Harvard University, is most significant:[99]

> Historians of science are familiar with this phenomenon (the nearly-complete resistance to drawing what would seem to be the inevitable conclusion from a vast amount of evidence); it happens almost invariably when new facts cast doubt on a generally accepted theory. The prevailing concepts, although more difficult to defend, have such a powerful hold over the thinking of all investigators, that they find it difficult, if not impossible, to free themselves of these ideas.

While it did not fit his argument in the article quoted, Mayr undoubtedly would acknowledge that historians of science are equally familiar with the antithesis of the described phenomenon — the *nearly-complete resistance to abandoning an attractive theory for which scientists have no data.* In both cases, scientists refuse to replace obviously outdated ideas with new ideas, and thereby they exhibit *dogmatism.*

With the almost unbelievable contribution to better living earlier credited to science, scientists would have to be superhuman not to indulge in a little romantic *reverie* about their achievements once in a while — even if it is another sign of aging.

But the ungrateful public does not always agree that past scientific accomplishments offer hope for the future. Consider this evaluation:[100]

> Such was their (scientists') success that many people became convinced that there were scientific or technological "fixes" for all the nation's problems, including its most serious social ills. Even as late as 1967, after Watts, Newark and Detroit had been engulfed in flames, the dean of M.I.T.'s College of Engineering, Gordon Brown, could be heard to proclaim: "I doubt if there is such a thing as an urban crisis, but if there were, M.I.T. would lick it in the same way we handled the Second World War."
>
> Such arrogant and naive optimism sounded questionable even then. Today it has a particularly hollow ring. For after years of sunny admiration, science suddenly finds itself in a shadow. No longer are scientists the public's great heroes or the beneficiaries of unlimited funding. Unemployment runs high in many scientific disciplines; the number of young people drawn to the laboratory in certain key areas has diminished significantly. Indifference to scientific achievement is the mood of the moment...The reversal is the result of a new mood of skepticism about the quantifying, objective methods of science.

So far, science has been shown to exhibit four signs of aging. What about the fifth — *insecurity*? Are scientists demanding that everyone agree with their ideas? Of course, this is the underlying demand behind all accusations of heresy that have been earlier elaborated. Perhaps one more specific example would confirm this phenomenon of aging.

Anthropologist Anthony Ostric, in an address to the Ninth International Congress of Anthropological and Ethnological Sciences in Chicago, sharply criticized his colleagues for declaring "as a fact" that man descended from ape-like creatures. Here is one account of his criticism:[101]

> The Darwinian evolutionary theory has been promoted by only a few leaders in anthropology and human biology, Ostric declared, but, he said, the vast body of professionals have fallen in behind them "for fear of not being declared serious scholars or of being rejected from serious academic circles."
>
> Ostric told the (attendees) that there was no evidence that man had not remained essentially the same since the first evidence of his appearance…"Man's unique biophysical and socio-cultural nature appears now to represent an unbridgeable abyss separating him from all other animals, even from his closest 'anthropoid relatives.' It is not possible to see how biological, social or cultural forces or processes could transform any kind of pre-human anthropoid or 'near-man' into homo sapiens."

Science evidently is exhibiting conclusive signs of old age. Is there anything that can be done to arrest this disturbing trend? Is the death of science inevitable? What are the forces at work that seem to be accelerating the aging?

Collectively, the aging forces or accelerators appear to lie on the frontier between scientific and Socratic truth. The impetus behind irrationalism — the New Dionysians as Holton calls them — and the increasingly-perceived social responsibility demanded of scientists are pushing hard for some kind of "value system" within science.

The dilemma, of course, is that there is no objective way to determine good and evil, proper and improper, right and wrong — all *moral* values. Therefore, the New Apollonians resist the onslaught of the New Dionysians.

A British interdisciplinary working party in 1972 sought some answers to the conflict over social responsibility of scientists. A summary of their work states:[102]

> The first thing to say is that we came to no radical conclusions. Those which have been proposed fall into three broad groups. First, that there

should be something in the nature of a Hippocratic Oath or code of ethics for all scientists, whereby they bind themselves not to take part in work which will have socially harmful consequences; second, given that we have an increasingly science-based civilization, that moral and political decisions about the social application of scientific work should be left increasingly to scientists themselves; and third, that there should be radical — if not revolutionary — reform of the whole social system. Of these, we think the first impracticable, the second dangerous, and the third beyond our competence.

The subject undertaken by this working group is obviously too large in scope to be discussed here. However, their second conclusion — that leaving moral decisions to scientists would be dangerous — is germane to the dilemma that I believe could kill science if not properly resolved.

Here's their reasoning:[103]

> We suspect that the idea that scientists should be left to decide how society should use their work is based on a tacit confusion. If decisions of this kind were "scientific" decisions, there might be something to be said for leaving them to scientists. But they are not; they are essentially political and moral decisions. We have no evidence to suggest that scientists are better qualified to take (sic) decisions of this kind than men trained in other disciplines, or indeed that scientists as a class are more rational, or more altruistic, or in any sense morally "better" than other classes of human beings. Neither their training, nor their experience in laboratories or academic institutions, particularly equip them to make important decisions for the community at large. Nor has past experience with different forms of government shown that our society would be better ruled by a small elite of specialists, not subject to the checks and balances designed to avoid too great a concentration of power among too few people, and not accountable to the public whom they serve.

There it lies. Scientific success has bred power, and, in turn, that power has bred the danger of death to science. If scientists properly recognize the twofold limitation of science, it may survive. If they fail to recognize it, Harvard biologist-historian Everett I. Mendelsohn may be correct in his observation, "Science as we know it has outlived its usefulness."[104]

DANGER ONE: DOGMATISM

Science will be killed if either of two power-bred dangers are ignored. The first danger to viability is *dogmatism.*

As you recall, the *Law of Scientific Dogmatism* was defined in Chapter 1. And dogmatism has been noted in this chapter as a sign of old age. However, both of these references to dogmatism have been limited in scope — *illustrative* rather than *substantive.*

At this point, we are discussing the *disease* of Dogmatism. This disease is caused by power — too much of it. You see, *anyone* can be dogmatic. Small children playing in a sandbox...soldiers in the barracks...old men around the pot-bellied stove in a country store.

But these are examples of dogmatism, not *Dogmatism.* The latter is possible only when the dogmatist perceives himself to be omnipotent, impregnable to questioning.

In the grip of Dogmatism, the victim becomes "increasingly entranced with his own navel." He gets disconnected from reality. To use a sexual analogy, this person commits incest — has intellectual intercourse exclusively with those whose ideas agree with his. This inbreeding only compounds his illusion of infallibility.

The death of science, if it occurs by Dogmatism, will progress in stages. First, there will be an elimination of the concept that something could be *scientifically unknowable.*

Next, unsubstantiated personal beliefs will be declared to be *scientific facts.*

Finally, science will be converted to the *religion of scientism.*

Let's examine these three stages in greater depth.

Murder of Mystery

Healthy science thrives on enigma, intrigue, wonderment and the unknown. As Wernher von Braun says in the Foreword, "The mainspring of science is curiosity." Great scientists have been those who were spurred on by what others thought were impossible puzzles or riddles.

Mystery is a frequent term in science. Like when rocket-borne instruments detected a new and extremely powerful source of X-rays in deep space trillions of miles beyond the Milky Way, astronomers in charge of this experiment said:[105]

> Quasars and radio galaxies have strained our conventional concepts of the generation of power by nuclear fusion — and the discovery of even greater galactic power radiated in X-rays has deepened the mystery of the basic energy source.

In an article about scientific achievements in the 1960's (dubbed the "Scientific Sixties"), here are some words used:[106] "Amazing, incredible, not understood, unexplainable, stumbled, ironically, peculiar, strange, unusual, puzzling, and surprising."

Who could fail to catch the thrill experienced by the physicist who exclaimed:[107]

> The suddenness of the discovery, coupled with the totally unexpected properties of the particle are what make it so exciting. It is not like the particles we know and must have some new kind of structure...(We) are working frantically to fit (the particle) into the framework of our present knowledge of the elementary particle...it must have some novel, as yet not understood properties...

Less than a week after that sensational discovery, a second "new and mysterious" particle was found. This provoked additional excitement.[108]

> "I don't know just what to think right now" said physicist Burton Richter, when asked about the significance of the burst of new activity in the world of physics...(The two discoveries) "leave us with still more questions to answer...and very little sleep for anyone."

> The discovery of the first psi or J particle did not fit into any of the theoretical schemes the physicists had constructed to explain the nature of the matter. The discovery of yet another particle is not yet understood, either. "There has been a population explosion of theorists trying to explain the new psi particles. It has created a madhouse..."

Astronomers also have fun puzzling over *conflicting* evidence. In a meeting of the American Astronomical Society, Kenneth I. Kellerman of the National Radio Astronomy Observatory described a whole *galaxy* that appears to be expanding faster than the speed of light — that supposed ultimate of all speed limits.[109] Kellerman's discoveries, observers said, "simply add to the puzzlement of the astronomical community."

Contrast all this mystery and intrigue with the closed-minded accounts in science textbooks of the origin of the universe, life, and man, as described in Chapter 1. There's no mystery left in those accounts. It's *all* known. What is left to discover?

In fact, don't you dare question the "facts" that the universe started with one big bang, that life had a soupy start, or that man's parents were ape-like. If you do, you'll be sorry...On those subjects, mystery has been murdered — first stage of "death by Dogmatism."

Facts or Beliefs?

In the opening chapter entitled, "Scientific Philosophy," of his outstanding book, *The Foundations of Metaphysics In Science*, Errol E. Harris discusses *empiricism.*

Empiricism can be described as the search for knowledge by observation and experiment, or the belief that sensory experience is the only source of knowledge. Empiricism forms the backbone of science. It is what freed the Middle Ages from superstition. It is what distinguishes "objective" from "subjective." The concept of empiricism *seems* simple, but it really isn't easy to even define! Consider Harris' description:[110]

> (Empiricism's) starting-point, tacitly or openly, is always the fortuitously occurring data of sense. These are assumed to be simple and irreducible particulars without mutual connection, though they are, of course, variously conjoined. It is the conjunction of these primary data that constitutes matters of fact and it is solely through our experience of them that we can and do acquire knowledge of the world. Matters of fact are, therefore, on this view, mutually independent. Though conjoined, there is no discoverable necessary connection between any two of them. The world is thus presumed to consist entirely of unconnected, mutually independent facts, described as atomic, any one of which could be other than it is without changing any of the others, or could remain the same despite alterations elsewhere. Similarly, propositions in which such facts are stated are atomic and are logically independent of one another. No factual proposition is deducible from any other by virtue of its factual content, though such deduction is sometimes possible in consequence of formal characters of the propositions.

"Wow," you say, "that is some definition!"

Perhaps we could summarize Harris' technical statement by stating that (1) you start with observations, (2) you do not allow the observations to influence one another, (3) you tie these independent observations together, (4) the observations thus tied together form a scientific fact, and (5) the resulting scientific facts constitute our knowledge of the world.

So you can see that scientific facts do not jump right out at you. Further, they are subject to debate among scientists.

More importantly, the layman must understand that, when the scientist starts to tie two independent observations of nature together, he must make an *assumption* or *judgment* that they belong together. This obviously cannot be a value-free conclusion.

Therefore, at the very foundation of science — deciding what is or is not a scientific fact — there is "a little bit of sawdust mixed into the hamburger." Harris quotes two observations of Bertrand Russell on this point:[111]

> Science is at no moment quite right, but it is seldom quite wrong, and has, as a rule, a better chance of being right than the theories of the unscientific.

> (It is a mistake) to begin with how we know and proceed afterwards to what we know...because knowing how we know is one small department of knowing what we know...for another reason: it tends to give to knowing a cosmic importance which it by no means deserves.

Many, if not most, scientists are not students in the *philosophy* of science. Because of their ignorance of this subject, they can easily skip over the difficulty of inherent, value judgments in so-called scientific facts.

However, I propose, at this point, that we do exactly the same. Let us set aside this admixture of observation and judgment. I feel that the amount of intrinsic subjectivity of basic scientific facts is insignificant when compared to *private beliefs* that scientists defend in the name of science.

"Do you mean to say that scientists pass off their private beliefs as though they were scientific facts?"

I do.

Go back and look at Figure 2 in the Prologue. The filter of scientific methodology simply doesn't catch all the "impurities" of belief.

One of the most amusing illustrations of scientists allowing their preconceptions to overrule facts occurred during the early Apollo flights to the Moon. When Neil Armstrong and Buzz Aldrin brought the first Moon material back on Apollo 11, the chief geological investigator of this material was a Caltech professor of geology, Eugene Shoemaker.

In an article dated 20 November 1969, Shoemaker was quoted as saying that there can no longer be any doubt that the vast plains of the Moon — the maria — were created by lava flows.[112] He further stated that that conclusion was the key to a new understanding of the Moon and the most important result of the Apollo 11 mission. He elaborated in great detail the significance that this conclusion carried. He ended his interview by saying, "Once and for all and forever, the stuff of the maria is lava."

Exactly ten days later — 1 December 1969 — the first report of investigation of Apollo 12 Moon material showed Shoemaker's bravado to be completely unwarranted. Here's how that report read:[113]

> After the Apollo 11 rock samples were examined, many scientists were describing in detail the formation of the moon's maria areas. Their conclusions for formation of these broad, flat and smooth lunar plains were based on what the Apollo 11 rocks showed apparently happened at the Sea of Tranquility.
>
> Their reasoning was that if one mare was formed in a particular way then almost certainly the moon's other maria were formed the same way. Apollo 12 rocks apparently have ended such ideas.

Something that was "once and for all and *forever* determined" was changed within ten days! What actually changed? *Facts* or *beliefs*? Obviously, the beliefs.

The *age* of the Moon is critical to beliefs that various geologists hold about its origin. In Chapter 1, two radically different beliefs about the origin of the Moon were mentioned. One way to settle the argument between beliefs about the origin of the Moon might be to settle the *age* of the Moon.

In other words, if Theory A proposed by Professor X says that the Moon was spun out of the earth, obviously he needs a date for the Moon much *younger* than the earth — if his belief is to be the "correct" one.

On the other hand, if Professor Y is pushing Theory B, which says that the earth and Moon were simultaneously formed, he needs a much *older* date than Professor X.

Do you think that these two professors have a totally "open" mind as they walk into the Lunar Laboratory to start dating Moon rocks? *This* "Dating Game" is not likely to end in "Moonlight and Roses!" Professional reputations are at stake...

Leon Silver, another Caltech geology professor, disagreed with his colleagues who made definite statements about the Moon based on the single "bucket of dust" returned on Apollo 11.

> "To make such announcements as we have heard would be like figuring the age of the earth from a few pieces of rock from the San Gabriel Mountains scooped up in a backyard in Pasadena."[114]

He pointed out that, in the dust and few rocks he had evaluated, he found materials ranging in age from 4.1 to 4.63 billion years! Not *too* precise — 530,000,000 years variation in one little area of the Moon.

By the time Apollo 14 returned to Earth with a different sample, the "antiquity antics" became even more amusing. A third Caltech professor, Gerald J. Wasserburg, confessed surprise that, instead of the Fra Mauro material being 4.4 billion years old as the experts had estimated, it was 3.85—3.95 billion.[115] Another small prediction error of 450-550,000,000 years. That's worse estimating than the garage mechanic does on your car — and he's not even a scientist!

Later, after Apollo 15 returned from the lunar Apennine Mountains with the so-called "genesis rock," a team of scientists from the State University of New York in Stony Brook used a technique called Argon 39-40 to date the rock at somewhere between 3950 and 4350 *million* years.[116] That narrowed the estimate to somewhere within a *400,000,000-year* range. Yet Argon 39-40 is called a sophisticated and consistent radioactive dating technique!

Meanwhile, Professor Wasserburg determined that the "Great Scott" rock also returned on Apollo 15 was only 3.36 billion years old.[117] So, if the "genesis rock" is the maximum age allowable, astronauts Dave Scott and Jim Irwin brought home one Moon rock that was 1,000,000,000 years older than another rock! It's beginning to look like they should find a better way to settle scientists' beliefs about the Moon's origin than using the *age* of Moon material...

This example of conflict between beliefs and facts about the Moon was selected because it was widely reported, involved disagreements about numeric (rather than semantic) values, and represented a huge expenditure of public funds on a scientific question. However, myriads of other examples of the fact-versus-belief conflict are found in scientific literature.

Perhaps the one that laymen most frequently recall is the Piltdown Man hoax that persisted from 1911 until 1953. Of course, this archeological hoax is not unique — there have been others.[118] The key to understanding these fossil hoaxes is in understanding the overriding belief in the amoeba-to-man idea. If one looks hard enough for something he believes *should* exist, he can actually *invent* it like was done with Piltdown Man.

This type of bogus behavior is certainly atypical of scientists, and by no means is it suggested that such practice is widespread. On the other hand, a much more likely result of the influence of private belief takes place in the *interpretation* of observed data.

It occurs when a person *believes*, prior to any evidence, that the earth cannot be any older than 4004 B.C. and thereby this person rejects all data indicating otherwise.

It obviously also occurs when, for example, the Biological Sciences Curriculum Study group coins a catchword, "A hundred years without Darwin were enough" and sets about to reform content of textbooks and promote the amoeba-to-man idea in public schools as they did in 1960.[119]

Unfounded beliefs, whether unique or widely-held, are most influential in scientific circles. To illustrate, if one *believes* without any evidence — just a hunch — that life exists elsewhere beside on Earth, this belief influences (1) the data you seek, (2) what you think about the data you find, (3) what data you accept or reject, (4) how and where you will continue searching for data, and (5) what arguments you generate to refute data that apparently contradict your belief.

One article, depicting this belief of life elsewhere, is filled with words that betray the strong conviction of blind belief:[120]

> Suggesting the possible...may provide...probably the first conclusive proof...sequence related to the scientific belief...is also believed to be...strongly suggests...hints...can imagine that...could be...

Or see if you can detect this biologist's belief system in the following report concerning the mythology of aging:[121]

> It is beguiling and attractive to think we shall live forever but evolution has made it an unrealizable wish...(that man may not exceed 70 years of age) is nonsense because man already has evolved from a short-lived to a longer-lived creature. In the course of becoming an animal that is capable

of dominating the earth, man's anthropoid ancestors had to evolve into longer-lived forms. No other primate lives as long as man, nor takes so long to mature. Man's pre-eminence is largely based on his ability to store and correlate information, and to act upon what he has learned. This is the product of long evolution...It (is) naive to think that all possible evolutionary adaptations favoring longer lives already have been incorporated into human genetic heritage.

Sometimes the beliefs that scientists hold will inhibit research. Until recently, cancer research was thwarted because molecular biologists believed that "genetic information in a living cell flows in one direction only — from DNA to RNA and from RNA to protein."[122] Then a series of experiments "produced conclusive proof that this dogma is wrong." Molecular biologist Howard Temin, for six years a heretic, was suddenly the father of the Teminism theory — by overthrowing a belief system!

The social sciences perhaps suffer more from unfounded beliefs than the physical sciences. The President of the American Psychological Association, Donald T. Campbell, addressing that organization's convention in Chicago in September 1975, criticized psychiatrists and psychologists who believe that "the human impulses provided by biological evolution are right" and "that repressive or inhibiting moral traditions" are not. He went on to say:[123]

"This assumption may now be regarded as scientifically wrong, in my judgment." He urged his listeners to revise their teaching of the young so as to remove "any arrogant scientistic certainty that psychology's current beliefs are the final truth on these matters," and even suggested that (those) who object to current school textbooks may have something on their side.

Perhaps no one has more succinctly summarized the problem of belief overriding facts in science than the Past President of the American Physiological Society, Allen C. Burton:[124]

Science is, at its core, not concerned with what "must be" and far less with what "ought to be" (this is the realm of morals and ethics), but with "what is." The true scientist is he who in the ultimate conflict of even the best of logical systems, and of prediction from what is already known, in a conflict with the results of experiments, acknowledges the primacy of the evidence. In this sense, science is almost anti-rational, anti-intellectual; and the 19th century identification of "science" and "reason" has lead many astray...

The facts must mold the theories, not the theories the facts...I am most critical of my biologist friends in this matter. It seems to me that they have allowed what is a most useful working hypothesis in a limited field of the whole of biology, to become "dogma," in their worship of the principle of natural selection as the only and sufficient operator in evolution. If they have done this, they no longer can act as true scientists when examining evidence that might not fit into this frame of concepts. If you do not believe me, try telling a biologist that, impartially judged along with other accepted theories of science, such as the theory of relativity, it seems to you that the theory of natural selection has a very uncertain, hypothetical status, and watch his reaction. I'll bet you that he gets red in the face. This is "religion" not "science" with him.

Science is struggling for survival of facts over belief. Far be it from us to propose that scientists should not believe. If they didn't have some idea what they were looking for, it is very unlikely that they would find anything worthwhile. If scientists just collected facts, without either an idea of what they *expected* to find or what they *wanted* to find there could be no science. As Barton says:[125]

> Prejudice, preconceived ideas and systems of ideas, enthusiasm for his own point of view — these are the working attributes of a productive scientist. Yet he must unerringly, in the crisis of a contradiction, by the data, of all he has cherished in theory or prediction, renounce Love for Duty (like the hero of a Victorian novel).

Failing to surrender his belief in the face of fact, the scientist subjects science to the second stage of "death by Dogmatism."

Siren of Scientism

I have held responsible management positions in three major American corporations — Litton Industries, Northrop Corporation and North American Aviation (later merged into Rockwell Corporation). Corporations, like organisms, are born — and many die. An old management adage regarding corporate survival says, "Discover those unique attributes which make the company succeed and then *nurture* them!"

As a manager, I often pondered the practicality of that maxim. I used to ask myself, "How can I, right in the viscera of the corporation, even hope to *know* what it is that has contributed to company success?"

Even if I *could* know them, are the "knowables" sufficient to explain why companies succeed? Are there not *intangibles* — public mood, tax structures, state of technology, government policies or world market values — that would completely override or mask whatever overt management actions that might be taken in response to known success factors? Can one really reduce success down into a few pat rules or attributes?

This idea, of reducing the complexities of a corporation down to a few principles, illustrates one aspect of what is known as "scientism." In fact, a dictionary definition of scientism is "the tendency to reduce all reality and experience to mathematical descriptions of physical and chemical phenomena."

Isidor Chein of New York University, lists four characteristics or attributes of scientism:[126]

1. A profound commitment to the "scientific method."
2. A doctrinaire or extremist view of the nature of science, including "proper" rules of scientific conduct and expression.
3. An emphasis, even above the value of semantic precision, on "respectable" technical elitist language.
4. A commitment to seek explanations of all phenomena in terms of a particular set of primitive terms and propositions.

Chein's discussion of these points is worthy of further study for those interested. For example, regarding the second characteristic he says:[127]

> By strict application of some of these rules, a considerable array of sciences, from anatomy to zoology, would be ruled out of the domain of science because they are, in the main, not experimental, not quantitative, not concerned with prediction, and/or not hypothetico-deductive in structure. A work like Darwin's *Origin of Species* would similarly not be expected to make the grade since it promulgates as a theory propositions that can only be applied on a post hoc basis and do not serve the ends of prediction. Most scientismists, however, would exempt those disciplines because they compensate by their highly technical vocabularies that cannot be confused with ordinary English; and Darwin is, by tradition, an examplar of the true scientist.

However, I would take the scope of scientism even a step further. Chein properly describes undisguised chauvinism about scientism — an arrogant, pompous, exclusionary elitism. What I would include, beyond this description, would be an almost *religious fervor* based on virtually identical confidence to that attributed to theologians of the Middle Ages.

Consider these statements by the noted paleontologist, George Gaylord Simpson:

> Adaptation (by natural selection) is real, and it is achieved by a progressive and directed process. This process is natural, and it is wholly mechanistic in its operation. This natural process achieves the aspect of purpose, *without the intervention of a purposer*, and it has produced a vast plan, *without the concurrent action of a planner.*[128] (emphasis added)

> Man is the result of a *purposeless* and *materialistic* process that did not have him in mind. He was *not planned.* He is a state of matter, a form of life, a sort of animal.[129] (emphasis added)

On what basis does Simpson know this to be *scientifically* true? Does he have access to a special channel of revealed truth? Hardly. This is scientism at its best — playing God.

Maybe you are a part of the public that has heard of the "new" biology — those scientists who talk about changing a baby's heredity or seeding tailor-made life forms on Mars. At the annual meeting of the American Society of Cell Biology, Stephen H. Howell of the University of California at San Diego said:[130]

> It really isn't as complicated as once thought...The awe which others have for living systems is based on the knowledge that life processes are very complex and we don't yet understand their interaction. But scientists are finding that what they see in a simple bacteria can be compounded to explain higher forms of life. It was not long ago that things which we now understand were looked upon as complicated processes. But they have yielded to investigation and the things we see as complications today will yield, too.

Howell, who was under 30 years of age when he made that statement, was opposed by Herbert Stern, the more mature Chairman of the Department of Biology at the University of California at San Diego. Stern made it clear that some scientists have been guilty of overstating the prospects of the new biology.

While agreeing with others that recent biological findings offer potential, he is far more skeptical that biologists will be able to master the intricacies of life processes in the omnipotent manner implied by some of them.

"They are uncertain what will be the results of the experiments they do in the lab — if they were certain of the outcome there would be no need to do the experiments. So how can they possibly be so confident about things on a cosmic scale?" he asked:[131]

Why does scientism arise to prostitute science? Why can't scientists be content with the probabilistic nature of science (where there is always a degree of uncertainty)? Probably because they are human. They gravitate toward power just like all other people.

Dogmatism, a danger bred by power, seems to grow in proportion to the threat of being overthrown. In the Middle Ages, the church recognized the threat of reason and became increasingly dogmatic. But they could hold off only so long against reason.

There is almost a complete reversal of roles today. Scientism has adopted the intolerance of medieval theology. The heretics are still scientists — just as they were then. But the authoritarianism has a different source. Instead of being based on the revealed truth of the church, it is now based on the self-established truth of the scientist.

The next time you meet a scientist infected with scientism, ask him to describe the conditions under which arguments, backed by demonstrable evidence that invalidate a currently established scientific truth, should be suppressed. His answer will either cure or kill him. Just as the medieval church ultimately capitulated to reason, scientism may also.

A beautiful sea goddess, part bird and part woman, used to lure sailors to their death on rocky coasts by seductive singing, according to Greek mythology. Such sirens are yet alive. One's name is Scientism. If she is successful, "death by Dogmatism" will have completed the third and final stage.

There is no posthumous science...

DANGER TWO: "OFF LIMITS"

If science escapes the power-bred danger of "death by Dogmatism," another threat to viability lies just around the corner. Power may induce "death by Aberration" because this disease, like the disease of Dogmatism, is caused by excessive power.

Whereas Dogmatism may be considered a vertical problem (too high and mighty), Aberration is a horizontal sickness, caused by wandering away from the central *causa causans* (causing cause).

If Dogmatism bespeaks *incest*, Aberration is like *adultery* — being untrue to one's pledge, forgetting the ground rules. Aberration is a departure from what is true, proper and correct.

Aberration in science is not due to stupid, dull or uninformed scientists. Instead, it more often besets scientists who are over-ambitious, excessively bright and impatient — those who feel that science is too limited in scope. These folk are convinced that, if scientists would open up the frontiers a little wider, science could gain the title, "Sole Source of Truth." It could put philosophy, metaphysics and religion out of business forever. The world could be conquered in the name of knowledge.

How to Why

The current ground swell of irrationalism, as exemplified by Holton's New Dionysians, represents a force in the direction of Aberration. Other scientists, disturbed by the public's recent disenchantment with science due to "depersonalizing" effects of technology, desire to enlarge the classical domain of science so as to include scientific recognition of *spiritual* phenomena.

In August 1971, six months after Apollo 14 returned from the Moon, I met privately for several hours with astronaut Ed Mitchell who made that flight. As you may recall, Mitchell conducted — without NASA approval, several extrasensory perception (ESP) experiments while on his way to and from the Moon, as well as while he was in the lunar module on the Moon.

He and I discussed those experiments at length, as well as his personal struggle to gain recognition in the scientific community for higher sensory perception phenomena. As a minimum, he hoped at that time to get scientists to acknowledge the reality of what is commonly called "the spiritual dimension of man."

While we did not necessarily agree on particulars, we were in agreement that there is observable evidence that man possesses qualities not reducible to strictly physical and chemical constituents, and further that scientists seem to be deliberately avoiding such evidence.

Mitchell, who earned a doctorate in astronautics at Massachusetts Institute of Technology, is sincere in his conviction that the limits of science have been drawn unnecessarily narrow — that the world of nature includes factors beyond physics and chemistry. He is further convinced that instrumentation could be built which would measure spiritual realities.

Since resigning from NASA, Mitchell has established an institute to sponsor basic research on human consciousness and conduct educational programs to enhance the climate of acceptance in the scientific community for the relationship of mind to body. His viewpoint is summarized in *Psychic Exploration: A Challenge for Science* that he recently edited with John White:[132]

> Psychic exploration had already put us in the position where science's basic concept of man and the universe must be revised. Clearly the universe has meaning and direction. It was not perceptible by the sensory organs, but it was there nevertheless — an unseen dimension behind the visible creation that gives it an intelligent design and that gives life purpose.

If Mitchell were successful in securing scientific respectability for concepts like the mind, soul or spirit of man, would this constitute "off limits" for science? I do not believe so. The key to determining whether power has carried scientists beyond rightful limits is in deciding whether science has stopped explaining *how* natural phenomena function and started explaining *why* they do.

"Why" relates to those ultimate questions — *meaning*, *purpose*, and *value*. If Mitchell were to believe that scientists could *determine* the "meaning and direction" that he attributes to the universe, obviously he would be "off limits" by my definition. Perhaps specific illustrations of scientists exhibiting the disease of Aberration or being "off limits" would help us make this distinction.

A classic example of Aberration is provided by a winner of the 1967 Nobel Prize in chemistry, Sir George Porter. Porter, who is Director of The Royal Institution in London, delivered an address to Caltech chemists in which he attempted to establish the relevance or usefulness of science by saying:[133]

> (Science) has been negative in the sense of destroying old conceptions... without providing a new positive philosophy and purpose...Is it not possible that our way to a new faith, a new purpose for life is through further knowledge and understanding of nature... It is, of course, quite possible that we can never understand, never discover a purpose, but we shall not succeed if we do not try...There is, then, one great purpose for man and for us today, and that is to try to *discover* man's purpose by every means in our power. That is the ultimate relevance of science...

Follow his logic. First, he admits that the scientific method has destroyed the idea of purpose (that he attributed to outmoded philosophy and religion). Second, he proposes that more scientific research might ultimately restore purpose (a new and different one, of course) back to mankind. However, restored purpose is not at all probable.

So, though there is *no* current purpose, Porter claims that we already *have* one great purpose — to try and *discover* our purpose! Are you confused? Don't feel inferior or lonely about it...

Porter's address was also published in the public media.[134] Reader response to this article was most perceptive — and *critical.* One reader quoted another knighted Englishman:[135]

> Sir F.M.R. Walshe had an answer for people like Porter back in 1952, to the effect that "it often is a cloistered scientist who knows least about men who is apt to pontificate most loudly and confidently about Man. Beware of him when he assures you that he knows all the answers about us, for too often he is one of those Peter Pans of science that every generation produces: a clever boy who hasn't grown up."

Another example of scientists wandering "off limits" occurs in the pontification about human values in the name of science. In a wide-ranging paper, the President of the American Academy of Arts and Sciences, Hudson Hoagland, proposed that since ethical and religious leaders have failed in the past, scientists should take over the job of determining values and purpose for mankind.

Sounding very much like earlier-quoted G. G. Simpson, he says:[136]

> Man himself and his behavior are an emergent product of purely fortuitous mutations and evolution by natural selection acting upon them. Nonpurposive natural selection has produced purposive human behavior...

Based on that illogical and inexplicable premise, Hoagland reaches several startling conclusions that should deter endorsement of his premise of scientifically-determined ethics and values for quite some time into the future. One such conclusion was:[137]

> Objection to population control by methods of contraception represent(s) value systems based on archaic and parochial notions at variance with what science has learned about the nature of human conduct necessary to advance cultural evolution in the nuclear age.

One respondent to Hoagland's conclusion pointed out that, even if population control were proven scientifically to be required, scientists would be in no position to decide whether contraception, rhythm, periodic continence, abortion, sterilization or any other method should be used. This person pointed out that choice is:[138]

> ...based on value judgments that rest outside of the scientific or biological realm, namely, on convenience, moral, philosophical, theological, or even trivial reasons. To reduce moral, philosophical, or theological reasoning to "archaic and parochial notions" reveals either a lack of the concept of true science or true humanism...

Maybe scientists could learn a lesson from economists about the problem of human values. One of the nation's best-known economists, Walter W. Heller, who served as chairman of the Council of Economic Advisors under Presidents Kennedy and Johnson once said:[139]

> Economists are human beings. And as such, we differ as sharply among ourselves on goals, values and priorities as those who read these lines disagree among themselves. For example, how much unemployment and recession are we willing to endure to beat back inflation? That answer can't come out of economic science. It comes out of people's hearts and souls, where differences abound...
>
> It would be helpful if economists were more meticulous; if they would "come clean" more often and say, "You've got to understand that here I'm not speaking as an economist — I'm speaking as a value-charged human being."

Regarding the pressure that power produces to tempt scientists to go "off limits" in order to address not only *how* but *why*, I find myself in basic agreement with Steven Weinberg, Higgins Professor of Physics at Harvard University:[140]

> Science cannot change (from how to why) without destroying itself, because however much human values are involved in the scientific process or are affected by the results of scientific research, there is an essential element in science that is cold, objective, and nonhuman...(the scientist) learns the boundaries of science, marking the class of phenomena which must be approached scientifically, not morally, aesthetically, or religiously. One of the lessons we have been taught in this way is that the laws of nature are as impersonal and free of human values as the rules of arithmetic. We didn't want it to come out this way, but it did.

Yet, at the same time, I sympathize with the despair of Theodore Roszak when he speaks of the "monster of meaninglessness" that has been produced by modern science – "the psychic malaise, the existential void where modern man searches in vain for his soul" brought about by scientists.[141]

My personal solution to this tension would be for scientists to (1) recognize and believe the limitations of science (perhaps as defined by Nielsen), (2) announce those limitations openly and widely to the entire world, and (3) point the world toward those who deal with Socratic truth — philosophers, metaphysicians and religionists.

Meanwhile, failing to take these rational steps, scientists (1) remain publicly mute, (2) accept all the public adoration (as though it were deserved), and (3) knowing full well (except for a few deluded ones) that they can't deliver meaning, purpose or values to mankind, allow science to be torn apart internally by divergent forces.

Science would be *strengthened* not diminished by a clarification of its limits for the non-scientific public. By honestly and openly admitting that they have no means to objectively determine all truths, scientists would gain great credibility with the public.

Just as America's Founding Fathers discovered the wisdom of dispersed power, the public would rediscover the virtue and richness of decentralized Truth.

Origins – Forbidden Fruit

Tempting as they may be, *origins* mark the terminal stage of the "disease of Aberration" for science.

Chapter 1 provides rationale for this statement. The desirability of being "The Answer Man" for such intriguing events makes the temptation almost irresistible — especially when one possesses the power that success has bestowed on science.

Why should science be content to remain "less than complete"? Why not "go for broke"? Science has eliminated all competitors for Truth, hasn't it? It neatly took care of philosophy, ethics, religion, metaphysics, and art — managing to thoroughly discredit those charlatans and depose them from power.

Science alone "proves" anything, so no one could possibly arise to *prove* science wrong about origins. Has the time not arrived for science to ascend the throne of omniscience?

Understand that there is an unseen but real payoff concerning origins — for the party that can make their claim stick. If they can convince the world that their version is, in fact, the authentic and only possible account, then they can claim the real clincher — assertion that they are able to know the ultimate *end* of all things as well.

After all, "what has been" can never compare with "what will be" in terms of popular appeal. If the origin of the universe, life, and man can be considered closed issues — no longer subject to question, then working from "knowns" to predict "unknowns" of the future fits ideally into the portfolio of science. Isn't prediction an inherent process in the scientific method?

In the minds of some scientists, the origin of the universe, life, and man are known quantities — despite the ambiguities discussed in Chapter 1. Leaving to their colleagues the job of selling the public on this mythology and assuming that the selling will be a success, they have already proceeded to work on defining the *future* of the universe, life, and man.

Typical of this futurism in the name of science are the earlier-quoted statements of Captain Jacques-Yves Cousteau and Isaac Asimov.

One more example, however, might illuminate this symptom of Aberration more clearly. The Vice President of Hewlett-Packard for research and development, Bernard M. Oliver, excited Caltech students not too long ago with a lecture that exemplifies the confident extrapolation into the future based on "myths-assumed-to-be-facts."

His lecture is well worth careful study and analysis in this regard. Note his (1) unqualified statements, (2) air of certainty and (3) cosmological assurance in these few excerpts:[142]

> Out of all of these (scientific) discoveries has emerged a new and grander picture of the world, one that may ultimately reshape man's entire philosophy. The picture we now behold is one of cosmic evolution...

> (By applying) the known laws of classical and nuclear physics, we can reconstruct what things must have been like about 15 billion years ago when it all began...
>
> We should look briefly at what did happen on earth...The young sun blasted most of the hydrogen and helium out of the inner part of the solar nebula, leaving behind only the dust — or perhaps the already accreted planets. It is out of this dust that you and I were made...On Earth the steam from volcanoes condensed into seas...(organic compounds) fell out of the early atmosphere into the early seas, turning these seas into a consomme or chicken soup, Literally. It was out of this broth that life began...
>
> Life invented sex...the geological record shows that it was right after the invention of sex that evolution really took off...
>
> Out of chaos the wonderful laws of nuclear physics and chemistry have produced the complexity of the brain. Only the expansion of the universe prevents this evolution from being a violation of the second law of thermodynamics.
>
> What I find so impressive is that the steps from fireball to life, from chaos at the beginning to the indescribable complexity of, say, the human brain, all happened through natural law. Surely this must be the greatest miracle of all — that the universe, out of the fire in its beginning, has evolved not only into stars and planets, but into living things that now have eyes and minds that can contemplate the universe that begat them. What greater miracle does man need?

Scientists, for their own good as well as the protection of science from Aberration, must avoid the forbidden fruit of origins. In the greed that power encourages, scientists risk repeating the mistake of the dog in Aesop's fable, who, with an ample piece of meat in its mouth, looked down at its reflection in a pool of water as it crossed the bridge, and desired the piece of meat that "the other dog" had also...

Set for Eruption

By the late 1960's, there had developed in the scientific community, particularly in America, a certain air of arrogance. There seemed to be no limit to what scientists could accomplish — either in terms of destructiveness or of extending life comforts to humanity. The projected curves of where science could take humanity seemed to be unlimited — in both height and duration.

Transplanted organs were successful, man had escaped from Earth and returned, labor-saving conveniences had proliferated, and most mysterious diseases had lost their mystery. All thanks to science.

"If you've got it, *flaunt* it," even became the battle cry for science.

Particularly within the *historical* sciences (paleontology, archaeology, and geology) and the *life* sciences (biology with all its subsets like zoology, biochemistry, anatomy, and embryology), a restive nature brewed.

Perhaps rightly so, some of these scientists had resented for decades the restraint that the residual inertia of Judeo-Christian beliefs in America seemed to impose on their allegiance to the amoeba-to-man idea. One of the directors of the Smithsonian Institution in Washington, who is a personal friend of mine, even confided to me that the reason that there is no exhibit in the Museum of Natural History on the amoeba-to-man concept is because a strong reaction from the American people is feared if such a display were to appear.

However, the time had arrived by 1969 to come right out into the open with their belief system — to "let the chips fall where they may."

With the armor of scientific expertise, the sword of respectable technical language, the shield of prestigious scientific officialdom, and the helmet of academic inbreeding, they were set to deliver the knockout punch in a war that many of them believed began with Copernicus.

Little did they realize that they were going to set off an eruption — rather than seal any fate...

Chapter 3

ENTER CONCERNED CITIZEN

"The causes of events are ever more interesting than the events themselves" – *Cicero*

I suffer from a guilt complex. And I have been unable to get rid of either the guilt or the addiction that causes me to feel guilty.

My addiction? Reading the *Los Angeles Times* every morning. The reason I suffer is that I feel that I should be using my time to better advantage.

With the wide variety of activities that I pursue in my personal and professional lives, I have become increasingly jealous of my time. There are never enough hours in a day. Like many, I keep a perpetual list on my desk consisting of at least 35 items that require my attention.

Therefore, reading the *Los Angeles Times* every morning delays the accomplishment of these activities and generates within me a nagging sense of guilt regarding priorities for my time.

This guilt, for me, is a cumulative phenomenon, beginning every morning. It starts at zero and builds gradually. My little daily ritual of physical conditioning precedes reading of the *Times*. I generally warm up with some calisthenics, and then run a relatively fast half-mile (at about the pace that I used to run the mile on a college track team).

As I return home, I pick up the *Times* off the front lawn. The guilt, at that point, is about to start. But glancing at the headlines on the front page is okay. "I *should* be aware of major events going on in the world," I have convinced myself.

I can generally read the entire front page while my pulse and breathing are returning to normalcy. Still no guilt. *But*, from there on, guilt sets in.

EDITORIAL EVOLUTION

On Tuesday morning, 14 October 1969, I hurriedly scanned the bulky *Los Angeles Times*. The Tuesday editions average 80-90 pages in five parts. You don't come to the editorial portion of the newspaper until the latter part of Part II. By that time, my conscience is hollering for me to stop and get to work.

That morning, the normal urge to get going for the day was there when my eyes fell upon an editorial entitled, "*Evolution and the State Board*" (Exhibit 1). As I read it, a sense of indignation fast replaced my priority for work. The editorial was so illogical as to be ridiculous – cluttered with contradiction, bias, and arrogance.

The editorial's message did not hit me completely without warning. Several days earlier, a front-page article in the *Times* had discussed a California State Board of Education meeting in which evolution was a topic. However, I read that article with only passing interest. If anything, I was *annoyed* that the issue had been raised.

Evolution is what S. I. Hayakawa, renowned semanticist turned United States Senator, calls "an unexamined key-word in our thought process" that can hinder and misdirect our thought "by creating the *illusion* of meaning where no clear-cut meaning exists" (emphasis added).[143]

Evolution is an illusionary word. Trying to get someone to succinctly define it is like trying to pick up a drop of mercury. It goes in all directions…

Hoping to reduce my ignorance of evolution, I once purchased a book with a very appropriate title, *Understanding Evolution.*[144] The first chapter, "Meaning of Evolution," seemed to be the place where I could find a precise definition. Expectantly, I searched. Finally, on the last page of the chapter, the only definition appeared:[145]

> This, then, is evolution — *changes in the genetic* composition *of a population with the passage of each generation.*

All this says — in an excess of words — is that offspring are not like forebears. Well, any child can see that. Is that what excites everyone — either definitely *for* or definitely *against* evolution? Of course not. Is this what the State Board of Education was upset about? No. How often would the average layman define evolution that way? Never.

I finally learned something about evolution that biologists have known all along but have failed to tell the public. There are two different phenomena that are called evolution! As Macbeth says:[146]

> Evolution has two aspects: one large and relatively easy, the other smaller and much more difficult. All biologists know this, but it is often forgotten.

The first aspect of evolution simply affirms that living things change over a period of time. That's obvious, and it was for many years prior to Charles Darwin. This type of evolution (Evolution I) doesn't disturb anyone at all. Everyone agrees that it's true. We'd probably all say that it is a *fact!*

Evolution II, the second aspect of evolution, is, as Macbeth says, "the modus operandi, the *how* and *why*" living things change. This is *not* obvious.

There is great disagreement about Evolution II. It is not an irrefutable fact. It involves a *belief system* that cannot be tested nor demonstrated right at those critical junctures where proof is needed most. This is the concept that we have been describing as the "amoeba-to-man" idea thus far.

The damage to communications by using one word to mean two distinctly different things — especially when one has universal agreement and the other is controversial and irresolvable — is both critical and inexcusable. I cannot understand why biologists do not do something to correct this practice — unless they hope to have Evolution II "piggy-back" on the universally-accepted Evolution I.

The reason, then, for my personal annoyance with evolution being discussed by the California State Board of Education was that I believed nothing fruitful could come from a discussion of a term that has two meanings. Even worse, in a sense, it is like trying to settle whether the South had a right to secede from the Union, or whether two-platoon football has ruined that sport.

The sides have been taken before the argument starts, and no one is ever converted to the other viewpoint.

Guilt Loses Out

The 14 October 1969 editorial, however, introduced a factor that even overruled my annoyance with evolution. Guilt-ridden or not, I could not let such offensive journalism stand without challenge.

Imagine the *Times* editor even wondering whether the State Board of Education "would impose their views on schools"! What did he think such boards were for?

Without attempting to be melodramatic about it all, I had a strange but gripping compulsion to tell that editor what I thought of his editorial. Undoubtedly, this compulsion is what provokes most "letters to the editor." And it's a healthy application of the First Amendment to our Constitution. However, I had never written a letter to an editor in my life.

Once more, my conscience surfaced and began to haunt me because writing a letter would take additional time. Somehow, the compulsion to write a letter won the tug-of-war with guilt.

The letter was brief (see Exhibit 2). Sure, I polished it more than I would have an ordinary business letter. But even after I had it all typed, I wasted some more time wondering what to do with the letter beyond simply sending it to the editor.

"Why not send it to the members of the State Board of Education just in case the *Los Angeles Times* doesn't publish it?" I pondered.

"But you don't even know the names and addresses of the Board members, and to find them would take a lot more time," came thundering into my head. After seesawing back and forth for a few minutes, I decided to "buy extra insurance" by sending the letter to all the Board members.

It was not easy to get all the names and addresses of the California State Board of Education — but I managed it. When I had finished this entire operation — reading the editorial, writing a letter to the editor, locating the names and addresses of the State Board of Education, reproducing copies for each Board member, as well as attaching a cover letter, and mailing these letters — I had invested approximately four hours of my time.

With all my hang-ups about spending my time wisely, I often make foolish mistakes. I frankly thought that after spending four hours on this subject that I'd made another mistake with my time. However, I did feel that I had properly responded to that rather strange impulse that I had to write to the editor.

To my shock, before the *Los Angeles Times* could have reprinted my letter (which they never did!), I had a long distance telephone call from Thomas G. Harward, a Needles, California physician and the Vice President of the California State Board of Education.

"First of all, I would like to thank you for the very nice letter that you sent to the *Los Angeles Times*," he said. "You expressed yourself so much better than I could on the subject, that I wonder whether you might be willing to appear before the State Board of Education at their November meeting on the 13th in Los Angeles."

"Well, I don't know how effective I might be, but I am concerned about this issue and would be glad to help you in any way that you feel might be appropriate."

Hurriedly, I glanced at my desk calendar and saw that I had no out-of-town assignment on the 13th.

"If you are willing to appear and express your views, I will notify the Department of Education in Sacramento so that they can list your name on the agenda and issue you an official invitation to appear before the Board," he said.

"All right, since my calendar is clear for November 13th, you may notify them that I will be pleased to be a witness at the hearing."

As I hung up the telephone and walked into the kitchen to share this news with my wife Phyllis, my heart was in my throat.

First of all, what did I *know* about evolution? Secondly, I had just accepted an appointment by Wernher von Braun to the NASA Safety

Advisory Group for Space Flight that was scheduled to meet soon. My other business was also pressing me at the time. So I would have no time to research the subject. Third, I hardly knew *where* to find information that I might present to the State Board.

In the brief cover letter to the State Board members that I enclosed with my letter to the editor, I had stated:

> I would be pleased to assist you, in any manner you feel appropriate, in your efforts to assure a truly scientific approach to the origin of life. We can ill afford, especially in the name of science, to allow personal bias to preclude the teaching of *all* theories on a given scientific subject.

I had no *particular* type of "assistance" in mind. It was just a passing phrase. Good etiquette — I simply wanted them to know that I too cared. Dr. Harward quickly followed up his telephone call to me.

On Friday, 17 October 1969, my phone was ringing again.

"Hello, this is Walf Oglesby from the Department of Education in Sacramento. Dr. Harward has asked me to officially invite you to address the State Board of Education at their next public hearing on Thursday, November 13th, at 10:00 A.M. in Room 1138 of the Junipero Serra Building in downtown Los Angeles. He said that you will be speaking in support of the proposed changes in the *Science Framework*. Is that your understanding?"

"Yes, I agreed to be present and lend my support to revisions in the *Framework*."

"Fine, I will confirm our conversation by a letter to you which assures you a place on the agenda for that meeting."

By that time, I knew that I was in for some *serious* preparation. Where could I turn for help?

Help!

In 1952, I joined an organization known as the American Scientific Affiliation (ASA). It is small for a professional society — only about 2,000 members. Despite being "dues poor" because of many professional societies I had joined, I kept my ASA membership current because it is an *interdisciplinary* group, consisting of scientists from all the sub-disciplines of science.

In this day of increasing specialization, I find it necessary to keep the balance that only a generalist organization can furnish. In addition, I thoroughly enjoy the ASA quarterly journals. From time to time, these journals have discussed evolution in depth — not only as a *scientific subject* but also as a *philosophy*.

At the time that I received Walf Oglesby's call, I had attended only 5 or 6 ASA meetings in my 17 years of membership. However, I had somehow been elected Secretary of the ASA Executive Council for Southern California.

An idea popped into my head! Perhaps the best place that I could look for help would be to have the Executive Council meet with me right away. The next scheduled Council meeting wasn't until December however.

I called Dave Sheriff, a physicist at Consolidated Electrodynamics Corporation and the current President of the Executive Council. Hurriedly, I explained to him my dilemma and asked whether he thought we could assemble some of the ASA Council members who might be interested in helping me.

"Well, let me give it a try!" he said.

In just a short time, Dave called back. "We managed to get 4 or 5 fellows to meet tomorrow afternoon (Saturday, 18 October 1969) at 5:30 at Ralph Winter's home in South Pasadena."

"Great!" I exclaimed.

"But what do I want these fellows to *do*, now that they've agreed to meet with me?"

I began to panic.

I had not seen one thing in writing except the *Times* editorial. "What is the *issue* that I am urging to be revised?" I asked myself. "How can I get a copy of the *Science Framework* in which the so-called objectionable language appeared?"

The State Board of Education member who lived nearest me was Donn Moomaw, the former All-American football player from UCLA who was now Governor Reagan's pastor at Bel Air Presbyterian Church in Los Angeles.

I located Moomaw after several telephone calls and asked if I could at least borrow his copy of the *Science Framework* for my meeting with the ASA Executive Council. He seemed a bit dubious about meeting with me — even after I told him that I had received an invitation to appear before the State Board. However, he agreed to meet me at 3:30 on Saturday afternoon at his church.

In Moomaw's office overlooking the whole San Fernando Valley atop Mulholland Drive, he went over the contested part of the *Science Framework* with me. As it turned out, Moomaw had not originated the objection to the language about the origin of life in the *Framework* during the October Board meeting.

The primary spokesmen for revision had been Harward and John R.

Ford, another physician from San Diego. Moomaw had serious doubts about any benefit I could bring by testifying before the State Board. If anything, he tried to persuade me from appearing in the November meeting. I believe that he feared that I (or anyone else) would only "muddy the waters."

Following the 9 October 1969 Board meeting, Moomaw had requested Richard J. Merrill, a curriculum consultant in the Mount Diablo Unified School District, to express his viewpoint in writing on this controversy. He handed me Merrill's letter (dated 14 October — the same as mine) and asked me to read it.

It was a lengthy and thoughtful proposal that the issue of creation and evolution was improper — one that should not be pursued by the State Board of Education. (Later in the struggle, I became personally acquainted with Dick Merrill and found him to be a very fair and reasonable person.)

However, neither Moomaw's reticence nor Merrill's written arguments were enough to dissuade me from proceeding with my plans to meet with the State Board of Education. I thought that I perceived an issue that neither of them did.

Although it was the only copy that he had, Moomaw gave me his *Science Framework* so that I could go directly that afternoon to my meeting with the ASA Executive Council. On such short notice, Dave Sheriff had done an excellent job of getting a diverse group of men together.

Ralph Winter, in whose home the meeting was held, had been a personal friend of mine for some time. He later became a major source of counsel and encouragement to me during the entire struggle. Without his help, I doubt whether I would have continued. Upon completing an engineering degree at Caltech and a masters degree in education at Columbia University, Winter had received a doctorate in cultural anthropology and structural linguistics at Cornell.

Truly an interdisciplinarian, Ralph is a prolific writer as well as a well-read scholar. He and his family lived in South America for a number of years, and this experience enriched his perspective for the long struggle that none of us could foresee.

Also attending this early evening meeting — in addition to Winter and Sheriff — were Bernard Ramm, a theologian who has written widely on the confluence of science and religion, Mark Biedebach, a biology professor at California State College in Long Beach, and Bill Iwan, Professor of Applied Mechanics at Caltech.

We had very little factual information to discuss at our meeting because I did not even yet have a letter of invitation from Sacramento.

But as we all studied the *Science Framework* together, the only place that there was any language dealing with evolution was in one of the appendices of the *Framework*. Even there, the treatment of evolution was oblique rather than direct. Nonetheless, it was quite dogmatic. We agreed that it should be softened considerably.

Several — if not most of those present — felt that there was very little that I could contribute of a substantive nature. Some felt it was not worthy of any effort on their part. There didn't seem to be anything that we could really sink our teeth into aside from "softening up" some of the dogmatic statements in the appendix to the *Framework*.

One or two of the men were concerned that the press would only malign and distort the issue even if I did attempt to do something constructive. (And they turned out to be good prophets!)

After extended discussion, there was a weak consensus that perhaps I should prepare a formal statement — attaching to it a revision of the pages in the appendix that appeared to be overly dogmatic. If I was enthusiastic about making a contribution to science education when I arrived at the Winter home that evening, I didn't leave that way. I could have thrown in the towel right then...

The week following this ASA Executive Council meeting afforded me little time to put my thoughts together for my upcoming appearance before the State Board of Education.

A prior commitment to prepare a 20-year summary of aerospace activity was due — and as yet unwritten. As a member of the System Effectiveness and Safety Technical Committee of the American Institute of Aeronautics and Astronautics, I had accepted this assignment. The summary was due in a meeting of the Committee at Disneyland on 22 October 1969.

Likewise, most of the next week was committed to work at the Air Force Space and Missiles Systems Organization in Inglewood, California. So October slipped away without any preparation for the State Board of Education.

The NASA Safety Advisory Group for Space Flight, to which Wernher von Braun had appointed me, was meeting in Seattle on 4-5 November 1969. We were to be guests of the Boeing Company where we would be briefed on space simulation at the Boeing Space Center. We also were scheduled to have a presentation of systems analyses of the 747 jumbojet and the SST (supersonic transport) to aid us in our preparation for upcoming Apollo and Skylab missions.

I had intended to fly to Seattle to attend this meeting. However, as I weighed the upcoming date before the State Board of Education and my lack of preparation for it, I modified my travel plans.

Richard H. Bube, Chairman of the Department of Materials Science and Engineering at Stanford University in Palo Alto, California, had for many years been a frequent contributor to the *Journal of the American Scientific Affiliation* as well as serving as its editor. I greatly admired his writing.

Although we had never met previously, I called Dick to see if I could stop on my way to Seattle and "pick his brain" about some worthwhile thoughts to present to the State Board of Education.

When he consented to see me, I elected to drive instead of fly. That gave me flexibility en route to Seattle while having a car to visit many friends while in Seattle, our former home.

Dick and I met in his Stanford office on 31 October 1969. It was a very pleasant meeting, and one that I had looked forward to for some time. I thought he would be excited about this opportunity to make science education more objective.

However, Dick was not overly enthusiastic about my appearance before the State Board of Education. He felt, as had several others two weeks earlier at Ralph Winter's home, that this was at best a very superficial issue. It was so fraught with details of controversial terminology as well as the potential for being misunderstood, that Dick's enthusiasm was noticeably absent. He agreed that he would review anything I wrote, but he had nothing of consequence to offer me.

Leaving Dick's office and driving on to Seattle, my spirits were certainly not soaring. My personal limitations and inadequacies were rapidly coming into focus. On top of my incompetence and inability to enlist expertise to help me, I had not really run into *anyone*, with the exception of Ralph Winter, who thought the issue was worth preparing for!

"Maybe I should drop the whole thing," I remember saying to myself.

I tend to be a procrastinator. Or you could say more positively that I work best under pressure. Either way, it wasn't until I returned from Seattle on 7 November that I began in earnest to prepare some thoughts for the State Board of Education.

Neither the telephone call nor the letter of invitation from the State Department of Education in Sacramento had indicated any time limits for my presentation, so I felt no restriction on the length of my address to the Board. After drafting my thoughts, I reviewed them with Ralph Winter.

During this review, Ralph reminded me of something that I had tentatively agreed to do at the meeting in his home. I was going to *rewrite* that part of the appendix of the *Science Framework* that seemed to be overly dogmatic. I had forgotten about it...

More Than Bellyache

I had concentrated on my oral presentation. But, as Ralph Winter reminded me, "What do you want the Board to *do* — after they've heard your speech?"

"Obviously, if they just *listen* to me — or even if they get excited by my message, I have accomplished nothing. They must take some overt action to change *status quo*."

I had to provide some words that they could easily adopt right into the *Framework* if I had persuaded them.

Most public hearings, like the upcoming one, are nothing more than a platform to allow folks to discharge some emotional feeling. Or, as is commonly said, it allows you to "vent your spleen." Citizens seldom present a *solution* to the governing body before whom they appear. They simply express their side of a dilemma.

But it is important to do something beside register a complaint. I was beginning to realize the wisdom of my colleagues in suggesting that I rewrite a portion of the *Framework*.

To provide the State Board of Education a defined alternative (instead of just a complaint), I decided to *revise* a small section — rather than edit individual words — consisting of two paragraphs in Appendix A of the preliminary draft of the *Science Framework*.

It seemed to be simpler to do it that way. The small section that I selected to rewrite was later described as:[147]

> In Appendix A, the (California State Advisory) Committee (on Science Education) listed seven major conceptual systems, with subdivisions, that "provide a foundation for the understanding of how certain facts are related" and "offer a perspective by means of which future discoveries may be correlated and understood." Conceptual system G was "Units of matter interact," and the subdivision G-2 was "Interdependence and interaction with the environment are universal relationships." Examples of inter-dependence and interaction were given from the fields of astronomy, physics, chemistry, and biology. The biological examples included mention of the integration and interaction that occur at the levels of the components of cells, entire cells, organisms, populations, communities, and ecosystems. It was noted that some of these interactions are cyclic — such as those involving food, water, and the stages of the individual's life. The final example of interactions was that of evolution, involving both living and non-living phenomena.

It was my intention that the State Board, if they approved of my approach, would simply replace the earlier version with my rewritten version *in toto*.

Admittedly, the latter was approximately twice the length of the former, but I felt that the extra length was justified because we were proposing an additional alternative concept; i.e., creation, as well as softening dogmatic statements about evolution. (Exhibit 5 contains both the original Appendix A version and my proposed alternative.)

The theme of what I thought I was trying to accomplish by rewriting those paragraphs was *reconciliation*. I visualized two forces that were polarized and highly-charged. Each of these camps of thought sought the *elimination* — not the coexistence — of the other. I saw merit in both forces. My thrust, I then reasoned, should be toward giving each view an opportunity to be presented — a sure cure for dogmatism.

As I rewrote the Appendix A paragraphs, I tried to be as fair with both points of view as possible. "Those who will wish to criticize me (and there were many who eventually did) may call me 'naive' but should not be able to criticize me on *scientific* grounds," I repeatedly said under my breath.

I was to learn later that, because of my naiveté, I *was* in error occasionally on scientific grounds. On those occasions, I was chagrined and sorry because I never had any intent of abusing or misusing science.

Here's an example of my naiveté forcing me right into the middle of crossfire. I honestly (but mistakenly) thought that educators and scientists wanted to be *objective* about origins. My role, I believed, was to point out some common ground that the two forces could jointly occupy comfortably — a goal that Henry Kissinger probably visualized in his concept of *detente*.

One possibility for common ground seemed to lie in pointing out that science is not based on *singularity* — requiring only *one* explanation for natural phenomena.

Dualism is well known in science. Gerald Holton, the Harvard physics professor and science historian quoted previously, even wrote an article that affirms dualism as an ingredient in the growth of scientific thought: "On the Duality and Growth of Physical Science."[148]

In fact, Holton cites two types of dualism — a dualism between *public* and *private* science and a dualism between the *conscious*, rational work of a scientist and the *unconscious*, irrational and even metaphysical creative drive of the scientist.

Here's how he describes their interrelationship:[149]

> As contrasted with his largely unconscious motivations and procedures, the intellectual discipline imposed upon the physical scientist is now quite as rigorously defined as the conventionalized form for research papers. This super-position of discipline and convention upon the results of free

creation represents a dualism in the work of the scientist which runs parallel to the dualism in the nature of science itself.

So in my search for a common ground, I wrote in my proposed version for Appendix A (see Exhibit 5):[150]

> All scientific evidence to date concerning the origin of life implies at least a dualism or the necessity to use several theories to fully explain relationships between established data points. This dualism is not unique to this field of study but is also appropriate in other scientific disciplines such as the physics of light.

Although I did not cite any specific reference in that document to support my contention of dualism, I did elaborate about dualism as well as pluralism in my oral presentation before the State Board of Education (see Exhibit 4). In that address, I quoted my own undergraduate physics text regarding dualism in the physics of light:[151]

> The present standpoint of physicists, in the face of apparently contradictory statements, is to accept the fact that light appears to be dualistic in nature. The phenomena of light propagation may best be explained by the electromagnetic wave theory, while the interaction of light with matter, in the processes of emission and absorption, is a corpuscular phenomenon.

The book's author, Francis W. Sears, Professor of Physics at Massachusetts Institute of Technology, undoubtedly had heard of Max Planck's quanta, Louis De Broglie's matter waves, and Niels Bohr's two integrating postulates about electron behavior. Even I had. Neither of us was trying to say that light could not and would not ever be explained by a single unifying theory. Rather, it was easier to work with two concepts of light — and it still is.

Later chapters will document the hysteria that this duality analogy generated. My obvious error was in assuming that either of the two opposing forces *desired* a common ground. They were not about to listen to reason.

One more account will describe how I caught brickbats from both opposing forces by proposing a commonality. In concluding my rewrite of Appendix A, I attempted to support the duality concept by specific examples from both camps:[152]

> Note that creation and evolutionary theories are not necessarily mutual exclusives. Some of the scientific data (e.g., the regular absence of transitional forms) may be best explained by a creation theory while other data (e.g., transmutation of species) substantiate a process of evolution.

Since these two examples later became of consequence in the struggle, I want to explain how I selected them. The first example, "the regular absence of transitional forms," was taken directly from the writing of G. G. Simpson — probably the most widely-read and quoted paleontologist on the subject of evolution. By no means was the phrase taken out of context. Exhibit 48 provides Simpson's setting for this phrase. The paragraph from which the phrase was taken reads:[153]

> This regular absence of transitional forms is not confined to mammals, but is an almost universal phenomenon, as has long been noted by paleontologists. It is true of almost all orders of all classes of animals, both vertebrate and invertebrate. A fortiori, it is also true of the classes, themselves, and of the major animal phyla, and it is apparently also true of analogous categories of plants. Among genera and species some apparent regularity of absence of transitional types is clearly a taxonomic artifact: artificial divisions between taxonomic units are for practical reasons established where random gaps exist.

The second example, "transmutation of species," was quoted from the writing of James O. Buswell III, an anthropology professor at St. John's University in Jamaica, New York, who is well known for his position as a creationist:[154]

> Although evolution does not consist of wholly distinct processes, and a fossil series, for example, can exhibit speciation, or splitting, in a phyletic pattern through geological time, nevertheless the well documented data of natural selection on these levels – the genetic and geological processes – may be abstracted from the overall theory of organic evolution. *The creationist may accept all of the facts within these two areas of consideration.* Thus he need have no quarrel with the transmutation of species or other taxonomic categories, and may fully accept the genetic explanation for variation.

Again, one can see that the phrase was not lifted out of context. Instead, I sought out experts from both poles of thought so as to avoid error or sensationalism.

Despite my efforts to be scientifically correct, I was later attacked by evolutionists for using Simpson's words (although I am rather certain that they were unaware that he had written them) and by creationists for using Buswell's phrase. Sometimes you can *never* win...

Another very difficult semantic problem I faced in my proposed revision of Appendix A involved the two sticky words — *evolution* and *creation*. Not only are there two major and different definitions or uses of evolution — Evolution I and Evolution II (as earlier described), there are also at least two radically different uses or meanings of the word *creation*.

Creation I might be defined as the means or mechanism by which something comes into being. Creation II is a cosmology that attributes origins and preservation of the universe to a supernatural Being (God). Creation I *is* a scientific use of creation while Creation II is *not.*

I was aware of the confusion that was potential in both of the terms *evolution* and *creation* as I set out on my task of rewriting Appendix A. However, I vastly underestimated the difficulty I would experience in trying to keep terminology straight in the minds of those directly involved in the struggle.

(I later even tried substituting "chance" for evolution and "design" for creation in order to clarify differences, but it was also unsuccessful for reasons to be explained later.)

When I used the word "creation" in my oral presentation and amendment to Appendix A (Exhibits 4 and 5), I had in mind Creation I (the *scientific* instead of the *religious* use of the word) as illustrated in the following report:[155]

> Scientists at the University of California San Diego believe the spectacular space shows put on by comets may be the actual process of creation going on before the eyes of man. The new theory, developed by Dr. D. Asoka Mendis, research physicist, holds that comets may be forming, disintegrating and regenerating continuously in a process basically similar to that which brought the earth and the solar system into being. The theory upsets the accepted view that comets came into being at the same time as all other matter in the solar system some 4.5 billion years ago. Mendis' work, soon to be published, was announced Monday by two colleagues, Dr. Hannes Alfven, Nobel laureate in physics, and Dr. Gustav Arrhenius, professor of marine geology at the Scripps Institution of Oceanography. The scientists agreed that if the theory holds up, comets may in the near future help unlock the mysteries of how the earth and the solar system were created. Alfven and Arrhenius are among 10 investigators serving on a NASA panel to examine the possibilities of sending unmanned missions to comets and asteroids before the end of the decade to seek clues to the processes of creation.

Creation, in scientific terminology, is even used by astronomers in connection with the Big Bang theory for the origin of the universe, as indicated by this article:[156]

> According to this (Big Bang) theory, the universe began as a super-dense ball of energy. The ball exploded – astronomers refer to this as the Creation Event – and the energy was scattered, gradually becoming matter as the universe constantly expanded.

Another example of the scientific meaning of creation is frequently used in genetics. A team of scientists at the University of Wisconsin, headed by the Nobel laureate, H. Gobind Khorana, announced that they had created a man-made gene. This gene, according to these experts, could lead ultimately to the "artificial creation of life itself:"[157]

> With this first artificial gene, the team of scientists...has taken a profound step toward correction of inherited diseases perhaps genetic "engineering" of improved humans and animals, and perhaps ultimately artificial creation of life itself.

Since one seldom hears of something *artificial* that does not have a corresponding *real* equivalent, scientists must be acknowledging an actual creation of life with a mechanism that they are attempting to duplicate. This, on the other hand, is not meant to imply that they are acknowledging a Creator.

They are only using creation as a description of the means by which life came into being from a previously non-existent state without known natural cause.

My purpose, then, was to show that the history of the universe was not necessarily a slow, gradual, uninterrupted process as Evolution II proposes. There is an alternative possibility that the history of the universe is like a series of distinct chapters, punctuated by discrete events representing discontinuities in a continuum.

Those points or events which appear as discontinuities or origins could be accounted for by (1) postulating some hypothetical (but not demonstrable) factor like G. G. Simpson's "quantum evolution" (see Exhibit 48) or, (2) by postulating the intervention of a Creator.

In describing these origins themselves, scientists would be using the word creation in a *scientific* — not religious — sense. On the other hand, in accounting for *why* they occurred or in attempting to postulate a mechanism for occurrence of origins, scientists would have to leave the domain of science in order to describe what they *believe* about these unique events.

The origin of life — that issue that had erupted in the State Board of Education — was simply one of those creative events. Undoubtedly, this subtle distinction failed to be recognized by Board Members, the news media, or the general public. In most of these people's minds, creation had an exclusively religious meaning. If *creation* was used, it automatically implied a *Creator.*

By such detailed explanation of why I amended Appendix A as I did, I wish to establish two premises. First, I had no deliberate intent to capitalize on the ambiguity of the two words *evolution* and *creation.* Secondly, I was sincere in proposing that creation is a word used in scientific circles quite apart from any religious connotation.

I was also aware, although I had not met any of the members of the

State Board of Education at the time I prepared my amendment, that much of the controversy really dealt with disagreement over Evolution II (the *belief* in the amoeba-to-man idea) and Creation II (the *belief* in a Creator).

Since neither one of these belief systems are *scientific* (because they actually deal with Socratic truth), I knew that they could not be resolved in terms of science. On the other hand, since Evolution II was already entrenched in the *Science Framework* under the guise of science, my intent was to introduce Creation I and, at the same time, hope to limit all discussion of evolution in science textbooks to Evolution I.

If I was successful (and it was a big gamble), the discussion about origins in science textbooks would be *cleansed* from private beliefs — in either *scientific naturalism* (the amoeba-to-man idea) or *religious creationism* (a Creator). At the same time, opportunity was being presented to the science student to *understand* two things: (1) neither of the two belief systems could be validated by the scientific method, and (2) either of the two belief systems is *equally valid* from a scientific viewpoint.

Because of my naiveté, this ambitious gamble failed...at least partially.

FRAMEWORK FRAY

The Scopes trial of 1925 was deliberately provoked. Before Martin Luther King, Jr. was born, his belief in "civil disobedience" was used by John Scopes' friends. As described earlier, Scopes was arrested willingly. A confrontation was *sought.*

Not so in 1969 in California. Confrontation caught up with people who were running away from it. Like many wars in the past, everything done to prevent it seemed to guarantee its inevitability. The focal point of confrontation was an instrument unique to the State of California — a *Science Framework.*

As in many states, California has boards of education at the city, county and state level. Without any question, this is a clumsy arrangement that many believe to have outgrown its usefulness – like many other bureaucracies. All the jurisdictional jealousy, administrative anarchy, political polemic and operational obtuseness attributed to government agencies in general are also evident in California's many boards of education.

The capstone of the educational hierarchy in California is the State Board of Education. Its 10 members are appointed by the Governor. By law, the State Board is responsible for establishing overall educational policy as well as approving all textbooks used in elementary and junior high schools. These two responsibilities merge, in a sense, in documents known as "frameworks."

Frameworks are just that. They are a loose set of guidelines – not detailed requirements. They stipulate broad, foundational aspects that should be taught rather than specific rules for doing so. Instead of definitive standards to be followed by educators and publishers, goals, operational objectives, and instructional strategies are stressed.

In 1969, the *Science Framework* wasn't *unique* either. Frameworks are prepared in *every* major subject — in music, mathematics, social science, language arts, as well as science. This is a continuous, on-going activity, with frameworks being prepared for several subjects simultaneously.

Some naive observers of the Science Textbook Struggle erroneously believed and reported that the *Science Framework* of 1969 was some type of rare, novel and monumental work.[158] *Science Frameworks*, like frameworks in all other subjects, are either updated or rewritten every three to four years.

Therefore, the confrontation that erupted during the approval of the 1969 *Science Framework* was purely accidental — instead of being deliberately concocted like the Scopes trial of 1925.

As you might guess, framework preparation is a long, tedious bureaucratic process. For anyone with an analytical mind who is accustomed to making clear-cut decisions, the process is completely frustrating. And it takes no profound insight to realize that, by the time a framework has been prepared, reviewed throughout the state by over 1000 educational groups, cliques and boards, and presented to the State Board of Education for adoption, it is not universally acclaimed. It has both friends and enemies. The saying, "A camel is a horse designed by a committee," certainly applies here.

Educational Enhancement

Providing frameworks for each academic subject in public schools is admirable — for several reasons. Frameworks are a practical tool for carrying out the legal requirement assigned to the State Board of Education to establish educational policy. By periodically rewriting or editing existing frameworks, educators are also obliged to incorporate material and update textbooks and programs to reflect progress in each subject. Most importantly, frameworks provide a periodic excuse to solicit expert counsel and to focus on the enhancement of a given subject.

California's leadership in educational innovation has long been recognized. Further, because 10% of all the nation's textbooks are purchased in California, whichever direction California takes pretty well determines which way the rest of the states will ultimately go. Frameworks form the backbone of California's leadership in primary and secondary education.

How are frameworks prepared? Obviously, the 10 people appointed by the Governor to the State Board of Education cannot write such documents. First of all, they are not qualified to do so. They are deliberately chosen to represent a cross-section of society – from business, homemaking, the professions, agriculture and labor. Secondly, they pursue full-time careers and serve on the Board only as unpaid public servants. Third, framework preparation is but one of their many wide-ranging functions as Board members.

The responsibility for framework preparation is delegated by the State Board of Education to one of several major commissions formed by the Board — the Curriculum Commission. This Commission, renamed in 1972 the Curriculum Development and Supplemental Materials Commission, has primary responsibility for securing and organizing expertise to write the frameworks (see Exhibit 15). In the mid-1960's, the Curriculum Commission appointed a framework-writing committee of 15 people known as the State Advisory Committee on Science Education.

This committee was then authorized by the State Board of Education to start writing the *Science Framework*. It took about four years for them to complete the 1969 version of the *Science Framework*. Without any doubt, when they finished their work, this Committee had produced an outstanding document.

However, as all of us who have worked with the democratic process are well aware, excellence is not the only criterion in public documents. In fact, excellence is seldom of great importance. Of far greater consequence is securing the official approval that gives a document public authority. The State Advisory Committee on Science Education could not foresee a snag ahead in the approval process that would offset the excellence of their work...

Publisher's Nightmare

While frameworks represent a noteworthy innovation in education, they are not all "peaches and cream." Publishers, for one, do not consider them to be a godsend. Certainly, it is valuable for publishers to have input (like a framework) from a group of experts that the State Advisory Committee on Science Education represented in 1969 on the subject of science. But publishing textbooks and other supplementary teaching materials such as audio-visuals, laboratory experiments, and workbooks is complex business that must integrate information and expertise from many diverse sources.

The California frameworks, *per se*, do not become contractual instruments between the state and the publisher. On the other hand, publishers cannot ignore a framework. The Curriculum Commission, which performs all screening and review of textbooks and materials, as well as recommending which of these shall be adopted by the State Board of Education, bases their opinion, in part, on the frameworks. Therefore, a framework might just as well be a contractual requirement for publishers.

Publishers must remain profit-making businesses or die. They are not altruistic. They feel no calling to re-invent American education or champion social reform. They must sell their materials.

California frameworks are simply an additional factor to complicate a publisher's marketing problems. Publishers must maintain awareness of what the framework writers are thinking — from the time they start until they finish. Since it took four years to prepare the 1969 California *Science Framework*, you can imagine the amount of dollars spent by every publisher to keep aware of the latest consensus of the State Advisory Committee on Science Education.

Textbooks also take a long time to prepare. It often requires several years between the time a publisher decides to issue a textbook and that textbook appears in a classroom. To compound the publisher/framework interface, it would only be coincidental if a publisher would start work on a new textbook precisely at the time a framework was completed. So textbooks and frameworks are generally out of phase.

Revising a book is expensive and almost prohibitive once a publisher decides to print a textbook. So any last-minute reversals or changes of conceptual direction expressed in a framework have predictable repercussions throughout the publishing world.

Earlier, it was pointed out that frameworks are only intended to be loose guidelines. But that intention also presents publishers with mortal danger — *proving compliance* with obtuse and vague concepts.

California, in addition to being America's most populous state, is also thought by many to be America's oddest state. It has earned the title of "Heretic Heaven" for its many religious cults and bizarre theologies. Politically, it defies prediction. It has been more than one pollster's nemesis. Educationally, California likewise has wandered all over the ballpark.

One little-known fact — even to Californians — is that publishers do not print the textbooks used in California public schools! They are printed by the Office of State Printing in Sacramento. This practice allows some very irregular and questionable procedures to be carried out. Most germane to the Science Textbook Struggle, it permits the state, through its State Board of Education, to *edit*, *excise*, *rewrite* or otherwise *modify* textbooks.

Major publishers seldom prepare textbooks for a *regional* market — even for the largest state in the Union. They aim at a *national* market in order to realize the greatest profit. Since California uses frameworks as a means of educational innovation as well as specifying unique requirements for the state, the fact that California prints its own textbooks makes it possible to have a *California* edition of a national textbook.

If you pause to consider the ramifications of this practice, several startling reactions can jump out at you. "Then, the state is using tax money to compete with free enterprise (the publishing industry)?" Precisely.

"Do you mean that the state can *dictate* what will be said about history, science or any other subject even though none of the State Board members possess expertise in these fields?" Yes, I do.

"Why would California want to become a textbook publisher in the first place?" you might ask. The answer is almost ridiculous. California stumbled into publishing by a series of unrelated events.

By the 1880's, the state had already invested over 2 million dollars — at that time, a large amount of money — in a state printing plant. About that same time, some textbook publishers got themselves involved in some price-fixing. This unethical activity caused anti-trust action to be initiated against them.

Popular sentiment against publishers prompted the people, in 1884, to amend the California constitution by the so-called Perry Amendment which called for a uniform series of textbooks throughout the state, to be compiled by the State Board of Education, *printed by the State Printing Office* and sold at cost of printing, publishing and distributing.[159] In 1885, legislation was enacted, appropriations made, and this plan put into operation.

The Perry Amendment also contained a "sleeper." It stipulated that *California* books and authors must be given preference in textbooks selected by the State Board of Education. This stipulation ran into high headwinds right from the start. By 1890, resolutions were being passed by educators throughout the state, demanding relief from this requirement because California authors could not provide high quality books.

So in 1903, the California Attorney General wrote an opinion that permitted the abandonment of local authorship and allowed the *leasing of plates* from regular publishing companies.[160] Instead of reexamining the *primary* problem of quality education and textbooks, the Attorney General addressed the problem of how to sustain the State Printing Office!

Then another event took place that further compounded state

printing of textbooks. Not many parents today realize that, at that time, pupils (or their parents) had to buy their own textbooks!

It seemed logical in 1912 to amend the state constitution once more — to provide *free* textbooks to pupils. This amendment was carried by a large majority of the voters. And the comedy of errors continued — always *avoiding* the quality education problem and *focusing* instead on the support of a state printing bureaucracy.

Thereby, California became and has remained a publishing firm. Of course, they consume their own product and do not compete outside California with other publishers.

However — and most significantly — this manufacturing monopoly obviously removed the choice of textbooks from the local level (where parents could have a voice) and centralized it in the hands of 10 powerful people — the State Board of Education.

Centralized power always has two sides. It can either be a *blessing* or a *curse.* It's great to have all decisions centralized if you wish to spread a theory or institute reform. It's so *efficient!* You simply convince the few decision-makers, and your ideas are implemented.

However, if the decision-makers decide to order a new approach or teach precepts which you oppose, there is virtually no way to stop them. It then becomes *frightening!*

California never went to the extreme of having only *one* textbook per subject, although there were attempts at statewide standardization (to clinch the rationale of state printing monopoly) in 1916 and again in 1927.[161] However, these efforts were both soundly defeated.

Over the years, there have generally been three or four books selected for each subject by the State Board of Education and printed by the state. However, the significant cultural, ethnic, geographic, economic and racial diversity throughout California recently introduced another new wrinkle to the textbook scene. It caused the State Legislature to tinker some more with the experiment of textbook acquisition. A bill was passed in 1972 that stipulated:[162]

> The (State) Board (of Education) shall adopt a minimum of five instructional materials and instructional materials systems per subject, per grade; there shall be no limit on the *maximum* number of materials on any list.

This law obviously opened up the number of options available to the local school districts for instructional materials. However, it virtually eliminated the original reason for California being in the textbook printing business!

It was no longer efficient to print relatively small numbers of textbooks. Do you think that state printing officials recognized this obvious change and abandoned the business? Of course not. I am unaware of any bureaucracy that ever voluntarily proposed a reduction in its own size.

Perhaps one of the reasons that California continues to hang onto textbook printing, despite its blatant inefficiency, is that the state can thereby maintain centralized control of curriculum. It is clear that the 1972 legislative action diffused the power that had previously existed as well as virtually returned *in toto* the choice for instructional materials to the local school districts. Yet, just as long as the state continues to print textbooks, the archaic system of reviewing and approving textbooks at the state level will limp along.

Appendix Apoplexy

John R. Ford, a physician from San Diego, was named by Governor Reagan to the California State Board of Education in 1968 to fill an unexpired term. He was appointed in 1969 to a regular four-year term on the Board.

In August 1969, Ford was Chairman of the Board's Policy and Programs Committee. All frameworks were presented to this Committee for review prior to being submitted to the full Board for adoption. It was a routine matter for the Curriculum Commission to present the 205-page draft of the *Science Framework*, written by the State Advisory Committee on Science Education, to Ford's committee during the August Board meeting.

"We were quite concerned with the *methodology* of teaching science at that time because many scientists, in proposing new ways of teaching science, felt that we ought not to tell the children the *facts* — but only let them ask *questions*," Ford recounted later. "My response, along with others, was that if they don't know the *facts*, how can they know what *questions* to ask?"

This disagreement over teaching strategies and the use of the inquiry process caused Ford's committee to ask the *Framework* writers to rework the document before the September Board meeting. The Curriculum Commission then brought a revised version of the Framework to the Board in September.

"It was evident that the Commission thought that the *Framework* would be adopted now without any problem at all," Ford said. But the one-month delay had allowed copies of the Framework to be reviewed by various private citizens, among them Mrs. Marilyn Angle, who had somehow secured a copy and had written a critique that she circulated to the Board members.

"As I was idly perusing the *Framework* and Mrs. Angle's memo while the *Framework* was being presented to the Board, my eyes struck a portion of her critique which said that the *Framework* dealt too much with evolution. She referred to some 20 pages that bothered her. So I began thumbing through those pages. All of a sudden, my eyes fell on these words:

> From the origin of the first living particle, the evolution of living organisms was probably directed by environmental conditions and changes occurring in them. A soup of amino acid-like molecules, formed in pools some 3 billion years ago, interacted with oxygen and other elemental constituents of the earth, probably giving rise to the first organization of matter which possessed the properties of life."

"This really struck me," Ford recalls. "Could I as a physician sit back and let this go through? Never."

As soon as the Curriculum Commission finished their presentation of the *Framework* to the Board, Ford said, "I feel that this guideline should not be accepted without at least alluding to creationism which is an accepted theory by scientists in this country. I think that we, as a State Board, would be remiss if we did not include the theory of creation along with the evolutionary theory of life."

The Board Vice President, Thomas G. Harward, also a physician, then spoke up: "You know, I agree with Dr. Ford. In fact, every time I deliver a baby, I think of the wonderful creative power of God. How could anyone believe in evolution who witnesses such an event?"

Board member Donn Moomaw added his support to Ford and Harward.

Without much more discussion, the Board passed a unanimous resolution to send the *Framework* back to the Curriculum Commission.

It is important to stress, once again, that the words that caught Ford's attention were *not* in the main portion of the *Science Framework*. They were in a rather obscure portion of Appendix A, one of three such appendages to the *Framework*.

Probably because of this relatively unimportant status of these words, the Curriculum Commission did not attach any great significance to what they considered rather routine action by the Board of sending the *Framework* back for a little cleanup of phraseology prior to adoption.

Instead, the Commission delegated to the State Department of Education, the Sacramento bureaucracy devoted to supporting the State Board of Education (but frequently usurping their authority), the task of drafting a few words to placate the obviously obscurant thinking of two non-scientific physicians.

How ironic that tiny threads of misjudgment could be forming a confrontation that no one wanted! Taken singly, the response of Ford, the action of the Board, the decision to delegate by the Curriculum Commission, and the rewriting by the State Department of Education were individually inconsequential. Yet together they were coagulating to produce the sudden paralysis of apoplexy.

No one could really argue with the brief statement that the State Department of Education drafted to "scoot across the shoals" that temporarily hung up the *Framework*:

> The origin of life and the evolutionary development of plants and animals have long been of interest to the layman and scientist alike. The oldest explanation is a religious one — that of special creation. Aristotle proposed the theory of spontaneous generation. In the 19th century, the concept of natural selection was proposed. This theory rests upon the idea of diversity among living organisms and the influence of the natural environment upon their survival. Fossil records indicate that hundreds of thousands of species of plants and animals have not been able to survive the conditions of a change in environment. More recently, efforts have been made to explain the origin of life in biochemical terms.

It was an admirable attempt to desensitize an emotional edge — where science and religion appear to conflict. It *didn't* use "evolution." It *did* acknowledge "creation." Quite conciliatory, to say the least. It avoided any judgment as to which explanation was the *best* (except perhaps by implying the most recent). It masked any implication or suggestion of the "amoeba-to-man" idea.

What else could you ask for? Nothing — except an understanding of what Ford, Harward and Moomaw had *meant* by their objection. The State Department of Education had failed to empathize with these men and thereby missed the target by a million miles!

When the State Board convened on 9 October 1969, the conciliatory paragraph was presented to them. John Ford scanned it — obviously looking for what was said regarding creation. There it was — "The oldest explanation is a religious one — that of special creation."

So *that* was the only bone they were throwing to him? He had said in September that *scientists* accepted creation as a theory. Yet this statement equated creation exclusively to *religion*!

"This explanation is entirely inadequate. I feel that we ought to have a hearing on it, let the public speak — coming to us with ideas on how we can present both creationism and evolutionism to the boys and girls of this state," Ford roared.

In short order, the Board concurred with Ford that a public hearing should be called for November. Battle lines were drawn. Forces could now be mobilized. The confrontation had become clearly inevitable. The Board had choked on one small sentence that had been meant to make swallowing easy...

"MEMBER OF THE AUDIENCE"

In complete ignorance of the State Board actions that have just been recounted, I had prepared my oral presentation to them (Exhibit 4) together with a proposed rewritten version of two pages of Appendix A in the *Science Framework* (Exhibit 5). Ralph Winter had reviewed and polished both of them for me.

But I was still nervous about them.

"Is there anything I could do to confirm that I am even on the right track?" I wondered.

"Why not call Dr. Harward, Vice President of the State Board and the person who recently invited me to speak?" entered my mind. I weighed the wisdom of the thought. Then reacting to this impulse, I called him in his office in Needles, California.

"Dr. Harward, this is Vern Grose. I am concerned about whether the talk I intend to give before the Board this week really meets the intent of your invitation to me. Would you possibly be coming into Los Angeles prior to the meeting Thursday morning so that we could meet together?"

"Yes, I will be driving in during the late afternoon on Wednesday (12 November) and registering at the Biltmore Hotel downtown. Why don't you come down after dinner — say around 8:00 o'clock?"

"Fine. I'll bring my speech so that you can review it. See you then!"

I hung up feeling a little bit relieved. At least, someone on the Board would see what I was saying before I made some horrible blunder. On the other hand, discussing it with Harward that late on the night before a meeting early the next morning really gave me no time to revise the 13 typewritten pages of my speech. So I suppose I was really seeking *confirmation* rather than *correction* from him.

Still seeking to assuage my insecurity and doubt about this up-coming venture, I thought about calling my neighbor, Chuck Welsh, to go with me to see Harward. Chuck was Dean of Hamilton High School in West Los Angeles. We had become close personal friends because his three boys and my two boys lived together almost constantly.

"Chuck, how would you like to meet the Vice President of the State Board of Education?" I half jokingly phoned him.

"I might as well peddle a little influence instead of revealing my insecurities," I said to myself.

"What's the deal?" he asked.

"Oh, I've been asked to make a little pitch to the State Board on Thursday morning, so I'm going down tonight to try it on for size with him ahead of time. Would you like to come along?"

"Oh, I don't know...Maybe. What time are you leaving?" He didn't sound enthused.

"Around 7:15."

"Oh, I suppose I could go. What's the reason again?"

"It's too involved to discuss now. I'll tell you about it while we drive into town."

Chuck is a low-key guy. He quizzed me as we drove the 30 miles into the Biltmore Hotel. I don't think I was very effective in convincing him that I had any business in the issue. If anything, he seemed to think it was incredible (a) that the State Board was spending *its* time on an issue like this and (b) that I was wasting *my* time getting involved in such a foolish controversy.

Parking in the lot beneath Pershing Square that the Biltmore faces, we walked through the hotel lobby to the desk, asked for Harward's room number, and then called his room. He invited us up.

Consonant Caucus

When we arrived in his room, Dr. Harward was busily unpacking a large briefcase full of papers to review for the Board meeting the next morning. Not wishing to waste any of his limited time, and yet hoping that he would read my 13-page speech, I immediately handed a copy to him after I had introduced him to Chuck. He rapidly scanned both the presentation and my proposed revision.

"I think it looks okay to me. May I keep these?"

"Certainly," I replied — with great relief.

"Fine. I'll get them reproduced for each Board member before we convene in the morning so they can follow what you say in writing."

Harward mentioned that the issue had stirred lots of reaction — including considerable mail. As it turned out, my letter to the State Board was not the only one provoked by the *Los Angeles Times.*

Among others that the Board also received was one written by N. H. Horowitz, a Caltech biology professor (see Exhibit 3). His letter was co-signed by 14 other biology professors at Caltech, including Max Delbruck, who won the 1969 Nobel Prize for physiology and medicine in the field of bacteriophage research — the specialized virus that destroys bacteria by taking over their genetic apparatus. I found their letter most confusing — and apparently so did the State Board. Only rarely can impressive credentials be substituted for common sense.

As Chuck and I drove home that evening from the Biltmore, I felt a wee bit more confident that my preparation was adequate. Our brief caucus had produced consonance for the morrow...

Raucous Reception

Anticipation mixed with apprehension ...I think that's what I felt as I walked into the Junipero Serra Building at First and Broadway in downtown Los Angeles on Thursday morning, 13 November 1969.

Public speaking doesn't bother me. I have delivered dozens of papers at international technical symposia and meetings. However, I had always known both my subject and my audience in those situations. But here ...I knew neither. Doubts crept through my mind.

"Maybe I've bitten off more than I can chew," I thought. A counter-argument raced in to replace that concern — "You have nothing to lose because there is nothing at stake!"

My only brother Jerry lives just a mile from us. Feeling the need for all the moral support I could muster, I had asked his wife Shirley to accompany Phyllis and me to the public hearing. The three of us looked for Room 1138.

There it was — with large double doors.

All my friends know that "punctuality" is my middle name. Today was no different. "Public hearings do not always come off as smoothly as the Board would like. I suggest you plan on being present from 10:00 a.m. until 12:00 noon," Walf Oglesby's letter of invitation from Sacramento had said. It was only 9:30 when we walked through the double doors.

Room 1138 is actually a very large auditorium with several hundred padded theatre-style seats. It was about one-third full as we walked in.

The Board was in session. Max Rafferty, State Superintendent of Public Instruction (an elected position) who also serves as Executive Secretary of the Board, was making his report.

Phyllis, Shirley and I found seats in the center section about four rows from the front. Except for the well-lighted semicircular arrangement of desks on the dais, the rest of the auditorium was quite dark.

A few moments after we were seated, Mark Biedebach, the biology professor who had attended that first planning session at Ralph Winter's home, slipped into the seat next to me. Was I ever glad to see him!

"If I get in over my head with the Board, I can at least call on Mark," I thought to myself.

After Max Rafferty finished his report, the Board turned to their agenda. Item 1 was a Preliminary Statewide Testing Report by Dr. James

Crandall, scheduled for 9:30 a.m. They were running a little behind, just like Oglesby had said they would. However, they dispensed with this report without too much delay.

About 10:15 a.m., Item 2, listed as "Science Framework for California Public Schools – Public Hearing," was called by Howard Day, State Board President.

Glenn F. Leslie, a professor of elementary education at Fresno State College and also the Chairman of the Science Committee within the Curriculum Commission, was listed on the agenda as the respondent to the Framework in the hearing. However, Leslie was not immediately called upon to speak.

"We will now open the public hearing on the *Science Framework*," intoned Day without emotion or emphasis. "Our first speaker will be Vernon L. Grose."

At the mention of my name, I was on my feet and moving down the aisle toward the microphone. Adrenalin was flowing. I had to squeeze by the press table and the TV cameras in order to find my way to the podium.

President Day used the time it took me to reach the podium to instruct all those who intended to speak in the public hearing that they would be limited to 5 minutes each. My heart jumped into my throat! I had read my speech aloud at home — just to time it, and it took me about 23 minutes!

"What if they cut me off before I have even made my points?" I worried.

"May I ask all of those who are going to participate in the public hearing, starting with Dr. Grose, to get right up next to that microphone and speak up good and loudly because the acoustics in here are lousy," Max Rafferty boomed in his rapid-fire way. Addressing me, he continued, "Prop that mike up so that you can talk right into it."

"Mr. Day, members of the Board, for my appearance today before the State Board of Education, I am deeply indebted to the *Los Angeles Times*..." I began. Out of the corner of my eye, I saw several at the press table bristle. Glancing from my prepared text occasionally, I could see that the TV crews were shaking their heads from side to side, almost in contempt. There were even audible groans from media personnel, as I continued.

In one way, I was pleased that I had "reached" the press by my comments. They had been most unfair in their reporting, so how else could I make my point? In retrospect, I realize that this was not wise — even if it were true. The media have deep bias — but they also have long memories.

Nearly 20 years earlier, I had stood in Concord, Massachusetts beside Daniel Chester French's first important work — the statue of "The Minute Man." On the pedestal of that famous sculpture is engraved the first stanza of Ralph Waldo Emerson's stirring "Concord Hymn:"

> By the rude bridge that arched the flood,
> Their flag to April's breeze unfurled,
> Here once the embattled farmer stood,
> And fired the shot heard round the world.

The salvo I fired at the very outset of my speech was symbolically analogous, in the Science Textbook Struggle, to that famous shot fired on 19 April 1775. The news media never let me forget it, either.

Less than a minute into my speech, I made a slip of the lip — "I myself believe in the various theories of creation…" I quickly caught myself, "I mean, of *evolution*…"

But I wasn't quick enough for Board member Mrs. Seymour Mathiesen, a farmer's wife from the San Joaquin Valley. She burst forth with a loud guffaw, that seemed to me to be a combination of derision and glee at my mistake.

"Quite a raucous reception," ran through my mind. Though I tried to ignore her uncouth response, it momentarily broke my concentration.

But I picked up the cadence and rolled on — concerned that the 5-minute limit could be imposed on me at any moment and yet not wishing to rush, thereby forfeiting the impact of what I had to say.

I suppose all public speakers secretly hope to spark a response among listeners — regardless of the sophistication of the audience or the official "appropriateness" of displaying response. I was no different. Of course, Mrs. Mathiesen had shown response — but hardly the type any speaker could desire.

Just once, during the 24 minutes I spoke, I felt that the audience "came alive" and joined me. It was when I asked the hearers to imagine the impact of being forced to amend Jefferson's inspiring words in the Declaration of Independence to read:

> We hold these truths to be self-evident, that all men arose as equals from a soup of amino acid-like molecules and that they, by virtue of this common molecular ancestry, are endowed with certain unalienable Rights such as Life, Liberty and the pursuit of Happiness.

There was no applause at that point because it is forbidden in public hearings, Yet, I knew that I had communicated with those whom I was addressing. It felt good…

Maybe because I never gave President Day an opportunity to

graciously interrupt me or possibly because the copies of my speech Harward had reproduced and distributed to all the Board members allowed them to see that it was more than a 5-minute speech, I was not stopped for the nearly 24 minutes it took for me to finish. My greatest relief was that I had been allowed to finish uninterrupted.

I started back to my seat in the auditorium. As I brushed by the press table, several correspondents jumped up and followed me back to my seat, bombarding me with questions:

"Do you have any extra copies of your speech?"

"Are you representing any organization?"

"What subjects do you teach, and where do you teach them?"

It was obvious to me that their early contempt and open ridicule of what I was saying had changed, at least to some degree. They now seemed to acknowledge that I had a serious message. Although I had no extra copies of my speech, I began to wish that I had.

Like so many hawks converging on a prey, the newsmen pressed me for a business card. Because I am frequently involved in several professional ventures at the same time, it is not unusual for me to carry several types of business cards.

I took a bunch of cards out of my pocket and handed them out. Evidently among my Tustin Institute of Technology cards, there was one from the University of Southern California.

(In 1967, I taught a graduate course in physics and chemistry for USC in Germany. Then, late in 1968, I returned to Europe for USC — this time to Spain, where I taught two other technical courses in the Graduate School. I had returned from Spain in May 1969, and the USC card was evidently misplaced among my other cards.)

Even though the second speaker in the public hearing had already begun to speak, bedlam with newsmen was still going on around my seat in the auditorium. The correspondent who received the USC business card must have asked me what subjects I teach because, within hours, I was being dubbed a USC physics professor on radio and television. This, of course, was not true. However, as I was to learn with distress, correcting errors made by news media is not unlike "trying to get feathers that have been released in a tornado all back into a bag."

The public hearing had attracted quite a few people who wished to express themselves on the *Science Framework*. The State Board has a procedure that requires all speakers wanting to appear in a public hearing to request permission in writing from Sacramento in advance of a meeting. The list was not short.

Without exception, all the remaining speakers wanted the *Science*

Framework revised. There were speakers from the Creation Research Society and several Christian colleges in Southern California. Mrs. Nell Segraves, who had been instrumental in securing an opinion from the California Attorney General in 1963 concerning the teaching of evolution, spoke as a private citizen. I knew none of these people.

All the speakers — except me — seemed to be calling for either a total rejection of the *Science Framework* or else a major rewriting of it. As testimony continued, I could see that I was standing more and more apart from all the other speakers. Among the differences were:

1. The Board had allowed me to speak nearly five times longer than anyone else (apparently because *they* had invited *me* to speak).
2. I was the only person to praise the *Science Framework* as a fine piece of work.
3. I alone said that only minor revision was needed to make the *Framework* acceptable.
4. To the exclusion of all others, I furnished the Board a specific revision in writing that would supposedly satisfy all parties to the dispute.

By the time the final speaker was finished, I had turned out to be a moderate voice of conciliation in the midst of strident voices of dissent. All other speakers had been in close agreement that the *Science Framework* was unacceptable without major action.

Vice President Harward was quick to sense the urgency of resolving the issue in this meeting. Delaying a decision could only be confusing to all.

"Mr. Chairman," Harward spoke up, "in order to conclude this (issue) in this session of the State Board, I would make a suggestion..."

No news reporting of the Board's deliberations, following the public hearing, came close to telling the truth. For that reason, I have included a verbatim transcription (Exhibit 6) of the Board's discussion that lasted approximately one-half hour. Those who read that exhibit will note the following facts (that were either erroneously reported or omitted from press reports):

(1) It was Board member Commons who overrode members Day, Ragle, Ford and Harward in their desire to return the *Science Framework* back to the Curriculum Commission and the State Advisory Committee on Science Education for additional

editing. He insisted that the Board had "every right to amend the guidelines as we see them right here and now."

(2) It was Board member Mathiesen who endorsed Commons' position that the Board amend the *Science Framework* without allowing it to be returned to the original authors by calling for the question to vote on immediate amendment and adoption by the Board.

The significance of these facts is that the press consistently reported that the Board, at the insistence of "conservative" Board members Ford and Harward, reacted to a suggestion from a "member of the audience" (meaning me) and overrode their own illustrious State Advisory Committee on Science Education. News reports always pointed out that Ford and Harward were Reagan appointees.

Interestingly, Commons and Mathiesen are both appointees of Democratic Governor Edmund G. Brown! They were lauded as "liberals" when they completed their terms in 1970.[163]

Why does the press refuse to tell the truth? What dangers would the news media incur by accurately reporting "liberal members of the State Board overrode the urging of at least four of their colleagues to allow the original authors to rewrite contested portions of the *Science Framework*?"

Wouldn't that be exciting, sensational news — because the Commons-Mathiesen action was a *reversal* of the traditional liberal position? But even a more basic question is, "Would newspapers fail to sell if they were *accurate*?"

The moniker "member of the audience" that Jack McCurdy, education writer for the *Los Angeles Times*, tagged on me the next day really stuck.[164] It appeared over and over again in public media as well as scientific publications.[165]

By that description, readers were led to believe that there was an emotional and irrational tirade carried on by religious fundamentalists on the Board that was only quelled when some person in the audience jumped to his feet and offered a few words that caught the imagination of the Board.

By this lengthy account of how I became involved, you now have an opportunity to judge the extent to which I was a "member of the audience." Exhibit 6 accurately recounts the resolution of Item 2 of the agenda: "Science Framework for California Public Schools – Public Hearing."

Ford's fateful decision to lift the 2nd and 8th paragraphs out of my proposed revision of Appendix A (see Exhibit 5) and insert them into the *Framework* did not particularly please me. Those paragraphs obviously were not meant to be consecutive and were taken out of context. Had I been given opportunity beforehand of knowing how those paragraphs would be adopted by the State Board, I probably would not have consented to it.

As the Board adjourned for lunch, Phyllis, Shirley, Mark Biedebach and I stood up, stretched and looked at each other. I had mixed emotions. Obviously, I was personally pleased that I had been a source of help to the Board. Yet, what could two small paragraphs, taken out of context, accomplish in a 205-page document?

People all over the auditorium seemed to be engaged in animated conversation as they left the auditorium. As we were walking out, we could see reporters clustered around Dr. Ford at the front of the auditorium.

All my letter writing, consultation, phone calls, and preparation had culminated in only two short paragraphs (127 words!) being accepted by the Board. Was it worth it? I had no idea…

Infamy or Inspiration?

As we filed out of the auditorium into the crowded hallway, something quite unusual occurred. A very attractive, middle-aged lady walked up to me, took my hand and said in a quiet voice — almost a whisper, "Isaac Newton would have been pleased today."

For a moment, I was somewhat startled by her comment. She quickly followed with, "I want to thank you for your preparation and speech before the Board this morning. You did an admirable job and the school children of California are certainly going to profit from it."

Permit me to delay introducing this lovely lady momentarily, in order to explain my stunned reaction to her observation.

On my first visit to London, a few years before this eventful November morning, I had toured Westminster Abbey. In my estimation, this marvelous shrine to honored citizens and patriots has no equal in any nation. Dating from the 11th century, it has marked the scene of many great events in English history. Except for Edward V and Edward VIII, all English rulers from the time of William the Conqueror have been crowned there.

Westminster Abbey, however, is more than just a world-famous church, one of the most beautiful in England. Either in its walls or beneath its floor, the bodies of many of England's greatest lie — kings, queens, statesmen, military heroes, and poets. One of the greatest honors England can bestow is to be buried in Westminster Abbey.

Not as many scientists are buried there as persons of other accomplishments — primarily because science arrived so recently in history. Therefore, I confess that most of the names that caught my eye as I wandered through the huge, ornate cathedral were those of people who had lived prior to the era of modern science.

Yet, as I turned and entered the North Choir Aisle, I came upon a large and beautiful memorial to Sir Isaac Newton, who is also buried in Westminster Abbey. He has always been one of my heroes — not solely for his scientific contributions but also because of his personal knowledge of the Creator.

It is not well known that Newton was a bachelor who actually spent very little of his time studying mathematics, physics and astronomy — subjects for which he is best known. He spent far more time on questions of theology and in Bible study. He, like nearly all the forefathers of modern science, believed that the role of science was to discover and expose the works of the Creator.

On that first visit to Westminster Abbey, as I finished admiring the memorial to Newton, I continued on the North Choir Aisle to where it intersects the North Nave Aisle, about 20 feet away. As I glanced at the floor at this intersection, there was a large slab of gray stone with the inscription "Charles Darwin." It marks his burial place beneath the floor of the Abbey!

I remember so distinctly standing directly on this inset tablet of stone and contemplating Darwin's impact — not only on science but on the thinking of the entire world. Since my rather odd attachment to burial places has already been described, it should not surprise you that I had many thoughts running through my head as I stood on Darwin's grave.

Turning back to look at Newton's memorial, I said to myself, "I wonder whether Darwin ever stood here in Westminster Abbey and contemplated Isaac Newton?" Since Newton died 82 years before Darwin was born, it was entirely possible!

"What a discussion those two would have had if they had been contemporaries!" I mused. Before departing this scene of contemplation — you guessed it — I took a picture!

The mention of Newton's name in the context of a debate over Darwin's claim to fame — evolution — had set my mind spinning with a flashback par excellence…

But now I am ready to introduce the gracious woman who said, "Isaac Newton would have been pleased today." She was Mrs. Joseph P. Bean, wife of the President of the Glendale School Board.

"I make it a practice to attend every meeting of the State Board of Education, and I can tell you that I have never heard a more thoughtful and inspiring address than you made this morning." With such flattery ringing in my ears, I asked her if she would join Phyllis, Shirley and me for lunch. She consented.

Mark Biedebach had to rush back to the University and was unable to have lunch with us. In the meantime, as we continued to be buffeted in the crowd rushing to get to lunch, two other friends spotted us. Dave Irwin, an interdisciplinary colleague of mine who was at that time completing some graduate work in anthropology after living several years in Malawi, was first to greet me. Right on his heels was John Gregory, an attorney with whom I had been in an investment club for many years. I was surprised to see both of these fellows, even though I knew that they were aware that I was going to speak in the public hearing.

"Care to join us for lunch?" I said to Dave and John.

"Why not?" they responded.

The six of us decided to leave the building to find food. Before I got outside, however, several of the other speakers in the public hearing who had been urging either total rejection or major revision of the *Science Framework* confronted me and said, "We know that you are probably happy about the Board's adoption of your words, but we think it is a disaster. You have compromised so much. What we've been fighting to gain for years appears now to be lost."

That took a lot of wind out of my sails, believe me.

I retreated back to the solace of my friends, to catch my breath again after that body blow. As they cheerfully chatted among themselves, I was wondering to myself, "Which way is it — did I accomplish anything today or did I actually destroy the opportunity for a much grander plan?"

"Order whatever you like for lunch, because this one is on me," John Gregory said firmly as we sat down and surveyed the menu. As we waited for our food, we listened intently as Mrs. Bean shared with us some of her experiences in attending Board meetings in the past.

Obviously a very intellectual and refined woman, she had insight to share with all of us that we had no other way of knowing. We were all neophytes, never having attended a meeting before.

Mrs. Bean attempted to put into a broader perspective what had happened that morning in Room 1138. Among the rather startling and thoughtful implications that Mrs. Bean noted in the decision by the State Board was the possibility of returning to the public school child a perspective of seeing himself as a creature of planned destiny.

In other words, she interpreted the exclusive teaching of evolution as indoctrination in a particular philosophy of life. It was *more* than science. It promoted the concept of a purposeless existence that even a child could perceive.

"When we teach our children that they are nothing more than a chance coincidence of sperm and egg which will proceed to grow into a blob of protoplasm weighing somewhere between 100 and 200 pounds, dying in approximately 70 years without purpose, direction, rationale or destiny, what can we expect of the human race?"

Quietly but firmly, she continued, "When people believed the words of Thomas Jefferson about man being a very unique creature specifically created by a Creator with purpose and plan in mind, the nation grew and was prosperous. In other cultures where this teaching has not been present, the same results have not come."

We sat and listened as Mrs. Bean completed her thought, "In America, one can readily see that there is a definite and visible correlation with disintegration of spirit and sense of destiny as we have become exclusively indoctrinated with the purposeless chance philosophy that evolution portrays."

"Very interesting observation," I thought to myself. So that was the reasoning, then, that had prompted her to say that Isaac Newton would have been pleased with what happened on 13 November 1969.

While we were not forcing children to believe in the *planned* existence of the human race, at least the opportunity was now going to be available for them to view this as an alternative to the *purposeless* existence demanded by Evolution II.

Certainly there are those who will argue with Mrs. Bean that evolution does not exclude the concept of the dignity of man. However, it is very difficult to imagine from whence the dignity would come if, in fact, George Gaylord Simpson's words are true:[166]

> Man is the result of a purposeless and materialistic process that did not have him in mind. He was not planned. He is a state of matter, a form of life, a sort of animal...

As soon as we finished our luncheon discussion, Phyllis, Shirley and I got in our Pontiac station wagon to drive home. As we turned on the car radio, it was tuned to KNX News Radio, Los Angeles' CBS affiliate. Startling the three of us, a staccato newsflash that was to become as repetitive as it was inaccurate was on the air.

"The California State Board of Education comes up with new guidelines, including the Bible's version of earth's creation as a part of the curriculum," the announcer declared in authoritative voice. Following up with details, he said:

> Evolution will be downgraded, and the Bible's version of earth's creation will be further incorporated into school curriculum. California's State Board of Education has revised their guidelines for indoctrination and teaching public school science in regard to the start of everything. The Vice President of Santa Barbara's Tustin Institute of Technology, Dr. Vernon Grose, testified that the Biblical interpretation is now needed to give the California student more freedom to make his own decision on creation and evolution..."

"That isn't what I said at all," I protested to Phyllis and Shirley. "In fact, I *denied* that we were talking about the Bible when we used the word creation."

It was no use. As we got home and listened to news reports throughout the afternoon and evening, we became increasingly disheartened at the distortion of what I had said. The 6 o'clock news on Channel 2, KNXT, carried an accurate exception, however, to the prevalent misreporting:

> State Board of Education members voted today to give the theories of evolution and creation equal footing in California's science classrooms. The Board, meeting in Los Angeles, approved a so-called *Science Framework* — a master guidebook for classroom instruction in science and adoption of textbooks. The *Framework* had already been revised to meet objections that it ignored the theory of creation in dealing with man's origin. But critics complained that even the revisions failed to recognize creation as any more than a religious conviction. So a parade of speakers appeared at today's hearing, asking for equal treatment of the creation theory along with the teaching of evolution. One speaker, USC physics professor Vernon Grose, proposed that curriculum include the fact that several origin theories have been advanced and that creation be considered a scientific, as well as a religious theory. The Board unanimously adopted the revised *Framework* with Grose's basic proposal included.

Later, Board member John Ford discussed the effect of the Board's action with Big Newsman Howard Gingold:

Ford: We are hoping that now, for the first time since the *Framework* has been adopted, that as texts are readied for this that the children of our state will be able to be taught the various theories of the origin of man — what has happened to him during the centuries — so that then they can choose whichever theory they decide that they want to believe.

Gingold: That's something that's not being done now in the public schools?

Ford: That is not being done at the present time in most of the public schools.

Gingold: How long will it be then before this filters down to the public school level — to the actual classroom level?

Ford: This we do not know. It depends upon the various teachers, and it will depend upon adoptions for science textbooks in the future.

Nonetheless, feathers of fiction had been released in tornadic winds, and there would be no way to get them back into a bag...

CITIZENS SPEAK OUT?

Voices — confused and confusing — are crying aloud about the role of the individual in today's society. "The day of an individual having any impact is long past," some claim. Others shout the battle cry, "Join the rebirth of citizen involvement!" Who would Ralph Nader be today if he had not championed the role of the individual citizen?

Yet, as successful as Nader's consumerism has become, it seems to be consequential only when the government and industry can afford to treat it as a novelty. When the chips are really down, consumerism, environmentalism and all other collective concerns are quickly overrun. Remember the collision between the energy crisis and environmentalists?

On the other hand and despite all pessimism concerning individual effort, I have a deep-down conviction that if I am willing to give up the important role that an individual can play in society, then I am also ready to give up the American form of government.

Without any question, there is a price paid by any individual who speaks out on issues in society. The price paid is not just in dollars and hours spent. The involved individual actually gives of *himself*. He can never again be the same after having divested himself of a certain part of his being in an issue.

But maybe that is only a proper and right way to look at society. We *should* be willing to give of ourselves. Most young men assume that they must sacrifice a portion of their life in military service. Politicians accept the obligation to give themselves to the nation in terms of public service. Perhaps every citizen should budget a part of himself for the ongoing sustenance of his civilization.

However, the involvement of individuals in issues is not likely to come from a rational decision that they owe a part of themselves to society. More than likely, it comes unexpectedly but subtly, as it did for me in the Science Textbook Struggle. And because involvement is not rationally entered into, it generally surprises the involved individual that he must divest a part of himself in an issue — never to return to being the person he was earlier.

Among those intangible parts of yourself that you forfeit by becoming involved is your *privacy.* This is not to say that you become famous and thereby suffer the fate of the famed. *Any* "private citizen involved in a public issue" is simply *not* a private citizen!

You become a focal point — a *target* for those opposed to you and a *torch-bearer* for those allied with your views. People whom you have never met feel no reticence in giving you instructions, telling you what you had better do. You receive mail with all kinds of orders "to do this or that."

Somehow, a person, upon becoming involved in a public issue, is supposed to automatically possess the gift of perceiving the "public interest" (whatever that is). Then, he is expected to "act in the public interest." This term "public interest" is about as indefinable as the "law-abiding public."

Is there a portion of the public that obeys all laws? How many laws can you *break* and still be a member of the "law-abiding" citizenry? Are policemen even "law-abiding?" The more you pursue these folk, the less certain you are that there is anyone there.

So it is with "public interest."

Plain old dollars are also required. After I had been involved in this struggle for about two years, I made a very modest estimate that I had invested at least $30,000 in telephone calls, correspondence, car mileage, reproduction and printing together with my own personal time. And there was no way to be reimbursed!

The only thing that kept me continually investing my time and energy was the conviction that I had to see the struggle resolved — I couldn't quit halfway through. Somehow, I was *suspicious* of those who, in the name of science, said that they alone knew how the origin of the universe, life, and man had occurred...

Guard the Guards?

There is a prevalent tendency among Americans to hold lightly the trust that they have in officials whom they have elected to public office. From time to time, incidents like the Watergate affair come along to reinforce the healthy suspicion that Americans have about their elected leadership. Because they have voluntarily surrendered some personal sovereignty to a representative body or individual, citizens feel obliged to verify the integrity of their proxy.

Nearly 2,000 years ago, Decimus Juvenal wisely said, *"Quis custodiet istos custodes* (who will guard the guards)?" Citizen involvement in public issues quite often stems from the inherent feeling that supposed experts need to be checked — even by non-experts. At least two reasons account for this feeling.

First, as discussed in Chapter 2, power can corrupt even the most well-intentioned. Second, the individual citizen holds an exalted position in American life — even above the highest elective or appointive office.

I confess that I felt that I had every right, and even obligation, to deflate the arrogance of anyone, however "expert" they considered themselves to be, who would attempt to suppress all other ideas than the one they had already decided to be a "fact." Call it "righteous indignation," if you like. I was certain that both the press and the so-called scientific community were in error...and I haven't changed my mind to this day.

Unavoidable Abuse

While a private citizen *forfeits* a part of himself by entering a public controversy, he also *receives*. However, much that is received, in no way, compensates for what he gave up.

Quite the opposite! I wasn't prepared for some of the things that came my way as the result of this struggle, and I don't think that I ever got used to having them happen to me.

One of the first things I began to receive was a large amount of mail. Most of it was "kooky." Perhaps because of the nature of the issue, with its possible religious overtone, many people were apparently stirred to write to me. Astrologers, people who had invented their own private religion, frustrated parents and those generally dissatisfied with the state of affairs in the entire world all wrote me letters. Some encouraged me while others criticized me harshly.

Many interpreted my action as being *against* evolution — rather than *for* additional theories. These people had further advice on how I could torpedo evolution.

Probably, no other subject could have inspired the wild variety of letters that the issue of origins did. One 3-page, single-spaced "treatise" was typical of mail that offered elaborate but most unorthodox descriptions of origins, apparently intended to encourage me in calling for alternative theories.

For example, about the origin of light, this letter said:

> Light was created in full extent, from source to the eye of man. It did not travel for any "light years" to reach our eyes. True, the photons

> apparently travel at a fixed measurable rate now (though it cannot be proved that it does so in the most distant reaches) but so does electricity AFTER you plug in the cord. God plugged in the light rays to Adam's retina, so to speak. Thus, the most distant stars are only as old as this earth, the moon and our sun — a little over 6,500 years (according to the Septuagint Translation of the literal family history book called Genesis.) The Deluge deposited all the fossils. Dinosaurs and every other so-called "prehistoric" animal, lived together on this planet up until they were smashed by the great canopy of water when it fell down from above the atmosphere. Before that fall, it was a solid piece of ice-glass, and the ancients were *correct* in describing it as the "crystalline firmament", which it was, keeping the planet in a beautiful, lush, life-sustaining estate.

Regarding the formation of oceans, the letter contained some real eye-openers. For example:

> The crust of the pre-Flood earth was hand made by angels, not accidentally formed by erosion, volcanic action or earthquakes, etc. It was arranged with a set of subterranian (sic) tubes heated by volcanic pockets which forced water upward and out onto the surface, to water the ground. The land mass was of one chunk, and may have had, actually, four corners — as that expression is of very great antiquity. The watery sphere was not quite as watery until after the Flood, when it was necessary to pull open the deep chasm we call the Atlantic Ocean, to make room for excess water to allow the continents to once again appear.

Apparently the writer of this letter expected the textbooks to include his interpretation of many ancient books for the origin and state of man. One excerpt was particularly interesting:

> The Jubilees, Genesis and Enoch accounts tell of the horrible pre-Flood civilization, enmeshed in starvation, rampant murder of beasts for food, angels copulating with womankind to produce hybrid monsters who even now are remembered in the Greek myths, the warnings of Enoch, who, as a token of the oncoming disaster, was allowed to return into the Paradise behind the flaming sword and after which, the city was taken up off of the planet out of sight. Then came to fruition the total evil of man and angels which brought the Flood. From that day to now, man has lived in a stark, water eroded environment, subject to the elements of natural whim.

One letter even caused me to turn it over to the police — not because I feared for my own safety, but because I hoped the author, who was probably in desperate need of psychiatric help, could be located. The letter was hand-printed, with every word retraced several times. The return address was an expired post office box.

A typical excerpt illustrates how it combined evolution and God in language that had no meaning:

> Darwin's theory of evolution could have been created by a Bible faction to which it becomes the positive or negative pole yet it keeps both in contents of energy storage of a battery. The recognition of a God now included in scientific is an over-simplification of spoken and written symbols in which any thoughts of desire, need, rationalist legalities, and other aspects of pinpointing things for conclusions or to whatever could follow — ad infitum (sic).

I was attacked with equal fury from all sides. There were letters from adamant creationists who were highly insulted by what I had done. One, from New York City, illustrates the bitterness that came from those I thought would be pleased with the State Board's action:

> Either wittingly or unwittingly you have not only overlooked the greatest challenge of science to religion but you have used it in your own arguments and apparently fostered it. I am referring, of course, to the Second Law of Thermodynamics which is THE scientific challenge to basic Christianity...This *cannot* be true for it is in no way supported by the word of GOD and is, indeed, denied by it.
>
> How you...can support such blasphemy is beyond me, unless you are an unconscious agent of the *Devil.* That you support and cite what has been inserted in science books to deny the very basics of our CHRISTIANITY makes you an unwitting tool of the Leninist/Marxist conspiracy! If you are concerned with the souls of our children as you pretend to be, only the denounciation (sic) of this insideous (sic) principle can SAVE YOUR SOUL!
>
> Perhaps you have fallen into the semantic trap of thinking that because it says "law" of thermodynamics, it is an established *fact*...It needs be *routed* out of school materials! What are you planning to do about it?

This was not just an initial surge of correspondence. It continued over several years. Books, pamphlets and magazines arrived by the bushel. Every time my name appeared in print, the postman knew it for sure — and was sorry! Fourth-class mail increased many-fold as I was placed on mailing lists for all types of organizations.

Frequently, because mail would demand an answer, I had trouble with my conscience. Public exposure imposes certain responsibilities, and I often pondered the duty I had to respond to mail requesting answers. Not being independently wealthy or having a secretary I could devote full time to correspondence, it bothered me to ignore mail that requested a reply.

Of course, not all the mail was answerable — some was either rhetorical or irrational. On the other hand, there were concerned parents who solicited advice on what they might do to open up objectivity in the science classroom. While I obviously was no expert on the subject, I did have some ideas that I thought could help them. At the same time, I had to continue making a living that meant that I could not afford the time to answer the mail.

Correspondence and telephone calls are obvious and *overt* aspects of unavoidable change for the "private-turned-public" citizen. However, there was also a *covert* dimension to the treatment that I received. Let me illustrate.

The *Los Angeles Times*, possibly because I attacked them at the outset for their role in the struggle, carried on what could only be interpreted as a campaign to discredit me as a person. They did little things — nothing individually consequential. Just cheap shots like misspelling my name. Or by casting doubt on my professional affiliation with the following phrase: "Vernon L. Grose, *who said he is* vice president of the Tustin Institute of Technology (emphasis added)." This is not unlike saying, "Grose, who said that his name is Vernon L. Grose." The only conclusion one can draw is that the *Times* does not believe that I was honest in describing myself — and *ipso facto*, all that I have done is likewise dishonest.

Or consider once more the descriptor "a member of the audience" that was assigned to me. Had the *Times* done any investigation on who I was, they would have easily found that I was invited, without solicitation, by the Vice President of the State Board to express my views. I was not a *spectator* at the hearing. Or if I was, then so was everyone else except the Board Members sitting on the dais.

In an apparent attempt to further drive a wedge between my credentials and the issue, the *Los Angeles Times* began to refer to me as "a Santa Barbara management consultant." It is true that, like many other educators, I derive income from consulting with government agencies and private industry. However, most of my income comes from teaching. The farther they could separate me from education, the better. So the *Times* would never acknowledge that I was primarily an educator.

It may be over-reactive to equate this kind of treatment by the media with "abuse." However, at least some of it is worthy of that description. Are these examples of innocent sloppiness in reporting or deliberate attempts to discredit? Why did the "errors" always happen to those opposed to the editorial position of the *Times*?

I do not wish to imply that I was a special target for such abuse. It is a normal product of a public exposure of a private citizen – particularly when the citizen takes a position contrary to that held by the media.

There is more editorializing than appears on the editorial page of a newspaper. The selection of news that will appear in the paper, together with its location in the paper and the size of print used for headlines are all reflective of the bias that the newspaper editors control and utilize.

Further examples of ill-will that came my way only because I took a public position will follow in later chapters. Scientists openly threatened me, accused me of deliberate misinformation, and impugned my intelligence and integrity. Some even wrote secretly to my personal friends and tried to get them to testify against me! Is it paranoid to call this "abuse?"

Ready-Made Labels

Communication has been called the world's biggest problem. Whether between husband and wife, parent and child, teacher and parent, or Congress and the Executive Branch, communication is difficult. Words, the basic substance of communication, are actually only *symbols*. They stand for or *represent* something real but in themselves are *not* real.

Some words represent reality in such a way that *everyone* recognizes that reality. Other words do just the opposite. They depict one real thing to one person and an entirely different real thing to another. In the worst extreme, a single word can raise a different picture in the mind of every single person who hears or reads it. Such a word can truly be called "ambiguous."

A "label" is a particular kind of word. It is used to conveniently classify a person, group or theory. It is a generalization — that is, it is a loose, vague description rather than a specific and precise one. Labels aid communication — up to a point. If you are labeled as "energetic," "brilliant," or "handsome," you probably *accept* it — without bothering to find out specifically or precisely why you won the label.

On the other hand, when labeled "lazy," "stupid," or "ugly," you are not even likely to demand details on why you deserve the label — you *reject* it outright.

Acceptance or rejection of a label by its recipient is obviously not the label's measure of merit. The point at which a label ceases to aid and begins to ruin communication is when its generalizations are *false*. Since generalizations are built up from specific instances, it follows that a label should be able to be proven false by precise, clear-cut exceptions to the generalizations.

The news media had labels "poised and waiting" for the Science Textbook Struggle. For anyone opposed to the *status quo* position that "the amoeba-to-man myth is a proven scientific fact," the labels of *fundamentalist*, *religionist*, *non-scientific*, *conservative*, *right-wing*, *creationist*, and *reactionary* were preprinted and gummed.

All they needed was to be licked and slapped on the unwitting beneficiary. Like shells in a cocked shotgun, all that remained was for someone to aim in the general direction and pull the trigger.

A barrage of labels splattered me. Because I lacked the thick skin that experienced politicians or others in public life must develop against such buckshot, it hurt — at first. It was especially painful when I knew that I could specifically prove that the label was false.

Bearing the stigma of inaccurate and unwarranted labels is another price a private citizen pays for becoming involved in a public issue, I no longer feel the need to get rid of the labels that the news media stuck on me. I accept them — not as valid but as *inevitable.*

On the other hand, it is critical to the future discussion of origins in a scientific context to *expose*, and if possible *eliminate*, the use of phony labels by the media. It is critical not because innocent people get falsely branded but because *issues* are shielded from view by a wall of fallacious and inflammatory catch-words.

To the end then of *exposure* rather than personal defense, I would point out that there were three types of labels used by the media in this issue — political, religious, and scientific. The primary political labels were *conservative*, *reactionary* and *right-wing*. The favorite religious label was *fundamentalist*, although *religionist* and *sectarian* were often slapped on indiscriminately. While they hated to admit that any bona fide scientist could acknowledge creation (even in a religious sense), the media generally pinned the label of *creationist* on any scientifically-trained individual who opposed their thinking.

Not just the popular press engaged in the labeling game. *Daedalus*, the Journal of the American Academy of Arts and Sciences, carried an article by a biology professor from the University of California at Riverside, John A. Moore, that engaged in the same name-calling rhetoric:[167]

> Within decades of the publication in 1859 of *On the Origin of Species*, the terms "evolutionist" and "creationist" began to be taken as synonymous with "scientist" and "fundamentalist"... Fundamentalist theologians, possibly less sure of their faith than the more enlightened nineteenth-century natural theologians, have continued to insist that Biblical statements about nature really mean what they seem to be saying...It is this fundamentalist strain in religion, so strong among certain Protestant sects, that has led to recent vitriolic conflicts with science.

Note carefully Moore's "unbiased" and "objective" approach to labeling. On what authority does he interchange labels? By what standard does he declare 19th century natural theologians to be "more enlightened?" Later in the same article, Moore continues to reveal a "careful and thoughtful attempt to be accurate and fair" by this statement:[168]

> I am not aware of any member of the Creation Research Society who is doing research that any biologist or geologist who is not also a member would regard as pertinent to either evolution or creation.

Please understand that I am not, and have never been, a member of the Creation Research Society. Therefore, I am not defending that organization. I simply want to point out what a monumental task has apparently been accomplished by Moore in order for him to make such a sweeping statement. Unless he made the statement in a most frivolous and irresponsible manner, I would be obliged to believe that a careful scientist, as he must be, would have done the following:

1. Completely investigate the 500-plus scientists (all holding MS or PhD degrees in science) who are members of the Creation Research Society to (a) select those conducting research and (b) determine the nature of research each of these selected scientists is performing.
2. Develop specific criteria for research that can be universally regarded by the entire scientific community as pertinent to either evolution or creation.
3. Submit for review against the criteria of step 2 by all non-CRS biologists (of which there may be perhaps thousands) and all non-CRS geologists (of which there must also be thousands) a complete summary of the research being done by each of the 500-plus scientists.
4. Receive from *every single non-CRS* biologist and geologist confirmation that *not one CRS* researcher meets the criteria of step 2.

Of course, the phrase, "I am not aware..." could be a wide-open escape hatch for an unethical person because, after all, *everyone* is unaware of anything they haven't studied. Though wishing to believe that Moore is a serious scholar, I became somewhat insecure in that belief when I read:[169]

> Although the creationists' antagonism toward organic evolution has not abated, their strategies for undermining the influence of evolutionary biology in the public schools have changed. Whereas the main thrust of the creationist movement of the 1920's, which culminated in the Scopes Trial, was to get evolutionary theory out of the schools, they are now demanding that their own theory be given equal time.

Several obvious errors exist in that brief statement:

1. As anyone who has studied the events leading to the Scopes Trial of 1925 well knows, the only "movement" was by those who desired to overthrow the teaching of creation as the sole explanation of origins. There was neither a "main thrust" nor a "strategy for undermining the influence of evolutionary biology" at that time for a very simple reason. Creation enjoyed a *monopoly*!

2. There was no evolutionary theory to *get out* of Tennessee schools — because it wasn't *in* the schools until John Scopes agreed, on a trumped-up charge, to say that it was.

3. There is absolutely no evidence that "equal time" was or is being demanded by anyone for a "movement's own theory" (whatever that might be) — at least in the Science Textbook Struggle.

4. Even if some other theory (whether taught on an "equal time" basis or otherwise) were to be taught side-by-side with evolution, there is no reason to say that evolutionary theory would be undermined. In fact, evolution should thereby be *strengthened*, if it is as overwhelmingly supported by scientific data as its advocates state, because its superiority can be shown more clearly by comparison with an obviously weak alternative than by standing alone. Is this not the traditional liberal argument in permitting Communists to teach communism in our universities?

Moore's argument — that when any other theory is permitted exposure alongside a monopolistic one, the coexistence is based on "a strategy for undermining the influence" of the first theory — sounds frightingly similar to that attributed to *conservatives* in the 1954 McCarthy era. Traditionally, academicians like Moore pride themselves on their liberal viewpoint. It is almost a mark of success in science to be a liberal:[170]

> The most eminent and successful scientists generally hold more liberal views than their less well-known colleagues.

The most frequent political label that the media chose for me in this struggle was *conservative.* Frankly, I became quite confused as to how they decided to dignify me with that one — especially after I observed the thinking of people like Moore. My confusion persisted as I read:[171]

> Political conservatives thought that science would dissolve the traditional ties of master and man and the religious beliefs that legitimated those ties. They feared it would introduce a rationalistic, experimental attitude toward social

and political institutions and would thereby fail to perceive what was essential in them...Political radicalism always regarded science as its great ally against the forces of clerical and worldly authority. Science and reason were at one in their implacable opposition to the traditional, the arbitrary...Progressives and liberals regarded science as their ally in the campaign to erode the superstitions of traditional beliefs in hierarchical institutions.

In the Science Textbook Struggle, I was appealing "against the forces of scientific authority, opposing the traditional and arbitrary." I regarded myself as allied with true science in a campaign to erode "the superstitions of traditional belief."

Those *opposing* me were the ones wanting no change — claiming that anyone not agreeing with them would surely bring Temple Science crashing to the ground. *They* were the ones who feared the dissolution of traditional ties and "legitimate authority." How could I be *conservative?*

Then, I read an article by an avowed liberal that began to disperse the liberal-conservative fog for me:[172]

(The liberal is) open-minded and unfrozen. The liberal ideology is to be non-ideological. But some self-styled liberals insist on being doctrinaire...I'm afraid it's true about many liberals; they have closed minds and locked ears.

We (liberals) continue to demand free speech for those who agree with us. We sometimes support free speech for those who disagree with as but don't frighten us. We are enraged when the principle of free speech is applied to those whose opinions we find dangerous.

This writer went on to explain a prejudice among liberals — that some of them think they know best what other people should be allowed to hear. This prejudice he attributed to "an ugly lack of trust in people." He summarized the liberal prejudice this way:[173]

There is a liberal bias, alas, absolutely unlike a liberal mind. It feeds on reinforcement and regurgitation. It secretly thrives on absolutes. And it tries to *discredit counter-theories by discrediting credentials instead of theory*. It is always easier to attack *men* than *ideas*. The greatest conspiracy on earth continues to be the denial of expression, a vulgar act. (emphasis added)

So there it was. Many so-called *liberals* are afflicted by a bias that makes them act in a manner that the media traditionally attribute to *conservatives*. And I, though branded by the press as conservative, was actually acting in the best tradition of liberalism (that I had thought I was all along).

I never ceased wondering, from the time of the first label the press conferred on me, *who* decided and *how* it was determined which label I deserved. Did they consult the intelligence community to determine my covert activities? Although I am listed in several "Who's Who" volumes, none of *them* give the kind of information required to assign labels like *conservative*, *fundamentalist*, or *creationist*. Do the media conduct a private investigation? How do I qualify for such a distinction?

Governor Ronald Reagan of California did appoint me to the California Council on Criminal Justice and the Board of Directors for the California Crime Technological Research Foundation. I also served him full-time for seven months on the Governor's Select Committee on Law Enforcement Problems.

But all these appointments were made after 1971, and I was being called a conservative in 1969! (It may surprise many to know that I am not a Republican, and have never registered as one. Further, I did not know Governor Reagan personally until after I became involved in the Science Textbook Struggle. Our acquaintance had absolutely nothing to do with my involvement in this struggle either.)

Had the media carefully examined my life, they would have known that I became heavily involved and committed to correcting what I considered inequities among racial minorities as early as 1949, while still a senior in college. Research for a thesis I prepared, "The Federal Role in Civil Rights," convinced me that civil rights in the United States were being violated on a large scale. This was long before there was any concerted or publicized effort — in or out of Congress — known as a "civil rights" movement.

While I was stationed with the Air Force in Texas during the Korean War (my first visit to the southern part of the United States), I witnessed such Jim Crowism as separate drinking fountains and toilet facilities. Phyllis and I were so disturbed by this practice that we deliberately rode in the back ("colored patrons only") section of city buses, hoping to be punished for it. Our social consciences seemed to demand it — without any organized campaign. (This was five years before Martin Luther King, Jr. became famous by organizing a similar practice.)

My conviction concerning equal rights for all Americans has never changed. It was disturbing, therefore, to be labeled or categorized with groups who opposed my convictions. What can an individual citizen do about this obviously one-sided, false and irresponsible practice by an overpowering press? Nothing...

Before leaving the subject of political labels among scientists, I wish to recount a rather unusual incident that occurred while I was a member of the Governor's Select Committee on Law Enforcement Problems.

Our five-man committee was researching, among other subjects, the classic question of punishment-versus-rehabilitation in criminal justice.

Herman Kahn, President of Hudson Institute and frequently called "one of the world's leading thinkers," had made some public pronouncements on this subject.

So I called him in New York, inviting him to Sacramento to meet with Governor Reagan and our Committee on 14 February 1973. This was the first time that Kahn had met Reagan, and as I introduced them to each other, I thought, "How unbelievable that the most prominent conservative politician in America and an outspoken liberal intellectual from the Eastern establishment should be meeting to discuss one of the most divisive issues separating conservatives and liberals!"

Kahn is a nuclear physicist — graduate of Caltech, who is considered synonymous with the epithet "Think Tank." One of his books, *On Thermonuclear War*, is reputed to have inspired the film *Dr. Strangelove* in 1964.

No one would accuse Kahn of being a right-wing, hard-line conservative. Yet, some of his views about liberals were most startling! Four points that Kahn made concerning intellectuals during his meeting with us that day bear on political labels. In brief, they are:

1. Most intellectuals consider themselves to be liberal.
2. Intellectuals can be defined as "people who deal almost exclusively with second-hand information."
3. Intellectuals frequently reverse their position on major issues within a short time span.
4. Liberal intellectuals control the three mouthpieces or outlets of public opinion in America – academia, courts, and the media.

Kahn's first point can be easily confirmed. To be liberal is to be fashionable, chic and suave among intellectuals. This is because to be liberal is to be successful! A 1971 study of politics of academic natural scientists revealed that, in all scientific disciplines, achievers are much more liberal than rank-and-file scientists.174 This conclusion was consistent with a long-established pattern in which achieving intellectuals are inherently the most critical of existing social institutions and practices.

A more recent study of the social origins of American scientists and scholars showed that scholarly doctorates (generally associated with intellectuals) also come disproportionately from religious groups having certain beliefs and values.[175] Those religious groups that are traditionally classified as theologically liberal (e.g., Unitarians, Quakers and Jews) produce far more scholars and intellectuals than theologically conservative groups (e.g., Roman Catholics, Lutherans and southern white Protestants).

The second point of Kahn's is verified in this conclusion:[176]

> Observers in a great variety of settings since the end of the Middle Ages have found intellectuals in a common posture — standing outside their own societies, acting as critics of them. As a group, intellectuals are people engaged in work that emphasizes the importance of creativity, originality, and innovation. Many writers have pointed out that inherent in the obligation to create is the tendency to reject the status quo, to oppose the existing or old as philistine. Intellectuals are more likely than others to be partisans of the ideal and thus to criticize reality from this standpoint. This pressure to reject the status quo is compatible with a conservative or right-wing position, as well as with a liberal or left-wing stance. But in the United States since the 1920's (and increasingly in other Western countries as well), intellectual politics have become left-wing politics.

Kahn elaborated on this point by noting that intellectuals seldom hunt or fish, replace the water pump on their car, or wallpaper a room in their home. These activities are considered, in a sense, *primitive.* They relate to the *lower* end of Maslow's "hierarchy of needs" model — that of just staying alive.[177]

Because intellectuals spend their time and energy on *idealism*, as noted earlier, they are seldom in touch with "the real world." They have little appreciation for "the nitty-gritty" of daily life. Instead, they become partisans of the ideal and criticize reality from that standpoint.

This vulnerable position then leads to Kahn's third point — that liberal intellectuals frequently reverse their position on a subject within a relatively short time span. He pointed out that liberals were the ones who convinced President Kennedy to become heavily involved in Vietnam via the "domino theory." Yet, within ten years, the liberals were denouncing American involvement there as "heinous and immoral!"

Likewise, the liberal community urged Kennedy to establish the space program as a noble, peaceful national goal. Less than a decade later, these same voices were crying with lamentation that money spent on this endeavor was "scandalous" in view of newer liberal goals like urban renewal or income redistribution.

One can get dizzy trying to follow such reversals. However, they are to be expected from people who only think of "what *ought* to be" and who have no concept of "what *is*." While the ground they occupy is lofty, it is also precarious for its instability.

An interesting incident that illustrates the impact of reality on idealism recently occurred in Florida. A criminology professor at Florida State University joined a police force, spent four months as a patrolman, and later worked as an undercover agent dealing with pimps, prostitutes and pushers. This experience radically revised his views on criminal justice — and his approach to teaching in the university. At the same time, it turned the faculty against him — and may ruin his career.[178] He was no longer idealistically "pure."

These sporadic liberal reversals have caused a blurring of distinction between liberal and conservative labels on certain subjects in recent years in America — to the point that many are abandoning political parties. Instead of a *two-party* system, we have drifted into a *four-party* system — liberal Democrat, liberal Republican, conservative Democrat and conservative Republican — which chooses the oddest times and circumstances to fall into loose, random and temporary coalitions on any given issue.

Many foresee a complete breakdown of traditional political parties — and an accompanying rise of independence (rejection of labels):[179]

> The failure of our political parties has been accompanied by a rise in the number of persons who register as Independents. In 1964, 22 per cent of the registered voters were Independents. Today this has grown to 32 per cent. More interestingly, the percentage of Independents increases with education. Among Americans with only a grade school education, 21 per cent are Independents, while 38 per cent of college graduates are Independents. The percentage of Independents is also higher among the upper income groups and among young people. For those under 30, 45 per cent are Independents as opposed to 40 per cent Democrats and 15 per cent Republicans.
>
> Being an Independent today may not so much reflect apathy, as a disgust with the alternatives. Independents may be like Jefferson who said, "If I could not go to Heaven but with a party, I would not go there at all."

Not too long ago, many thought it easy to spot either a liberal or a conservative by what they said. Not so today. Liberals, recently making another of their classic turnabouts — this time on fiscal responsibility of government, now talk of themselves as "new progressives" in order to be differentiated from the conservatives whose party line they have adopted *in toto.*[180]

Take, for example, California's Governor Edmund G. (Jerry) Brown, Jr. He completely confused his liberal colleagues in fields like education, welfare and transportation by being more "conservative" than his conservative predecessor Ronald Reagan. His reversal of traditional liberalism is, at once, both lauded and criticized by his colleagues:[181]

> Brown, elected as a liberal in 1974, has hit upon the fatal flaw of American liberalism, the fact that social welfare measures, like everything else in this country, were founded on a premise of unlimited resources

and a set of curves — in production, in population, even in what used to be called "the moral condition" — which were expected to continue to rise forever. That era, he is telling us, is over — "that time is no longer"...

One of the possible consequences of limited growth is a reversion to a social Darwinism, a rapacious survival of the fittest in which we all struggle to carry off as many as possible of the goods that are still available.

Kahn's fourth point — that academia, courts and the media, who control the mouthpiece of America, are liberal — is corroborated by ABC Newscaster Howard K. Smith:[182]

Networks, says Mr, Smith, are almost exclusively staffed by liberals. "It evolved from the time when liberalism was a good thing, and most intellectuals became highly liberal. Most reporters are in an intellectual occupation." Secondly, he declares that liberals, virtually by definition, have a "strong leftward bias." "Our tradition, since FDR, has been leftward."

Not only the TV networks have an acknowledged leftward bias associated with liberal views. So do nearly all other news media. Therefore, when the media desire to castigate someone, they stigmatize them with the most despicable label in their repertoire – *conservative.*

It is my personal conviction that the reason I was labeled *conservative* was not because anyone had researched my views and aligned them with traditional conservative dogma. Instead, I was declared their "Public Enemy No. 1" (which is a conservative).

Believing that political labels tend to be what Hayakawa calls "unexamined key-words in our thought processes," I have managed to find definitions for both conservative and liberal that I can subscribe to as my own:

Conservative doctrine holds that, to the greatest degree possible consistent with a demonstrably overriding public interest, the individual citizen should be both free and responsible — free to do what he pleases, responsible for the harm he himself may suffer.[183]

Liberalism is a tolerance of views differing from one's own. It favors reform or progress which tends toward democracy and personal freedom for the individual.[184]

One more category of label needs mention — the *religious.* Right from the beginning, reporters tried to find out if I went to church and, if so, which one. I could see no good — only much harm — that could result from another label, especially a *religious* one. Struggling as I was to separate creation from its religious connotation in order to have scientists consider it in their context, this would be going in the wrong direction.

Newsmen were persistent. I became correspondingly irritated by their persistence. On 5 January 1970, the telephone rang.

"Vernon Grose, please. Long distance calling."

"Speaking," I responded.

"Hello, this is Rebecca Larsen of the *Christian Century* in San Francisco. I am preparing an article on the creation controversy in the public schools. May I interview you right now?"

"That's fine. Go ahead."

"Among the things we would like to know is what religious beliefs you have. Are you a member of any church?"

"Yes, but I don't see its relevance to this issue. I'll admit that it has been frequently asked of me since November 13th, but I think it is no more appropriate than asking about my political affiliation or my patterns of sexual behavior."

There was a short pause. "I see," she finally replied.

Realizing that she was not convinced, I asked her, "Are you intending to ask those who are opposed to my position the same question? After all, you would not want to be biased. If my religious affiliation influences my view of science, then the same must be true for all involved in the controversy."

In a follow-up letter that she requested, I included a statement from a widely-distributed "collective reply" by Ralph Gerard, one of the members of the State Advisory Committee on Science Education and the chief spokesman in opposing the inclusion of my words in the *Science Framework*. This statement confirmed my position that private religious convictions were *irrelevant* to the issue:[185]

> Although it is really irrelevant, for your information I am an agnostic (which means one who is not certain about the existence of God) although I certainly lean towards the atheistic rather than the theistic position. I do not accept the Bible as the unfiltered word of God...and I would be surprised if, after having considered and discussed this matter earnestly over some seven decades, I would now find some argument to convince me of the divine origin of this volume.

Despite this documented evidence that I sent her, Rebecca Larson wrote an extensive article that presented Gerard in a very favorable and responsible light — though erroneously listing him as head of a curriculum committee:[186]

> One alienated scientist is Dr. Ralph Gerard, a biologist on the faculty of the University of California,who views the amendment as equivalent to telling children that babies are brought by the stork. After asking a

> rhetorical question — "Did it require the Apollo 11 mission to prove the moon is not made of green cheese?" — Gerard went on to say: "In the case of evolution, I know of no responsible person who has examined the evidence who questions that species arose by a continuing series of changes from ancestral ones."

No mention of Gerard's religious convictions...*He* is quoted exclusively as a scientist. *He* is objective. *He* is permitted to ridicule those opposed to him by using obviously ridiculous analogies.

Contrast Gerard's treatment with mine (bearing in mind that she knew Gerard's beliefs but did not mention them). Recounting the 13 November 1969 meeting, she wrote:[187]

> Representatives of several fundamentalist groups appeared, among them members of the Bible Science Association and the Creation Research Society, a Michigan-based organization of scientists who are engaged in writing a text that incorporates the creation theory. According to Grose, he is not affiliated with any religious group of this type. He also has refused to state his church affiliation, commenting: "I'm a church member, and I believe in God. But I'm not representing my church in this matter, and I don't want to be categorized on the basis of religious affiliation." (Several persons at the hearing reported that Grose is a Christian Scientist.)

She had to have the last word. There was no way that she would let me get away without a religious label stuck on me. Why? There was an insatiable passion by the media — even by religion journals — to pin a religious label on anyone opposed to the "amoeba-to-man" myth. If you resisted, they would manufacture a phony one for you — even as Rebecca Larsen did.

So much for labels — political, scientific or religious. Ostensibly designed to aid public understanding of an issue in science education, they really served quite another purpose. They became the skeletons of "straw men"...

Chapter 4

RAISE A STRAW MAN

"Men often applaud an imitation, and hiss the real thing." —*Aesop*

By the time I went to bed on Thursday night, 13 November 1969, truth had been dealt a heavy blow.

Deliberately or accidentally, the news media had wrecked what I had tried to accomplish that morning. While I suspected that the news coverage was purposely inaccurate, I had no proof. So there was still hope…

I reasoned to myself, "Since both my speech and my suggested revision (including the two paragraphs that the State Board adopted) are in *written* form, I'll be able to find a reporter or newscaster who will tell the truth — especially after I show it to them in *writing!*"

You see, I was naive enough at that time to believe that errors in newspapers, magazines and telecasts were inadvertent. I suppose that I accepted at face value editorial comments like this one from the *Los Angeles Times*:[188]

> We, as a medium of communication, strive constantly to evaluate our product. We seek facts. On what we sincerely judge to be the facts, we then express our editorial judgments.

"If they are seeking facts, I *have* the facts. All I have to do is to present the facts to them and they will set the record straight," I muttered under my breath, just before dropping off to sleep.

ALARMING APPARITION

I awoke Friday morning then with hope — hope that I could correct the erroneous radio and TV reports by talking to *newspapers* and *magazines*, showing them in writing what I had said and what the State Board had actually adopted. But my hope was short-lived…

In fact, it *died* when I picked up the 118-page *Los Angeles Times* off my front lawn and spread it open.

Dead-center on the front page, a double-column article jumped right out at me. STATE EDUCATION BOARD BACKS TEACHING OF EVOLUTION AS THEORY, the headlines read.

"Not *too* bad," I said quietly as I walked toward the house. But then, I read the secondary headline – REJECTION OF HANDLING SUBJECT AS FACT IS COUPLED WITH INCLUSION OF DIVINE CREATION

IDEA IN TEXTBOOKS GUIDE. (Major articles on the *Times* front page sometimes carry double headlines — one in large bold print and the other in slightly smaller, less bold copy.)

In the very lengthy article (over 40 column-inches of newscopy) that accompanied these headlines, numerous errors appeared, among them these:[189]

> The (science) framework also will include the Biblical version of God's creation of life. No other religious scriptures, except for the Bible, are mentioned in the adopted framework.
>
> The compromise amendments were suggested by Dr. John Ford, a Board member, after two colleagues, Dr. Thomas Harward and Eugene Ragle, proposed postponement of adoption.

The unbiased reader of that detailed account would assume that (1) Biblical accounts of God's creation were mandated by the State Board to appear in California science textbooks, (2) no other religious accounts of creation would be allowed in science textbooks because the Board favored exclusive rights for the Bible and (3) Dr. Ford had ignored several attempts by his colleagues to postpone adoption by proposing amendments to the *Science Framework*.

All three of these assumptions are *absolutely false*. Instead of publishing what had happened, the *Times* had *invented* news. It was not a series of simple errors. The errors were collective in one direction. They did not cancel one another out — as should occur with random error.

I was beginning to detect the truth of what I often heard later: "The news media do not *report* the news — they *make* the news." And they make it *their* way…

The dim outline, almost ghost-like, of a "straw man" was beginning to form in my mind's eye. It was an alarming apparition. For one thing, ghosts are not respectable in an intellectual environment. You can't prove their existence. They are almost exclusively reserved for those who are mentally deranged — the paranoid or psychoneurotics.

Nonetheless, I could detect the phantom of a straw man named "Religion Masquerading as Science." He was invented by the news media. His purpose, like all scarecrows, was to avoid a *real* issue — to divert attention by substituting a phony for the genuine.

Surprisingly, this straw man wasn't the only one to appear in this struggle. Others came later…

Harassing Headlines

About the time I finished reading the *Los Angeles Times* article that morning, the phone rang.

"Do you get the *Valley News*?" It was our neighbor across the street, Robbie Newman.

"No, we don't take it."

"Well, you should see the headlines this morning! I'll bring it right over."

She was knocking at the door almost as soon as she hung up. She excitedly thrust the front page toward me.

"Look at this!" The top story on the front page of this newspaper with a circulation of 260,000 carried a double-row headline in letters one inch high: EVOLUTION FOES WIN REVISION OF SCIENCE TEXTS IN SCHOOLS.

"Well, that's really interesting. But I'm not an evolution foe..." I started explaining to her. And it was only the first of hundreds of explanations I've given since because headlines have been in error.

Headlines are handy. They enable you to read a newspaper much faster than if the print were all the same size. All of us make reading decisions by means of headlines. Assuming that they reflect or summarize the content of a news item, we decide by a headline whether we should read further about that item...At least, that's the assumption I always made about headlines.

After November 1969, I changed my mind about headlines. I discovered that news editors had additional uses for them. They didn't always have my *reading efficiency* in mind. Headlines could also *influence* people who read no more than the headlines. Even worse, headlines can be employed to deliberately mislead or harass the public.

How does harassment occur by headline?

Consider this example. On 14 November 1969 — the same morning that the earlier-mentioned headlines appeared in Los Angeles papers — the *Palo Alto Times* carried large headlines reading, DIVINE CREATION BELIEF ADDED TO SCHOOL BOOKS.

Those headlines aroused so much reader reaction that, two weeks later on 1 December, the Palo Alto Unified School District became the first district in the state to take official action against the State Board for their framework revision. By a 4 to 1 vote, these trustees demanded that the Board rescind its ruling (to incorporate my words) or they would join with other districts to "seek legislative and/or injunctive relief."

The next morning after that unique and emotional action, the *Palo Alto Times* headlines read, PALO ALTO TRUSTEES CHALLENGE STATE BIBLE RULING. By this time, a totally fallacious issue had been created by the press. They had *invented* and *perpetrated* a lie — that Biblical creation would be taught as science.

To be fair, all the blame cannot be heaped on the news media.

Although they deliberately created the falsehood and gave it unwarranted prominence, the school trustees are also guilty of irresponsible behavior.

They took action based exclusively on press reports. None of them bothered to find out whether the reports were true. They didn't talk to State Board members or to me. No source material, such as the minutes of the meeting, was examined. If they had shown any kind of mature response, this whole fracas would have collapsed of its own sheer weight. Instead, it continued to snowball.

It was as though the newspapers wanted to whip people into a frenzy of absurdity. Consider this very emotional reporting of a media-created crisis:[190]

> The state board decree that science textbooks should include with the Darwinian theory of evolution such "competing theories" as that in the Book of Genesis is attacked by the (Palo Alto) resolution on the grounds that it is religious teaching.
>
> "The teaching of religious beliefs in the public schools of this nation has been declared unconstitutional," it adds, saying that the "doctrine of equal validity" favored by the state board "would lead to chaos in our science curricula."

This article further stated that the Palo Alto Board was "extremely disturbed." One could presume that they might be driven to drastic action! The headlines had harassed them into irrationality, like a stampeding herd of wild beasts. These desperate and disturbed people had a warning — no, more like an *ultimatum*:[191]

> "If the State Board of Education does not comply with the above request, this district will join with any and all other districts of this state to seek legislative and/or injunctive relief from any and all actions that would force the teaching in our schools of religious beliefs or their presentation on an equal basis with scientific knowledge."

In addition to the formal resolution, several Palo Alto trustees sent a joint letter to the State Board that gave clear evidence of their agitated mental state:[192]

> If the "competing theories" doctrine were carried "only a little further," the letter stated: "We will then have to teach that disease may not be caused by microbes but may be due to evil spirits invading the body (in which case, burn feathers to ward them off); that lightning may be electrical discharges between charged clouds or it may be the wrath of a vengeful God...The state board has not yet suggested that the schools teach communism and fascism on an equal footing with representative democracy, yet how can it avoid such a recommendation if it is to be consistent?"

Who can we thank for such a thoughtful monologue in the name of sophisticated education? The news media. In fact, how can anyone differentiate what the media did in this case from someone hollering "Fire!" in a crowded theatre?

But this isn't the end of the story. As they had threatened to do, the Palo Alto board introduced a resolution later that week at the Annual Conference of the California School Board Association held in San Francisco. The delegates to this conference showed some temperance by rejecting the threat of legislative or court action against the State Board, as proposed by the Palo Alto resolution. However, the 100 delegates then passed a resolution by a margin of 2 to 1, that addressed the completely phony issue invented by the press:[193]

> Delegates to the CSBA's annual conference in San Francisco asked the state board "to reconsider the inclusion of the Genesis theory in science textbooks." The resolution also expressed "concern about the introduction of religious beliefs in the scientific curricula in contrast to the study of religion in other proper areas of the curriculum."...Dr. Brooks Lockhart, CSBA president, said the association in its resolution has the effect of asking the state board to reverse its action on evolution.

Without any doubt, the news media had won a great victory — if their intent was to mislead the public. Believing Joseph Goebbels' old adage, "If you say it often enough, the people will accept it as truth," the *Los Angeles Times*, on 9 December 1969, apparently felt secure enough in the success of their campaign to make it official.

In a lengthy editorial the *Times* addressed what they called *the issue*:[194] "Should theology be mixed with science in guidelines for technical textbooks to be used in the state's public schools?" They could have just as appropriately asked, "Should physical education programs include music theory as a requirement in the state's public schools?"

Setting up this ridiculous issue, they then proceeded to show how ridiculous it was — by being ridiculous! The editor of the *Los Angeles Times* said that the State Board approved "a fundamentalist statement proposed by a member of the audience at its November meeting." What made my two paragraphs *fundamentalist?* They actually *deny* that the Biblical use of creation is proper in science textbooks! My only reason in citing the Bible was to assure that it *wasn't* used.

Headlines had not only created a straw man, but they were the prime weapons being used in the battle to shoot down the straw man (see Figure 4). The news media had resurrected God from oblivion and Darwin from the dead.

They were now pitted against one another...

God Versus Darwin

On 21 December 1969, a nationwide United Press International news release was carried in the Santa Barbara, California *News-Press* under the following headline: GENESIS-DARWIN TEACHING DECISION STIRS ARGUMENTS.[195]

"To err is human..." is a saying all of us have heard. Further, we all know its truth. However, there is both *deliberate* and *inadvertent* error. Which kind of error do you think occurred in this UPI story?[196]

> The state of California is attempting to give the Biblical theory of creation equal time with Darwin in the public schools. The state Board of Education has adopted a new science framework to present the version of Genesis as an "alternative" to Darwin's theory of evolution accepted by most scientists...The Board approved the document (*Science Framework*), but with a three-paragraph amendment proposed by one of a dozen people in the audience speaking in its favor. The action was a victory for some fundamentalists who began a campaign seven years ago against the teaching of Darwin's theory. They said the school shouldn't teach children ideas that contradict what their parents tell them at home. The fundamentalist population in California is large.

Details of the events leading to the State Board's adoption have been documented in Chapter 3 and can be verified for their accuracy. Those details clearly contradict this UPI account.

Consider one error that illustrates bias to support the "God versus Darwin" phantom. No one "in the audience" proposed the amendment. Board member John Ford did. Further, his amendment was drafted and offered *after* all speakers had spoken! While the Board's action may have caused some fundamentalists to cheer, many other groups viewed it as a good move, too. It certainly benefited, did it not, the news media — by providing an exciting source of headlines!

Why, then, single out "fundamentalist" reaction? How did the UPI determine that "the fundamentalist population in California is large"? Does the almanac carry that information? Maybe the Census Bureau now asks, "Are *you* a fundamentalist?"

As much as newsmen might have wished or believed it, the scientific community has never been atheistic. Going back in time, the Greeks deified nature — attributing to it independent powers and capacities. At the rise of modern science in the 16th and 17th centuries, however, scientists recognized that nature was distinct from God, holding that it was wholly dependent upon Him. Greek and medieval rationalism gave way to a thorough-going empiricism at that time.

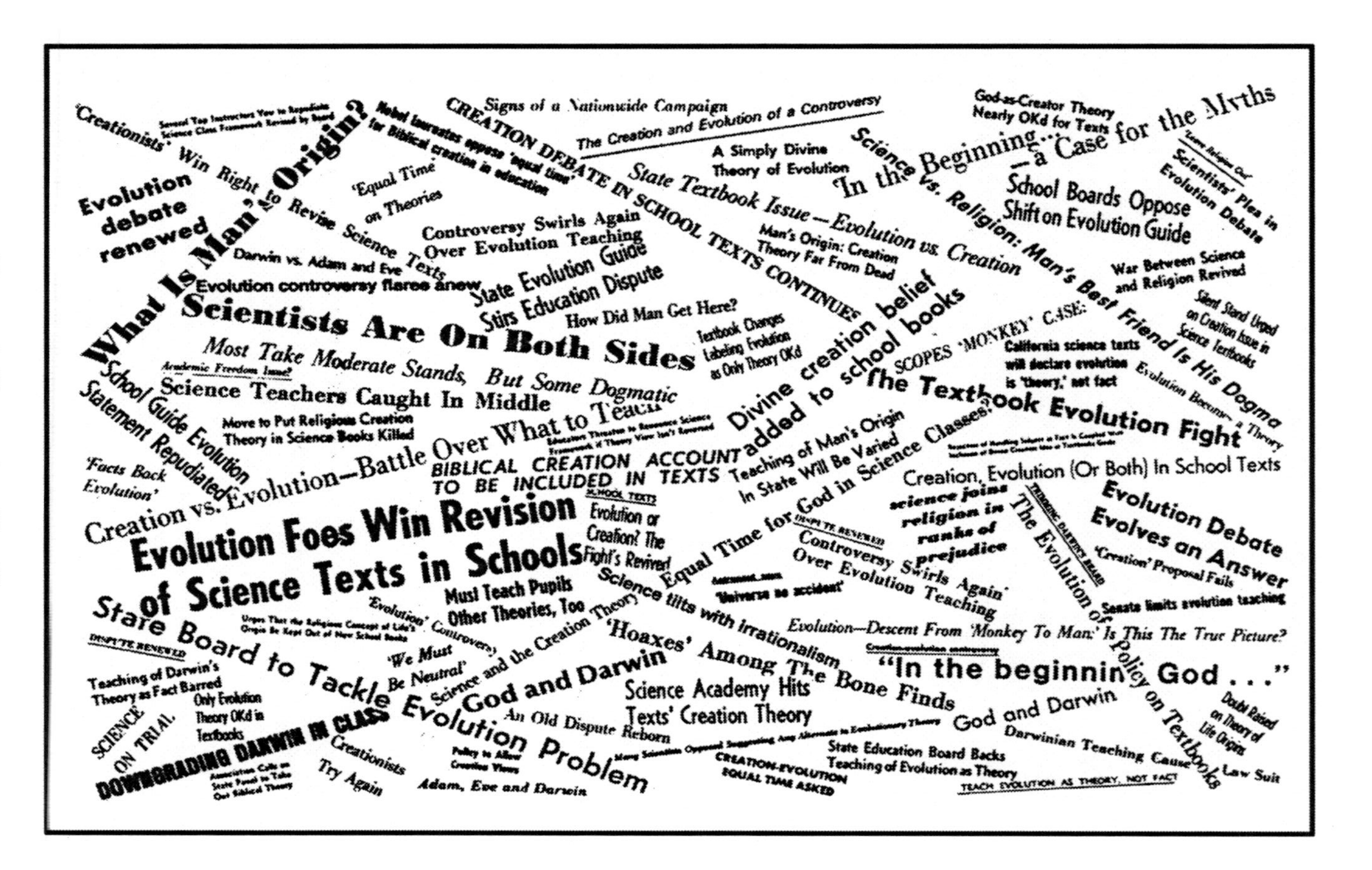

Figure 4. Harrassing Headlines

Recognizing that God could have created the universe any way He pleased, scientists, at the rise of modern science, set out to learn how He *did* create it — by observation and manipulation. Moreover, the recognized giants of modern science — men like Johannes Kepler, Galileo Galilei, Blaise Pascal, Isaac Newton, Michael Faraday and James Clerk Maxwell — all experienced a personal relationship with God. Faraday even preached sermons every Sunday for over 20 years! These men viewed their work essentially as uncovering or exposing the works of God in nature.

Among historians of science, there is a large body that goes even further and attributes the very *cause* of the birth of modern science to Christianity. A Dutch science historian at the University of Utrecht, R. Hooykaas, recently published a learned and erudite historical analysis of this viewpoint.[197]

As distasteful as it might be to newsmen and others who wish to divide and polarize, history bears out the fact that modern science arose and flourished only in a monotheistic culture — one in which there was only *one* God instead of *many* (as the ancient Greeks and most Asian and African cultures believe). Belief in only one consistent and unchanging God led to the term "*uni*-verse," as contrasted with the obvious "*multi*-verse" that would result from acknowledging many gods — each of whom had his own idiosyncrasies.

The extent to which the monotheistic beliefs of these early men of science influenced their scientific work, of course, will always remain a debatable and unresolved subject. However, it cannot be denied that they had totally integrated the concept of God the Creator with their scientific work. There was neither division nor antagonism between God and science in their minds. Therefore, the false, media-derived issue of "God versus Darwin" is historically incredible as well as without documented support in today's world.

But what about this phony flimflam fostered free from fact? Sure, it's comical and cute to have God and Darwin jousting with each other. However, is there a deeper significance to this caricature that the media desire to transmit? Are they implying that if you believe Darwin's ideas, you must deny God's existence? Or, if you believe God created everything (including Darwin), you are required to deny everything Darwin ever wrote? After all, Darwin acknowledged God's existence in his famous book:[198]

> Have we any right to assume that the Creator works by intellectual powers like those of man?...May we not believe that a living optical

instrument might thus be formed as superior to one of glass, as the works of the Creator are to those of man?

Therefore, how could you really "vote" between God and Darwin? While those voting for Darwin might believe that Darwin was simply an advanced piece of protoplasm in a long chain of living things that started by itself without a Creator, Darwin obviously didn't believe that. From Darwin's viewpoint, the fact that *he* existed pointed to the concurrent existence of God the Creator. Conversely, those voting for God have an obvious explanation of Darwin's existence. God and Darwin simply cannot be mutual exclusives! Voting for one in no way excludes the other. So what is the true *meaning* of the media's sophistry — God versus Darwin?

Perhaps the meaning could be ferreted out by a process of elimination. Let's examine some possible reasons why the press decided to avoid the real issue and focus on a straw man instead. While we won't be able to conclusively *prove* why they did so, listing alternative possibilities could help those in the future who are faced with newsmen who insist on inventing news instead of reporting it.

Here are some conceivable motives for false reporting:

1. *While the press understood the real issue, they felt that it was either too mundane or too complex, and thereby not newsworthy. So they decided to "spice it up a bit" with their own version (which they considered to be reasonably accurate) in order to sell the news.*

The real issue *may* have been too mundane. That's more likely than being too *complex*. Either way, what could be exciting about revealing the fact that science has no way to determine origins? Or that *all* explanations whether scientific, religious or philosophic are equally valid because *none* of them can be verified?

Since the news media are profit-making organizations, they must sell their news. There *is* a remote possibility that the public would not buy as many newspapers or listen to television and radio news if the truth were told in an unembellished form. So what could be wrong with "spicing it up" by over-simplification — particularly if the basic issue is still reasonably accurate?

2. *The press honestly overlooked the real issue, believing sincerely that what they wrote was accurate.*

It is very easy to listen selectively. All of us remember times when we "heard only what we wanted to hear" or were so preoccupied with what we *believed* someone was saying that we didn't actually *hear* what he or she did say. This possibility for error is very likely whenever the reporter has a presupposition already in his mind.

3. *The press understood the real issue as stated, but believed that it was a cover-up for another issue (the one that they invented and published).*

As ABC's Howard K. Smith was quoted in Chapter 3, the news media represent an intellectual community. Intellectuals frequently feel, as Seidenbaum earlier said, that they know best what we should all hear. Further, the news media often believe that they have an unwritten call and mission to be the final arbiters of truth in society.

Seeming to subscribe to the doctrine of original sin, they suspect everyone at all times of not telling the truth. Elected officials, judges — almost everyone except athletes — are, in the eyes of the press, guilty of deception and untruth until proven innocent. A ranking Congressman recently said, "Almost anything one does is subject to challenge on the grounds that maybe he is doing it for some ulterior motive, or seeking to evade the law."

This possibility suggests that the State Board of Education and all the public speakers in the hearing were suspect of an unspoken and immoral motive that the media felt impelled to ferret out. The press apparently knew more than the people involved — even to the point of knowing unconscious motivation. It was necessary, therefore, for the news media to invent an issue that would act as a catalyst to reveal the ulterior motives of all parties involved.

4. *The press honestly and inadvertently misunderstood the real issue, sincerely conceiving the issue to be a disguised attempt by religionists to discredit science. So they knowingly distorted the issue to compensate for this ostensibly elicit maneuver, thereby protecting the purity of science.*

This possibility, though similar to the third one, is different because the news media took on the role of champions of what they considered a vulnerable body of truth — science. Considering themselves excellent, self-appointed detectives of religious bias, they became convinced that they had discovered a horrible demon lurking in the shadows of this issue. They felt fully justified in their minds in distorting the issue because sinister religionists were doing the same thing. They fought fire with fire.

As earlier acknowledged, there is no way to determine which of these four possibilities (or some other possibility) caused the news media's distortion of the truth. It may not be critical that we ever know the cause. The critical focus must remain on the *withholding* of truth and the further *perpetration* of a fallacious issue upon the public — in the name of *news* yet!

"Why don't you sue newspapers that print accounts that are obviously in error?" you could ask. Good question. Libel need not be personal. It is defined as "anything that gives a damaging picture of the subject with which it is dealing." However, since libel suits must allege *monetary* damage resulting from libelous statements, it would be difficult to either estimate or prove financial damage in the Science Textbook Struggle.

In 1974, the United States Supreme Court established some guidelines for libel suits that do bear on the responsibility of the news media. The Court made a crucial distinction between public figures and truly private citizens; i.e., courts judge inaccuracy of news accounts by two *different* standards:[199]

> For a *public* figure to win a libel suit, he must prove that the erroneous story was published with "actual malice" — that is, that editors knew it was false or showed "reckless disregard" for the truth. By contrast, a *private* individual must show only that the press made a negligent mistake and that he suffered actual damage. (emphasis added)

The Supreme Court went on to explain that a person becomes a public figure only if he thrusts himself into the forefront of particular public controversies in order to influence the resolution of the issues involved. Obviously, the Science Textbook Struggle is a public issue and undoubtedly would be tried under the same rules that apply to a public person. The difficulty of conclusively proving "actual malice" on the part of the media has already been shown by listing and discussing various alternative motivations.

The Supreme Court further ruled in 1976 that publishers could be held liable for a reporter's *misinterpretation* — even if the news story were a rational interpretation of an ambiguous document![200] To reach this decision, the Court overruled the argument of newspapers and magazines that the First Amendment guaranteed them the right to say without restraint virtually anything they wished to say.

This decision had the effect of applying pressure on the news media to be both accurate and accountable for what they say. The courts seem to be making slow progress toward media responsibility.

Most importantly, "God versus Darwin" is not and never was the issue in the Science Textbook Struggle. Neither was it "science versus religion." Yet this is what the public, the scientific community, and educators throughout the United States were led to believe by the news media.

Therefore, forces began to mobilize around the straw man. Some religious periodicals announced the confrontation with great joy:[201]

> SACRAMENTO, CALIF — Perhaps one of the most significant educational events of our time transpired here recently without fanfare or ceremony. In fact, its effect may be felt in schools across the nation for generations to come.
>
> The California State Board of Education voted to include the Biblical creation story in the official state textbooks...

Other religious bodies, such as the Presbytery of San Francisco, representing some 44,000 United Presbyterians in that area, reacted violently by unanimously approving a resolution which read in part:[202]

> (The Board's action) completely misunderstands the significance of the Genesis account and raises serious questions, both about the teaching of religion in the public schools and about the integrity of the educative process.
>
> As Christians we find no conflict between the emphasis in Genesis that God is the creator and the objective findings of science that creation has occurred by means of an unimaginably long and complex evolutionary process. It is unfortunate that this issue, which has been settled in most of the Church for at least a generation or two, should now be raised in a secular body which is responsible for the intellectual development of the youth of our state.

The author of that resolution, Dr. Robert Bulkley, pastor of Portalhurst Presbyterian Church in San Francisco, was further quoted:[203]

> Once again we're making opposites out of science and religion and out of liberal and conservative members of the church. These are questions that should have been settled generations ago.

Although he is a professional religionist, Bulkley also detected *political* overtones in what the Board had done. Alluding to a long-standing antipathy between northern and southern Californians, Bulkley impetuously threw a wild punch:[204]

> According to Bulkley, the amendment is another attempt by southern California conservatives to crack the whip over the rest of the state. "It was just one more way for them to get in their slaps at liberals, the university, and liberal churches."

The respected journal, *Christianity Today*, showed unusual insight by pointing out that Bulkley's vehement reaction was based on misinformation and "inaccurate news accounts."[205] It chided Bulkley for reacting to news media that had "mistakenly reported that the Board had mandated the Genesis account of creation for inclusion in the state's science curriculum."

In contrast with the religious community that had shown immediate but mixed response to the phony "God versus Darwin" issue raised by the media, science journals were noticeably mute.

Some *individual scientists* however, reacted with either disgust, disdain, or derision. One of the most vocal was Ralph W. Gerard, Dean of the Graduate Division of the University of California at Irvine and one of the world's leading neuro-physiologists. A few selected words from his letter to the *Los Angeles Times*, that was printed on 6 December 1969, were widely quoted in both the secular and scientific press:[206]

> Should a scientific course on reproduction also mention the stork "theory"? Did it require the Apollo 11 mission to prove that the moon is not made of green cheese? Is the soothsaying of astrology really to be considered along with the precise content of astronomy that allowed man to plan his unerring trip to and from the moon?

Not all scientists were unhappy, of course. A research physicist at the University of California in San Diego objected to the arrogance shown by Gerard and other members of the State Advisory Committee on Science Education. While rebuking them, he also endorsed, not *religion* (as the press had pictured the issue) but the concept of alternative views concerning origins (that he was able to perceive as being the true issue):[207]

> As a physicist I must object to the attitude of the State Advisory Committee of Science Education. Their reluctance to allow the presentation of alternative views of the origin of life amounts to a religious faith in purely naturalistic science, which is poor science indeed. Furthermore it places students in the unenviable position of having only one side of a many sided problem presented to them before letting them come to their own decision.
>
> "The very essence of science" which the (committee) feels is offended by the introduction of the fact that there is at least one alternative to the theory of organic evolution is not science at all but scientific dogma which cannot stand to be challenged by other views.

So the news media really had success! They were able, with their straw man, to get some theologians to lose their cool while others were in ecstasy. Likewise, they simultaneously ruffled and smoothed the feathers of scientists. Some feat. In fairness, I must give the media an even greater compliment. They didn't just provoke an *ordinary* fight — their straw man incited a form of gallantry unknown since the knights of old!

Spokesmen from both camps came rushing to the rescue of their *opponents*. A professor of sacred scripture at Loyola University in Los Angeles was impelled by the phony straw man to protect science from the intrusion of religion:[208]

> As an active member of the Society of Biblical Literature and of the Catholic Biblical Society of America, may I voice an unofficial plea in the name of America's two most learned Biblical societies which represent Protestants-Jews-Catholics?...The (State) Board's manner of Biblical interpretation permeating the two added paragraphs, in the basic outline for new textbooks, is blatantly fundamentalist. The board's dichotomy of Bible vs. science is antiquated and unnecessary.

Another wholly unnecessary and inaccurate concern! Thanks to the news media, this Bible scholar believed that the Bible had been pitted against science, so he went to the defense of science. (Please, before reading further, go back and read once more Exhibit 5 — particularly the 2nd and 8th paragraphs of my revision that were adopted by the State Board — and see something "blatantly fundamentalist" if you can.)

Equally gallant scientists went to the rescue of religion — to keep it from being polluted by science! Typical of these was Ralph Gerard, earlier quoted, who said:[209]

> The State Board of Education has voted to have inserted into a long and meticulously prepared document on the teaching of science an utterly irrelevant statement concerning the Biblical version of man's origin...As a newspaper article put it, "God to get equal time." What an undignified approach to God!

You might even be persuaded that Gerard was a religious man, by the concern he showed for God. His earlier-quoted atheistic position would tend to cast doubt on the sincerity of his solicitude for God's reputation, however.

But there you have it. A totally false issue had resulted in getting everyone upset — one way or the other. Without any doubt, confusion reigned supreme at that point. What could be better for the news media? They had created a headline-producer that would last for several years...

Bait for Battle

In creating a ruckus, newspapers were the key. They were far more important than radio or TV newscasts. Their influence was many times that of magazines or the wire services. Why? Because all other news media built their stories from newspaper accounts. If the newspaper reporters had published the *facts*, the straw man "Religion Masquerading as Science" simply could not have been built.

Newspaper *headlines* obviously focused the issue. A few headlines spoke of the Science Textbook Struggle in subdued tones — as a *debate*, for instance. Most headlines, however, painted the issue in combative terms — like *dispute*, *confrontation*, *campaign*, or *trial*. Typical banners screamed, "CONTROVERSY FLARES ANEW!" or "EVOLUTION OR CREATION? THE FIGHT'S REVIVED." Inflammatory words like *win* or *killed* frequently punctuated captions as shown in Figure 4.

Perhaps the most descriptive word appeared in a huge headline carried by the *Los Angeles Times* in their Sunday "Opinion" section.[210] This headline introduced an impressive collection of 8 separate articles on the issue and read: "CREATION VS. EVOLUTION — BATTLE OVER WHAT TO TEACH."

It *was* a battle. Newspapers *wanted* a battle! It was good for their business. Besides, it was a happening of their own creation.

A battle requires two or more combatants. One army, person or theory must be opposed to another army, person or theory. Further, there must be a starting event or incident — a provoking *impulse* that sets off the fighting.

My brother Jerry is six years younger than I. Today he is bigger than I am, but as we were growing up, he was considerably smaller. One of his delights was to watch a fistfight — particularly one in which I was involved. On many occasions, he would attack someone much bigger than he, turn and run toward me as his victim retaliated, and then holler for me to protect him. Often I had to finish a fight he had started — while he cheered me on.

The news media in the Science Textbook Struggle reminded me of Jerry. How they must have delighted in seeing the irrationality they had generated! As provocateurs, they had no equal.

Having set the stage for war, the news media needed two opponents — two parties to be locked in mortal combat over the straw man. "The scarecrow must be shot down conclusively by one overwhelming opponent, while the other opponent is made to look like a simpering fool for defending a dummy," they plotted. "We need some bait for battle."

As any fisherman knows, bait is important. The idea is to lure, tempt or entice fish to bite. Much of the sport of fishing revolves around selection of the right bait. Therefore, fishermen often carry several kinds of bait in their tackle kits — along with leaders, sinkers, hooks and floats.

However, while bait is something you *use*, it is also something you *do*. If you bait someone, you goad or torment them, often with insulting or provocative remarks. The news media apparently decided to "bait by using bait." As foolish as that analogy might sound, this is essentially what happened.

A CREATION I	B EVOLUTION I
• ***Scientific theories*** **that account for something coming into being from a previously non-existing state** • **Subject to experimental verification** • **Examples:** **Abiogenesis (*novo* or *de novo*)** **Laplace's solar nebula theory for planets** **Woolfson's solar system creation model**	• ***Scientific theories*** **that account for variability and change over time among like plants and organisms** • **Subject to experimental verification** • **Examples:** **Hardy-Weinberg Law** **Darwin's theory of natural selection** **Mendel's laws of genetics**
C CREATION II	**D EVOLUTION II**
• ***Cosmological belief*** **that a supernatural Creator originated and continues to sustain the universe, life and various types or categories of living beings, including man** • **Accepted exclusively by faith (non-verifiable)** • **Examples:** **Genesis account in the Bible** **Islam's Koran account** **Zoraster's account of Ahura Mazdah**	• ***Cosmological belief*** **that the origins of the universe, life and various types or categories of living beings, including man, are interlocked and sequential events that happened without activity or intelligence external to the universe.** • **Accepted exclusively by faith (non-verifiable)** • **Examples:** **Quantum evolution** **Macro or mega-evolution** **"Amoeba-to-man" continuum belief**

Figure 5. Bait Boxes

Remember the brief discussion in Chapter 3 about two meanings of *evolution* and two uses of *creation*? Whether or not the news media realized it, those four terms — Evolution I, Evolution II, Creation I, and Creation II — were used as bait to bait combatants in the Science Textbook Struggle.

"That's too far-fetched for me," you might protest.

"Wait a minute. Look at Figure 5. It's just like the fisherman's tackle kit with four individual boxes for different bait!"

Let's examine the bait in that figure. This admittedly abbreviated and simplified display hopefully provides graphic "shorthand" to reduce some of the semantic difficulties inherent in exposing what has been called the *real issue* in the Science Textbook Struggle.

Box A in Figure 5, entitled *Creation I*, contains a definition and explanation of the *scientific* use of the word creation. Extensive research and theoretical modeling by mathematicians, astronomers, biochemists, and physicists to derive explanations of how previously non-existing materials or phenomena came into being are represented in this box.

Several examples of scientific creation theories are listed in Box A. One phenomenon for which scientists frequently propose theories is *abiogenesis*, the conversion of non-living material into living material. Scientific theories have been proposed for how life might have arisen spontaneously over and over again. (*de novo*) as well as uniquely only once (*novo*).

Louis Pasteur's experiments were an early part of this scientific work. Other examples might be Laplace's solar nebular theory that proposes that individual planets were created in a nebula revolving about the sun.[211] Similarly, Woolfson has proposed a model for the origin of the solar system wherein all our planets were created by being thrown from the sun.[212]

The important thing to remember about Box A is that these theories of creation are *strictly scientific* and are subject to demonstration or validation to verify their repeatability and reliability for prediction of future findings; e.g., creating life in the laboratory.

Box B in Figure 5, entitled *Evolution I*, describes and illustrates the *scientific* use of the word evolution. This box represents a vast amount of research in scientific laboratories by biologists, biochemists and geneticists to explain how living things change over a period of time. In contrast to the scientific theories in Box A, it is not likely that *originating* mechanisms would be of interest here. Instead, origins are presumed to have already occurred.

A typical example of scientific evolution theories would be the Hardy-Weinberg Law, that says that gene frequency will remain constant providing no mutations occur, all mating is random, and the population remains large. Another obvious example is Darwin's theory of natural selection. Likewise, Mendel's laws of genetics illustrate Evolution I.

Again, as with the scientific theories in Box A, these theories can experimentally be verified and provide information for predicting future findings.

Box C in Figure 5, entitled *Creation II*, retains all the various *beliefs* in supernatural causes for origins. Note that there is absolutely *nothing* scientific about the beliefs in Box C. These beliefs are independent of scientific evidence, and they can neither be proven nor disproven by the scientific method. They involve Socratic truth as defined in Chapter 2. Since they cannot be verified scientifically (note that this does not necessarily mean that they *conflict* with scientific evidence), they are accepted by faith.

While the Bible's Genesis story is most familiar to western civilization as a source for these beliefs, there are many other cosmologies based on a supernatural Creator, including Islam's account of creation in the Koran, and the story of Ahura Mazdah in the writings of Zoroaster.

Box D in Figure 5, entitled *Evolution II*, possesses those *beliefs* that origins can be wholly explained without referring to supernatural causes. These believers are committed to faith in *scientific naturalism* instead of faith in a *supernatural being* as occurs in Box C. As with the beliefs in Box C, these beliefs are outside the domain of scientific proof or disproof. Therefore, they are accepted by faith rather than upon scientific evidence.

George Gaylord Simpson's *belief* in "quantum evolution" (see Exhibit 48) as the origin of taxonomic units like families, orders and classes is typical of the beliefs in Box D. Exhibit 48 also contains Simpson's description of his *belief* in "mega-evolution" to explain origins without referring to supernatural causes. *Belief* in the "amoeba-to-man" continuum likewise falls in Box D because it can neither be proven nor disproven scientifically.

Now that we've examined the bait in each of the four boxes in Figure 5, we can draw some conclusions:

1. Boxes A and B alone contain *scientific theories.*

2. Boxes C and D contain *philosophic*, *metaphysical*, and *religious beliefs* rather than anything scientific.

3. Box A contains those scientific theories which deal with "non-existent becoming existent" phenomena in nature while the scientific theories in Box B "assume the existent." Therefore, Boxes A and B are interdependent and *complimentary* rather than competitive.

4. The beliefs in Boxes C and D tend to be mutually exclusive; i.e., if you believe in scientific naturalism, you will not likely believe in a supernatural Creator, and vice versa. Therefore, Boxes C and D are *competitive* rather than complimentary.

5. The beliefs in Boxes C and D are not necessarily anti-scientific or antithetical to science, although the possibility is always present. More frequently, from a historical perspective, the beliefs in these two boxes have *aided* scientific investigation rather than *hindered* it.

6. Because Box A (science) and Box C (belief) have identical titles — *creation*, confusion of the two is highly probable. The same confusion is likely for *evolution* in Boxes B and D.

Earlier, the news media were said to want "candidates to bait for battle." It is obvious now that they had four different baits to use as candidates. In fairness, it must be admitted that, prior to my 13 November 1969 speech, they might have thought they had only *two* — evolution and creation. But, if they had listened to what I said to the State Board of Education, they would have easily seen that two different types of thought are both called "creation."

Although I did not personally emphasize the difference between Evolution I and Evolution II in my remarks that day, documented proof exists that the media were informed of that distinction. Here's how the *Los Angeles Times* quoted Richard A. Smith, chairman of the Natural Science Department at San Jose State College as well as chairman of the State Advisory Committee on Science Education, immediately following that November meeting:[213]

> Smith said it is correct to distinguish between evolution as fact based on present-day evidence and evolution as theory in explaining the origin of the world.
>
> He said he believes the original framework made the distinction adequately, although it was misinterpreted.
>
> L. F. Mann, coordinator of the framework-writing project in the State Department of Education, also believes that the two aspects of evolution have been confused.

Therefore, there seems to be little excuse for the news media's failure to recognize all four boxes in Figure 5. Using those four boxes, the press could have logically pitted either A and B (science) against C and D (belief), A and C (creation) against B and D (evolution) or C (one belief), against D (another belief). Instead, they selected two illogical baits as opponents — B (science of *change*) against C (belief about *origins*).

To aid in explaining this choice the press made among boxes in Figure 5, consult Figure 6. The same relationship of boxes used in Figure 5 is followed in Figure 6.

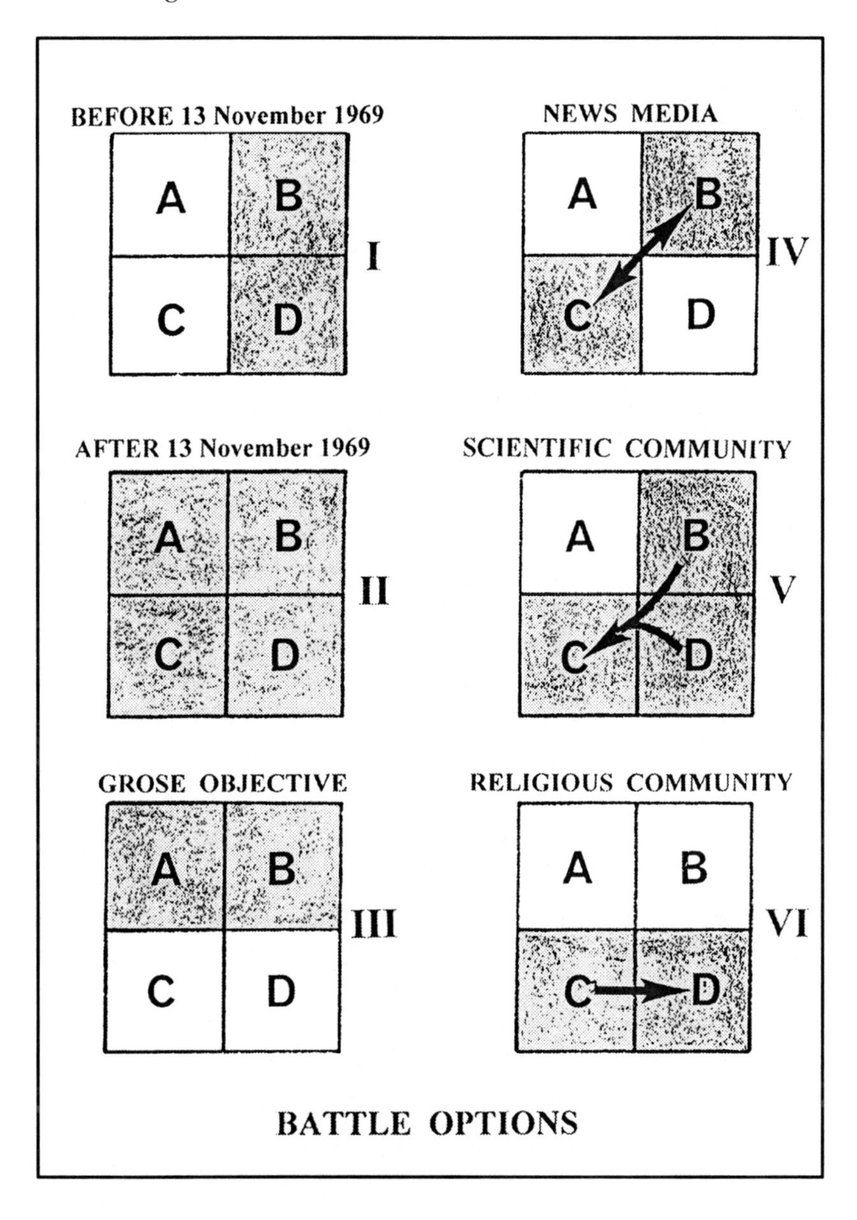

Figure 6.

Before 13 November 1969, the shaded areas in Group I of Figure 6 depict that evolution was being taught both as science (proper) and belief (improper if it is the *sole* belief offered). Further, science textbooks were absolutely *silent* about creation – either as science or belief. This situation was correctly diagnosed as requiring correction by the State Board of Education, as well as by concerned parents and citizens – myself included.

Immediately *after* 13 November 1969, the State Board of Education exposed to the public — by means of the *Science Framework* — the message that all four options must exist for a balanced science program (Group II in Figure 6).

Because I am convinced, along with Ralph Gerard, that "some things should not be mixed"[214] and that science classes should discuss only science, I personally favor cleansing science textbooks of *all* belief systems as shown in Group III of Figure 6. My speech to the State Board of Education in November 1972 (Exhibit 47) specifically and emphatically makes this point.

Group IV in Figure 6 illustrates the irrationality of the news media's choice of B to oppose C. About the only way that such a choice could be rational is if news media personnel, as individuals, were personally committed to the *beliefs* of D. They might then want to *kill*, once and for all time, the beliefs of C by making them look ridiculously indefensible in the light of *scientific* evidence for B. Group IV also shows that the news media never acknowledged the existence of A or D.

How and why the news media failed to get a full-fledged battle organized comes later. Briefly, they failed because their selected opponents viewed the battle from different perspectives.

Take a look again at Figure 6. Group V shows that the "official" scientific community marshaled both the *science* and *belief* of evolution (Boxes B and D) in an attack on the creation *belief* (Box C). Shielding the scientifically indefensible belief of D within the hard shell of scientific respectability of B, this elitist community sought to "out-Darrow Darrow" in killing belief in a Creator.

Group VI of Figure 6, on the other hand, depicts most of the religious community viewing *their* combat role in the media-created battle as having to attack the scientific naturalism of Box D. They generally did not mix science into the cosmological battle, however, as their opponents had done.

Obvious exceptions to these simplified illustrations occurred. For one, a group of scientists known as the Creation Research Society attempted to "mirror-image" the scientific community battle plan. They tended to ignore Box B just as the scientific elitists ignored Box A. On the whole, these pictures accurately reflect the major outline of battle.

Not a bit like the press reported, is it?

REPAIR BY REASON?

It was hopeless. True, I'd still had hope of getting the truth out when I had only heard the radio and TV reports. But then the *newspapers* began repeating the same false message. News coverage was not only profuse — it was also *unanimous!* What could I do about a wide-open breach in the dike, with a floodtide of error pouring through it?

Then, mid-morning on Friday, Ralph Winter called. "Man, did the *LA Times* ever misread your intent!"

"I know – and I'm sick about it," I moaned.

Ever the optimist, Ralph sensing my despair, bounced back with "Listen, Vern, don't give up yet. Dan Thrapp, the Religion Editor at the *Los Angeles Times* is a personal friend of mine. Let *me* call him and express *my* disappointment with the obvious error in today's article."

"Well, if he wants factual material to prove what was *actually* done yesterday, I'll be glad to furnish him with a copy of my speech and the *Framework* revision so that he can read the truth for himself," I offered.

"Let me see what I can do," Ralph said assuringly. "I think he should be more responsive to *me* because I am a disinterested third party in the issue, as well as a good friend of his."

A tiny spark of hope flickered within me once more as I thanked Ralph and hung up. Frankly, I stayed close to the phone the rest of Friday, expecting to hear either from Ralph or directly from the *Times*. But no call came...

Electronic Explanation

Later on Friday, as I waited for Ralph's response, I got another call. It was from KHOF-FM in Los Angeles, wondering if I would appear on a talk show called "Sharing Time," that night at 9:00 P.M. They wanted to interview me about the State Board's decision the previous day.

While I realized that the damage done in *writing* by newspapers could not be reversed by a *spoken* interview on radio, I felt that I could hardly lose by trying to set the record straight. So I accepted the invitation.

The talk show host, Cam Wilson, jumped right into the issue by reading aloud the headlines in that morning's *Valley News*: EVOLUTION FOES WIN REVISION OF SCIENCE TEXTS IN SCHOOLS. He went on to introduce the interview by saying: "As I read this front-page article and then went to the interior of the paper, I discovered that

Thursday, November 13, 1969, is likely to go down as a landmark day in public education for the schools in California."

Wilson's first question was: "What did you believe needed *changing* in textbooks and in classrooms of science in California schools? What was *wrong* with the way they were teaching?"

My answer was short and blunt. "They were teaching only one view about origins, and teaching it as *fact!*"

Moments later in this radio interview, Wilson posed another question: "We noticed an extensive article on the front page of the *Los Angeles Times* this morning, and I'm quoting from it, if you don't mind: 'The framework for the new textbooks also will include the Biblical version of God's creation of life.' Is this *really* true?"

"It really is *not* true...I was careful to show that it was not a *religious* issue that we were introducing into this *Science Framework*, but that creation is viable *scientific* theory — incidentally, endorsed by people who have absolutely no belief in God."

Wilson rebounded, "Then, I think what you're hinting at is that you felt yesterday morning the issue was *not* primarily a religious one. Is that correct?"

"Absolutely!" I said with every ounce of conviction I could throw behind it.

As I walked out of the radio station that evening, I felt a little better than I had when I first saw the morning headlines. Yet, I knew that a very small percentage of the people who read the headlines had heard me on radio.

I wasn't the only one to use the electronic media to rally public thought on the issue. Around dinner time on 4 December, the telephone rang, "Have you heard from KHJ-TV, Channel 9 yet?" Ed Shevick asked me rather excitedly.

"No."

"Well, you will be shortly. Baxter Ward, moderator of the TEMPO Show wants to discuss this textbook thing on his show tomorrow. He got hold of our Committee today while we were meeting in San Francisco. We elected Ralph Gerard to represent us, and your name was proposed as Gerard's opponent."

Shevick, a junior high science teacher and neighbor of mine, was a member of the State Advisory Committee on Science Education that had written the *Science Framework*. We had not known one another until my amendment was adopted. I spotted his name in the membership list of the Committee, discovered he lived nearby, and called him to introduce myself.

Shortly after that, we met in his home for an evening and discussed what I was trying to do versus what the press was reporting. We became good friends, and Ed supported my position for *objectivity* concerning origins while remaining opposed to the version of my words being reported by the press.

Somehow, Baxter Ward never did reach me to invite my participation in the debate. Instead, he interviewed Ralph Gerard on his TEMPO Show on 5 December 1969. I will be thankful to my dying day that Ward never located me, although I made no effort to avoid an invitation.

Baxter Ward had been a well-known Los Angeles television personality for several years prior to running for public office in December 1972, when he was elected a County Supervisor. Undoubtedly, his television exposure aided him greatly in his election to public office. There are only five Supervisors in Los Angeles County, and no other local official in the United States is assigned responsibility of the breadth and scale of those afforded a Los Angeles County Supervisor.[215] Ward took over as Chairman of the Board of Supervisors in December 1975, and today he wields great political power.

However, on the day he interviewed Gerard on his TEMPO Show, no one would have suspected how the interview would turn out — least of all Gerard himself.

By this time, Gerard had established himself in the Science Textbook Struggle as the incisive, articulate, self-appointed spokesman for the 15-man Committee. His letter to the editor of the *Los Angeles Times* had been widely quoted and re-quoted.[216] He had a commanding, almost arrogant air about him.

To say that Ward completely unzipped and humiliated Gerard that day on TEMPO would be an understatement. (Exhibit 9 is a verbatim transcript of the telecast.) Without much difficulty, Ward was able to make this brilliant, noteworthy scientist appear confused, picayune, and rattled. Like a cat playing with a tired mouse, Ward taunted Gerard to the point that Gerard lost his temper. At one point, Gerard appeared to become irrational, accusing the State Board of being "sincerely stupid." When Ward later reminded him of this accusation, Gerard *denied* it!

In a way, Gerard, for all his scientific reputation, was pitiful. Completely losing his composure, he said that my two adopted paragraphs had "assassinated" the *Framework*. "The changes, though small in extent, had the effect of entirely undercutting the thrust of the 205-page document," he declared. (I didn't know that I could write anything that consequential!)

If Gerard intended to clarify the issue by appearing on television, he failed miserably. The most unfortunate aspect was that Gerard simply had not done his homework. He had not consulted the original source material, or he would not have made the many inaccurate statements he made on television. Even eminent scientists fail at times to be careful, thoughtful and objective before taking a position.

Although I had hoped that radio and television might be the means to eliminating the straw man, I soon gave up that hope.

Journalistic Justification

Neither Ralph Winter nor anyone from the *Los Angeles Times* ever called back on Friday, 14 November, as I had anticipated. So, not knowing whether the *Times* would ever respond, I thought I'd take a stab at getting *national* news coverage to correct the error. Such coverage would be particularly important if the *Los Angeles Times* refused to listen to me.

On Monday morning, 17 November, I called the *U. S. News and World Report* office in Los Angeles. Susan Thornton in that office was cool about the possibility of a personal meeting. Typical stalling tactics sometimes used by secretaries convinced me that I needn't bother to go in person to their office.

On the other hand, when I contacted Sandy Burton in *Time* magazine's Beverly Hills office, she agreed to meet me that very day! Beside taking specific written exhibits for her, I decided to also take a witness. Dave Irwin, who had been present at the State Board hearing on 13 November 1969, agreed to meet me at *Time's* office.

Dave and I had a very enjoyable discussion with Miss Burton. We handed her a copy of my speech and my proposed *Science Framework* revision, with the two paragraphs that the State Board adopted marked in red. She listened most attentively as we showed her what the *Los Angeles Times* was saying about the issue. She appeared to understand exactly what our concern was. We urged her to have *Time* "scoop" the newspapers by exposing the straw man.

She agreed with us that this was a good opportunity for *Time*. However, she had a natural "escape hatch" — the magazine is produced in New York!

"I promise you that I will forward these materials to New York right away. Obviously, I can't guarantee that they will use them. However, it sounds logical that the story should be corrected."

Dave and I left the *Time* office reasonably hopeful that this would do the trick. "If they want to report the *truth*," Dave said, "they've got an excellent chance of doing so!"

It didn't take long for those hopes to be dashed. On 12 December 1969, in a lead article entitled EQUAL TIME FOR EDEN, *Time* showed their high regard for accuracy and truth by saying:[217]

> (California Public Instruction Superintendent Max) Rafferty and his fellow fundamentalists want equal time for the Garden of Eden and the rest of the biblical account of creation so that the children can decide for themselves between the *Book of Genesis* and Darwin's Revised Standard Version of creation.

Beyond the actual words of that article, any serious and thoughtful person would see that *Time* intended to do far more than report some news about California. Their facetious twisting of Darwin and Revised Standard Version (a well-known version of the Bible) had every mark of derision.

Further, I often wondered what went through that reporter's mind as he wrote something completely contrary to the written source documents I had furnished. I wrote to the *Time* editor after the article appeared, listed five specific errors, and noted that I had "furnished all materials necessary to preclude these errors from being reported." My closing statement had a needle in it, "You must have a penchant for distortion or an unseen ax to grind. Which is it?" (They never told me!)

I had been a consultant to several divisions of IBM since 1967. During the week following the State Board's action, I had to fly to IBM's General Systems Division in Rochester, Minnesota. Being out of California at this time was critical, of course, in my effort of correcting the news media error before it became "set in concrete."

Upon my return from Minnesota, I found a letter from Ralph Winter awaiting me. It said that Dan Thrapp, Religion Editor of the *Los Angeles Times*, had finally returned Ralph's call after a five-day delay. In their telephone conversation, Thrapp asked Ralph to have me send him a detailed letter, noting specific errors so that they could be examined. Ralph's letter was complete with instructions, even as to how to start my letter to Thrapp:[218]

> I told him as well as I could very briefly over the phone what the situation was, that I was concerned that the "Times" was going to get in trouble for misrepresenting this case because religious people would be overly optimistic and anti-religious people would be overly enraged...I finally suggested that I would have you send him a complete copy of what you actually presented, and then also — he suggested this — a list of the specific errors in the presentation of the article...I think this is the kind of request from the "Times" to you that you can legitimately act on, and your letter should begin by saying "Our mutual friend...has passed on to me your request that I send you a copy of what was presented..."

The dying embers of hope were fanned into flame by Ralph's intercession on my behalf. I could now write to the *Times* at *their* request, rather than submitting an unsolicited protest.

I followed Ralph's specific instructions faithfully. Exhibit 8 contains the full text of my letter to Thrapp. Carefully detailing four errors in the 14 November 1969 front-page article, I pointed out that all those errors were repeated in addition to others in the editorial of 16 November 1969 (Exhibit 7). More errors in their article of 23 November 1969 were also listed. The errors were summarized with a note that if there was an outburst at the State Advisory Committee on Science Education meeting scheduled on 4 December 1969 in San Francisco, I would personally hold the *Times* responsible.

(Of course, this meeting turned into a melee, with a wild and irresponsible resolution being passed by the Committee.)[219]

Believing that there was still hope to "repair by reason" all the damage that had been done to this issue, I concluded my letter to Thrapp as follows: "Via this letter, the *Times* now has the facts, I trust that they, in fairness, will have the grace to clarify and if necessary renounce their previous position."

Well, you'd better chalk up another one to naiveté! As you might have guessed, I never heard a word back from Thrapp or the *Los Angeles Times* in response to my extensive letter. Even though *they* requested the information, they lacked even the courtesy of acknowledging receipt of it. That whole effort could have been just as well poured down a rat-hole.

Not only was I concerned about what the written news media could whip up on its own. I also feared what individual members of the State Board of Education might say publicly that would add fuel to the media's fire. Therefore, on 1 December 1969, I wrote a personal and confidential letter to Vice President Harward of the Board.

At that time, Gerard — on behalf of the entire Committee — was threatening to disavow authorship of the *Science Framework* if my two paragraphs were not withdrawn. This reaction was exactly what the news media evidently had encouraged by their headlines. Accordingly, in my letter to Harward, I summarized what I could see happening to the issue.

One of the key points in my letter said, "We cannot expect to have public school teachers attempting to teach, under the subject of science, that God is the creator... I was sincere in my presentation to the Board that we should teach creation as a viable *scientific* concept, rather than one that requires a belief in God." Building off that statement, I made the following historically significant points:

1. We must be most careful to clearly differentiate between religious beliefs and scientific theories.
2. We must guard against well-meaning but less-than-scientific people who, in their enthusiasm to right many wrongs, could ruin the cause which has been so nicely started.

Meanwhile, the conflagration assumed ever-increasing dimensions. Day by day, articles appeared in newspapers, completely distorting the issue. With urgency, I wrote to Max Rafferty on 6 December 1969 (see Exhibit 10). I summarized my personal position as being to "specifically *exclude* any religious connotation regarding the concept of creation. The mention of the Bible in the adopted paragraph was in the sense of *denying* the Genesis account." I went on to explain, "I did this for the very purpose of offsetting the presuppositions of religionists, on one hand, who would claim, in error, a great victory for God and the Bible, and my scientific colleagues, on the other hand, who would falsely read Biblical creation (which is believed only by faith) into the *Science Framework*."

I explained to Rafferty my frustration experienced while watching Gerard on the TEMPO Show. "As I watched this telecast, I was agreeing with Dr. Gerard in his insistence that religion had no place in a science framework. Yet, I was the author of what he viewed as an anathema!"

Feeling an obligation to both Rafferty and the State Board for action that I felt at least partially responsible in creating, I prepared a statement in the form of a press release that Rafferty might wish to issue. Two paragraphs in that press release are critical to the argument that I was not attempting to introduce religion in the science classroom:

> The State Board hereby wishes to reaffirm that there is no intention to include religious instruction under the guise of science education. Neither the Biblical account of creation nor mention of God was intended nor implied in the *Science Framework*.
>
> The primary purpose of amendment was to broaden the scope of science education to allow the teaching of *all* theories, not just the theories of evolution and creation, which provide rational explanation for scientific observations of natural phenomena.

On 11 December 1969, State Board President Howard Day released the statement I had written. The *Los Angeles Times* immediately published a long article under the headline, MISCONCEPTIONS BLAMED IN FURORE ON EVOLUTION.[220] (For anyone interested in examining how written statements could be taken and twisted completely out of context into another story, I would suggest that they read this article.)

One statement in that article is most interesting when compared to the battle options in Figure 6:[221]

> This is what the scientists who authored the guidelines are objecting to — that the scientific "fact" of evolution based on concrete evidence is being reduced to the level of theory and that other theories such as religious explanations are being elevated to the level of scientific propositions. The scientists say that they have no argument with the board about evolution being a theory as far as the origin of life.

I certainly understood the old saying, "There are none so deaf as those who will not hear." The only conclusion you could draw was that the media were not going to let this issue be settled. At least not in written form. My journalistic efforts to justify my intent and repair the breach by means of reason had come to naught...

Personal Persuasion

The day after Ralph Gerard appeared on the TEMPO Show, I placed a long-distance telephone call to his office at the University of California at Irvine.

He betrayed quite a bit of shock in his voice as I revealed who I was. I briefly explained that I believed *I* was as interested in keeping science pure as *he* was. Further, I acknowledged that I was very disturbed, as I knew he was from his television appearance and his letters to various editors, about the possible introduction of religion into the science classroom.

Not easily convinced, he said he doubted that I fully understood what he was complaining about, but he would be willing to meet with me in person to discuss it further. Inasmuch as he was taking a trip to Africa almost immediately and would not return until after the first of the year, he requested that we set a date in January to meet personally. In the meantime, I assured him that I would write to Rafferty and all members of the State Board, explaining our conversation and reconciliation (see Exhibit 10).

I was enthused by my telephone conversation with Gerard. Even though my hopes to correct journalism, radio and television reports had appeared to fail, I felt that if there could be a cease-fire between the two supposed forces represented by the straw man, there would be nothing for the news media to distort.

In a letter to me on 8 December 1969, Gerard showed a warmth that I appreciated:[222]

> I was very disheartened by the TV appearance on KHJ, since the M.C., presumably trying to see that both sides of the issue were aired,

> succeeded in preventing any of the real issues from being discussed effectively. Rather, he degraded the period into a seeming cry of aggrievance by the Committee and myself. I am sure that, had you been my "antagonist," the public would have received some education rather than mere titillation over a "fight."

I really looked forward to our proposed meeting in January!

As might be expected, the *religious* community also misread my intent — thanks to the news media. Therefore, I received frequent requests to address religious groups. All the ingredients for becoming a folk hero to religious groups were available to me. However, I was adamant in my rejection of what they believed, based on what they read in the newspaper — that Genesis was now going to appear in science textbooks.

On 5 January 1970, I was invited to address the monthly meeting of the West San Fernando Valley Ministerial Association in Los Angeles. The newspaper account of this speech correctly interpreted my position:[223]

> "The teaching of creation in California public schools has been misunderstood by many churchmen," the West Valley Ministerial Association was told at their monthly meeting on January 5, 1970...Vernon L. Grose, Vice President of Tustin Institute of Technology in Santa Barbara, explained that some ministers have made the same error made by certain educators and scientists concerning the recent decision of the State Board of Education to include other theories of origins beside evolution in science teaching. They have thought that "creation" meant "Biblical creation." Neither the Bible, the Genesis account of creation, nor God will be mentioned in science texts.

As I look back over my frantic efforts to destroy the straw man created by the news media, I realize how ineffective *reason* is at times like that. While science is a field supposedly based upon reason, scientists themselves can be just as rapidly drawn into a whirlpool of irrationality as anyone else.

With 20-20 hindsight, people may now blame me for not having succeeded in communicating the truth of this issue. But I felt like I had given it "a college try." I had tried radio, the written word and personal speeches — all to no avail. Like an exhausted swimmer clutching desperately onto a slippery rock in the roaring rapids, I finally released my grip...

CAUTION TO THE WIND

Once I realized that I was no match for the news media and surrendered to their onrushing force, I began to see my efforts in a different light. True, I'd clearly been outnumbered and overwhelmed. The news media had been able to say, "We don't care how you approach this — by radio, television, in writing, or in personal speech — we will not tell the public what you've said." I had exhausted every means of public communication that I knew, trying to correct an obvious error. However, as I licked my wounds and reviewed my performance, I could see that everything I did to correct the direction the news media had chosen for this issue actually was *counterproductive.*

Every move I made to correct the record seemed to convince the news media that I had an ulterior motive in what I was doing. Their initial conviction — that this was a heavily-financed, well-oiled diabolical scheme to resurrect and rerun the Scopes trial of 1925 — became fixed in their mind. I was a decoy. Some big machine had set me up to take the heat while others were slipping in the back door to destroy science — and particularly Darwinian evolution. So the news media threw caution to the wind!

Error Sells!

What would *you* do if your tax consultant consistently made mistakes, resulting in your being fined or penalized for paying the wrong amount of taxes? I think you'd fire him. How long would you keep a secretary who could not spell, punctuate, or type without strikeovers? Not long. If the same filling kept falling out of one of your teeth, would you change dentists? Without a doubt. What would happen to a shortstop that bobbled the ball every time it came to him? He'd be replaced.

Behold I show you a mystery! The news media are *immune* to this universal law of mistakes. The nation continues to watch the 6:00 o'clock television news every night, whether or not they make mistakes. Newspapers and magazines continue to sell, even when they are filled with untruths.

Now, I am not trying to suggest that *all* news media mistakes are deliberate. Nor am I trying to propose that they don't expend great effort trying to be correct. Just as the accountant, the secretary, the dentist, and the baseball shortstop may *try* to do a good job, so do the news media. They place no premium on being wrong. But when the news media make the same mistakes the others make, they do not suffer. Due to a quirk in human nature, error often sells *more* newspapers!

There is apparently less profit to be made by *correcting* error than there is for *producing* error. Therefore, error always has greater prominence in the news than its corresponding correction. People are more willing to read about something that is *wrong* than they are to read about the *correction* of that wrong.

This line of reasoning, of course, could lead to a person becoming paranoid about the news media. That is not my intent. Rather, by calling attention to this quirk, it might stimulate thought about how it might be overcome or corrected.

Certainly, there is no meaningful legal recourse in the case of news media error. The courts have long perpetuated the practice of accepting retraction on page 37 of a page 1 error as being compensatory.

So there is little hope of accomplishing much through court action.

Likewise, radio, television and newspapers have such a monopolistic status in society that massing a meaningful boycott is virtually impossible. Therefore, the public cannot hurt the pocketbook of the news media as a lever of overcoming error. Error continues to sell.

Making a moral appeal for the news media to have responsibility for what they print is also fruitless. As logical as it might seem that a liberal news community should have open-mindedness and objective reporting, the moral appeal falls on deaf ears when the media are in a profit-making situation. The altruistic argument is no competitor for dollars.

What is left — short of revolution? Nothing except accommodation. News media error must be accepted as a "given" datum in life's equation. It really makes little difference whether the media make *deliberate* or *inadvertent* errors. They have the same effect. While corrective action might differ for these two types of error, there are no sanctions that can be levied against the news media to bring action toward correction for either one.

Some of the error in the Science Textbook Struggle was even *amusing*. It resulted in rather bizarre treatment of my words — certainly far afield from any effect I could have predicted.

"Hey, Vern, I read about you in the barbershop the other day!" Jack Mansfield, Director of Continuing Engineering Education at The George Washington University where I teach, greeted me with a sly smile as I walked in his office one morning.

"You're kidding," was my immediate reaction.

"No, I was sitting there waiting for a haircut, and I picked up the latest *Saga* magazine — you know, one of those men's magazines. There was some far-out article about life coming in from outer space, and you were mentioned right at the start of the article for what you had said to the California State Board of Education."

I was able to buy a copy of the magazine soon afterward at a local newsstand. The article to which Jack referred was entitled, "Is Man a Hybrid 'Developed' by a Super Space Civilization?" Sure enough, the authors of this article had managed to weave what I had written into a tapestry of wild concepts of man's origin:[224]

> In November 1969, the California State Board of Education declared that new textbooks must include all other theories of the origin of man, including that of the Biblical Creation; and children will also be allowed to learn the spontaneous generation concept of Aristotle, the panspermia (spores from space) theory of Svante Arrhenius, and others ... Some of the testimony that the California School Board based its decision on came from a physicist, Vernon L. Grose, who said: "Creation and evolutionary theories are not necessarily mutual exclusives, some of the scientific data may best be explained by a creation theory, while other data substantiated a process of evolution."

This article went on to propose that man is a hybrid that was developed and transplanted on Earth by people from outer space, rather than having come up through the "amoeba-to-man" continuum. Naturally, because it was in a girlie magazine, much of the article was devoted to sexual traits that the author felt pointed to an extraterrestrial origin for man.

(This article evidently was quite popular because it was reprinted in *Saga* magazine's *1972 Annual* two years later.[225])

Apparently because two of the State Board members were physicians and had been vocal about alternative theories concerning origins, *Medical World News* published an error-ridden article entitled "Downgrading Darwin in Class." Even the title is fallacious, because there were no negative statements made about evolution or Darwin in the words adopted by the State Board — unless the idea of permitting alternatives amounts to "downgrading."

It can be assumed that this article was addressed to a narrow segment of society — professional medical experts, people with above-average education. Yet, you would have thought the article was written to stir the basest emotions of the public. One could hardly help but wonder where the author gained, for example, the *objective data* to write:[226]

> While most MDs *believe* that *Homo sapiens* ascended, step by step, from protozoa by way of an ape-man, a few *cling to faith* in the divine creation of each species, despite persuasive biological evidence to the contrary and the blow dealt to fundamentalism by the Scopes trial 45 years ago. (emphasis added)

What is the difference between "believe" and "cling to faith"? Or where is the "persuasive biological evidence" that would rule out divine creation? This article likewise had the sequence of many events in the issue completely out of order.

For example, it had the press release that I had drafted for President Howard Day (that was released on 11 December 1969) being issued in response to a footnote to the *Science Framework* that wasn't added until

February 1970. It further had the State Board of Education recommending to the Curriculum Commission, when the Curriculum Commission is subordinate to the State Board of Education, with all recommendations coming *from* the Commission to the Board — not vice versa. Numerous other errors permeated this article.

One normally thinks of the *New York Times* as the king of all American newspapers. Its reputation has supposedly come from being both thorough and accurate. Yet it reported that the State Advisory Committee on Science Education had resigned in a body when the State Board attempted to dictate changes in the Committee's proposals.[227]

I was not the only person involved in the issue who could see error in the news. Another person — one who could be possibly considered opposed to my position — was well aware of the error committed by the news media.

On 9 December 1969, L. Frank Mann from the Bureau of Elementary and Secondary Education in Sacramento wrote to me, "I thought your presentation to the Board was very well done. It is most regrettable that some of your statements taken out of context have come to mean something quite different from what you intended. This is particularly true of your reference to the Bible."

The chairman of the Science Committee in the California Curriculum Commission, Glenn F. Leslie, who was frequently quoted in the press as opposing me, saw error too. His concern on 13 November 1969 was a very legitimate one (see Exhibit 6). He and I developed a warm personal relationship in the months following the adoption of my revision. An extremely fair-minded person, I learned to appreciate him as the issue progressed.

On 9 January 1970, Leslie wrote. "I am sorry that all the confusion, misrepresentations and adverse publicity resulted because of the amendment. However, I believe in the long run that the air will be cleared and the science program in California will be better because of the controversy. Perhaps you and I had to bear an unnecessary part of the criticism but I'm sure we can survive it." Undoubtedly, he had a more mature viewpoint than I did from which to judge the twisting by the news media.

Church periodicals were as guilty of error as the secular press. A vested interest was obvious in headlines that read, DARWIN GAINS IN BIBLE FIGHT.[228] However, one Palo Alto, California church published an extensive treatise entitled, TEACH CREATION IN THE SCHOOLS?[229] This lengthy article was remarkable because it so clearly pointed out that the issue raised by the news media was fallacious and that the Bible was never intended to be introduced into the science classroom. It is unfortunate that this article did not receive wide publication.

The Public as Judge?

One unique peculiarity concerning news media error has continued to intrigue me. And it has very little to do with the intent of the news media. Rather, it involves that portion of the scientific community that attempted (and succeeded quite well) in using the news media as a vehicle to appeal to the general public.

On one hand, consider the statement by G. Ledyard Stebbins, a member of the National Academy of Sciences, regarding those who should be allowed to talk about science. Stebbins defines a scientist as "a person who does experiments, gathers original observations, reads firsthand papers in which experiments and observations are described, and goes to scientific meetings or conferences to discuss problems with his colleagues." (See Exhibit 26) He goes even further by stating that scientific opinion can be considered *informed* only when the person has acquired firsthand information about the subject being discussed.

So, one could expect that scientists alone would be able to determine what represented scientific truth. This is a very small group of people percentage-wise in the world. If there were an issue of peril to science, it seems unlikely that scientists would be expected to become aware of it through the newspapers or television. Rather, it would appear much more likely that the learned journals of science would be the vehicles for exposing and correcting error in science.

One of the amazing aspects of the Science Textbook Struggle was that some of the more prestigious voices for the scientific community — like the National Academy of Sciences or the American Association for the Advancement of Science — apparently mobilized the news media to propagate their message. Aside from the fact that these learned bodies were concerned about a straw man rather than a real issue, it is puzzling indeed as to why they would appeal to the *public* to judge a *scientific* issue!

They had already set the ground rules that they would not listen to anyone who failed to have firsthand information on a scientific subject, so what was the urgency or even the rationale in appealing to the public?

The only reason that I could perceive for the scientific community insisting that *they alone* should judge and yet simultaneously appealing to the *public* to judge was that the scientific community is not unlike a state-supported church.

Certainly the vast majority of dollars for scientific research comes out of public taxes. Therefore, such support is in constant jeopardy, due to the fickleness of public opinion.

On the other hand, the scientific community is like a theological hierarchy. They have their own restricted definition of scientific truth that must be protected from heresy. Like every other organizational structure that man has ever built, the scientific power structure must retain its power and guarantee its perpetuity.

One obvious reason why some scientists thought Darwin was being persecuted, evolution downgraded, and science itself under a massive assault by the Science Textbook Struggle was that something the scientific barons had not approved might be compared side by side with their elitist truth.

For example, Max Rafferty, California Superintendent of Public Instruction, was widely quoted on 13 November 1969, following the State Board's adoption of my suggested revision, as proposing a supposedly outrageous thing — that school children would have a *choice* in what they believed about origins. Radio, television and newspapers quoted Rafferty as saying:[230]

> Evolution is to be taught in the schools of California. It is to be taught as a major milestone of scientific thought, but it is to be taught in the context that there are many people who disagree with it — and that there are other theories including Aristotle's theory of spontaneous generation, the biochemical theory for the origin of life, the creationism theory which many scientists as well as lay people hold to...These are all to be described to children with the scientific evidence for or against them adduced as evenhandedly as we can. And let our young people draw their own conclusions from the evidence presented.

Rafferty's statement was projected in preposterous proportions by the press. He was "quartered and drawn" for having made such a barbaric proposition as letting children make up their own minds. Was Rafferty *really* proposing preposterous possibilities?

The Science Textbook Struggle is so often compared with the Scopes trial of 1925 as though it was a rerun of a Charlie Chaplin movie. Not many people — even scientists — know a great deal about the Scopes trial. Sure, they know about Clarence Darrow beating William Jennings Bryan so badly that Bryan died within days after the end of the trial. They also know that science was finally freed from its shackles of religion by that noteworthy event...

However, there is an interesting bit of testimony in the Scopes trial that sheds light on Max Rafferty's preposterous statement. Clarence

Darrow wasn't the only attorney who defended John Scopes in that trial. Another prominent attorney — Dudley Field Malone — figured very prominently in the trial.

At one point, Malone delivered a stinging argument in behalf of children in the schoolroom:[231]

> I would like to say something for the children of the country. We have no fears about the young people of America. They are a pretty smart generation. Any teacher who teaches the boys or the girls of today an incredible theory — we need not worry about those children of this generation paying much attention to it. The children of this generation are pretty wise...The least that this generation can do, Your Honor, is to give the next generation all the facts, all the available data, all the theories, all the information that learning, that study, that observation has produced; give it to the children in the hope to heaven that they will make a better world of this than we have been able to make of it.

This impassioned plea by Dudley Field Malone in the Scopes trial might well be what Max Rafferty had in mind by his statement. In fact, it sounds almost *identical!* Yet, for some unknown reason, the scientific community had great fears of allowing this option to children (now that *they* had control). Maybe Malone's further argument should be considered as well:[232]

> For God's sake, let the children have their minds kept open — close no doors to their knowledge; shut no doors from them. Make the distinction between theology and science. Let them have both. Let them both be taught. Let them both live. Let them be revered.

The ambiguity of the scientific community's position — on one hand wanting to reserve the right to speak about science exclusively for the scientific elite and, on the other, appealing passionately for public pressure to be brought to bear on those who disagreed with what the scientific barons had already established as the only acceptable truth – is an enigma.

Of course, many imponderable questions arise from this bipolar hysteria. This book is not the appropriate place to discuss those questions, but the underlying issue of using the public, on one hand, to accomplish their purposes and then slamming the door shut on the public when the public has questions about the beliefs of scientists does, indeed, need exposure.

If the public is to be the arbiter, the referee, the judge, then the scientific elite must be prepared to lose some and win some. The only other alternative is scientific tyranny...

Irrevocable Fissure

Once the news media threw caution to the wind and decided to ignore the truth in order to exploit a straw man, there was no turning back. For any one — including me. The fissure between fact and fiction was fathomless.

Without any doubt, the news media had found a winner. They had polarized two forces that would never be able to be reconciled. They had driven a wedge into public opinion in such an amazing way that controversy would go on unabated.

However, the news media did not just divide the population into two camps. They chose sides. They picked the side that they wanted to win, and played their news coverage in such a lopsided way that it would always favor their team.

They had several means of aiding their "chosen ones." The *wording* of headlines, the *size* of headlines, the *location* in the newspaper of the headlines, *length* of articles and the use of *favor* or *derision* in their articles were typical means that were employed. One example will illustrate the bias in favor of one side over the other.

The *Los Angeles Times* ran two articles regarding the Science Textbook Struggle on the same page of their newspaper. They were not equal in length, in size of type, or in location on the page. The first article was located in the most prominent corner with the headline, TEXTBOOK FIGHT PARTLY LAID TO BIBLE SCHOLARS.[233] This article also carried a secondary headline, PRIEST SAYS EXPERTS HAVE NOT PUT ACROSS ASSUMPTION THAT GENESIS STORY IS A MYTH.

This article quoted a "Bible expert" to put down a primary thesis of the religious community — that Genesis offered something scientifically relevant to origins.

The second article was approximately one-third as long as the first. It was in a lower part of the page. Its headlines were printed in type about one-third as large as the other article and read, NEW VOICES HEARD AGAINST EVOLUTION.

Although this article discussed the same straw man issue, it did so in exclusively religious terms while the more prominent article had brought in scientific discussion. Interestingly, it contained a rather amazing aspect to the issue that had not previously been disclosed — that religious bodies not normally known as fundamentalist were concerned about the dogmatism of evolution on the subject of origins. Bishop Meletios, Greek Orthodox prelate for the Western states was quoted as saying that evolutionary theory, "is absolutely unacceptable to the logical and faithful Christian intellect."[234]

The Greek Orthodox church has never been accused of being fundamentalist. Is that a newsworthy revelation? Would not widespread public knowledge of Bishop Meletios' position change the character of the straw man issue into a more respectable intellectual debate?

One can only ask rhetorically why the news media, if they wish to be as fair as they openly claim to be, do not treat two sides of a controversy *equally*? Why do they "load the dice" in favor of one side over the other? Do they have special insight into which side *should* win that is unavailable to the average layman?

While an irrevocable fissure had been created by the news media, a statement — that John Scopes himself made long after his trial — continued to haunt me.

Scopes said shortly before he died, "The basic freedoms defended at Dayton are not so distantly removed; each generation, each person must defend these freedoms or risk losing them forever."[235]

I could see that roles had reversed since Scopes had his trial. Those who believed in the "amoeba-to-man" myth had the seat previously occupied by those who believed in the Genesis story — and vice versa. But the principle Scopes espoused was still valid. What could *I* do to carry out Scopes' concern about tyranny?

While the fissure might be irrevocable, there was still the possibility of bridges being built over the fissure. I determined that I would seek to be a bridge builder...

Chapter 5

SUPPRESS BY THREAT

The partisan, when he is engaged in a dispute, cares nothing about the rights of the question." — *Socrates*

Bridges are built — not from one point to another — but between two points.

But how do you build a bridge between two supposed enemies? How do you even *start*?

It may surprise many, but I never considered myself a partisan on the subject of origins. That there was *dogmatism* about origins in the science classroom I could plainly see. That science textbooks were *biased* in favor of one belief concerning origins — when such bias was *not* based on science — I had no difficulty spotting.

However, I perceived from the beginning of my involvement that the struggle was for retaining *objectivity* concerning origins, rather than for promoting *one belief* over another — particularly in the name of science.

When the news media produced the straw man, "Religion Masquerading as Science," I was greatly discouraged. This changed the entire character of my involvement because there was now a *fissure* in what had been solid ground. Instead of just *expanding a frontier*, it was now necessary for me to become a *bridge builder* — or else find myself on only one side of an artificial dispute.

BRIDGE TO RECONCILIATION

Before I even got a chance to do so, some tried to either spoof or correct the breach. One of the most clever attempts to reconcile (or perhaps to ridicule?) the combatants in the "God-versus-Darwin" battle occurred soon after the fight started. And it appeared in 75 different newspapers across the country!

On 30 December 1969, Frank Interlandi, who has a syndicated cartoon series "Below Olympus," needled evolutionists, environmentalists, and creationists — all in one cartoon (see Figure 7). His famous "little old right-wing, super-reactionary lady" was shown picketing an ocean beach and warning that, since evolutionists believe that "life began in the sea," those evolutionists who pollute water are polluting their mother!

Figure 7. Origin of Life?

Others also tried early to defuse the confusion by writing letters to editors. One of these was Dick Bube at Stanford, whom you may recall was not enthusiastic about my involvement in the first place. He had *warned* me that I would be misunderstood! Nonetheless, Dick wrote a thoughtful, low-key letter that was published in the *Palo Alto Times* (see Exhibit 12).

His noble but fruitless endeavor contained some notable points: (1) Genesis was not even suggested by my adopted paragraphs, (2) the theory of evolution is untestable at critical points, (3) origins are unique and outside science, and (4) defense of neither the Bible nor of evolution was necessary.

His letter accomplished no more, however, than had my many attempts to set the record straight. The fissure remained...

Before building a bridge, the bridge designer must decide which two points he is going to interconnect. This decision was not easy for me.

On the so-called "Darwin" side of the controversy, I could not understand why anyone felt threatened. After all, there was no attempt, implied or otherwise, to limit, restrict or modify any evolutionary teaching. Surely, the mere *comparing* of other theories with an impregnable theory like evolution couldn't be intimidating!

So I determined to find out why some scientists were upset with having an alternative viewpoint alongside the "amoeba-to-man" idea in textbooks...

Whence the Menace?

Threats can be real or imagined. However, to the person threatened, the threat is *always* real!

If I was going to build a successful bridge between adversaries in the "God versus Darwin" straw man issue, I would have to find out whether the violent reaction shown by some scientists to the State Board's action was based on *real* or *imaginary* threat.

To start with, was there any truth to Gerard's claim that my 127 words that had been adopted by the State Board could *destroy* a 205-page *Science Framework* that took 15 eminent authorities over four years to write? Which of those 127 words did the trick? Was I *brilliant* or the Framework *vulnerable*?

I was forced to believe that the 205-page document must have been so defenseless that *anyone* who modified it would have demolished it. Much as I would have liked to believe that I was "mighty with the pen," I knew better. Therefore, what was it about the original document that made it so open to "assassination?"

Going back one more time to Clarence Darrow who argued the case of John Scopes in 1925, I studied his arguments to find a possible clue. Darrow argued most persuasively that Tennessee's Constitution gave men the right to differ about religion because it stipulated that no act shall be passed to interfere with religious liberty and that everyone had a right to worship according to his own conscience. He eloquently pointed out that to teach biology only according to the fundamentalist conscience was to deny religious liberty to all non-fundamentalists. And therefore, it was unconstitutional.

Darrow's finishing remarks were pithy indeed — especially if I was to determine whether the threat to the Darwin camp was *real* or *imagined*.[236]

> He concluded with a plea for tolerance, intellectual freedom, and the right to be different. "To think is to differ," he said. If you discourage difference, you discourage thought. If today you make it a crime to teach evolution, tomorrow you may ban books and newspapers.

Tolerance, *intellectual freedom*, and the *right to be different*. Those were exactly the issues that the "Darwin advocates" found threatening or dangerous in the Science Textbook Struggle. Yet in 1925, their plea was based on these very issues! Today, *they* would make it a crime to teach *creation!*

"Now, wait a minute," the Darwinians can be heard to say. "We do not object to alternative theories to our chosen amoeba-to-man one. However, they must be *scientific* — not *religious*."

"Oh, I see. You want the power to decide whether the *right* alternatives are proposed before you will grant tolerance or intellectual freedom?"

"Not exactly. We Darwinians have special ground rules that determine what may be compared with what. If we don't enforce those rules, there will be confusion."

"Who will be confused — you or non-Darwinians?"

"Oh, the non-Darwinians, of course. We cannot be confused — you see, we made up the rules!"

Therein lies the answer. Whence the menace? In upsetting an entrenched *status quo*. And the threat to the Darwin disciples was *real* — not imaginary!

Now, do not mistake what this means. There was a strong possibility (especially after the news media inflamed the public on a false issue) that the Genesis account could have been included — under political, religious or popular pressure — in science textbooks. And that would have been *most inappropriate* — for both science and religion. Crazy things happen when the public is seduced by the mass media.

However, the Darwinians were not protecting Dame Science from such unchastity as "Genesis infection." Nothing close to that noble intention! Their motives sprang from *beliefs* — not the rules of science. Their passion originated in cosmological *faith* — not the rationality of objectivity.

On what basis can the motives of "Darwin's defenders" thus be lowered from lofty levels? How can we say that they were impelled by pique instead of principle? Consider these clues:

1. Darwinians reacted to second-hand *interpretations* instead of seeking first-hand *facts*. Thereby, the news media had them "tilting at windmills."

2. Darwinians failed to differentiate between Evolution I (science) and Evolution II (belief). Their "amoeba-to-man" belief, in terms of scientific validity, was just as vulnerable as belief in a supernatural Creator.

3. The Darwinians concluded that Darwinism was under attack. Yet the State Board of Education had not issued one word of limitation, restriction or modification of Darwinism. Nothing concerning evolution had changed, and Darwinians were free to teach anything they wanted.

4. Darwinians ignored a ruling by the California Attorney General that teaching of evolution was constitutional only "if it did not involve indoctrination of the idea."[237] This ruling, in effect, *required* an alternative to Darwinism. Without such an alternative, how could indoctrination otherwise be *avoided?*

5. Darwinians rejected the plea for "tolerance, intellectual freedom, and the right to be different" (that they had used in 1925) and adopted instead a *totalitarian* stance.

Even though the news media can be blamed for the false report that triggered irrational Darwinian reaction, a nagging question persists, "What menace did the Darwinians perceive that justified such an *unscientific* outburst?"

Astronomers, for example, do not lash out emotionally, threatening legal action if a currently-popular theory of x-ray production is compared to contradictory ones, do they? Of course not. In fact, it often *delights* them![238]

Then what makes Darwinianism so uniquely sensitive to comparison? Could not Darwinists be gracious (and scientific) enough to (a) *listen* to details of alternative ideas, (b) *compare* point-by-point the differences between theories, and (c) *acknowledge* the right of the science student to evaluate the comparison? There must be a reason...

Earlier, we concluded that the perceived menace for Darwinians was "an upsetting of *status quo*." Particularly one in which they had total domination. And we can all identify with their reticence to share their monopoly.

But, the menace had to be greater than "having to move over and share the spotlight." Very deep within the recesses of private belief lies, for everyone, a cosmological conviction concerning ultimate causation. It is a binary conviction...0 or 1, black or white, true or false, yes or no. Either there *is* a supernatural Creator who causes all things to exist, or there is *not.* No ground in between. No room for agnostics. Not *maybe.* Nor *possibly.* Not even *perhaps...*

To ferret out the *real* menace that Darwinists recognized in the straw man issue, maybe the following logic would help:

1. Scientific naturalism (upon which Evolution II is built) is openly and clearly based on the premise that there is not and has never been supernatural activity in the universe.

2. The scientific validity of Evolution I has been subtly but illegitimately extended to embrace Evolution II, so that the public has been misled into confusing the two as being one.

3. The anti-theistic character of Evolution II can remain *incognito*, passing as "science" by its semantic association with Evolution I, *unless* it is compared with a theistic alternative.

4. Once a theistic concept is placed side-by-side with anti-theistic Evolution II, the cloak of scientific respectability is stripped from Evolution II — because *both* are thereby exposed as lying outside the domain of scientific verification or disproof.

If these four points are valid (and I believe they are), the path to the real menace for Darwinians is clear.

First, anti-theism is unpopular with the public — at least in a Judeo-Christian culture like the United States. No one running for national public office would, in his right mind, stand up and openly declare himself to be an atheist.

Secondly, modern science exists today only with the taxpayer's support. Without question, public funding would be cut off if the anti-theism of Evolution II were to be flushed out into the open.

In the long run, Darwinians could hope to convert all young minds (and thereby, the general public) to the anti-theism of Evolution II — particularly if they continued to enjoy a monopoly in the science textbooks. However, this possibility of indoctrination or brainwashing of children's minds was the very reason that the California Attorney General ruled that teaching of evolution was constitutional *only if it did not involve indoctrination* of the idea.

So Darwinians in 1969 were apparently trapped between two difficult options:

1. If they were consistent with Darrow's liberalism (tolerance, intellectual freedom and the right to be different) by which they themselves had achieved a voice, creationism or the belief in a supernatural Creator would be introduced into science classrooms and thereby "blow the lid" off Evolution II — revealing both its anti-theism and its unscientific character.
2. If they were reactionary and totalitarian, they could hold off any theistic comparison with Evolution II until such time as they had indoctrinated all young minds with the anti-theism of Evolution II. By then, such revelation would make no difference.

It is ironic and rather sad to realize that the first option was totally fallacious — postulated on the straw man created by news media. However, because the Darwinists failed to do their homework and get the real facts in the issue, they fell for the straw man. They *had* to elect the second option.

The menace they were fighting was *real*, all right. But it was within *themselves*. How tragic that they had to violate their own liberal ethics and deny others the right to be heard. How unnecessary their arrogance and boorish behavior. If they had only bothered to read my 127 words that the State Board had adopted.

Creation Without God?

While the Darwinists had been taunted into indiscretion on their side of the "God Versus Darwin" fissure, the Godly also were guilty of succumbing to press reports. They didn't bother to read my 127 words either!

If I was to build a bridge between these two forces, I also needed to clarify what I meant by Creation I, the *scientific* use of the word.

Returning momentarily to the battle options of Figure 6 in Chapter 4, you can see that my personal objective in the Science Textbook Struggle, depicted as Group III of Figure 6, was different than that of both the scientific and the religious communities.

The Darwinists had merged Evolution I and II in an assault on Creation II, as depicted by Group V. However, the religious community, for the most part, misread the battle line to be drawn between Creation II (a Creator) and Evolution II (anti-theism).

As Figure 6 illustrates, I actually had a more difficult communication problem with the religious community than I did the scientific.

In the weeks immediately following the State Board's action on 13 November 1969, I was contacted by many representatives of the religious community. Most of them wished either to congratulate or to encourage me.

I found myself in the awkward position of frequently having to explain, "I may not be your champion after all." No one likes to disappoint someone, and yet I found that that was what I often did. Many were puzzled by my refusal to accept their congratulations. Others undoubtedly felt that I was some sort of religious renegade.

At this stage, I confess to becoming perplexed. It was far easier to tell everyone (whether they were opposed to or in favor of what I had done) what I had *not* meant, than to explain what I *did* mean — especially regarding the *scientific* meaning of creation (Creation I).

I had had so little time to quietly ponder and resolve my own thinking on this subject, especially after the State Board decision. I honestly felt that I *knew* what I had meant. Yet, my mind was becoming clouded with *doubts*. Meanwhile, my increasingly unstable position was beginning to be bombarded from all sides by those who were sure *they* knew what I had meant!

Once more, I turned in desperation to the American Scientific Affiliation — this time to the editor of the *ASA News* who was a biochemistry professor at Iowa State University. Strictly on the strength of things he had written in ASA publications and without personally consulting him, I had listed Walter R. Hearn, in my address to the State Board on 13 November 1969, as one who endorsed my position that creation was a scientific term as well as a religious one.

I was totally unknown to Walt Hearn. So I really had *three* reasons to write to him — to introduce myself, to confess that I had used his name without consulting him, and most importantly, to seek help from him.

Early in December, I prepared a collection of items — my 13 November 1969 speech, newspaper articles and letters to key people — and mailed them to Hearn. In that mailing, I proposed a meeting on 29 December in Los Angeles to prepare some material on scientific use or meaning of creation. I asked Hearn if he could personally list some critical scientific references to creation as well as enlist ASA members to attend the meeting.

My letter caught him at a very bad time — in the midst of final exams. On 17 December, he responded with a long, detailed letter. It was loaded with disturbing news for me. Though couched in the most apologetic and kind terms, his disagreement with what I had done was irreconcilable.

First, Hearn was not sure that I had correctly stated *his* views about creation in my State Board address. Secondly, he felt that the *scientific* use of creation bears so little resemblance to the *Biblical* concept that there is no use of trying to co-opt it.

"I think scientists who have used the term have weakened the word by their use and I'm all for letting it drop out of the scientific vocabulary!" he said.

Next, Hearn said he couldn't "see how *science* can handle anything *but* cause and effect relationships, so 'scientific creationism' doesn't make sense to me."

Then, he observed that the California controversy seemed to limit the discussion of creation to instantaneous events, and this only boils down to a "fight for the possibility of miracles *within* science."

Walt Hearn's letter did have one encouraging thought for me: "I am with you all the way in trying to take the dogmatism out of science and make it perfectly respectable to say 'I don't know and what's more I don't think we can find out' as a scientist or as a textbook writer."

But that encouraging thought could hardly compensate for all the other things he had to say. His concluding summary was perhaps the most disturbing of all to me. It had a ring of prophetic authority: "My experience...leads me to think your greatest headache of all will not be the *L. A. Times* but the Christians willing to wreck science to support miracle *within* science."

So there I was — shattered. Asking for help and encouragement, I had gotten "shot down in flames." Nicely, of course. In a gentle manner, without a doubt. But shot down, nonetheless.

As I tried to recoup my sense of balance, I slowly realized that most of what Hearn had said was absolutely correct. He *had* misread a few things — like the correlation of creation with the instantaneous. Or the implication that I was proposing "scientific creationism" as a term, process or belief. Neither of these were correct. But I couldn't refute the rest of his arguments.

Hearn later sent my letter and materials to another Iowa State colleague, Donald S. Robertson, Professor of Genetics. Robertson, in a letter to me on 23 December 1969, was just as unable as Hearn to help me. He expressed genuine sympathy and interest in what I was attempting to accomplish in my 29 December meeting.

"If it is any help, you can list me as one who believes in creation," confirmed his evident support. Yet the heart of his letter was a bone-crusher:

> As I understand, the objective of the meeting is to collect scientific data in support of creation. I interpret this to mean that you are seeking

positive scientific evidence supporting creation and not just evidence against evolution. To be perfectly frank, I am not aware of the existence of the kind of evidence you are seeking.

As you might have guessed by now, the meeting on 29 December 1969 turned out much different than I intended. Several other scientists whom I invited to the meeting had either found convenient excuses or openly said that they "did not feel it would be appropriate to attend."

I was beginning to feel like the music had just stopped in the game of musical chairs, and I alone had no chair! Quite lonely, to say the least.

This loneliness forced me to begin a desperate search within my own mind for a rational explanation of what I intuitively felt was still a valid issue. Somehow I knew, with an impelling inner conviction, that the struggle was legitimate and must not die.

That conviction was balanced by another — that I personally did not have the proper education or experience to successfully persuade the scientific community to change its mind on anything.

Those two convictions had caused me to solicit help from those whose credentials and views I was sure would be acceptable to scientists in general. But, I had been "turned down flat." Where else could I go for help?

Up to this point in time, I had steadfastly maintained a strict *neutrality* between those Darwinists who were upset and those people favoring recognition of creation who were overjoyed. The latter group was a mixture of scientists and non-scientists, while the former was comprised almost exclusively of academic biologists.

I wanted to keep that neutrality in the worst way. It was critical, I felt, to my role of securing *objectivity* concerning origins. I dared not take sides.

Yet, I was fast being boxed in. My two paragraphs were being branded "rhetoric without scientific basis." I had a personal meeting scheduled for 16 January with Ralph Gerard — and I was ill-equipped to discuss the scientific aspects of creation with a scientist of his stature.

If I was going to go ahead with the 29 December meeting, I would have to find someone to attend who could help me — even at the expense of temporarily abandoning my desired neutrality.

It was this dilemma that caused me to decide to turn to some of those who favored the recognition of creation in science texts and who had called or written to me following the adoption on 13 November. Many of these folk had asked for an opportunity to meet privately with me, and I had refused. Circumstances now had pushed me to the point of accepting their invitation.

Douglas Dean, a biology professor at Pepperdine University, had offered his home for such a meeting, so we met on 29 December in his lovely Glendale home. About a dozen people attended, representing an amazing diversity of thought about creation being taught in the public school classroom.

In addition to Dean and his wife who is a physician, the group that met included professional scientists and educators like Robert C. Frost, an experimental biologist who had been chairman of natural sciences and professor of biology at three colleges and universities, R. Clyde McCone, an anthropology professor at California State in Long Beach, Mark Biedebach, the biology professor at Cal State in Long Beach mentioned earlier, and Richard C. Spurney, a philosophy professor at Long Beach City College.

Among those present at this meeting who had no scientific training or background was Mrs. Nell Segraves and her son, Kelly. Mrs. Segraves, a housewife who had been instrumental in securing a ruling in 1963 by the California Attorney General about the constitutionality of teaching evolution, had been one of the speakers on the day my words were adopted by the State Board. Her son, at the time of our meeting, was developing a series of taped interviews on topics related to science and creation for release on radio. Their interest in attending the meeting was obviously peripheral to my purpose — seeking scientific evidence or basis for the term "creation."

With the exception of Bob Frost and Mark Biedebach, this was the first time that I had met any of the people present that day. The meeting started with a review — from my viewpoint — of events preceding and following the 13 November decision by the State Board.

Due to the diverse backgrounds of the attendees, my account was frequently interrupted. Several thought that they knew much more about the sequel than I did — and in some cases, they were right.

It also became immediately evident to me that I had not been the first person to have taken action toward getting alternatives to evolution regarding origins taught in science classes. In fact, the same resentment that had been expressed to me immediately following the State Board meeting — that I had ruined the work of many others who had been seeking a creation alternative for years — flared again in the meeting. Several were unhappy with my adopted words.

For one thing, it meant that their various individual efforts would now have to be funneled or combined with what I was going to do from then on. Various State Board members had advised these people to work with me as a focal point. Frankly, I did not agree with much of what some of them had to say that day. This created some undesirable tension in the meeting.

Another factor had arisen by this time. I had become aware that the *Science Framework* was not really a contractual document for publishers. Instead, the *Framework* is a background reference attached to a contractual statement of work, "Call For Bids For Textbooks and Reusable Educational Materials in Science."

Therefore, we had a dual objective for our meeting. First, we needed possible specific criteria regarding creation to be included in the "Call for Bids." Then, we also needed general backup information and references of a scientific nature concerning creation.

After quite heated and lengthy discussion, a consensus was reached. Each person was to provide — in writing to me within a week — data or references that could be considered meaningful and proper for either inclusion as specific criteria in the "Call for Bids" or that would aid me in preparing a background statement on the subject. This latter statement would have *multiple* uses or applications — such as being the basis for my discussion with Ralph Gerard on 16 January.

The first written response I got was from a biologist, who had received his PhD from Johns Hopkins University. He had been vocal in our meeting in opposing what he *thought* I was attempting — a synthesis of evolution and creation, perhaps something akin to "theistic evolution" or "evolutionary creation."

Although I repeatedly denied this synthesis and firmly stated that I wanted *alternatives* instead, this biologist remained unconvinced. Because he didn't believe me, he was not going to help or support me in my efforts.

His letter also contained, like a parting shot, a statement very similar to earlier ones I had received from Hearn and Robertson at Iowa State:

> I see no point in trying to make creation appear scientific. I do not think there are any grounds for trying to make a case for a scientific creation. It just admittedly is not scientific.

Some unforeseen problems arose from this meeting almost immediately. A few attendees were apparently impressed with the vast amount of newspaper publicity that had persisted since 13 November. Perhaps they attended the meeting in order to become publicly identified with an issue that was attracting great interest. Or they may have hoped to steer the issue in a different direction than it had taken to that point.

Regardless of his motive, Dick Spurney called me on 7 January to inform me that he had drafted an eight-page statement that he was going to discuss with D. Russell Parks, Chairman of the Curriculum Commission on 9 January. His reason for meeting with Parks was to see if his statement was in the proper format for curriculum criteria.

Spurney had been among those at the meeting who were unhappy with what I was proposing. Although I didn't know it at the time, and didn't find out until months later, Spurney was going to "go it alone."

Since the meeting in the Dean home, he had met with Mark Biedebach and George E. Massey, a professor of philosophy at Cal State in Long Beach, proposing that they all jointly write a book on the philosophy of science — particularly concerning origins.

The draft he intended to discuss with Parks was evidently the introduction to their proposed book. He wasn't calling to get my *permission* to go directly to Parks, he was *informing* me that he was going to do so. His independence made it clear to me that he was not going to abide by the consensus agreement we had reached when we adjourned on 29 December.

I tried diplomatically to remind him that everyone had agreed to forward his or her thoughts to me for compilation prior to another meeting on 10 January. His impatience was not disguised. He did not feel bound to any consensus agreement.

After considerable discussion (much of which was pretty testy from both sides), he finally agreed to not "rock the boat" by going to Parks.

Spurney never did send me his eight-page draft, however. While I couldn't blame him for wanting to capitalize on the issue's high public visibility at the moment by trying to get a part of his book "blessed" by the Curriculum Commission, he had nearly sacrificed our collective approach for personal gain. Though I had successfully stopped him this time, I began to worry.

"How many others are going to 'do their own thing' without even bothering to let me know?" I wondered. "At least, Spurney was kind enough to *notify* me of his abortive intentions!"

My goal of establishing "creation without God" was in danger if everyone acted like Spurney and went their own way. But I had to do more than wring my hands. That would never build a bridge.

Peace Proposal

Once I had the real menace to the Darwinians ferreted out on one side of the fissure and, at the same time, hopefully had convinced the Godly that I was only interested in "creation without God," I was ready to build a bridge.

It was not going to be easy though, because neither the Darwinians nor the Godly wanted to believe that the two points between which I was going to build a bridge really existed!

The disagreement about where the bridge would go reminded me of another bridge-building squabble that occurred while we lived in Seattle, Washington from 1952 to 1959. Seattle is bounded on the east by Lake Washington — a body of fresh water about 18 miles long and only a few miles wide. A rather famous floating bridge was built in 1940 between Seattle and the eastern shore of Lake Washington. Within a decade, however, this single bridge was absolutely inadequate for the traffic load. So the city fathers began planning a second lake bridge.

Engineering study contracts were awarded. Surveying began on both sides of the lake to determine the optimum location for the bridge terminals. Everyone was happy. There would be a new bridge. Excitement continued to build until the two selected points for the bridge were announced.

Then a storm of protest broke out. Over the necessity for a second bridge? No. On where the bridge was to be *located!* The protest was so heavy that a new engineering study contract was awarded. Again, the happiness index went up because there was going to be a second bridge. How long did the happiness last? Until they announced the selected terminal points for the bridge. Though these two points were different from the two selected in the first engineering study, the storm of protest was equal to the first storm!

And so the ping-pong match continued. For more than a decade, the same pattern was repeated. New engineering contracts were awarded. As soon as selected bridge terminals were announced, protest exploded. And the bridge was never built — while we lived there.

Years after we moved from Seattle, a second bridge was built. I've never heard how the city fathers finally settled on the bridge terminal locations. Maybe they just didn't announce the final location decisions and surreptitiously began building the bridge one dark, moonless night...

I suppose that the *location* of a bridge may be as important a parameter as its style, length or load capacity. It was obvious to me that I certainly wasn't going to build a bridge between two points of *agreement.*

Fortunately, I didn't need the approval of either party. On the other hand, I ran a real risk of building a bridge that could not be connected to either side!

On 7 January 1970, after blocking Spurney's abortive attempt to go directly to the Curriculum Commission, I had decided to "head off everyone else at the pass" by calling Parks myself.

Not having his telephone number, I thought that Ed Shevick might have it. Shevick, you will recall, is my neighbor who was a member of the State Advisory Committee on Science Education that had written the *Framework.*

"I really don't have Dr. Parks' number, Vern," Shevick said when I called him. "But I do have some information that I think will interest you."

"What's that?"

"Well, Dr. Smith, chairman of our committee, has just sent out a letter to all members of the Curriculum Commission's Science Committee — as well as to State Board President Day — saying that our committee is still unhappy that the State Board hasn't withdrawn your two paragraphs."

I felt immediate knots in my stomach. I was glad to know about it — in a way. However, when you know something but can do nothing about it, you are often better off in not even knowing.

Shevick continued, "You know, everyone on the committee feels as though the State Board is giving us 'the silent treatment.' They haven't communicated with us since they modified our *Framework* in November. Because the committee knows that you and I are neighbors and friends, some of them have asked me to get you to agree to join in a committee rewrite of your paragraphs."

"Ed, you know that I wasn't happy with the two paragraphs the State Board chose to adopt. Further, I think what the Board wants could probably be said better than the way the paragraphs now read. However, I don't feel it's my prerogative to do any rewriting — with your committee or anyone else — without being directed by the Board to do so."

He was less than happy with my response. But I was sincere. The State Board was still the primary authority. On what basis could I embarrass the Board in order to placate the pique of an *ad hoc* committee that reports to that Board? Especially when that committee hadn't really paid attention to what I had written and had gone off half-cocked over press interpretations of my words?

Out of courtesy, I felt obliged to notify State Board President Day of my conversation with Shevick — especially about the committee's hurt feelings. Day wasn't in when I called, but he returned my call around midnight on 7 January.

"I hate to bother you, but I thought that you might want to know that the *Framework* committee feels as though the State Board is ignoring them," I explained as to why I had called him. "You'll be getting a letter soon from them which expresses their gripe."

"Well, I can only tell you that the State Board has no intention of revising those two paragraphs — under any circumstances," Day said firmly. "That clarifying statement of yours that I released on December 11th should have cleared up their confusion, as far as I'm concerned.

Maybe they haven't read it yet."

I was impressed and encouraged by his resolute response. Our further conversation dealt with what type of information would be required for the "Call for Bids" in terms of specific criteria to carry out the philosophy of my paragraphs. He suggested that I call Dr. Parks, the chairman of the Curriculum Commission, right away and gave me his telephone number.

Before 8 o'clock the next morning, I got in touch with Parks at his home. After introducing myself, I said, "Howard Day suggested that I call you to find out what type of criteria I might prepare to support the two paragraphs of mine that were adopted by the State Board."

"Before I answer that question," Parks replied, "permit me to compliment you on the fine presentation you made on November 13th. I think that it was an excellent compromise between all the forces that were present that day. I can assure you that the Curriculum Commission has no qualms about accepting and working with the State Board's amendment. The unfortunate thing is that the press has so badly maligned what you actually said."

"Thank you very much for your compliments. I too have been disturbed by the apparently deliberate twisting of my words."

"Now regarding your question about criteria, I suggest that you get in touch with Glenn Leslie in Fresno, just as soon as you can. He will help you get your thoughts translated into the criteria correctly. The Curriculum Commission will be meeting in San Francisco on 11-13 February. We'll be taking up the science criteria at that meeting."

Just as soon as I hung up, I called Glenn Leslie. Our conversation was very friendly, and he complimented me once more on my presentation of November 13th.

"I must confess that I had considerable fears, due to the applause from the audience when the Ford amendment was adopted by the Board. I was afraid that you had ordered the Bible inserted into the science textbooks, especially after I heard comments by some Creation Research Society folks to the effect that their goal of 7 years had been realized by the adoption of your words. However, after I read what you actually said, I was very happy with the wording."

"Well, you are one of the few people who've actually bothered to read what I said instead of becoming alarmed over how the press reported it," I responded.

"To let you know that I was comfortable with what you did, I will tell you that I convened the Science Committee shortly after that meeting and suggested that they ask you to participate with them. However, most of them thought that they should communicate directly with the Board member who had made the motion, Dr. Ford. Ford has never responded to their invitation to meet with them, but they have interpreted the press release of December 11th as the answer to their note of invitation to him. In fact, we believe that the Science Committee's request probably initiated that press release."

I didn't have the heart to tell him that I had written the press release myself, without knowing that the Science Committee was trying to meet with Ford.

Leslie continued, "I'll tell you what I'll do. Rather than explain the ground rules for the criteria, why don't I just send you the original working draft of the criteria as well as the changes our Committee made after your words were adopted? This is on a confidential basis, please understand. You can either approve the revisions they made or make further suggestions for additions in writing. In that way, our Science Committee can present a unified position to the Curriculum Commission for approval."

"I certainly appreciate your helpfulness in this matter. You can be assured that I will get the material back to you just as soon as possible," I said.

"Most of the criteria were written by the *Framework* committee rather than by our Science Committee. I will need to have your contribution before our meeting on 11 February...Incidentally, before I hang up, I want you to know that I have tried numerous times to defend your true position to the press. But I have been continually misquoted! I don't really know what we can do about it."

"Did you know that I called Ralph Gerard a week or so ago and attempted to settle any differences we had?"

"No, but that sounds like the right thing to do. The press release of December 11th should also take most of the steam out of those who think they are opposed to what you have done."

Probably the thing that forced me to actually start building the bridge was my fast-approaching meeting with Ralph Gerard. Having observed him on television with Baxter Ward, I recognized that he was a man of very quick mind and fast tongue. I knew that if I did not have my thoughts well collected prior to sitting down with him, there would be no chance during our conversation. So I began to draft some basic thoughts about the relationship of science to the subject of origins.

My intent was to offend neither of the two poles in the straw man issue — God nor Darwin. One of the primary purposes was to show the influence of philosophy and presupposition on scientific investigation.

While this bridge that I built represented my best thinking at that time, I have come to change my mind about certain things I said in it.

However, I think that it is important for the reader to recognize that my own thinking was evolving throughout the entire Science Textbook Struggle.

By 9 January 1970, I had worked and reworked this bridging proposal into a statement, "Origin of the Universe, Matter, Life, and Man" (Exhibit 11). Some of those attending the 29 December meeting in Glendale had made suggestions that I incorporated.

The following day, I met with Doug Dean and Clyde McCone in Dean's office at Pepperdine to discuss my final version of the bridge. I wanted those who were committed to the introduction of teaching about creation in science textbooks to fully understand the meaning of the bridge. Dean and McCone were meeting with me for that purpose. Importantly, this statement was not meant to be a subtle or tricky way of introducing the Genesis account into science textbooks. Instead, it was a device to clearly identify both the religious and scientific connotations of evolution and creation.

The meeting in Dean's office was not a short one. The three of us worked together well into the afternoon. Our discussions were heated at times. By no means did anyone acquiesce to the other one's viewpoint without thorough, in-depth analysis.

I left Dean's office with a feeling of confidence that we had worked out a good statement. More importantly, I felt assured that I had a meaningful base for rational discussion of the issue with Ralph Gerard. The bridge was at least anchored on one side of the fissure, and all I needed to do was to anchor it on the Darwinist side...

INTIMATE INTIMIDATION

By the time I was 24 years old, my wife and I had driven our automobile into every one of the 48 states (neither Alaska nor Hawaii were yet states). We had even visited 40 of the 48 state capitols by that time. By traveling that extensively throughout the United States, I had made up my mind that I could live *anywhere* in the country — with one exception: Los Angeles.

My list against Los Angeles was long. Los Angeles was the very definition of *urban sprawl* — a *non-city* — *smoggy* — *artificial* — *earthquaky* — *high-speed* — *transient* — and on it went...

Yet, I was only 31 years of age when I moved to Los Angeles — and I've been here ever since.

From time to time, my mind flashes back to that long list of reasons why I shouldn't live in Los Angeles. It happened again on 16 January 1970, as I drove from my home in the western San Fernando Valley 70 miles across town to visit Ralph Gerard at the University of California in Irvine.

For one thing, it began to rain after I left home — *hard*. Los Angeles has many different climates within its city limits. It annoyed me that I hadn't brought an umbrella. Then, the monotony of driving through 70 miles of urban sprawl hit me. It was like being on a conveyor belt, going by the same type of architecture for more than an hour.

Maybe these were just excuses to take my mind off my meeting with Gerard — something that I feared, down deep, the closer the hour came.

State higher education in California fares very well. The University of California system consists of 9 major campuses. These are not only prestigious institutions scholastically. They are also beautiful to the eye. (There are also 19 additional campuses in the California State College and University system.)

One of the newest University of California campuses is located at Irvine in Orange County. The buildings at Irvine are of striking architecture and widely-spaced to take advantage of a beautiful setting on gently rolling hills. The spacing of the buildings is not for esthetic reasons alone, however. Large parking lots on campus are also a necessity, especially in California.

As I drove toward Irvine, the intensity of the rain increased. Charlotte Bell, Gerard's secretary, had given me specific instructions on how to find their office. I began to glance more apprehensively at her directions that I had written down as the rain got heavier.

Most campus buildings are not marked for outsiders. Irvine is no exception. The few existing signs are both small and designed to be obscure so that they do not ruin the open, park-like atmosphere of the campus. I began to wonder how far the parking lot was going to be from Gerard's office.

At last, I spotted the parking lot that Mrs. Bell had described. As I stepped out of the car in a downpour and began to run toward the building I thought she had meant, I stepped right into a big puddle.

I ran up a flight of stairs from the parking lot toward the building, and when I got there, it was the wrong one. Securing new directions from a student, I resumed my mad dash in order not to be late for my 11:30 AM appointment with Gerard.

I was a wet mess — and just a few minutes late — when I finally walked into the office and greeted Mrs. Bell. I started to apologize for being a little late, but she could understand by just looking at me. Without wasting any time, she ushered me into Gerard's office.

"Into My Parlor..."

As I entered Gerard's office, he rose from his desk. "May I take your raincoat for you?" was his first comment.

"Oh, thank you very much. I'm sorry that I'm just a little bit late, but I went to the wrong building first."

"Think nothing of it," he said in a fatherly way. "We have a few minutes before we drive over to the dining room for lunch, anyhow." Taking my raincoat, he draped it over a hanger and hung it on a coat tree.

I easily recognized him, of course, because I had seen him only weeks earlier on the TEMPO Show. Although I am a little over 6'2", it seemed to me as though I was a full *foot* taller than Gerard!

He immediately urged me to be seated, and just that fast, we were engaged in rapid-fire conversation. Two or three minutes into our discussion, he asked about my background and education. I honestly cannot recall how I responded to him because he had already overpowered me with his credentials.

I'm sure that Gerard had no intent of overwhelming me with his accomplishments. However, I was supersensitive. When he casually mentioned that he had received his *first* doctorate at 21 (he held both a PhD in physiology and MD degrees), it did not slide by me unnoticed. Low-key, perhaps, but also devastating. I was well aware that I was not in his league by any stretch of the imagination.

Dean of the Graduate Division, Professor of Biology, and Director of Special Studies at Irvine, Ralph Gerard was one of the world's foremost physiologists. Internationally known for his pioneer work on the chemical and electrical activity of nerve and brain, he had lectured on six continents, served as professor at 8 major universities, and received a number of honorary degrees as well as foreign medals and awards.

On top of that, he had published nine books and around 500 articles on a wide variety of scientific subjects. He had been the editor of a dozen scientific journals, and in 1955, he was elected to the National Academy of Sciences, the leading American scientific body.

If I felt at all secure before entering Gerard's office, I was clearly on the defensive by the time he looked at his watch and said, "Well, why don't we drive over to the faculty dining room now?"

The ensuing scene in his office — the two of us helping each other with our topcoats — was a bit incongruous. Try to imagine this short, gracious, 69-year-old genius struggling to help me get my long arms into a still-soaked coat. But we made it.

"We'll drive over in my car because the weather is so bad," he said matter-of-factly.

"We could take my car just as well," I offered.

"No, mine is parked much closer — beside, you are *my* guest today."

His charm was disarming. He had not raised one possible point of controversy.

As we walked outdoors, the rain had nearly stopped. But the skies were still threatening. We got into his new Chevrolet sedan, and as we slowly drove to the faculty dining room, he pointed out features of the campus. Still no discussion of my two paragraphs...

I recalled that old saying, "Come into my parlor, said the spider to the fly." The spider was spinning such an attractive web. How were we ever going to disagree?

Sage and Student

As we entered the dining room, Gerard carefully explained the buffet luncheon options. The room was quite large, but the tables seated only four people. The deference and respect paid Gerard by other faculty members was obvious as he and I walked through the buffet line together.

I followed him to a table almost in the center of the dining room. It was so quiet in there — more like a library than a place to eat and chat. Everyone talked in subdued tones.

Gerard and I discussed our likes and dislikes in various foods. I inquired about his vacation trip to Africa from which he had just returned. But the small talk couldn't last forever.

"I was pleased that you called last month to explain your true position on the evolution issue," he said softly. "I've been looking forward to discussing it further with you in person." My pulse quickened.

"So here we go," I thought. "He hasn't forgotten why I'm here after all!"

Trying to appear nonchalant, I answered, "Well, despite what the press has been saying, I don't feel that you and I are in actual disagreement."

There, the ball was back in his court — and I hoped it would stay there for a while. I had determined beforehand that I would try to *listen* rather than *talk*. At least two things would result from that approach: (a) I would learn a whole lot, and (b) he couldn't exploit my inabilities.

That strategy was actually based on an old proverb, "A fool utters all his mind: but a wise man keeps it in till afterwards."[239]

Gerard's response was interesting. "So few people really understand science. Laymen, especially those with no training in science, fail to realize the difference between how *they* seek truth and how the *scientist* determines truth."

From that point of departure, Gerard began to deliver a virtually tutorial lecture on the scientific method. The relaxed, soft gentleman fast disappeared.

In his place appeared an intense, driving personality. Words, clipped and precise, poured forth almost like a volley of machine-gun fire. He had a mind like a steel trap — incisive and quick. No one would believe him to be nearly 70 years old.

By this time, I had finished my meal and was giving him 110% of my attention. I honestly felt that I was a fledgling student sitting at the feet of a master, drinking in knowledge. Neither one of us was any longer aware of our surroundings. We had true rapport. While he wasn't necessarily saying anything I hadn't heard before, I admired his organized and logical arguments.

At several points in this rapid-fire review, I had a notion to break in and question him just a little, but I held my tongue. On one hand, I felt that a little dialogue would let him know that he was communicating with me. Yet, I didn't want him to view me as an adversary.

Finally, however, I overrode my reticence. Gerard was on a line of discourse that included the phrase, "...what scientists *believe* about their observations..."

"Pardon me a moment, Dr. Gerard. When you just said that 'scientists *believe* something,' what do you really *mean* by that?"

He paused only momentarily. "Well, I certainly don't mean it in the same sense that *religious* people believe something!"

His eyes betrayed him — I had struck a sensitive nerve. Not once had religion, faith or belief been mentioned in all our conversations to that point. *He* introduced it. But with his raising it came almost a defiant belligerence.

I quickly assured him that I wasn't thinking of religious belief at all. "I simply found it interesting that you used the verb *believe*, instead of *know*, *assume* or *suppose*."

His anxiety appeared to subside slowly as he continued his expounding. Yet, he left me a little puzzled. Why was he so sensitive to religion? What threat did he see that caused him to jump so suddenly and separate science from religion? Did he think that *I* was seeking a bridge between science and religion?

With my puzzlement still unresolved, I turned my full attention back to his reasoning.

After several more minutes during which I listened intently without interrupting him, Gerard began to describe how scientists go about amassing the evidence to support a scientific theory. I found this particularly interesting because it had occurred to me that a person's biases, prejudices, and presuppositions could influence the gathering of evidence, and even the design of experiments that were conducted to validate scientific evidence.

Remembering his earlier emotional reaction, I was hesitant to break in again. Yet, I wanted to check out his matter-of-fact explanation of scientific evidence in a setting that possibly could also answer my lingering puzzlement about his sensitivity to religion.

So I took a wild chance…

"Dr. Gerard, if I may ask another question — what *scientific* evidence would you require to confirm the existence of God?"

He stopped cold. That mind that had seemed to have no limit was momentarily frozen.

Although it may have been only five or ten seconds, his silence seemed embarrassingly long to me. I had not intended to be facetious or cute — or to try and stump him. Still, he appeared stumped!

With far more deliberation and at a much slower rate, he said, "You have asked a very good question." He paused again. Then quietly, he continued, "… and I have no answer."

In a way, I was stunned and hurt. It seemed a shame to me that such a brilliant man would be forced to make that admission.

Instead of feeling pleased that I had postulated a question he couldn't answer, I felt quite the opposite. After all, this intelligent scientist had widely distributed the following statement after the adoption of my words in the *Science Framework*:[240]

> Although it is really irrelevant, for your information I am an agnostic (which means one who is not certain about the existence of God) although I certainly lean towards the atheistic rather than the theistic position. I do not accept the Bible as the unfiltered word of God…and I would be surprised if, after having considered and discussed this matter earnestly over some seven decades, I would now find some argument to convince me of the divine origin of this volume.

If he had "considered and discussed" matters concerning God for nearly seven decades, what *criteria* had he been using? While questions like these were still racing through my mind, Gerard began once more to speak — still quite slowly and subdued.

"The best response I can offer you to that question is that, while I was a young teenage boy, I went out one night into the middle of a plowed field during a violent electrical storm. I held out my arms, looked up to the sky and said, 'If there be a God, strike me dead.' And I wasn't struck dead!"

Lest the reader mistake Gerard's attitude in recounting this story, let me assure you that he was deadly serious. It was no pun. He was not joking. I believed him when he said that it was the best that he could offer.

I almost wished that he had not offered that explanatory story. It wasn't worthy of a man with his brains, experience and accomplishments. Much of my *respect* for him suddenly changed to *pity*. The sage had disappointed his student...

Thundering Threat

Gerard's chain of thought seemed to have been broken by my question about "evidence for God." In fact, he didn't seem to regain his enthusiasm with which he had been discussing science during the previous 20 or more minutes. Instead, he slowly and cautiously began to discuss specific aspects of the State Board's decision — including what they apparently had in mind.

Undoubtedly, he was searching to find out how much influence I had exercised with the Board as well as what my personal objectives were in the issue.

I was naturally wary of his probing. I had a clear-cut objective in driving 70 miles to see him — to close the fissure that the news media had created. Therefore, I was seeking *peace* — not war.

Though I had a copy of my "bridging proposal" (Exhibit 11) with me, there wasn't an opportune time to expose and discuss it with him. So we continued to "shadow box" for a few minutes.

Then Gerard began to bring some heavy artillery to bear.

"If you are in as much agreement with me as you have indicated, why don't you get the Board to remove your paragraphs? They have obviously been misunderstood from what you had intended."

"I agree that there have been erroneous conclusions reached as a result of their decision, but I blame the press for this."

"But how can you stand by and let an excellent document be subjected to all types of irrational comment when you could clear it up by withdrawing your own words?"

That question probably had several objectives.

First, I think he was testing me to see if I had sufficient influence to get the Board to reverse itself.

Second, he was checking my reaction as to how much damage, if any, I felt my paragraphs had done to the *Science Framework*.

Third, he was trying to determine whether I felt that my words still belonged in the *Framework*.

I decided to take a different tack instead of answering his question directly. For the next few minutes, I tried to summarize my conviction that *origins* lie outside scientific resolution.

Gerard either didn't listen to me or else deliberately dodged the issue of origins. I made several attempts to explain to him that I felt that there was certainly more than one explanation that could be given for the origin of the universe, life, and man — especially from a *scientific* viewpoint. After all, no one could verify how all these events had actually occurred.

In retrospect, I have decided that it was my unwillingness to succumb to his pressure for withdrawal — together with my unspoken but evident conviction that there was something *science could not do* (i.e., determine origins) that provoked Gerard to the point of explosion.

Before describing his violent outburst, let me recount the setting once more.

We were in the faculty dining room. It is a very well-appointed room — rugs, drapes and small tables. Most everyone had finished eating by this time, and the conversations — though animated and widespread — were not much above the whisper level. Our table was virtually in the center of the room.

Gerard's exasperation arose with no apparent warning. I had shown no disrespect, not raised my voice or confronted him in any way.

Suddenly, he slammed his fist down on the table, rattling our dishes, and, with his large eyes virtually bugged out of their sockets, screamed, "You'd God damn well better learn a hell of a lot more than you know now if you expect to stay in this issue!"

This unexpected eruption caused everyone to jump at the sound of crashing dishes. They all turned to see what was going on. I am sure that I blushed. What could I reply to that thundering threat?

I just kept my eyes riveted on him. Although I waited, he did not follow up with any consequences I could expect if I didn't comply with his threat.

Gerard came to his senses about as fast as he had lost control of himself. Rather sheepishly, he suggested that we had probably explored the issue adequately.

As we drove back to his office, I apologized for anything I might have said that disturbed him, insisting that my motives were still to seek reconciliation rather than confrontation. To that end, I asked him if he would be kind enough to review and comment on my "bridging proposal."

I also suggested that G. A. Kerkut's *Implications of Evolution* had been helpful to me on the subject of evolution. He said he had never heard of the book and asked again who the author was.

"He's a physiologist at the University of Southampton in England, I believe," I explained.

He retorted, "I know every physiologist of consequence in the world, and I've never heard of him. But I will get his book and read it. In return, I will send you a list of my papers that deal with the limitation and scope of science, as well as some that address the role of science in ethics, values and aesthetics."

Although our parting was amicable, I could not easily forget Gerard's belligerent intimidation. Did he *really* expect me to capitulate right on the spot? Was his goal to suppress the issue by *threat?* Why had he *abandoned* the rational, objective scientific approach?

Gerard was very prompt in sending me his comments on my statement, "Origin of the Universe, Matter, Life, and Man" (Exhibit 11), as he had promised to do. In his cover letter to those comments, dated 19 January 1970, he said, "I enjoyed meeting and talking with you on Friday, and hope that you were not too distressed to find that our area of disagreement is considerably larger than you had thought."

One note that Gerard made on my bridging statement convinced me that I had at least made him think:

> You asked me what evidence I could think of that would prove or disprove the existence of God and I found it difficult to offer any. The idea of God is bad as a scientific hypothesis, since it probably cannot be disproved, even less proved. It is therefore accepted on faith or rejected, but it leads to no objective and testable outcomes.

Of course, Gerard hardly speaks for *all* scientists in his observations about God. Because I respected his right to reach his own conclusion, I resisted the temptation of sending him another observation concerning the existence of God that was made 2,000 years before his:[241]

> Since the beginning of the world the invisible attributes of God, for example, His eternal power and divinity, have been plainly discernible through things which He has made and which are commonly seen and known, thus leaving men without a rag of excuse. They knew all the time that there is a God, yet they refuse to acknowledge Him as such, or to thank Him for what He is or does...Men deliberately forfeited the truth of God and accepted a lie, paying homage and giving service to the creature instead of to the Creator.

Obviously, evidence to one person is not evidence to another — whether they be scientists or not...

FOOTNOTE FURY

Screaming unsuccess! That's what I evidently had in my reconciliation attempt with Ralph Gerard on 16 January 1970.

While two letters he wrote to me subsequent to our personal meeting were cordial and friendly, he was in no way reconciled to the State Board's amendment of the *Science Framework*.

Within a few days of our meeting, I got another telephone call from Vice President Harward of the State Board of Education.

"Ralph Gerard and his committee are demanding that we withdraw the amendment to the *Science Framework*," Harward said. "They are threatening to disavow authorship of the entire document unless we do so." He sounded very disturbed.

"Well, what do you intend to do? Isn't that blackmail?" I protested.

"Of course it is. I'm certain that the Board wouldn't consider withdrawal. But we are going to have to do *something*...We are thinking of meeting with Gerard privately to perhaps convince him that we are only interested in science — not religion."

"Did you know that I had a private meeting with Gerard just a few days ago for this very purpose?"

"No. How did it go?"

"Well, I think that he was hopeful of convincing me to voluntarily ask the Board to rescind my two paragraphs. When I held firm, he blew up and started screaming. It was some scene."

Obviously, I wasn't optimistic that another personal meeting with Gerard would produce anything but emotion. Further, Gerard was able to get favorable press coverage while the State Board couldn't even get newspapers to print what they released in writing! Nonetheless, Harward was firm about the need to meet Gerard.

"I've talked to Howard Day about meeting with Gerard, and we think that perhaps an unofficial breakfast meeting with him and possibly Dr. Smith, the committee chairman, could be arranged with John Ford, Day and myself during the February Board meeting in San Francisco. Is there any chance that you could join us for such a breakfast on Thursday, the 12th?"

I grabbed my desk calendar to check the date. Oh no! I was scheduled to be teaching an on-site Tustin Institute course in environmental simulation at Lockheed Propulsion Company in Redlands, California that entire week. There just seemed to be no way I could make it.

"Are you *sure* you can't fly up the night before and get back in time?" Harward sounded anxious.

"The trouble is that Redlands is so far from LAX airport. It would take me about two hours just to drive from there to Redlands, and I start teaching at 10 every morning," I explained. "Let me see what I can work out, though. I'll let you know."

Bellicose Bluster

As late as the end of January, I was still uncertain whether I could rearrange my Redlands teaching assignment so as to attend the breakfast meeting with Gerard in San Francisco on 12 February. However, my first-hand experience with Gerard's volatile invective made me want to protect anyone else from experiencing it.

On 29 January 1970, I made a long-distance call to Glenn Leslie in Fresno. He was amazingly reassuring about Gerard.

"I honestly think Gerard's more upset over the fact that his committee didn't get to review your two paragraphs than he is about their *content*," he said. "I assume that there's not the slightest chance that the State Board will change their mind about the amendment — regardless of how much noise Gerard makes. After all, the Committee was appointed by the Board to write something the Board could endorse — not vice versa!"

"That makes sense to me," I admitted. "But when Gerard gets excited, he becomes very irrational."

"Well, I wouldn't be upset by him…Incidentally, is there any chance that you would be attending the State Board's February meeting in San Francisco on the 12th and 13th? If so, our Science Committee will be meeting on Wednesday afternoon, just preceding the Board meeting. We'd be delighted if you could meet with us to explain your position further."

"I'm flattered by your invitation. Dr. Harward has already asked me to have breakfast with him on Thursday, but right now, it looks like I won't be able to be in San Francisco due to a teaching assignment I have." (I didn't share with Leslie the purpose of the breakfast.)

"If your plans change, let me know," he said cordially.

When I hung up, I wrote a long letter to Harward. Mentioning Leslie's calm response to the hysteria generated by Gerard, I suggested some strategy to Harward for the breakfast meeting:

> Keep pushing Gerard in order to force him to state what it is exactly that he disagrees with. The more than you can force Gerard to talk, the better…The best policy for you and the other Board members at the breakfast would seem to be to simply hear what Gerard has to say but commit to no action until later. Avoid any specific scientific wrestling with him because he is caustic and pulls his eminence of

> repute on anyone who crosses him. He is not bashful about how brilliant he is. He has asked to be heard, so *hear* him. Since the full Board will not be present, you cannot commit to any action.

I had earlier sent the State Board members a written transcript of Baxter Ward's TEMPO Show of 5 December 1969 when he interviewed Gerard (see Exhibit 9). Therefore, they were aware that Gerard had called them, among many other terms, "sincerely stupid."

At the time I sent it to them, I had no idea that they would be meeting him in person. However, with that possibility now imminent, I offered Harward further advice about Gerard, "If he gets particularly unruly, his TV remarks concerning the Board could be quoted back to him, and the suggestion that they hardly set the stage for rational and thoughtful discourse could be proferred."

In retrospect, I am embarrassed by the brashness and boldness of this unsolicited counsel to Harward. It was probably a product of naiveté coupled with high spirits provoked by Gerard's bellicose bluster. How else can I explain my concluding observation of Gerard in this letter to Harward?

> There is no doubt in my mind that this man is desperately defending a "religion" of evolution. At his age, he no longer has the time remaining to be calm, rational, reasoning or scientific about his feelings. He is so committed to evolution, not only in science but also in culture, ethics, behaviorism, and society that any threat to the concept knocks his whole house of cards down. With the great intellectual capacity he possesses, he cannot conceive of having been wrong for 70 years on a basic premise like this one! Therefore, he is fighting like a wildcat.

The renowned anthropologist Margaret Mead described men like Gerard a little bit different than I did — but with the same conclusion. While pointing out that the concept of human evolution often is mistakenly taught in schools as fact rather than theory, she once said, "Some scientists are as dogmatic about evolution as some preachers are about religion."[242]

After considerable juggling of my teaching schedule at Redlands and the discovery of a commuter flight from Ontario International Airport (only 30 miles from Redlands) to San Francisco, I called Harward back and confirmed that I could be present for the breakfast meeting on 12 February. He then had Eugene Gonzales, Associate Superintendent of Education in Sacramento, send me an official letter of invitation, placing me on consultant status to the State Board of Education that allowed me to be reimbursed for my travel expenses.

Arriving in San Francisco on Wednesday afternoon, 11 February 1970, I arranged to meet privately with Harward early that evening at the Clift Hotel. We reviewed the arguments and threats that Gerard had made and that had prompted this unusual meeting to be scheduled.

The purpose of meeting was to reduce further inflammation of the false issue — that the Bible was going to be introduced into science textbooks. The key seemed to lie in forcing Gerard to be *specific* by what he thought had ruined the *Science Framework*. Harward had considerable respect for Gerard's incisive mind. He had been a student under Gerard in years past.

I brought Harward up-to-date on my most recent conversation with Leslie, Science Committee chairman, and Parks, Chairman of the Curriculum Commission. Harward was pleased and encouraged that both of these key individuals viewed the *Framework* amendment as constructive and scientifically proper.

But I sensed that he remained quite apprehensive about confronting Gerard head-on, as we sat together in the hotel lobby. He was not looking forward to breakfast in the morning, as I bade him goodnight.

Later that evening, I called John Ford, the State Board member who had authored the amendment to the *Framework*, in his room. He invited me up. Ford, in preparation for the breakfast meeting, had also invited two biology professors from Loma Linda University (where he had completed his medical education) to attend the next morning. Mrs. Ford had accompanied her husband on this trip and quietly listened as Ford and I discussed — with Professors Ariel A. Roth and Berney R. Neufeld — potential problems we might have at breakfast.

That was the last time that I was to see Mrs. Ford. On Mother's Day 1972, her life was taken, along with two others, in a tragic head-on automobile crash near San Diego in which Dr. Ford alone survived. He was so severely injured that he remained on crutches until late in 1972.

Both Ford and I felt bolstered to know that two professional biologists would be available to aid us if Gerard tried any fancy footwork that he thought was outside our expertise.

It was late when I left Ford's hotel room and headed across the street for the Sheraton-Palace where I was staying.

Command Confrontation

I woke on 12 February 1970 with mixed feelings. In one sense, I was looking forward to settling the dispute with Gerard. On the other hand, I was uptight because I knew how volatile Gerard could be.

He might possibly scare Day, Harward and Ford into some unwise compromise. Without a doubt, the meeting would represent a watershed in the issue of science teaching regarding origins.

Because I was going to have to leave the breakfast meeting immediately to catch a flight back to Ontario, California, I hurriedly packed and checked out of the Sheraton-Palace. Walking across the street to the Clift Hotel, I found the Redwood Room where we were to meet.

By the scheduled 0730 hour, everyone was there — Gerard, Richard Smith, chairman of the State Advisory Committee on Science Education, Eugene Gonzales, Associate Superintendent of Education, Day, Harward, Ford, Roth, Neufeld and me.

The hostess found us a corner booth — sliding an additional table together with ours — so that all nine of us could be together. The seating was rather interesting. It even *looked* like a "command confrontation."

Gerard had center stage. He sat on the open side of the table. Flanked by Smith on his left and Gonzales on his right, he faced the rest of us seated on a continuous padded bench that curved to form the booth. I was crowded into the curved corner, representing the left end of the row of we six who were to be confronted. Harward was squeezed tight against me on my right, followed by Day, Ford, Roth and Neufeld.

It was far from comfortable, as far as eating breakfast was concerned. Yet, the togetherness seemed comforting, at least to me.

Details of how or what we ordered — or even how we managed to eat what we ordered — fade into obscurity when compared to the tension that existed.

Gerard used his great capacity to be gracious and almost obsequious. In fact, it seemed initially as though everyone was bending over backward to be nice to everyone else. However, because we were all on a tight time schedule — for example, the State Board was scheduled to convene at 0900 — the confrontation could not be delayed long.

Howard Day, State Board President, is a soft-spoken man. While we were yet eating, he quietly asked Gerard if he would like to express his reason for requesting the meeting.

Gerard was ready. Starting with a "sweetness and light" approach, he reviewed the development of the *Science Framework* over a four-year period by the State Advisory Committee on Science Education — noting the high caliber of the 15 members as well as their hard and diligent work.

Instead of a frontal attack, he made an oblique thrust that spoke of "disappointment that the State Board had not found the *Framework* totally acceptable." That point then allowed him to remark that the committee felt that they should have had the right to modify the document, instead of the Board doing it themselves. After all, the Board was not qualified in the field of science.

Gerard then went on to say that the amendment was totally irrelevant to the theme of the *Framework*. It further violated the intent of what the committee had intended.

Gerard kept on belaboring *generalities*. Not once did he get to what was *specifically* wrong with my two paragraphs. As his diatribe evolved into a tirade, he made a statement about the successes of microbiology in synthesizing life in the laboratory.

At this point, Professor Neufeld, a microbiologist himself, broke in with a comment that, in great measure, challenged Gerard's assertion as being a bit broader in scope than really warranted.

Right on Neufeld's heels, Professor Roth joined in suggesting the need to be a bit more cautious about the achievement of laboratory scientists. And they used obviously correct technical terminology in making their points.

Surprisingly, Gerard accepted their correction. In fact, he not only acquiesced — he paid them a compliment by saying, "You know something? You talk like *scientists*." From Gerard, that was some admission.

As Gerard resumed his dissertation, he once again began decrying the unilateral Board action in modifying the *Framework*. Finally, Ford had listened to enough.

"You know, Dr. Gerard, we all know that you think we are stupid and..."

Gerard broke in. "I never said or intimated such a thing."

"Oh yes, you did!" Ford retorted. "You accused us of having the *stupidity*, *audacity* and *temerity* to modify what you had written."

The transcription of the TEMPO Show on TV that I had sent to the Board members obviously had not escaped Ford. Now he was feeding Gerard's words right back to him, eyeball to eyeball. And Gerard was backpedaling — *fast*!

"I don't ever recall anything of the sort..." he protested.

"You even said it on *television*!" Ford had obviously taken the offensive at this point. "And let me tell *you* something, we do not intend to remove the two paragraphs that have offended you and the committee."

Gerard was incensed — but also cowed — by Ford's forceful counterattack. Having sent Gerard reeling, however, Ford had the presence of mind not to overkill. With excellent timing, he shifted into a more conciliatory mood.

"Now we all have pride of authorship. In no sense, do we wish to force you and the committee to accept *our* words as being *yours*." This line of reasoning obviously began to relax the tension that had peaked as Ford had choked Gerard — with his own words.

There then ensued a period of lighter discussion between several others around the table that tended to shift the focus away from the two prime combatants. Then Eugene Gonzales spoke up.

"In view of the apparently fixed positions that have been expressed — on one hand, that the Committee does not want to sign their names to the contested amendment, and on the other, that the State Board does not intend to remove the amendment — may I suggest a possible compromise?"

Everyone seemed relieved that Gonzales had not only summarized the controversy succinctly, but that he also might have a workable compromise.

It appeared to me (although I couldn't see the piece of paper from which Gonzales was reading) that his compromise had been drafted and typewritten prior to the breakfast. I never bothered to confirm it, but it seemed that there was no way that he could have prepared the compromise while Gerard and others were talking. The confrontation was so absorbing that no one could have concentrated long enough to prepare anything in writing. He evidently had guessed ahead of time what might happen, and was prepared.

Gonzales continued, "I have a statement that could be inserted as a footnote in the *Science Framework* concerning the two contested paragraphs. Here is how it reads:

> This statement was prepared by the State Board of Education as an explanation of its position and inserted in this copy in lieu of two sentences which were deleted. This statement does not meet with the approval of the State Advisory Committee on Science Education, nor does its inclusion in this manuscript have the approval of the Committee."

Gonzales looked up expectantly. Day looked at Gerard and said, "Would that satisfy your objections?"

Gerard seemed obviously pleased and responded, "I think that's an excellent statement of the true facts of the matter."

Day looked around the table and queried, "Does anyone else have any objection to this footnote being inserted?"

There was no dissent. Gonzales had drafted a masterpiece! I was personally surprised that such a simple footnote could resolve the high emotion that had existed only moments before.

Though I have never verified it with Harward or Ford, I do not think that they were aware, ahead of time, that Gonzales would attempt such a tactic. Therefore, it caught everyone by surprise. Was the war ended? It certainly looked like it, as everyone shook hands and prepared to depart...

Spotlight – Not Secession

Some human events defy analysis. Just when you think you understand what they mean, you find out that they meant something entirely different.

One such event was the insertion of a footnote in the *Science Framework* at the breakfast meeting in San Francisco on 12 February 1970. The footnote was obviously a compromise — between the State Advisory Committee on Science Education that *wrote* the *Framework* and the State Board of Education that is *responsible* for the *Framework*. Neither of these two parties could have guessed, on that day when they agreed to the footnote, what it would *really* mean later, however.

Recall the original position of both parties. The Committee wanted the two paragraphs *removed* forever — destroyed so that no publisher would ever be able to read them. The Board wanted the two paragraphs to *remain* forever — as an indigenous, unidentifiable element of the *Science Framework*.

What did the compromise footnote do? It permitted the two paragraphs to remain forever in the *Science Framework* (a victory for the Board). However, it clearly relieved the Committee of any responsibility for the two paragraphs (victory for the Committee). Both sides won victories! Were they equal or offsetting victories?

By no means.

With only brief insight, anyone can easily see that the *victory* that Gerard and his committee won was actually a *defeat*. They suffered a serious setback by having the footnote inserted. If they had simply accepted the paragraphs as their own, they would have been far *ahead* of where they ended up.

This conclusion is based on the following reasons:

1. The contested paragraphs would have remained buried unmarked among hundreds of other paragraphs. No one would have been able to find them. Now the only footnotes in the entire document (aside from a few bibliographic references) would guide the reader straight to the controversial paragraphs.

2. The Committee evidently thought they carried more weight with the publishing world than did the State Board of Education. Therefore, the Committee thought that if they disavowed the two paragraphs, the publishing world would likewise ignore them. Just the opposite was inevitable. The State Board had, in effect, overridden a veto by the Committee and announced that, despite the Committee's view to the contrary, they were insisting that these were important paragraphs. Since the State Board had all the power with the publishers and the Committee had exactly none, whom did the Committee expect the publishers to heed?

Many, with 20/20 hindsight, have wondered why a brilliant man like Gerard got "trapped" into accepting the footnote. I am not ready to say that he was trapped. Having had three major exposures to Gerard (the TEMPO Show on TV, my private meeting with him at Irvine, and the San Francisco breakfast), I am firm in my belief that one of two possibilities caused Gerard to accede to the footnote compromise as the best available alternative in his programmed harassment of the State Board:

1. He had never actually read what my two paragraphs said. Therefore, he got caught up by the press reports that religion was being mixed with science and "tilted at that windmill." Yet, when backed against the wall for reasons why the two paragraphs should be removed, he had no *specifics* to bring to bear. He didn't realize this void until the breakfast, sized up the situation, and selected the footnote as the route of honor.

2. He finally read what I said, but not until he had already gone public with his "hue and cry" and decided to play the belly-aching to the hilt. After all, there was always some possibility that the State Board would capitulate. When he saw Ford's firmness, he realized that he had no *specific* complaints with my words. Sticking with generalities, his only remaining complaint was that the Committee should have been consulted prior to amending the *Framework*. The footnote obviously took care of that complaint.

In actuality, the Committee had sealed its own fate in this matter long before the footnote was inserted. It was doomed to defeat from the time it childishly objected to the very minor revision by the State Board of Education — that body which was, after all, totally responsible for the *Science Framework* being written! The Committee had absolutely no function or existence apart from that delegated to it by the State Board.

Remember the arrogance of the Committee's 4 December 1969 widely-disseminated resolution? First, they took "strong exception to the alteration of content" by the Board. Then, they accused the Board of asking for changes to the *Framework* "essentially for injecting some extraneous religious or metaphysical views." Complaining that "the changes (introduced by the Board), though small in extent, have the effect of entirely undercutting the thrust" of the *Framework*, they charged that the Board "had thus offended the very essence of science, if not also that of religion."

Prefacing them with "it is our intent to make a prompt public release of this full statement" they then presented the State Board with the following list of demands, in order of their preference:[243]

> (la) Restore the *Framework* to the form submitted for its November 13 meeting.
>
> (lb) Restore the Framework as in (1a) but add the statement on creation as a clearly separate item over the name of the Board.
>
> (1c) Incorporate into the Framework as now altered a statement prepared by the Committee, explicitly repudiating the objectionable passage, with reasons. Further, the Committee states its intent to recommend against teaching materials in science that follow the disputed guideline.
>
> (2) If the Board refuses to take any of the actions indicated above, the Committee members decline, individually and collectively, to have their names associated with the document in any way. We also suggest that others whose names are mentioned as contributors be offered the opportunity to withdraw their names.

So, by alerting the news media, hurling out childish epithets, threatening to disavow authorship, and making irresponsible demands of the State Board, the Committee had deliberately "flung *down* the gauntlet."

Having reached the apex of wide publicity for their charges, any direction from there was down — with one and only one exception. If the Committee had been able to *destroy* the two paragraphs, they would have held their high ground. However, *anything less than destruction was automatically a defeat.*

The footnote obviously did not destroy the paragraphs — it raised them to the level of highest visibility! Instead of the two contested paragraphs *seceding* from the *Science Framework*, they were ushered right into the *spotlight* — where they remained.

CRITICAL CRITERIA

A few days before I met with Gerard in his office at Irvine, I received the working draft of the "curriculum criteria" that Glenn Leslie had promised to send me.

As mentioned earlier, these criteria are the *contractual requirements* for publishers of textbooks. While the frameworks are philosophic documents, the criteria are brief and succinct. For example, the final printed version of the *Science Framework* had 148 pages, while the criteria consisted of only 5 pages! Frameworks provide only explanatory background for the criteria. The publishers focus on (and must comply with) the *criteria* — not the frameworks. Textbooks are purchased only when they can be proven to meet the *criteria* — not the frameworks.

This subordinate role for the frameworks makes all the more ridiculous the hullabaloo that the State Advisory Committee on Science Education raised through their spokesman, Ralph Gerard, when the State Board inserted two small paragraphs in one obscure framework appendix. Regardless of what the *Science Framework* said, only the *criteria* would govern textbook selection!

Seeing the curriculum criteria for the first time, I immediately realized that I had a tough job ahead.

The sensitive words, *evolution* and *creation*, did not appear at all in the criteria. Neither did anything close to the subject of *origins*. If I proposed a criterion with those words in it, the State Advisory Committee on Science Education, the Curriculum Commission, and the State Board of Education would all focus on it — more on it, perhaps, than all the other criteria combined.

I also knew that, aside from the emotional sensitivity of my contribution, there would be resistance to *any* additions to the criteria because the list had been whittled down — with painful compromise — from a much longer list. *Modification* of existing criteria would be much more salable than trying to *add* new ones.

Finally, I felt constrained to be very brief. The criteria were expressed in *phrases* — not even complete sentences! Therefore, every word I used would have to be essential.

Hemmed in by these critical restraints, I searched the criteria for the appropriate section to suggest a revision. The criteria were listed in four major sections: (a) teacher materials, (b) learner materials, (c) instructional program, and (d) rationale for the criteria.

Where should I put rules for teaching about origins? What should I say? Should creation and evolution be specifically mentioned?

Upon returning to my office the day that I met with Gerard at Irvine,

I drafted three proposed curriculum criteria to carry out *contractually* what I, in a *philosophic* sense, had said in the *Science Framework*:

> *Suggested Addition No. 1:* The teacher materials shall "describe how science interfaces with other endeavors such as philosophy, particularly when considering origins, since theories like evolution and creation both have scientific adherents."
>
> *Suggested Addition No. 2:* The learner materials shall "provide a listing of unknowns in current scientific knowledge as a background for stimulating and focusing inquiry processes."
>
> *Suggested Addition No. 3:* The science program shall provide opportunities for the learner to "recognize the limitations of scientific modes of inquiry and the need for additional, quite different approaches to the quest for reality or for answers to questions like the origin of the universe, matter, life and even man himself."

Mailing these criteria to Glenn Leslie on 16 January, I explained my reasoning for each of the three criteria in a cover letter. Referring to *creation* and *evolution* as "magic words" since the press had maligned them, I said about my first suggested criterion:

> I agonized and wrestled with a possible way in which the magic words could appear (to placate those who will be looking for theories other than evolution now to be taught) and yet retain the truly minor character that they deserve (especially in the minds of the State Advisory Committee on Science Education). I believe that my Suggested Addition No. 1 to the criteria should meet these two objectives.

My second proposed criterion did not deal with origins, creation, or evolution. Instead, it was directed toward stimulating the student's inquiry processes. I explained my reasoning for its inclusion to Leslie as:

> Suggested Addition No. 2 is offered to support pages 57-60 of the *Science Framework* as well as one of the four major points I made to the State Board in my oral presentation on 13 November. This is only indirectly related to the amendment, but I feel it is vital to the overall objectives of the *Science Framework*.

The third criterion that I suggested was an expanded version of an existing one. Again, I offered an explanation for my modification to Leslie:

> Suggested Addition No. 3 is really only an extension of your Point 1.4 under "Attitudes of Science." This is offered to help desensitize this emotional issue of "evolution-versus-creation" and to point out that the question of origins really lies outside the legitimate domain of scientific quest.

In my letter to Leslie, I also offered to help him secure endorsement from the Curriculum Commission and the State Board of Education for my three new criteria. While I was not offering this help as a condition for their adoption by his Science Committee, I reminded Leslie that the three criteria were much milder than many of those who had wished for an extensive revision of the *Framework* — even its complete rewriting — expected. So I was trying once more to bridge a serious schism.

About two weeks after I submitted these criteria to Leslie, I called him to get his reaction. He suggested a few minor revisions — nothing serious. But he was non-committal about their adoption potential.

Since he was mailing my criteria to all members of his Science Committee, he urged me to try and meet in person with the Science Committee at the Sheraton-Palace in San Francisco prior to the February meeting of the State Board.

Leslie dropped one very interesting tidbit during our telephone conversation. He said that, in addition to those he received from me, one other person had submitted some criteria on the origins issue. These, Leslie said, were much too long and detailed.

"By any chance, was the author of that submittal named Spurney?" I asked, trying not to exceed propriety.

"As a matter of fact it was!" Leslie affirmed.

So Spurney had gone ahead on his own after all. Even though he'd agreed not to submit separate criteria when we discussed it on 7 January, he evidently reversed that decision without letting me know.

"Spurney and I had conversation a couple of weeks ago, and he agreed with me that you should receive only one set of criteria on the origins question. But I guess he changed his mind," I explained.

"Well frankly, I sensed some ulterior motive in what he submitted anyhow. It looked like he was after some personal endorsement or something. I would have been obliged to reject his material on that basis, regardless of whether I had gotten anything from you."

As recounted earlier, I was able to rearrange my teaching schedule and be in San Francisco on the afternoon of 11 February 1970. Accepting Leslie's kind invitation, I attended the scheduled meeting of the Science Committee at the Sheraton-Palace Hotel.

Dr. Richard Smith, chairman of the State Advisory Committee on

Science Education and Frank Mann from the Department of Education in Sacramento also sat with the Science Committee that day. We reviewed my involvement in the State Board amendment, as well as discussed my three criteria. The meeting was cordial and stimulating, with considerable interchange of ideas.

Before the Science Committee could come to the point of voting on my three criteria, two of the members had to leave to attend another meeting on adoption of social science textbooks. This disbanded the quorum, and the meeting ended officially without any action being taken on my criteria. However, discussion continued for quite some time with those who remained. I was pleased with the response to what I had written.

As the now "unofficial" meeting of the Science Committee finally adjourned, Leslie asked me to meet privately with him in his hotel room which I did. Together, we discussed in depth the pros and cons of the straw man issue raised by the news media. I also agreed, at his request, to rework my three criteria that evening, by incorporating suggestions made during the discussion in the afternoon meeting of the Science Committee. These revised versions of my criteria I assured him would be in his hands for the Science Committee meeting scheduled the next afternoon.

"While I can't speak for the entire Science Committee, I want you to know that I not only fully understand your objectives — I also endorse them. I am certain that California school children will benefit from the effort you have expended," Leslie said warmly, as we shook hands in parting.

Recalling as much of the afternoon's discussion as possible, I returned to my hotel room and revised the three criteria. Since I didn't have a typewriter available, I wrote out two copies in longhand on hotel stationery, keeping one for myself and putting the other in an envelope addressed to Leslie. Exhibit 13 contains this final version.

Comparing it with the original wording that I had sent Leslie on 16 January, you can see that the gist of all three criteria remained the same, although I managed to cut out some extraneous wording.

I felt good. The criteria had been thoroughly discussed by the Curriculum Commission's Science Committee. Also, the chairman of the State Advisory Committee on Science Education had been present to hear and participate in the discussion. Finally, their comments had been incorporated. Surely, the criteria should have no difficulty in being approved — first by the Science Committee, then the Curriculum Commission, and finally the State Board of Education.

I went to dinner that evening, breathing easy...

Closet Caucus

The breakfast hadn't ended any too soon. Despite the unanimity about the footnote, our meeting with Gerard and Smith on Thursday morning, 12 February 1970, ran late. Looking at my watch as we all shook hands, I realized that I wouldn't have time to run across the street to the Sheraton-Palace to deliver my revised curriculum criteria to Leslie.

In fact, I would barely catch my flight back to Southern California if I left immediately. What was I going to do about the criteria?

As I helped Gerard into his topcoat, I was desperately racking my brain on how I could get the envelope to Leslie. Suddenly, an idea popped into my head. Turning to Smith who was standing beside both of us, I said, "Dr. Smith, are you by any chance going to be seeing Glenn Leslie today?"

"Yes, we have a joint meeting of our two committees this afternoon," he said.

"I certainly hate to bother you, but I revised my three criteria last night to incorporate the consensus expressed in yesterday afternoon's meeting of the Science Committee that you attended. I promised Glenn that I would get them to him today, but I must run now or I will miss my plane. Would you please see that he gets them?"

"Oh, I'd be *delighted* to carry them to him," Smith responded — almost *too* willing, it seemed to me. Without a doubt, there weren't two people in the whole world who could have been more opposed to the survival of my criteria than Smith and Gerard. If those criteria were adopted, it really didn't matter what the *Science Framework* said about origins! And both of them knew it...

But what other choice did I have — especially after I had asked the favor of him? With a slight hesitation, I handed the envelope to Smith. As I did, I noticed that it was *unsealed.*

All the way to the San Francisco International Airport in the limousine, I was kicking myself for not at least sealing the envelope. Occasionally, I would counter my self-flagellation by reminding myself that Smith and Gerard, after all, were really honorable men who would not violate the privacy of communication — even of an adversary. The worst of all my thoughts that day, of course, was that the two of them might deliberately lose or destroy the criteria prior to meeting Leslie. I finally ceased my fuming and decided to "wait and see."

I didn't have long to wait. By the next evening, I was home from my teaching assignment in Redlands. After dinner, the phone rang. Ed Shevick, my neighbor on the State Advisory Committee on Science Education, was on the line.

"Hi, Vern. I thought you might be interested in what we did with your criteria yesterday in San Francisco."

He sounded so excited that I felt hope rising within me. "Sure would," I replied with expectation.

"Well, I guess you must have given your latest version of curriculum criteria to Smith and Gerard...at least they had a copy of them. Smith decided to caucus our committee prior to our meeting with Leslie's Science Committee. We had a long discussion of your criteria and finally prepared a unanimous resolution to the Science Committee recommending that they not be adopted. This resolution was then handed to Leslie at the same time that he received your proposed criteria."

Wasn't that nice? Talk about loaded dice! With that kind of fair play and objectivity, who really needs enemies?

Scientists have projected to the lay public the image of always dealing with facts and thereby being "clean as a clam" when it comes to skullduggery. Politicians, on the other hand, are automatically suspected of backroom maneuvering and shady deals. But you can be assured that they have no monopoly on such pranks. Scientists are just as good at "dirty pool" as anyone else.

Smith's closet caucus was no different than the political "bartering in smoke-filled rooms."

Fuming Frustration

I felt like I had been kicked in the groin! In fact, I was so stunned by Shevick's report of this obvious backstabbing that I wasn't able to say anything to him for a few moments.

"Ed, I guess that I should thank you for letting me know, at least." I was still reeling. "That certainly wasn't why I handed the criteria to Smith yesterday morning. I had no choice if I was going to catch my flight out of San Francisco."

Beginning to recover my senses, I continued, "But I can tell you one thing for sure. I don't really care *what* the Science Committee does with my criteria. If they accept your recommendation and throw my criteria out, I'll go right around them to the State Board!"

By now, my Irish temper was beginning to heat up...

Shevick exploded, "Where in the hell do you get your power? Do you know the Governor or somebody?"

"No, I don't, but..."

He cut me off, "You wanna know something? You *frustrate* us! Here we are — 15 qualified individuals who have worked together for four years, putting together a pretty nice piece of work. You come along out of nowhere, with no qualifications, and roll right over us. We can't figure out what you're up to! Why don't you go on back to engineering or whatever you were doing before?"

My first reaction to Shevick's outburst of frustration was to be amused.

"How could I really frustrate such an illustrious group?" I pondered. But as he went on to explain, it became more evident to me that he was not expressing just his own private opinion — it was undoubtedly the *collective* feeling of the Committee.

If you try to analyze *why* they were frustrated, answers don't jump right out at you. Sure, they were insulted by a group of laymen — none of them qualified scientists — who messed up their finely-honed masterpiece. But does insult equate with frustration? No.

Then, they also went to great pain (and expense — reimbursed by the State Board) to meet on 4 December 1969 in San Francisco to draft a self-serving "Magna Carta" for themselves. They successfully got wide press coverage for that pontifical piece...and waited...and waited — for weeks! — with no acknowledgement from the State Board that they had even read it.

Insulting again, perhaps...but not really frustrating.

And other events that the Committee legitimately could have considered insulting also occurred. However, it still holds that insults alone — however long the list — do not constitute frustration. To convert insults into frustration, it may be necessary to recognize a personal dimension in this issue — the one Shevick put his finger on.

Without any question, the State Board of Education treated *me*, the ignoramus, quite differently than they did *them*, the illuminati.

The Director of the Biological Sciences Curriculum Study, William V. Mayer, became irate over the same issue that Shevick raised — why I was apparently afforded different credibility by the State Board than their own Advisory Committee on Science Education. Mayer wrote an article that was less than complimentary to the State Board:[244]

> An unprecedented and almost unbelievable event occurred. Vernon L. Grose...prepared a thirteen-page statement of his personal viewpoint and presented it to the California State Board of Education on November 13, 1969. That this one man's opinion was accepted by the California State Board of Education over the considered recommendations of sixteen leading scientists and science educators who had labored for over a year to form them is unthinkable. It makes me believe that the organizational philosophy of the California State Board of Education must be like that of the Mad Hatter's Tea Party.

Mayer is hopefully a better biologist than he is a journalist because his article contains many unfortunate errors that he then compounds into even greater error. However, he does pinpoint what I believe translated *insult* into *frustration* for Shevick and his co-workers:[245]

> At the suggestion of one member of the audience, Vernon L. Grose, the Board inserted the two paragraphs containing a reference to the Bible. This certainly must be the most flagrant disregard for logic and reason to be recorded in modern educational annals. The State Board's repudiation of its own committee in favor of a lay opinion from the audience should ultimately become a classic example in textbooks on school administration of how *not* to proceed with the development of standards. The fact that the opinion of one man could sway the California State Board of Education against its own prestigious committee is either a tribute to the eloquence of Grose or an indication that the board had no faith in its committee in the first place.

The key question then emerges, "*Why* did the State Board treat the ignoramus differently than the illuminati?"

Of course, the Board members alone can answer that question. However, careful study of what *actually* happened may reveal that the Committee, through their arrogant yet childish behavior, single-handedly provided all the necessary ingredients for their fuming frustration...

Remarkable Reversal

The Smith-Gerard "slick trick" with the Science Committee on 12 February 1970 had me worried. And I stayed worried for about a week — with no word of what had happened to my criteria.

A letter from Glenn Leslie, dated 18 February 1970, brought wonderful news. Not only had the Science Committee ignored, by *unanimous* vote, the unanimous resolution to destroy my criteria that Smith and Gerard had masterminded. The parent Curriculum Commission had also *unanimously* adopted my three criteria!

The final version of the three criteria now read:

1. The teacher materials shall "recognize the relationship of science to other disciplines, particularly when considering unresolved theories, such as origins, evolution, creation and so forth."
2. The learner materials shall "provide examples of unresolved questions in science as a means of stimulating the inquiry process."
3. The science program shall provide opportunities for the learner to "recognize the limitations of scientific modes of inquiry and the need for additional, quite different approaches to the quest for reality,

including the search for answers to questions like the origin of the universe, matter, life, and man."

Leslie wasn't sure that the State Board would be totally convinced that my criteria were strong enough on the origins issue.

"If you feel that we have come to grips with the problem fairly and effectively, your approval will be welcomed," Leslie wrote. I gladly obliged him with a letter to the State Board.

The State Board of Education, during their March 1970 meeting in Sacramento, approved the final science textbook criteria, with my three firmly locked into the document.

As would be expected, the *Los Angeles Times* continued to twist the issue by a secondary headline for a major article on the event,[246] "BIBLICAL AND EVOLUTIONIST THEORIES ON ORIGIN OF LIFE MENTIONED ONLY INDIRECTLY."

Once more, I suggest that you read my three criteria and find anything that alludes to Biblical theories — *directly or indirectly.*

The State Board of Education had now been joined by the Curriculum Commission and its Science Committee in overriding unanimous recommendations of the State Advisory Committee on Science Education. It's most unfortunate that Mayer selected only the State Board for his vitriolic venom.[247] Had he been objective, he would have pointed out that *no one* in the entire process had paid attention to the State Advisory Committee on Science Education.

Seeking answers as to why *everyone* ignored the committee might have been quite productive for future ventures undertaken by obviously qualified scientists.

The adoption of my criteria evidently prompted another letter from Walf Oglesby, the person in the State Department of Education in Sacramento who had originally sent me the invitation to appear on 13 November 1969 as an expert witness. One part of his 24 March 1970 letter said:

> We plan to keep you in our file of expert witnesses. Do not be surprised to hear from the California State Board of Education again.

I accepted that statement as a compliment. However, I was looking forward to a long-needed rest.

Almost six months had passed since I was provoked by that *Los Angeles Times* editorial. Hundreds of hours and dollars had been expended in opposing the onslaught of an overwhelming news campaign committed to a straw man. However, it had been worth it all.

On 2 April 1970, I wrote a letter of appreciation to President Day of the State Board of Education that summed up my feelings: "I want to thank you sincerely for your leadership, courage, and objectivity in the face of a concerted effort by the news media to distort the meaning of the adoption of my words into the *Framework*."

The tide had turned. There had truly been a remarkable reversal...

Chapter 6

BACK TO SLEEP

"Veneer'd with sanctimonious theory." — *Tennyson*

The crisis had passed.

Regardless on which side of the battle you stood, there was a truce. Each side thought they knew the other's position. It was much like the early part of World War II, right after Hitler finished off Poland in 1939.

Remember the so-called "Phony War" that existed between the fall of 1939 and the spring of 1940? There *was* a war — and yet there *wasn't.*

So it was in the "God versus Darwin" straw man issue, after the footnote was inserted in the *Science Framework* and the curriculum criteria were adopted by the State Board of Education in March 1970.

The Darwinians figured that they had successfully gotten my two paragraphs relegated to absurdity (by means of the footnote in the *Framework*) and had cleansed the curriculum criteria of any mention of the Bible or religion.

The Godly were satisfied that the *Framework* still contained my two paragraphs (with the added endorsement by the State Board) and there were at least two specific curriculum criteria that addressed origins.

The most likely way that two opposing sides of a dispute can fall into a false sense of security with respect to each other is by failing to understand the opposition. Hitler and the Allies misunderstood each other during 1939-40. So did the Godly and the Darwinians in 1970. "Out of sight, out of mind" describes both sides in the origins issue at that time.

As explained earlier, the textbook acquisition cycle in California is a long process — approximately seven years elapse from the time a framework on a given subject is initiated until textbooks in that subject appear in the classroom. Obviously, there are key events or milestones that occur during that seven-year cycle. Those milestones are often given public visibility through the news media. However, with the exception of these several milestones, nothing newsworthy is occurring.

Therefore, the process moves silently and inexorably along like a centipede. And, as occurs also with centipedes, the *path* between points of focused attention; e.g., milestones in the textbook process, can vary greatly from that expected by observers.

What are the major milestones in the textbook acquisition process? There are probably five. Two have already been discussed — (1) the adoption of a framework on a given subject by the State Board of Education, and (2) the adoption of specific criteria for publishers to meet that are based on the philosophy of the framework. The remaining milestones that generally attract public interest are:

3. The public display of teaching materials submitted by publishers in compliance with the curriculum criteria. This is the first glimpse of the actual product of the entire process. Dozens of display points are scattered throughout the entire state. Teachers, students, parents as well as Curriculum Commissioners all review the materials.

4. The selection of specific textbooks and materials to be recommended by the Curriculum Commission to the State Board of Education for adoption. The viewpoints expressed by those who reviewed the textbooks and materials while on display are collected, digested and funneled to this decision point — a most critical procedure because millions of dollars are at stake by the resulting decisions. Selection by the Curriculum Commission, with rare exceptions, is tantamount to adoption by the State Board of Education.

5. The actual adoption of textbooks by the State Board of Education. At this time, the Board may excise from, add to, or edit textbooks in order to bring them into compliance with their requirements.

In the Science Textbook Struggle, about two years elapsed between the second and third milestones just described. During these two years, both sides of the "God-versus-Darwin" battle went to sleep. Why did this happen? Everyone was probably pacified by a potion of presupposition.

POTION OF PRESUPPOSITION

Each of us has an internal set of beliefs that we do not openly acknowledge. We may not even recognize that they exist. Buried in our subconscious, they influence much of what we do or think.

One group that *should* be free of beliefs about their work is scientists. Yet they are plagued with preconceptions and bias. Harvard physics professor Gerald Holton says, "Science without a metaphysical substructure has never been possible."[248]

It is safe to say that most beliefs or philosophic commitments are non-rational. However, even among scientists, these subliminal presuppositions are more often helpful than hurtful. Holton points out, for example, that Max Planck, in laying the foundations of quantum theory, was motivated by metaphysical belief in the Absolute.

While there is little, if any, explicit acknowledgement among scientists of their private philosophic persuasions, such beliefs are definitely a foundational facet of all scientific work. Holton affirms this in the following way:[249]

> The personal metaphysical tenets of scientists, although sometimes very strong, are in a free society generally so varied, so vague, in fact technically so inept that in a sense they cancel out...It may be argued that our science will be healthy only so long as our scientists remain poor metaphysicians. In only those places where one codified, generally accepted set of dogmas exists — as in some of the old scholastic universities or in modern totalitarian states — can the extraneous and ultimately detrimental idea in a scientific publication still survive the scrutiny of colleagues.

According to Holton, the metaphysics of scientists are harmless as long as they remain *heterogeneous*. To state it conversely, if the metaphysics of scientists become *homogenous* or unanimous, science will become sick. Curiously, Holton overlooked another place where "one codified, generally accepted set of dogmas exists" — the biological sciences. The renowned entomologist, a Fellow of the Royal Society, W. R. Thompson expressed it this way:[250]

> The development of Science, as an autonomous discipline, seems to entail the rigorous elimination of philosophical notions...(Yet) evolutionary speculation is (full) of philosophical principles and suppositions. The concept of organic evolution is very highly prized by biologists, for many of whom it is an object of genuinely religious devotion, because they regard it as a supreme integrative principle. This is probably the reason why the severe methodological criticism employed in other departments of biology has not yet been brought to bear against evolutionary speculation.

Undoubtedly, Darwinians were lulled to sleep in 1970 because they thought that Darwinism was, once and for all time, a "codified, generally-accepted set of dogma." They committed intellectual incest — talking only to those who agreed with them. Asking any of them to list evidence against evolution would be comparable to asking the clergy for proof of the non-existence of God. To elicit evidence against a theory, one must not only *reexamine* the theory but seriously *question* it. That day has long since passed for many biologists regarding Darwinian evolution.

Just as historians change their minds over a period of time, not only about *facts* but about *interpretation* of those facts, scientists also change their minds. This is true with respect to scientific observations as well as to the interrelationship of those observations. In that sense, science is no more "true" than history is. Error is just as likely to occur in science as it is in history, music, art or politics.

In fact, science, due to its probabilistic nature, theoretically says *nothing* with absolute certainty. There is *always* the possibility that an exception may be found. The likelihood is ever present that some later person may make an observation that upsets or modifies what has been commonly accepted as true by scientists for many years.

Yet, an arrogant attitude of smugness emerged among those who thought they had humbled the California Board of Education by means of a footnote.

Where did the smugness come from? How could Darwinians fall asleep in 1970? Could it have been the product of tacit belief? All of us are subject to accepting myths as fact. A recent book has been compiled on "facts we know that are not so."[251] It points out that chop suey is not Chinese, sheet lightning is identical to bolt lightning, and India ink never came from India.

Yet a majority persists in believing otherwise. With respect to what even scientists hold to be "fact," there is danger in becoming smug or secure in believing that such a fact never again needs reexamination.

If we concentrate exclusively on *what* is a fact and thereby ignore *how* we know it is a fact, we can easily become misled — even in science. Theodore Roszak expresses his concern about such peril this way:[252]

> There are *styles* of knowledge as well as *bodies* of knowledge. Besides *what* we know, there is *how* we know it — how wisely, how gracefully, how life-enhancingly. The life of the mind is a constant dialogue between knowing and being, each shaping the other. This is what makes it possible to raise a question which, at first sight, is apt to appear odd in the extreme. *Can we be sure that what science gives us is indeed knowledge?*

But Darwinians weren't alone in falling asleep. The Godly also drifted into slumber. Their negligence in napping may have been due to a sedative of smugness, similar to that of the Darwinians. And the smugness probably could be attributed, just as it was with the Darwinians, to tacit belief.

The belief itself, obviously, was different than the one that induced drowsiness in Darwinians. The Godly became smug in their belief that the cosmology of a Creator had been successfully and permanently reinstated in public school science curricula.

Therefore, both the Godly and the Darwinians became drowsy in

1970 because of what they believed. Their beliefs, though poles apart, dealt primarily with origins.

In fact, much of the presupposition that underlies scientific work today involves *ultimate causation* — concern about why nature is the way it is. One of two major philosophic predispositions seems to dominate or influence a scientist as he observes natural phenomena.

One philosophy, *mechanism*, says that everything in nature can be described and explained in terms of physics and chemistry. The other, *teleology*, maintains that something beyond mechanical causes is required to explain the obvious design and purpose in nature. Harris describes the tension between these two types of presupposition:[253]

> There are still two schools of thought. There are still those who stoutly maintain their faith in the possibility of reducing all biological laws to those of physics and chemistry, who point to the statistical laws of physics and assert that from the randomness of mutations order may be extracted by similar mathematical means...The view has a certain plausibility so long as one forgets the odds against the actual production by any such process of the observed outcome of vital activity. If the statistical laws of thermodynamics were consistently applied this result would be virtually beyond expectation.
>
> Not surprisingly, therefore, we find a large and influential body of biological opinion still supporting teleological or quasi-teleological views.

It can reasonably be assumed then that the reason for "God-versus-Darwin" opponents going back to sleep in 1970 involved this classic, historic tension between *mechanism* and *teleology*.

Mechanistic Mania

Science neither *is* nor *represents* reality. At best, science is a method or way of *interpreting* reality. So far as origins are concerned, the events themselves are not influenced one whit by what scientists may decide about their date or means of occurrence.

What happened, happened. Nothing concerning those events will be influenced or revised by the findings of scientists. Of all the phenomena of nature, scientists should speak about origins — whether of the universe, life, species or man himself — with the greatest humility and hesitancy.

Origins are real events. They did occur. They were not observed. They are not repeatable. Whether or not an outside intelligence was involved cannot be proven scientifically. Just because a group of people — even a *large* and *sophisticated* group of people — should collectively agree on what must have happened at the beginning, there is no *a priori* reason to put faith in that agreement.

Should this group convince the whole world to believe their story and even upgrade the story to the status of a "scientific fact," there may be absolutely no resemblance between the event itself and the agreed story. While this may sound trite, it is critical to understanding the limitation of science in discussing origins.

It is correct to say, then, that scientists start guessing about the origin of the universe, life, and man with exactly as many clues as anyone else. Zero. If they are believers in the mechanism school, what impels them to that belief? Facts? No, because there are no *facts* available — just historical artifacts.

One possible reason why scientists choose to be mechanists might be to avoid reference to anything supernatural. This is quite a proper reason since scientific criteria for supernatural phenomena would be difficult, if not impossible, to establish.

On the other hand, there could also be some scientists who do not wish to acknowledge an intelligence greater than their own and therefore commit themselves to proving that the entire universe can be satisfactorily explained without the need to attribute anything to a higher intelligence.

However, whether due to objective or personal reasons, science has traditionally avoided the supernatural. As Harris says, "It is alien to science to attribute phenomena to supernatural causes."[254]

The important point to remember is that scientists choose to believe in mechanistic explanations *prior* to examining natural phenomena — not as the *result* of doing so. A theoretical physicist, Steven G. Brush, has pointed out that the public is often encouraged to believe in Francis Bacon's version of science that he describes as:[255]

> Collect a large number of observations and experiments about a subject and arrange them systematically. One may then propose a hypothesis to explain the data, but it must be tested by further experimentation. Any hypothesis or theory which is disproved by experiment must, of course, be discarded immediately.

Yet, as Brush explains, both Mendel and Einstein approached science quite differently — almost in conflict with what Bacon had advocated:[256]

> They knew that 'raw data' never lead directly to a single theory or confirm it precisely. What counts in science is not the accumulation of data but the brilliant insight that reveals the regularity lying hidden beneath the chaos of superficial appearances...If science does progress, it must be because of the special genius and effort of particular scientists, not because there is a foolproof scientific method which anyone can use.

Therefore, the mere gathering of historical artifacts or bits of primordial evidence by a scientist does not automatically lead him to conclude how origins occurred. No clear, irrefutable picture of origins jumps out at him from the bits of data.

Instead, he *furnishes a picture out of his mind.* This picture is a preconceived notion of how these bits might have fitted together into a cohesive pattern. It is a *guess, hunch* or *assumption.*

However, just any old guess or hunch will not suffice. For one thing, scientific guesses or hunches are given a new name — *hypotheses.* And they generally conform to some rules that have gained acceptance among other scientists over a long period of time. For example, Hoagland lists two tacit assumptions that govern or influence hypotheses:[257]

> A scientist operates under the tacit assumption that there is order underlying all phenomena that he studies. Otherwise his work would be pointless. He hopes to find the nature of this order. He also assumes that all forms of order are determined — that is to say, are caused — and his job is to discover these determinants or causes.

Yet, the mechanistic scientist is tacitly committed to the premise that the *order* and the associated *causes* of that order are totally *natural* — not supernatural. This tacit assumption closes — not opens — his mind.

Since hypotheses are not an automatic product of experiments, what a scientist *believes*, as he examines, selects, and assembles data, is obviously critical to the conclusions that he reaches. One renowned philosopher who has influenced many who seek mechanistic answers to the problems of origins is Bertrand Russell.

Try to imagine how scientists who have been swayed by Russell's philosophy would approach the question of whether or not a Creator was involved in the origin of the universe, life, and man. Here is Russell's view of science:[258]

> In this world we can now begin a little to understand things, and a little to master them by the help of science, which has forced its way step by step against the Christian religion, against the churches, and against the opposition of all the old precepts. Science can help us to get over this craven fear in which mankind has lived for so many generations. Science can teach us, and I think our own hearts can teach us, no longer to look around for imaginary support, no longer to invent allies in the sky, but rather to look to our own efforts here below to make this world a good place to live in, instead of the sort of place that all the churches in all these centuries have made it.

Is that statement biased? Would scientists who agree with Russell be likely to openly consider evidence for a Creator? Would they favor a purely mechanistic explanation of origins — even when they could not support the explanation with physical evidence? Probably.

In fact, anti-supernatural belief or bias is probably the most common factor in mechanistic thinking. Darwinism, for example, is frequently believed by biologists not because it can be scientifically verified but because the only other alternative would invoke the necessity of a Creator.

Kerkut, in his introduction to *Implications of Evolution*, recounts his years of teaching biology students to examine evolution on the basis of *evidence* rather than *blind faith*. His persistent questioning of students about the evidence *against* evolution frustrated students who had been taught only the points in *favor* of evolution while being warned that the sole alternative was a religious idea. Kerkut recalls student frustration this way:[259]

> The student would look at me as if I was playing a very unfair game. It would be clearly quite against the rules to ask for evidence against a theory when he had learnt up everything in favour of the theory. He also would take it rather badly when I suggest that he is not being very scientific in his outlook if he swallows the latest scientific dogma and, when questioned, just repeats parrot fashion the view of the current Archbishop of Evolution. In fact he would be behaving like certain of those religious students he affects to despise. He would be taking on faith what he could not intellectually understand and when questioned would appeal to authority, the authority of a "good book" which in this case was *The Origin of Species*.

Of course, there may be other reasons — beyond a bias against the supernatural — that cause scientists to believe *a priori* in the mechanistic approach. Another reason may be that it is a simpler approach. Mechanistic thinking eliminates the need to refer to forces other than physical and chemical analogies.

Mechanistic thinking can be illustrated in layman's language in a variety of ways. The American literary giant, Nathaniel Hawthorne, wrote a short story entitled, "The Birthmark." It is a story of the faith of a man who tries to control nature through science — a man who believes that science can accomplish more than it actually can. It is one of Hawthorne's many excellent short stories and portrays how a mechanistic look at life can ruin all that is meaningful. Many would see in this story the ultimate result of mechanistic thinking.

Hudson Hoagland, quoted earlier, wrestles with the problem of apparent *purposiveness* in nature even while subscribing to the mechanistic approach. Using computers as an example of a remarkable complex of feedback processes, including utilization of information storage and its appropriate retrieval, he compares them to the memory and recall capability of man. He proposes that computers could be called *purposive mechanisms.* Yet he seems to be perplexed with the problem of purpose:[260]

> Objection may well be raised to calling such mechanisms purposive, since their purpose has been built into them by man. But man himself and his behavior are an emergent product of purely fortuitous mutations and evolution by natural selection acting upon them. Non-purposive natural selection has produced purposive human behavior, which in turn has produced purposive behavior in the computers.

His perplexity with the problem of purpose illustrates well the dilemma of the mechanism school of thought. However, Hoagland is not alone in seeing difficulties with mechanism as a total explanation for origins. The British brain physiologist, Donald M. MacKay, describes an even more subtle result or illustration of mechanistic thinking.

MacKay speaks of the mechanism approach as being "ontological reductionism." But in terms of layman language he suggests it could be called "nothing-buttery."

Nothing-buttery to MacKay is characterized by the idea that if any phenomenon is reduced to its components, you not only explain it, but *explain it away.* In other words, for centuries God was a kind of *alternative* explanation that was used whenever you couldn't explain something in physical terms. As physics and chemistry advanced, God — as an explanation — was forced to retreat.

Further, a scientist apparently can "debunk love, or bravery, or sin for that matter, by finding the psychological or physiological mechanisms underlying the behavior in question."[261] So, in MacKay's eyes, the mechanistic approach can be used as a "God eradicator" through scientists claiming that soon, if not now, everything in the universe can be reduced to "nothing but" certain physical and chemical mechanisms.

The former editor of *Punch* and for many years an outspoken Socialist, Malcolm Muggeridge, has another view that carries out the mechanistic idea to its obvious impact on society and politics. He suggests that mechanistic thinking can affect whole civilizations in a deleterious manner:[262]

> Men can become decadent. Men can lose their sense of what life is about. Men can lose their religious faith, which is an enormously important thing in them being able to live together. They can lose these things. They can go after what are called false gods. They can think that something like science, which produces prodigies, can also produce a basis for living, and they can be wrong...If you could point to one particular thing (which led to the rise of Socialism) it really was belief in science. It was the sort of faith in science, first in the intellect, in explaining what life was about and then in science, in putting at men's disposal the power to carry out their ideas and produce their kind of

> world. It's been, of course, a total failure...I would say that from, for instance, Darwin, you see, there has been a trend in our society in the direction that I've been speaking about and that all our way of life, our intellectual fashions, even our art has shaped itself around this trend. And this trend is a disaster course, a Gadarene course.

Obviously, not everyone is as pessimistic about the extension of mechanistic thinking as Muggeridge. Some scientists are even enthusiastic that the mechanistic process will lead quite naturally to "scientific ethics." But even that belief leads to disturbing conclusions at times. Not all of mankind's behavior fits nicely into the idea of being ethical. The late Theodosious Dobzhansky, for example, could not ignore the obvious and ugly problem of inhumanity of man to man. Where such behavior rightly should fit in an ethical construct based on mechanistic thinking puzzled him:[263]

> Suffice it to mention the so-called Social Darwinism, which often sought to justify the inhumanity of man to man, and the biological racism which furnished a fraudulent scientific sanction for the atrocities committed in Hitler's Germany and elsewhere.

Some mechanistic thinking does not always lead to satisfying answers. In fact, it leads to a number of definite problems. These problems seldom show up in scientific literature — let alone in science textbooks in the public school classroom. In enumerating the problems produced by mechanistic thinking, I am not trying to discredit such thought. Rather, I am hopeful that more serious questioning of the mechanistic concept may occur. To that end, let's consider just a few of its problems.

A professor at the Institute for Enzyme Research at the University of Wisconsin, Garret Vanderkooi, is concerned about recent developments in molecular biology that should cause a revolutionary reconsideration of the mechanism idea:[264]

> There is no theory in existence today that even begins to explain the origin of life by natural means. The individual molecules in a living cell are extremely complicated, precisely made, and arranged in a varied but highly ordered network. Both the structure of these molecules and their cellular organization (and thus life itself) are passed on from generation to generation. To think that such a system could ever have come into being by itself is unbelievable.

Vanderkooi goes on to discuss the problems associated with our concept of *time*. Somehow, most scientists have simply accepted the idea that time is a steadily flowing phenomenon in the universe. Yet Sir James Jeans once said:[265]

> As I see it, we are unlikely to reach any definite conclusions on these questions (determinism and causation) until we have a better understanding of the true nature of time. The fundamental laws of nature, in so far as we are at present acquainted with them, give no reason why time should flow steadily on: they are equally prepared to consider the possibility of time standing still or flowing backwards. The steady onward flow of time, which is the essence of the cause-effect relation, is something which we superimpose on to the ascertained laws of nature out of our own experience; whether or not it is inherent in the nature of time, we simply do not know...it is always the puzzle of the nature of time that brings our thoughts to a standstill.

But even if there were a satisfactory overall mechanistic explanation for origins, there are physical problems *within* such a model. Two prominent biochemists, D. E. Green and R. F. Goldberger, described one of the major problems with the mechanistic explanation for the origin of life like this:[266]

> There is one step (in the mechanistic concept) that far outweighs the others in enormity: the step from macro-molecules to cells. All the other steps can be accounted for on theoretical grounds – if not correctly, at least elegantly. However, the macromolecule to cell transition is a jump of fantastic dimensions, which lies beyond the range of testable hypothesis. In this area, all is conjecture. The available facts do not provide a basis for postulation that cells arose on this planet. This is not to say that some paraphysical forces were at work. We simply wish to point out that there is no scientific evidence.

Of course, there are conceptual difficulties as well. Take for example the subject of mutation being a random process. It is frequently difficult to believe that biologists are not using circular logic on the subject of mutation. Consider the following statement:[267]

> The process of mutation ultimately furnishes the materials for adaptation to changing environments. Genetic variations which increase the reproductive fitness of the population to its environment are preserved and multiplied by natural selection.

Now compare that statement with one by C. H. Waddington:[268]

> It remains true to say that we know of no way other than random mutation by which new hereditary variation comes into being.

One can easily become confused as to whether the process of random mutation is proposed because it is *scientifically verifiable* or whether it is required as an *explanatory crutch*.

This book is not the proper place to list and discuss all the mechanistic problems. A few of these problems have been mentioned only to propose that the mechanistic approach may not be a total answer. Meanwhile, there seemed to be in 1970 a mechanistic mania maintained in many minds that lulled them to sleep.

Tenacious Teleology

While the philosophy of *mechanism* proposes that all of nature can be reduced to physics and chemistry, *teleology* proposes that something in addition to physics and chemistry is required to explain origins. In that sense, teleology *embraces* mechanistic explanations but also *adds* to them.

One way to view these two philosophies might be to consider that teleology forms a skeleton on which the flesh of mechanism is distributed in order to complete a whole being.

The tension that exists between *mechanism* and *teleology* often revolves about several questions concerning the origin of life. Harris proposes three such questions:[269]

1. Is the origin of life simply an accidental outcome of a process of random shuffling?
2. Is the origin of life something determined by inexorable laws such that from sufficient data the end can be calculated from a knowledge of its beginnings?
3. Is the origin of life a process in which the end is implicit from the start and molds its course throughout?

In proposing an answer to the first question, Harris is firm in his rejection of a mechanistic explanation for the origin of life:[270]

> The facts do not permit us to hold that the high degree of improbability generated by natural selection could have been the result of mere random shuffling and accidental changes. These may have played their part but cannot in their nature provide the integrating and organizing principles.

In his opinion, the answer to the first question is a resounding "No!" That being true, he then proposes that the latter two questions really introduce the need for some teleological viewpoint. While declaring that there is a large and influential body of biological opinion still supporting teleological or quasi-teleological views including such authorities as Oparin, Sherrington, Haldane and others, Harris openly admits that teleological explanations are usually *rejected* in science. He gives four possible reasons for the rejection of teleology:[271]

1. Teleology presumes or implies some extraneous agency (e.g. God) directing and arranging living processes to fulfill its own deliberate purposes.

2. Teleology is anthropomorphic in crediting lower organisms with purposive activity similar to our own.

3. Teleology, tacitly or openly, attributes to living things the consciousness necessary to pursue purposes in the absence of any sufficient evidence of its existence.

4. Teleology attempts explanation *obscuri per obscurius* by postulating as a cause of present activity an event which must occur in the future.

Harris admits that these objections are legitimate so far as they apply. However, he proposes that if teleology is defined as "the dynamics of organized wholeness," it is possible to pursue a *scientific explanation* of these dynamics without resorting to *supernatural causes.*

On the other hand, teleology would not necessarily be forced to reject the acknowledgement of a supernatural force. To state this argument in layman's terms, both Darwinism (the amoeba-to-man concept for which there are no demonstrable mechanisms) as well as creationism (the belief that a supernatural Creator was involved in the process of originating life) could *equally* be classified as teleological.

Likewise, both of these belief systems could be considered to be supra-mechanistic. Harris offers persuasive argumentation that there are at least four compelling reasons why biological development and evolution are teleological. Those interested in these arguments would be well advised to read his complete work on the subject.[272]

Harris is not alone in proposing that Darwinism is a teleology. George Cabot Lodge, a professor in the Harvard Business School, has detected Darwinian teleology in the business practice of specialization and fragmentation.

The American space program exploded the myth that if experts and specialists would attend to the parts, the whole would take care of itself. From this realization (that the "wholeness" of a space system needed specific attention) came the emphasis now known as "the systems approach" — a focus on systems science, which is my own area of professional expertise.

Lodge relates the belief that man has the will to acquire power — i.e., to control external events, property, nature, the economy, politics, or whatever — to an evolutionary concept. He notes that the presence of this will in the human psyche means the guarantee of progress through competition, notably when combined with the Darwinian notion that the inexorable processes of evolution are constantly working to improve on nature.[273] That Darwinian notion to which Lodge alludes is obviously not mechanistic but *teleological.* Furthermore, recent decades of human history — whether economic, social or political — cast shadows of doubt on the validity of such teleology. Yet there is no scientific way to validate whether or not the processes of evolution, since they exist if at all only in the teleological sense, are actually improving man's lot in the universe.

Inherent in the tension between mechanists and those believing in teleology is the persistent realization among biologists that the parts of a living thing are not so nearly *material structures* as they are *functional processes.* As Harris says, "We must at once acknowledge that the distinguishable factors which contribute to (a living organism's) organized make-up are utterly inseparable, are completely interfused and mutually constitutive in a thoroughly inextricable manner."[274]

While a biologist can distinguish and give separate attention to these factors, *he may forget their indissoluble interdependence.* If that happens, he is liable to gross misinterpretation of experimental findings.[275] Despite the continuity that exists within living organisms, the properties of the inorganic and the living are so divergent that they must be separately recognized.

In more practical and less philosophic terms, the presuppositions that lulled everyone involved in the Science Textbook Struggle to sleep in 1970 were rather simple. Some felt that everything could be totally explained by physics and chemistry. Others felt that something else was needed.

The Darwinians, combining the mechanism of Evolution I with the teleology of Evolution II, ignored the teleology of the Godly. Conversely, the Godly overlooked the mechanistic elements of Creation I and became exclusively entranced with the teleology of Creation II.

Let's consider these two types of error and their ramifications, First, the Darwinians evidently ignored the widespread conviction, religious or otherwise, that much more than a mechanistic explanation is required to explain the existence of the universe, life, and man. More than one hundred years prior to Darwin, the great English satirist, Jonathan Swift (who incidentally was most critical of religious activities) had made the famous saying:[276]

> That the universe was formed by a fortuitous concourse of atoms, I will no more believe than that the accidental jumbling of the alphabet would fall into a most ingenious treatise of philosophy.

That is not a "fundamentalist ranting," as Darwinians are so ready to classify any statement that runs counter to their beliefs. Many non-religious people still agree with Jonathan Swift.

A major front-page story confirming strong public opinion regarding origins appeared in the *Los Angeles Times* late in 1969.[277] Darwinians apparently overlooked its theme — that the public at large had not believed the idea of purely physical and chemical explanations for origins.

It is possible that Darwinians honestly felt that another *Los Angeles Times* article better represented the true position of the Godly — that the evolution-creation battle was "an indictment of Bible scholarship."[278] If that article were accurate, then it was only a matter of time until the Godly would either *wake* up or *give* up.

Yet, as Professor Viner of Princeton has pointed out, the Greeks had even appealed, as early as the 7th century B. C., to the order that reigns throughout the universe as evidence of design and therefore the existence of the gods when arguing against the doctrine that the universe was a product of chance.[279] So, the argument from design for demonstrating the involvement of a Creator was indeed deep-seated and long-lived. Perhaps the Darwinians should have paid more attention to the magnitude of that conviction before drifting off in slumber.

But in addition to the obviously religious cosmologies that the Darwinians overlooked as they fell to sleep in 1970, there were also other groups — none of them religious in character — that had not been convinced by the mechanism of Darwinian argument.

For example, the "meaninglessness" of scientific mechanism which people like Theodore Roszak decry is one form of skepticism.[280]

> I have another monster in mind that troubles me as much as all the others — one who is nobody's child but the scientist's own and whose taming is no political task. I mean an invisible demon who works by subtle poison, not upon the flesh and bone, but upon the spirit. I refer to the monster of meaninglessness. The psychic malaise. The existential void where modern man searches in vain for his soul.

Likewise, those whom Holton has labeled as "New Dionysians" were ignored by Darwinians in 1970. And a third possibility is that the Darwinians may have placed their hope in the relatively rapid emergence of such a compromise as Rene Dubos proposes — a *scientific theology* evolving from an ecological view of the earth:[281]

> From the beginning of time and all over the world, man's relationship to nature has transcended the simple direct experience of objective reality. Primitive people are inclined to endow creatures, places, and even objects with mysterious powers; they see gods or goddesses everywhere. Eventually, man came to believe that the appearances of reality were the local or specialized expressions of a universal force; from belief in gods he moved up to belief in God. Both polytheism and monotheism are losing their ancient power in the modern world, and for this reason it is commonly assumed that the present age is irreligious. But we may instead be moving to a higher level of religion. Science is at present evolving from the description of concrete objects and events to the study of

> relationships as observed in complex systems. We may be about to recapture an experience of harmony, an intimation of the divine, from our scientific knowledge of the processes through which the earth became prepared for human life, and of the mechanisms through which man relates to the universe as a whole. A truly ecological view of the world has religious overtones.

Some may ask, "Well, what about *theistic evolution* — the idea that God, as a Creator, utilized evolutionary processes to bring everything into being? Isn't that an integrative principle that eliminates the need to disagree?"

Much as all of us would like to eliminate conflict and reach agreement, neither the Godly nor the Darwinians accept theistic evolution as an adequate explanation for origins. The Godly are reluctant to do so because it obviously reduces the scope and the intimacy of God's creative involvement in nature. But Darwinians aren't crazy about a compromise like theistic evolution that lets God have a piece of the action either.

Harvard zoologist Ernst Mayr, undoubtedly speaking for the majority of Darwinians, discussed the nature of what he called "the Darwinian revolution" in an address to the 1971 meeting of the AAAS.[282] According to Mayr, the main reason why Darwinism has made such slow progress from its introduction in 1859 is that one entire *weltanschauung* has to be replaced by a radically different one. In particular, he pinpointed three existing "metascientific credos" that must be overthrown in order to accept Darwinism. The first and primary of these credos is that of a Creator, and Mayr was adamant about a *total rejection of creationism.* In his eyes at least, there is no room for a compromise known as theistic evolution.

Not even Mayr would be surprised to learn that Darwinism has not been a successful revolution. The "metascientific credos" have not yet been overthrown. As one example, the Commencement Oration to the graduates of Northwestern College in 1973 was entitled, "Evolution or Creation."[283] This address by David A. Kriehn quotes an evolutionary myth about how life supposedly began on earth (which is carried in California textbooks — see Exhibit 66). Kriehn called this story "evolution pollution" and expressed shock that it was used in public schools.

In fairness and in contrast to Mayr's dogmatic position, it must be recalled that Darwin himself was concerned about possible "God-versus-Darwin" conflict. Almost in an apologetic plea he wrote in his *The Origin of Species*:[284]

> I see no good reason why the views given in this volume should shock the religious feelings of anyone...A celebrated author and divine has written to me that "he has gradually learnt to see that it is just as noble a conception of the Deity to believe that He created a few original forms capable of self-development into other and needful forms, as to believe that He required a fresh act of creation to supply the voids caused by the action of His laws."

Yet Darwin, for all his scientific brilliance, was exceedingly naive about the inevitable teleological — not scientific — conflict that would arise from his book. Hardly a decade had passed before a "Doctrine of Descent" based on Darwin's work was published by Oscar Schmidt, a professor in the University of Strasburg. In the preface to his book, Schmidt splits the Godly and the Darwinians into two mutually exclusive camps with these words:[285]

> No sphere of thought agitates the educated classes of our day so profoundly as the doctrine of descent. On both subjects the cry is, "Avow your colours!" We have, therefore, endeavoured to define our standpoint sharply in the introduction, and to preserve it rigidly throughout the work. This is, indeed, a case in which, as Theodor Fechner has recently said, a definite decision has to be made between two fundamental alternatives.

With this absolutist premise, dated 18 October 1873, Schmidt then sets out the rigid dogma that he promised throughout his book. Here are three typical examples of his hard-line doctrinal position:

> The parallelism of the palaeontological and the systematic series is either a miracle, or it may be accounted for by the doctrine of Descent. There is no other alternative.[286]

> (The lowest living beings') origin from inorganic matter, as we have set forth above, is a postulate of sound human understanding. To this beginning we are led, not, as the opponents of the doctrine of Descent are wont to say, by a dogmatic after-philosophy, but by the unprejudiced consideration and computation of the facts of individual development.[287]

> Such remains as we have of this oldest man known to us, display a high grade of development, and certainly belong to the period at which man had already found in language the implement wherewith gradually to free himself from the dross of his lowly origin. Whether the primitive man be found or not, *his origin is certain.* (Emphasis added)[288]

The irony of Schmidt's book is not its message but its *date.* The dogmatic and almost theological fervor with which he seeks to persuade his readers that man has descended from lowly forms of life via a wholly unexplainable continuum of changes bespeaks a conviction stronger than scientific data (including those accumulated in the 80 years since he formed his doctrine) would justify.

He rightly calls it a "doctrine." Could it be that the 14 years between publication of Darwin's theory and Schmidt's doctrine was sufficient to produce such an elaborate doctrine, or was the doctrine "waiting in the wings" for some work like Darwin's to come along and give it scientific respectability?

At the very least, it should be evident that the teleological tenets of Darwinism (e.g., that the origin of life can be mechanistically explained, that man is simply a natural inevitability, or that all of nature can be accounted for in totally naturalistic terms) did not spring from scientific observations of nature. Those tenets *preceded* such observations. Beliefs — *a priori* beliefs — form the framework upon which scientific data were later placed.

While it is obvious that some scientists (including Kepler, Newton, Faraday as well as those of today) who are popularly called *creationists* have had a presupposition of the intervention of a personal and supernatural Creator in nature, it has not always been obvious that Darwinians also start with an equally consequential presupposition or belief.

In that sense, Evolution II and Creation II are teleological equivalents. Neither one is scientific, in the sense of being demonstrable by scientific means.

Therefore, it is not difficult to realize why both the Godly and the Darwinians went to sleep in 1970. They were both tangled in tenacious teleology.

SUBTLE SHIFT

If we're really *alert*, we spend *only two-thirds of our lives awake!* We are asleep the other third. While we may not say so, we tend to believe that the important events in our lives all occur while we're awake. In other words, we often act as though the world stands still while we sleep.

But of course, that's not true. And it wasn't true for the Godly and the Darwinians as they slept during 1970, 1971 and 1972. The inexorable process of textbook procurement went grinding on. Day by day, the scenario was changing.

One major change that occurred during this time was an enlargement of the confrontation. It no longer was limited to California. Other states, duly alerted by massive news coverage across the nation, began to wonder whether they should also revise their science textbooks. Their concern probably was twofold.

First, since California purchases approximately ten percent of all textbooks in the nation, their influence would undoubtedly swing the content of all science textbooks in their direction. No other state could possibly match California's impact, so other states would automatically inherit California's version.

Secondly, since California was not only the most populous state but also a leader in both educational innovation and scientific advancement in the nation, other states wanted to re-examine their own position relative to California's.

Nevada, California's immediate neighbor to the east, was the first to contact me regarding their science curriculum. I met with Dr. Kenny C. Guinn and his cabinet in the fall of 1970. Guinn, Superintendent of Clark County School District that embraces Las Vegas and is Nevada's largest school district (they do not set criteria at the state level in Nevada), requested that I present the approach we had taken in California so that they might possibly adopt the same.

Texas was the next state to ask my counsel on science curriculum regarding origins. Other states continued to examine the issue of origins until the practice became fairly widespread by 1972. At one point, over a dozen states were revising curricula and educational media to reflect the concern first exhibited by California in 1969.

In addition to out-of-state interest in the Science Textbook Struggle, there were major changes occurring *within* California during the slumber of the Godly and the Darwinians. Most significantly, the changes involved neither science nor education. Instead, they were *political!*

Political Promotion

The California State Board of Education, as described earlier, is comprised of gubernatorial appointees. Therefore, Board members tend to reflect the political viewpoint of the Governor who appoints them. Normal Board appointments are for four years. The Board's collective political position obviously changes much slower than the politics of the Governor's office. In any given year (barring deaths or resignations), the Governor can appoint a maximum of three Board members. So it takes four years for any Governor to totally convert the State Board of Education to his political persuasion.

Ronald Reagan was first elected Governor of California in 1966. Three years later, in November 1969 when the State Board of Education unanimously adopted my two paragraphs into the *Science Framework*, the Board was not yet an all-Reagan Board.

Three members — Dorman L. Commons, Mrs. Seymour Mathiesen, and Miguel Montes — had been named to the Board by Reagan's predecessor, Democrat Edmund G. (Pat) Brown. While many citizens might believe that education should be free from politics, it simply is not. Voters influence education, for example, through their votes regarding taxation. The political process also affords citizens at least a partial voice in determining the major thrust of educational philosophy via elected representatives.

In addition to the influence on education that voters exercise by their selection of a governor (and thereby his appointments to the State Board of Education), there is a more direct means available to California citizens for expressing their educational desires — the statewide election of a Superintendent of Public Instruction. It is possible for this person to receive more votes than the Governor!

In 1962, Max Rafferty was elected Superintendent of Public Instruction. He was overwhelmingly reelected in 1966 in the same election that put Ronald Reagan into office for the first time. Without any question, Rafferty was a political conservative. In fact, his political views were generally considered to be more conservative than those of Reagan.

Among the duties assigned by law to the office of Superintendent of Public Instruction is that of Executive Secretary to the State Board of Education. Although the Superintendent theoretically has no influence on the Governor's selection of appointees to the State Board of Education, it is interesting to note that Rafferty's personal physician, Thomas G. Harward who was Vice President of the State Board when the *Science Framework* was modified and adopted in 1969, was appointed to the State Board in 1967 by Governor Reagan.

Rafferty's statewide elections in 1962 and 1966 (together with the election of the conservative Reagan in 1966) evidently persuaded Rafferty that the political mood in California was receptive to his conservative views. California's Republican Senator Thomas H. Kuchel, the minority whip in the United States Senate, was much more liberal than Rafferty.

So, in 1968, Rafferty decided to challenge Kuchel for his seat in the Senate. He was successful in wresting the Republican nomination away from Kuchel for the Senate in a primary election during which Kuchel only polled 46.9%. However, even though he received 3,329,148 votes, Rafferty was defeated by Alan Cranston, the state controller, in the 1968 general election.

Though he lost this Senatorial election, Rafferty still retained his position as Superintendent of Public Instruction and was present at the November 1969 meeting when the State Board modified and approved the *Science Framework*.

The No. 2 man under Rafferty in the Department of Education was Wilson Riles. Having served as Deputy Superintendent of Public Instruction since 1965, Riles had developed some ideas of his own about education that conflicted with Rafferty's.

Even though Rafferty had achieved considerable public exposure by running for statewide office three times in six years, Riles felt that he had a good opportunity in 1970 to challenge Rafferty for the post of Superintendent of Public Instruction. He sensed that the mood in California had moved somewhat from the conservatism that had propelled Rafferty into public view in 1962, 1966 and 1968. Furthermore, Riles was black. By 1970, being black had converted politically from a detriment to an asset.

Although the office of Superintendent of Public Instruction is ostensibly non-partisan, the battle between Rafferty and Riles in 1970 became as heated and controversial as any partisan contest. Riles campaigned on the issue that Rafferty's conservatism was out of tune with the demands for social action which was at that time being reinforced by campus violence and general societal unrest.

Undoubtedly, Rafferty believed that his two previous elections plus his widely publicized yet unsuccessful candidacy for the United States Senate would more than off-set any impact that Riles, a political unknown, could muster against him — especially since Riles was his subordinate. Although Rafferty received 66% of the vote in the 1970 primary, he was forced into a runoff with Riles in the 1970 general election.

In a most remarkable turnaround, Riles amassed 3,254,365 votes (54.1%) and soundly trounced Rafferty in the general election. (Governor Reagan was reelected with 3,439,664 votes in the same election.)

Obviously, this stunning upset by Riles was to ultimately impact the Science Textbook Struggle. Rafferty had not personally led (or even supported) a campaign for objectivity concerning origins in science textbooks (despite news media accusations of doing so). However, his defeat and replacement by Riles was a portent of change for the future in this issue.

That the official position of the Department of Education over which Riles now presided would shift was certain. Only the degree of shift remained unknown.

On 25 May 1971, California's Attorney General Evelle J. Younger appointed me to a Task Force on the Judiciary Process in Criminal Justice. This appointment necessitated frequent trips to Sacramento, and on one such occasion, I arranged to meet privately with Wilson Riles in his office to discuss my position on the Science Textbook Struggle.

Riles is both charming and impressive in person. Tall (6'4"), lithe and handsome, he possesses that charisma required of leaders. Yet, his low-key, soft-spoken manner belies the obviously political sense that attracted for him more votes than any other black man in America.

During our first face-to-face meeting in his Sacramento office on 1 June 1971, Riles was courteous but noncommittal. He calmly puffed on his pipe, nodding knowingly as I recounted the 13 November 1969 meeting of the State Board of Education in Los Angeles. He acknowledged that he had been briefed in depth concerning the Board's action, although he had not been present at the meeting.

Obviously, I was probing for his personal convictions about the importance of the Board's decision concerning origins. Just as obviously, he was determined to withhold his views. For about one hour, we discussed the issue of origins and the requirement for more than one position to be presented in science textbooks.

At one point, I said, "I am convinced that this requirement could result in the greatest step forward in science education in 100 years." Riles listened — but his attitude expressed toleration rather than endorsement.

In the months just previous to our meeting, I had been in contact with several science textbook publishers, furnishing them with details of what I had meant rather than what the press reported my position to be. I felt obliged to share this activity with Riles. Again, there was neither surprise, rebuke nor endorsement.

As I left Riles' office, I felt relieved that I had been able to communicate my true views on the issue. However, I lacked any sense of conviction that Riles was going to support the State Board's position. His political promotion had apparently freed him from any obligation to endorse any past activities in state education. He may have even taken his sweeping electoral vote as a mandate to *undo* the past!

Publishers - Perplexed or Predisposed?

My first contact with a science textbook publisher occurred in a left-handed and ironic way. Due to the wide exposure in the press of the State Board's decision to adopt my words, I received several invitations to share my views about the role of science in origins on college and university campuses.

One such invitation to lecture was at California State University in Northridge, California. Handbills announcing my lecture on 26 February 1970 were apparently widely distributed on the campus.

Carl F. Loeper, educational representative in the Los Angeles area for The Macmillan Company, was making a business call at the University a few days prior to my visit. He was discussing Macmillan's latest high school geography textbooks in the University's education school. Needing desperately to make a few notes at the conclusion of his discussion and having no paper handy, he looked for some paper as he left the office. On the hallway floor, he saw a blank yellow piece of paper. Picking it up, he hastily sketched his notes on it and then turned it over. It was one of the handbills announcing my scheduled lecture.

Loeper had already become aware of both the issue of origins and of my name through press reports. His interest was piqued to return to the University to hear my lecture. Thus we met on 26 February 1970.

There was a flurry of questions at the conclusion of my lecture that day. Among those who stood and asked a question was Loeper. "Could you tell me whether or not the State Board of Education will incorporate any of your ideas in the specific curriculum criteria that publishers will receive?" he asked.

Flushed with the victory of having my three criteria unanimously adopted by the Curriculum Commission and its Science Committee only days earlier (over the unanimous resolution of the Smith/Gerard committee to reject them), I gave a forceful and emphatic answer that such criteria were already a *fait accompli!*

My cocksure response provoked Loeper to remain behind until everyone had left the lecture hall. I then shared with him my best recollection of what the criteria said. This information was forwarded, ultimately reaching Macmillan executives in Texas and New York. Within months, I flew to New York City and met with top management personnel of the School Division of Macmillan to explain the scientific significance of the State Board's action.

Some later contacts with publishers came as a result of referral from members of the State Board of Education. For example, American Book Company (a division of Litton Industries with whom I had previously been an executive) approached me in the summer of 1971 at the suggestion of John R. Ford, who by this time was Vice President of the State Board.

The extensive discussions that I had with publishing executives during this period convinced me that the State Board's decision had presented publishers with a major dilemma. As described earlier, publishers are not necessarily in phase with California's textbook acquisition process. Even though 10% of all textbooks are sold in California, publishers cannot focus exclusively on the California market. Therefore, there were science textbooks in all stages of development from concept to completion at the time the State Board issued its criteria requiring more than one account of how everything began.

Publishers appeared to be caught between two opposing predicaments. As I explained my real intent and contrasted it with press reports, publishers seemed to be *confused* or *perplexed.* While they appeared to want to believe what I was saying about my own philosophy, yet the vast coverage in the news media had indicated something quite different.

Whom were they to believe? How could one person be so overwhelmingly misunderstood by the press? Further, I had no official relationship to the State Board of Education. I was simply a private citizen. It probably seemed to some publishers that the Board might have used me only as a catalyst — without actually adopting my ideas. If the publishers went my direction, they might be taking off on a wild goose chase. Further, since the press had persistently painted a preposterous picture of the issue, publishers were naturally frightened concerning the rest of the American market.

The other predicament was that most publishers already had their own versions printed and ready for sale. They were *predisposed.* Even revising page numbers was too expensive an alternative for them to consider. The pressure for profits obviously was overriding any philosophic truth or scientific accuracy.

While textbook publishers are particularly sensitive about the possibility of anti-trust action against them, there was still a powerful yet unspoken possibility that if the publishers would collectively ignore the apparently minor modification that the State Board had made, there would be no reasonable way for the Board to completely disavow every single publisher's submittal.

Of course, the hope of maintaining a solid front among all publishers is always clouded by the fear that one expedient member of that community may decide to bolt and go after the business by meeting the most ridiculous of demands.

As the textbook submittal date neared, publishers saw only three possible avenues to pursue relative to my criteria that had been adopted by the State Board. First, they could completely ignore the criteria as being an insignificant element in the total textbook acquisition process. Secondly, they could patch or repair their existing textbooks. Third, they could rewrite their books to comply with the issue of origins.

The publisher's game of "wait and see" continued well into 1971. Whether the publishers were perplexed by the prejudice of the press or predisposed to previously prepared editions of their textbooks, they gave every sign of being what my father used to call "mugwumps" — people with their mug on one side of the fence and their wump on the other.

Need You Be Reminded?

Starting with my meeting in New York City with Macmillan executives, I began to see a clearer picture of the dilemma that publishers were facing concerning the issue of origins in science textbooks. If publishers were going to be moved to take any action, the motivation would have to come from those charged with *official* responsibility — the State Board of Education. To state it in the converse, I was positive that no publisher would take action strictly on what I — a private citizen — had to say to them.

To compound the reticence of publishers to "move off dead center," the news media kept up their same old tired melody — that any comparison of theories about origins was part of a scurrilous plot against Darwin.

One nationwide release, datelined in New York, was typical of this theme. It carried the headline, "Minority Still Hoping to Discredit Darwin." Consider this curious and inaccurate statement from the release:[289]

> Creationists frequently argue that their ideas should be taught alongside evolution in public schools. Most of their efforts have been in vain, but last year the California regents granted permission for inclusion of such views, but at the behest of the (Bible-Science) association the plan has not been implemented, however.

Obviously, no California *regents* (whoever was meant by this term) took action on creation, no *permission* was granted for inclusion of creation, and no association *persuaded* anyone to not implement the non-existing plan.

So much again for the accuracy of the news. Rather than see my investment of time and money in the issue dissipated through publishers ignoring the criteria, I decided to take the bull by the horns. I called John Ford and requested to meet him privately on 11 May 1971 in his San Diego office.

Ford has a large and successful medical practice. Due to his heavy schedule, he arranged to see me early in the morning. My purpose was not only to brief him on my contact with publishers, but also to warn him that there was a high probability that publishers would ignore both the *Science Framework* addition and the three criteria that had been adopted to carry out the idea of multiple discussions about origins in science textbooks. Ford is a man of quick and incisive mind, and he easily understood my concern. In fact, he had already reached many of my conclusions independently.

During our conversation, he said that he intended to issue a policy statement to the Curriculum Commission to emphasize the importance of the subject of origins. I was encouraged by Ford's forthright attitude of not being cowed by press reports or any other voice of public opinion. Ford also seemed pleased that I would be seeing Wilson Riles within the next few days on the issue.

On 14 June 1971, I wrote to Ford, reporting on my visit with Riles in Sacramento. In that letter, I also mentioned to him that I had flown to Fresno, California on 2 June to visit for a few hours with Glenn Leslie, the Curriculum Commission's Science Committee chairman.

In my meeting with Leslie, we discussed how inflammatory the two words, "evolution" and "creation" had become. It occurred to me during our discussion that we might be able to substitute *chance* for "evolution" and *design* for "creation" to dissipate the hysteria that evolution and creation had raised.

Leslie was enthused about this suggestion. He confided to me that he had already warned all the textbook evaluators in the state that the position taken by the State Board was unanimous and that particular attention should be paid to evaluating science textbooks on the subject of origins.

I concluded my letter of 14 June 1971 to Ford by reminding him of the policy statement he said he was going to prepare for the Curriculum Commission. I proposed that if he had not yet drafted such a policy statement, a suggested version for him was enclosed in the letter. My proposed policy statement for Ford mentioned for the first time the terms, "chance" and "design" as substitutes for evolution and creation.

I did not learn of Ford's response to my suggested policy statement for nearly three months — until our family was driving back from a summer vacation in Washington State. I went to a phone booth while the gas tank was being filled in Fresno and called Glenn Leslie. Among the things that Leslie told me was that he had just received in the mail a copy of a statement concerning origins that John Ford had made at the July 1971 meeting of the State Board of Education in San Francisco. Leslie said it was a very strong statement regarding the necessity of publishers to specifically address the issue of origins in science textbooks.

Upon returning home, I wrote to Ford and asked if I could receive a copy of his statement (see Exhibit 14). *Ford had released verbatim the statement that I had written for him.* If it did nothing else, this statement certainly removed any ambiguity about the State Board's position for the publishers.

However, a new danger now arose. The textbook adoption process had been grinding on with the calendar. By now, it was obviously getting to be too late for publishers to do anything significant about the issue, since they had to submit text materials for evaluation in September 1971.

In a letter dated 23 September 1971, I wrote Ford that his July policy statement had essentially slammed the door on *all* publishers. No publisher would be able to meet the State Board's criteria because "they had eaten up the clock." I could foresee the news media headlines saying, "The State Board's pet theory is so far out that not one publisher chose to even *acknowledge* it. Is the State Board going to have to write its *own* science textbooks?" It was certainly a disturbing possibility.

In this letter, I outlined and analyzed three options that appeared possible for the State Board at that point in time:

1. The Board could swallow their pride and accept the best of the books even though creation is not even acknowledged.

2. The Board could extend the submittal date for some specified time period, perhaps 6-8 months, so that the publishers could prepare acceptable material.

3. The Board could engage some persons or groups to prepare acceptable material on creation to be integrated in the textbook selection resulting from the first option. This action would obviously raise the price of the books, since no publisher would have had typeset this particular material previously and would have to repaginate and reset all their textbooks.

I knew that from Ford's personal position, none of these options were desirable. In discussing each of them, I proposed that the first one would be politically disastrous because it would loose the news media like hound dogs in harassment of the Board on any controversial issue in the future. I felt that the second option, while reasonable, carried a potential danger that after six or eight months, there still might not be any publisher who had put anything together on the subject. That would put the Board back in the spot they were already in — only with a *double* slap in the face. The third option, though involving increased cost of textbooks, seemed to be the least offensive.

If the Board could have available for public display material that they found acceptable, they could offset the impact of the news media's campaign to side with all the publishers. However, the key to the success of the third option would be its *timing*. Such materials would have to be available while the textbooks were still being evaluated.

As it turned out, the Board elected none of the three options at that time. Instead, they delayed the confrontation until the books were brought to them for final adoption.

Regardless of when the confrontation was to occur, however, the State Board had at least reminded the publishing world that they were serious in their intent to have more than one story concerning origins in science textbooks.

ASTONISHING APPOINTMENT

I had almost forgotten that there was a Science Textbook Struggle. Many months had passed since I last gave it a thought. Perhaps I, too, had gone to sleep like the Godly and the Darwinians. But it was to be a short-lived siesta.

I was jarred out of lethargy on our 21st wedding anniversary. Phyllis and I celebrated this event on 14 April 1972 as guests of three couples at a Japanese restaurant in Encino, in the San Fernando Valley area of Los Angeles. Actually, it was a "double-purpose" evening since the couples — Ron and Beverly Ketchum, Bob and Della Schaeffer, and Jack and Doris Rediger — were putting the finishing touches on a vacation the six of them were taking in Europe. Because of our previous European travels, I was assisting them in planning an itinerary for their trip.

We drove to Rediger's home following dinner to work out final travel plans. The telephone rang, and it was for me. John Ford, Vice President of the State Board of Education, was on the line.

"Have you heard any news on the television or radio this evening yet?" he asked.

"No, I haven't."

"Well, you've just been appointed a charter Commissioner on the Curriculum Development Commission!" Ford sounded excited.

"You don't say!"

My reaction could probably be described best as stunned. The appointment had come completely by surprise, and my low-key response may have disappointed Ford. He went on to explain how it had happened.

"As you might imagine, most of the Commissioners must be selected from among teachers and school administrators," he explained. "However, there was one position open for a generalist or interdisciplinarian. There were 43 candidates for that one position, and you were selected!"

I was beginning to understand why Ford was excited about my appointment. Expressing my gratitude to him for his support in nominating me, I voiced hope that I could do an acceptable job in the position.

Upon returning to the living room, I briefly announced the news to Phyllis and our friends. We immediately returned to the discussion of European travel plans for the rest of the evening. There was no way at that time that I could have foreseen the impact of John Ford's telephone call on my life...

Inaugurated in Absentia

As a Commissioner, I started right off with a problem. I was unable to attend my own inauguration! Wilson Riles, in his role as Executive Secretary to the State Board of Education, sent the official letter of appointment and congratulations dated 17 April 1972. His letter also contained the dates of 25-26 April as the inaugural meeting of the Commission.

Checking my calendar, I had a sinking feeling...

Several months prior to my appointment, I had accepted an invitation by the Technischen Uberwachungs-Verein (an agency of the West German government) to participate in a colloquium to be held 24-27 April 1972 in Cologne. I was one of four people with backgrounds in the American space program that were invited to share in this effort to transfer space technology to the civil sector of the German economy.

There was no way that I could cancel my involvement at this late date. Further, I was scheduled to teach a Tustin Institute of Technology course in Cologne following the colloquium.

I was going to miss more than just my personal inauguration. A whole new Commission was going to be empowered, The California State Legislature passed a law in 1972 (see Exhibit 15) disbanding the previous Curriculum Commission that had been involved in preparation of the *Science Framework* and its associated criteria. Five Commissioners were carried over from the old Curriculum Commission to the succeeding Curriculum Development and Supplementary Materials Commission to which I was appointed. The rest of us were brand new.

The inauguration of the new Commission occurred on the evening of 25 April 1972 in the Department of Education Building in Sacramento. I was the only Commissioner absent that evening.

Wilson Riles, as Superintendent of Public Instruction, was present both to congratulate the Commission and to deliver an introductory statement. Among his comments, he specifically focused on the importance of our Commission:

> I came tonight to add my word of congratulations to you on your appointment to this group, to impress upon you, if you do not already realize it, the importance of the Commission.

> I would like to review in very brief terms why this is a new Commission. We have had a proliferation of commissions in California for a long time that have grown over the years. I went to the State Board early in 1971 and asked if the Board would support a program to consolidate some of the commissions and redefine their functions and eliminate some that were no longer needed. The Board studied the proposal carefully, endorsed it, and AB-2800 was the vehicle which moved through the Legislature and was signed by the Governor.

Riles was followed by Newton Steward, President of the State Board of Education. Steward had succeeded Howard Day, who had been President of the Board during the adoption of the *Science Framework* and the criteria. Steward's greetings included the statement:

> I was impressed by the number of vitas received on people who were willing to serve on this Commission and the qualifications and capabilities of these people. Your selection as members of this Commission was a long, hard task but we of the Board are satisfied that we have chosen the best. We are anticipating great things from you because this is a time of change.

Virla Krotz, a member of the State Board of Education who had been present when my words were adopted in November 1969, was named by the State Board to be their liaison member to our new Commission. After Board President Steward had finished his greeting, she delivered the inaugural Charge to the Commission. It was a lengthy statement couched in legal language. With the delivery of the Charge, the Commission was officially and duly constituted.

Only one other thing happened on that evening ... and it was prophetic of the months ahead for the Commission. Elaine Stowe, an educational specialist in the State Department of Education, briefed the Commission on the issue of *origins in science textbooks*. She particularly emphasized that the Commission should read John Ford's statement of 8 July 1971, calling for science textbooks to discuss at least two major contrasting theories for the origins of the universe and man.

She apparently touched some raw nerves. Lively discussion exploded. There were questions as to whether the dual theories would have to be in every book. Some of the new Commissioners felt that the Board should give further direction as to what they wanted.

However, President Steward replied that since the Board was now composed of different members than had passed the *Science Framework* in 1969, he could not say what their desire would be. Instead, he directed the Commission to continue the same trend that the previous Commission had taken. It was obvious that origins would be the primary issue in the days ahead.

Would you believe that I slept through Wilson Riles' stirring charge to the newly-formed Curriculum Development and Supplementary Materials Commission? Even Newton Steward's words of wisdom failed to raise me from slumber!

If you consider, however, that I was 6,000 miles away in Germany where it was pre-dawn, maybe I can be forgiven of otherwise boorish behavior. More than likely, I was savoring the memories of a wonderful evening spent on the River Rhine only a few hours earlier.

Because of an eight-hour time differential between Cologne and Sacramento, the inauguration of the Commission did not occur until the wee hours of the morning of 26 April 1972 in Germany. However, the 25th had been a very exciting day for me.

Although I had previously taught for the University of Southern California in Germany, this visit to Cologne was much more pleasurable. The German officials really rolled out the red carpet...receptions, press conferences, tours, banquets — the works. 25 April 1972 was not a day that I would soon forget.

The Institut Fur Unfallforschung (Institute for Accident Research) Colloquium-1972 took place in the Congress Center of the Cologne Fair located on the east or right bank of the Rhine, directly across from the magnificent twin-spired cathedral, one of the world's finest examples of Gothic art.

I was the first of the four American guest lecturers to deliver a technical address to the Colloquium. Although I studied German in college, I am not fluent in the language. Nonetheless, I wanted desperately to bridge the cultural gap between the audience of 300-400 German professionals and myself. Therefore, I made a few introductory remarks in German before delivering my address in English.

German audiences are very businesslike and impassive. There was not one smile, nod or sign of acknowledgement among the attendees as I shifted from German to English. I was not to learn the true reaction of my feeble attempts to speak German until later that evening.

During a delightful cruise arranged in our honor on the Rhine aboard the double-decked motorship "Rheingold", we had dinner. Only then did several of my dinner partners remark privately that they appreciated my gesture of trying to speak their language.

Another unforgettable experience occurred while we cruised on the Rhine after dark that evening. Cologne, the fourth largest city in Germany, straddles a gentle bend in the Rhine. Lining the river are many burgher's houses, some dating to the 14th century. In the summer months, it is customary to floodlight these unique buildings so that they may be seen from the hundreds of boats that ply the Rhine. Since our Colloquium was held too early in the year to see the lighted houses, arrangements were made with city officials to turn on the lights just for us!

So you can see that many memories of that day danced through my head as I went to sleep at a late hour. The beautiful weather, the music and jovial laughter during dinner aboard ship, the inspiring view of the cathedral from the river, and the excellent food all combined to offset my absence from my inauguration in Sacramento.

The Commission meeting that evening adjourned about 10 PM (about the time I was awaking in Germany). However, there was evidently some feverish activity among some Commissioners in hotel rooms during the night.

When the Commission reconvened at 9 AM the next morning, Commissioner Junji Kumamoto had an elaborate motion prepared.

Picking up where they had left off the night before, the first item of business for the Commission was how to comply with my two paragraphs and their three associated criteria. Undoubtedly hoping to scuttle the entire issue with one fell swoop, Kumamoto introduced the following resolution:

> WHEREAS there is some confusion on how the requirement for a necessity of having more than one theory for the origin of the universe relates to the doctrine of the separation of Church and State;
>
> WHEREAS there is even more confusion as to why science writers are demanded to do something that possibly theologians are better equipped to do; and
>
> WHEREAS there appears to be a need to provide a forum where the science authors are afforded some recourse to adversary proceedings,
>
> BE IT RESOLVED that this Commission request the Board of Education to clarify this issue by passing a formal resolution on the matter.

As might be expected, this resolution provoked some extended and heated discussion. Obviously, there was a schedule that had been followed for years that would have to be reset if the Commission had to wait for a resolution by the State Board.

Further, President Steward who had stayed for the second day of the Commission's meeting pointed out that the adoption process on science textbooks had gone a considerable distance already. If the Commission were to approve such a resolution, the science textbooks could not be adopted under the existing schedule.

Ruth Howard, one of the Commissioners reappointed from the previous Commission, called attention to the action taken at the final meeting of the old Curriculum Commission of which she had been a member. That Commission had recommended, prior to its dissolution, that one of the recommendations that should be attached to the science textbook adoption read as follows:

> Whatever books are recommended must conform or must be adjusted to conform to the requirements of the Board regarding theories of life origins in the appropriate subject matter area.

Ellsworth Chunn, Chief of the Bureau of Textbooks in Sacramento, further pointed out that there were legal ramifications in upsetting what the Board and the previous Commission had already started. He recalled the fact that the Board had the legal authority to change, alter, omit, or introduce into textbooks any changes that they might require. Chunn's arguments were immediately followed by a roll call vote. Kumamoto's efforts to kill the issue failed as the vote went 9 to 4 against him.

Right on the heels of this defeat of Kumamoto's resolution, another motion was introduced, seconded and carried that the new Curriculum Development Commission accept the recommendations of the former Commission and move ahead on the same schedule of adoption as previously instituted. Kumamoto was thus prevented from blocking the inevitable confrontation over the issue of origins. The next meeting of the Commission in May was thus destined to awaken all those who had been asleep...

Meanwhile, I remained oblivious of all this mad maneuvering in Sacramento. (Distance can be wonderful!) At the moment that the Curriculum Commission meeting was being called to order on the morning of 26 April 1972 and the big power play to block any discussion of origins in science textbooks was being executed, I was just arriving at the Cologne Rathaus (City Hall) to attend a reception with the Burgermeister (mayor) in honor of the four of us who had come from the United States to participate in the Colloquium.

The reception, that included a buffet supper, was held in a unique place — the sub-basement of the Rathaus. One of the rare blessings of the Allied bombing of Cologne during World War II, in which about 90% of the central city was reduced to rubble, was the discovery of Roman ruins deep below the foundations of destroyed buildings.

When the foundation for the new Rathaus was being laid in 1953, the ruins of a Roman Praetorium, dating from the 1st century, were uncovered. These remains of the Roman governor's palace have been excavated, cleaned and transformed into a beautiful gallery of Roman relics, remains of structures, and other mementoes of ancient civilization.

It was in this rare and novel setting that the Burgermeister, Herr Jacobs, met with us to present each of us a beautiful autographed book of colored photographs of Cologne. Jerry Lederer, NASA's first Director of Safety, is shown in Figure 8 responding on behalf of the American contingent to Burgermeister Jacobs' greeting in the Praetorium reception. Some of the Roman artifacts can be seen in the background.

Following this reception, we were given a private, guided tour of the Cologne Stadt-Museum by its director, Dr. Gunther Albrecht. For anyone with an appreciation of medieval history, this two-hour exposure was an education in itself. As we left the museum late that evening, the Curriculum Development and Supplementary Materials Commission was just adjourning in Sacramento.

While absence is supposed to "make the heart grow fonder," my absence at the initial meeting of the Commission failed to do that. Instead, my German experiences later seemed to be compensation for the confrontation that was looming dead ahead...

Booked Books

Telegrams, letters and phone calls poured in as my appointment to the Curriculum Development Commission was announced in the press. I received many even before I left for Germany. Most of the congratulations were identical messages that were sent to all 14 Commissioners. However, a few of them acknowledged that they recognized my personal role in the issue of origins in science textbooks.

"I wonder what they think of me *now*?"' I mused as I read some of the letters.

Those publishers who had contacted me during 1970-71 undoubtedly viewed my appointment to the Commission with some apprehension. They had ignored my *advice* (that was all that I could offer at the time). Now I was in a position to officially enforce my views.

Several publishers had decided to "patch in" a few references in their science books to Genesis, Moses or Michaelangelo's artwork on creation — evidently to placate the sensitivities of those State Board members whom they believed to be interested in a religious reference. Some even inserted direct quotations from the first two chapters of Genesis. But not one publisher made a serious attempt to address the *scientific* questions about origins.

Although Macmillan, the first publisher to contact me in 1970, engaged me to prepare written material on origins, they ultimately did not make any modification of their textbooks in response to the new criteria.

Figure 8. Burgermeister's Reception in Cologne
Jerry Lederer (facing camera) responds to Burgermeister Jacobs – V. L. Grose in rear

The original Curriculum Commission, now disbanded, had already eliminated Macmillan as a contender in the science adoption (not, however, for non-compliance with the origins criteria) by the time our new Curriculum Development Commission had been appointed in 1972.

Though John Ford in July 1971 had reminded — and even *warned* — all publishers to pay attention to the criteria regarding origins, his tipoff had been ignored.

Did they believe that there was safety in numbers?

Did they consider Ford's warning a hollow threat?

Did they trust the news media to rescue them from non-compliance?

Did they really care about children's minds and the need to truly educate?

We probably will never know the answers to these questions. However, the questions make me reflect on a recent statement by one of the most prominent radical educators in the United States, Jonathan Kozol:[290]

> It's the essence of a free enterprise society to allow free and open competition for businessmen, for doctors, for lawyers, for professionals of every kind. Why then should we be scared of a free and open market of ideas? The public schools, in their present form, have established a monopoly of ideology. There's one consensus viewpoint. It's hammered down through every textbook, no matter what the brand name and the binding. We don't have twenty thousand different school systems in America; we really have one school system. The schools have a monopoly on our imagination at the present time. They sell a single product in a single, homogenized consensus of moderate views — no dissent. When we must deal with people such as Helen Keller or Thoreau — who did dissent — we render them innocuous by a process of intellectual emasculation.

Without doubt, the publishing community opted for the easy route. They simply submitted materials that were already written prior to adoption of the *Science Framework* and its associated criteria.

In fairness, this would seem to be the rational business decision, from a resource allocation viewpoint. After all, since the press had made the State Board's position appear as a *wild religious aberration*, what could a publisher possibly write that would satisfy religious fanatics and still be scientifically acceptable? Yet, can textbooks — even *science* textbooks — be marketed in a strictly business environment? Publishers at least hoped so.

Perhaps they had even become drowsy like everyone else...

Chapter 7

AWAKEN WITH ALARM

"A thing that nobody believes cannot be proved too often." — *George Bernard Shaw*

"Our route this afternoon will take us over the Irish Sea, past the west coast of Scotland where we'll over-fly Ireland near Belfast — proceeding just south of Iceland over southern Greenland and then enter Canada over Baffin Island. We will cross Hudson Bay, then Manitoba and Saskatchewan before entering the United States over Montana..."

The TWA 747 Captain continued his description of our nonstop flight to Los Angeles as we climbed out of London's Heathrow Airport.

"We are scheduled to arrive in Los Angeles at 3:30 PM."

I glanced at my watch. It was 1:10 PM London time. In the next *ten* hours, my watch would be advanced only *two* hours!

"This is going to be a long afternoon," I mused. Pondering further, it seemed incredible that I was flying one-fourth the way around the world in only two hours clock time.

Although I had worked on the design and development of the inertial navigation system that now guided this 747 unerringly home, I was still amazed when we touched down at Los Angeles International Airport within *three minutes* of the estimated time of arrival which the Captain had given us more than ten hours earlier over London. It was like a reverie...

Later, this return flight from the Cologne Colloquium in May 1972 seemed like that last fleeting dream one has just before awaking.

If I had fallen asleep with the Godly and the Darwinians in 1970-1971, I was going to be wide awake shortly after landing in Los Angeles. In fact, the jolt back to reality was rudely sudden.

About the first thing I did upon arriving home was to hurriedly sort through the three-week accumulation of mail. A letter from Wilson Riles was one of the first I spotted. Quickly opening it, I read his announcement for another Curriculum Development Commission meeting in Sacramento the following week!

BACK ON TRACK

While I had missed the inaugural meeting of the Curriculum Development Commission in April 1972, I was even late arriving for the *second* one. All Commission meetings are scheduled for 9 AM. The earliest flight I could get from Los Angeles to Sacramento did not *arrive* until 9 AM. By the time I got a cab from the airport to the El Rancho Motor Hotel where this meeting was being held, it was 9:30 AM.

Embarrassed, I sheepishly walked into the hotel's Sutter Room. The Commission was already in session. I was sure no one recognized me — even though the room was crowded with 150-200 people.

Glancing quickly at the U-shaped table around which the Commission was seated, I spotted a vacant chair on the left side. Trying to appear confident that I knew where I was going, I threaded my way through the crowded room toward the vacant chair. As I approached this table, I noticed that nameplates were in place at every chair.

The empty chair bore my nameplate, much to my relief. So I eased into my chair as unobtrusively as one can under the circumstances. Even so, I felt as though every eye — not only of my fellow Commissioners, but also the publishers, press and public in the audience — was riveted on me.

I nodded and smiled at the apparent Chairman, Charles S. Terrell, Jr., sitting at the center of the table. He winked and nodded in return.

Not having received any information from the Commission, I knew none of the names of other Commissioners. Even the minutes of the first meeting had not reached me. It was tough trying to act as though I was a full-fledged member of this important body.

At that moment, I was yet an outsider.

But that didn't last long. Right in front of me was a thick, black three-ringed notebook with the title of the Commission in large gold block letters. My name was also printed in gold: "Vernon L. Grose, Commissioner."

As the discussion of a calendar for the coming year proceeded in stilted parliamentary manner, I flipped the black notebook open to the minutes of the meeting held while I was in Germany. Scanning swiftly, my eyes fell upon a startling sequence. From the moment the inauguration ceremonies were completed in the April meeting, the *issue of origins in science textbooks* dominated the entire proceedings. It was incredible!

The magnitude of the issue was far greater than I had imagined. Coming to the May meeting, I anticipated that I might have to *introduce* the issue to the Commission — on the assumption that it had died. Instead, it already occupied center-stage.

However, though the Commission's attention was riveted on the Science Textbook Struggle, this exclusive focus in April was due to *confusion*. It was a frustrating *puzzle*.

On one hand, the July 1971 statement of John Ford, Vice-President of the State Board of Education, was firm and explicit (see Exhibit 14). No ambiguity to muddle in. No uncertainty as to what he intended.

On the other hand, the crux of Ford's concern — presenting at least two major contrasting theories for origins — was an enigma to the Commission. It had apparently never been done before. There was no model to copy. It seemed that a "general theory of creation" would certainly have to be *religious* rather than *scientific.*

My pulse quickened as I read on.

It was obvious that the Kumamoto resolution, had it passed in April, would have effectively killed the issue. By sending the vexing question back to the State Board, the resolution would have nicely set everything at zero.

Only 3 of the ten original members who had unanimously approved the origins requirement in November 1969 were on the State Board in April 1972. Major public debates would have then been inevitable, giving the news media one more golden opportunity to deliver the death knell.

Perhaps I was paranoid by the time I finished reading the April minutes. My Irish temperament often causes me to overreact. In many cases, I'd rather fight than eat. But whatever the source of my agitation, I was itching for combat by this time.

Clearly to me, the issue had been *derailed.* There was undoubtedly strategy to complete its destruction already planned for this May meeting. My mission seemed critical — I had to set everything *back on track!*

Later, I was to learn that my role was not as singularly crucial as I had feared it to be. Teachers and administrators throughout the state had already been urging that the State Board's November 1969 position be enforced and carried through to completion. Exhibit 16 is one typical example of such prodding.

Not realizing that many others were already demonstrating concern — and thinking I perhaps alone perceived the precarious status of the issue, I reluctantly rejoined (at least with eye-contact) my fellow Commissioners in their tedious discussion of parliamentary procedures. Mentally, however, I was preparing for a head-on confrontation — if necessary, a *provoked* confrontation...

"We are pleased to have Vernon Grose with us today for the first time. He was unable to attend our April meeting due to a previous commitment."

Chuck Terrell, interim Chairman of the Commission, graciously surprised me and the rest of the Commission by this introduction at a logical break in the agenda. "Welcome aboard!"

"Thank you, I'm glad to be here."

The next item of business was another routine detail — committee assignments. Much of the Commission's work, as is the case with most public bodies, is accomplished by committees and ratified by the parent body.

Twelve committees devoted to specific teaching subjects like health, mathematics, foreign languages, and science had to be appointed. A slate of assignments to these committees was proposed without consulting any of us Commissioners. I was assigned to the Science Committee, as well as named Vice-Chairman of the Mathematics Committee.

Obviously, I was sensitive to the membership of the Science Committee. Junji Kumamoto, who had made the unsuccessful attempt to scuttle the origins issue in April, was proposed as Chairman of the Science Committee. Since he was a research chemist at the University of California in Riverside, this proposal seemed logical. However, the proposed Vice Chairman of the Science Committee was Commissioner Alpha Quincy.

Having noticed that there was a biographical section in the black notebook that contained resumes for each Commissioner, I flipped back to read Mrs. Quincy's qualifications. She had no education or experience in science. The other two proposed members of the committee likewise lacked any involvement related to science.

Naively assuming that the assignments should be as closely related to qualifications as possible, and realizing that none of us had been consulted about the arbitrary assignments, I quietly slipped out of my chair and walked over to Mrs. Quincy as discussion continued in the Commission.

"Hi, I'm Vern Grose," I whispered to her. "I note that you are proposed as Vice Chairman of the Science Committee. Would you consider swapping as Vice Chairman of the Mathematics Committee? My personal interest and experience in science much exceeds that of mathematics. I noticed that you have neither education nor experience in science. I would gladly make the switch if you are willing."

I knew right away that I had bombed it. Her smile froze. She curtly replied, "I'm *very much* interested in science, and under no conditions would I consider the switch."

A bit shocked but yet undaunted, I tiptoed back to my seat. The pros and cons of taking my proposal before the full Commission raced through my mind.

On one hand, I hated to propose a "personality contest" as my first action on the Commission (especially since I would be in the underdog position due to my absence from the first meeting). I realized that everyone detests having to choose between two people with whom they are going to work on a long-term basis. Further, the odds would be even higher against me since I would be opposing a *woman*.

On the other hand, there were *two* factors that favored a resolution by the entire Commission if Alpha Quincy would not yield to my personal request.

First, I was the only Science Committee nominee beside Kumamoto who had any qualifications in science. Logic seemed to heavily favor my request on that basis alone.

Secondly, since Kumamoto had clearly established in April his *antagonism* to the origins issue, I felt that it required that some stature or recognition — albeit less than Kumamoto's — should be granted to a person representing a *favorable* position on the issue.

Logic won out. I decided to "go for broke." And "broke" it turned out to be...

"Mr. Chairman, I would like to request a change in my proposed assignments," I began. "I have just had an opportunity to briefly scan the minutes of the April meeting for the first time today.

"In fairness, I should acknowledge that I am the author of the two paragraphs in the *Science Framework* — as well as three criteria unanimously adopted in support of those paragraphs — that apparently monopolized the Commission's time in the April meeting. Because of that involvement spanning over two years in this issue, it should be evident that my interest in the science textbook adoption is more than superficial."

I had everyone's attention by now. Later I was to learn that my revelation of authorship was really no surprise to most of the Commissioners, as members of the former Commission had spread the word of my identity with the issue.

However, I had no way to know this.

"Further, I have carefully examined the biographical material on all the Science Committee nominees and discovered that only Dr. Kumamoto and I have any educational qualifications in science. Therefore, I would like to request — even to the extent of forfeiting the vice-chairmanship on the Mathematics Committee — that I be named Vice Chairman of the Science Committee."

"Would you care to make that request in the form of a motion?" Terrell queried.

"I so move," I replied without hesitation.

My motion was quickly seconded by Commissioner Bud Creighton, and debate immediately ensued. Some discounted the importance of the Vice Chairman on any committee. Others admitted the importance of being professionally qualified in the subject matter. Another asked Alpha Quincy how she felt about trading positions with me. Rather than addressing her lack of qualifications, she reiterated her personal interest in science, saying she wished to remain as Vice Chairman.

Sensing that continued debate could only degenerate into personal bickering and division, Terrell wisely called for the question. All voting was by roll call — three yes, four no and eight abstentions. My bid had failed.

Kumamoto then moved that there be *two* Vice Chairmen for the Science Committee. It failed — six yes, eight no, with Alpha Quincy abstaining.

It was obvious that I had been heard by the Commission. Yet they didn't know how to resolve the dilemma I had posed. Very shortly, Commissioner Bob Rangel moved that the position of Vice Chairman on all committees be *abolished!* The motion carried on a voice vote.

The dispute over vice chairmen was not important in itself. However, it proved to be the catalyst in establishing that the origins issue would not die. The Commission could no longer claim ignorance of what the State Board had intended. They had the *source* of the Board's action as a colleague. And I had gotten their attention.

But the action was yet to get hotter — and fast! As soon as the committee assignments were approved (now without vice chairmen), the Commission recessed to break up into committee meetings.

Each committee met separately at various locations scattered around the large meeting room. Though I was also a member of two other committees by now, I obviously met with the Science Committee...And it was a disaster scene.

Evidently, my forthright comments regarding my role in the origins issue "lit the candle" of every publisher in the house. They converged on the Science Committee meeting like vultures after prey. It was a near riot.

Because we had no sound system available, publishers crowded closer and closer. This only compounded the bedlam because the Committee members remained seated, while dozens of publishing personnel stood, encircling us at least ten deep. A few had tape recorders that were thrust through the crowd to catch our discussion.

Whatever Kumamoto, as Chairman, had in mind to discuss in committee, there was only one subject on everyone else's mind — what to do about *origins*! Sensing that Kumamoto was overwhelmed — not only by the mob-like disarray surrounding him but also by my emphatic attitude, I asked him if I could make some clarifying remarks on the origins issue. He readily acquiesced.

"At the outset, let me establish some basic premises," I said loudly enough to be heard by all.

"Regardless of any confusion that might have existed in the minds of either individual Commissioners or publishers after the April meeting, the issue of opening up options for origins in science textbooks is very much alive. I say that on the authority of members of the State Board of Education with whom I have regular contact."

I felt that I might just as well get the cards out on the table right away.

"Further, it is a *scientific* — not a *religious* issue. Those who are attempting to pit *evolution as a scientific theory* against *creation as a religious belief* have been listening to the news media too long."

At that point, I pulled out of my briefcase the rationale that I had prepared two years earlier for my meeting with Ralph Gerard (see Exhibit 11). Using it as a framework, I delivered a rapid-fire summary of the struggle to be resolved.

Kumamoto seemed to be dumbfounded and offered no rebuttal to my comments. Instead, he proposed that I should probably repeat this same exposition to the full Commission when they reconvened. A strong vocal endorsement burst forth from the publishers, some of whom had probably not heard all my remarks.

Upon reassembling, the Commission heard reports from each of the committees. Kumamoto called on me to present the same background on origins that I had given to the Science Committee.

At last, I had the platform and audience that I needed. And the chief *opponent* to my position was asking me to give my views! The confrontation that I had sought had come about in an ironic way — better than if I had *planned* it. Was it *provoked* or not?

The very air seemed electric as I reached for the table microphone. If I ever felt like I was "on stage," it was now. Later, the Commission would pass a resolution of complaint about the "deplorable conditions" including inadequate lighting and sound amplification at the El Rancho Motor Hotel. But there was no difficulty in *me* being heard at this time. I am sure that everyone in that large hall heard every word that I said.

Taking advantage of this unique and rapt attention, I drove home critical points concerning the nature of science — particularly the need for *objectivity* and *conditional statements* concerning events that were never observed. Again basing my comments on the previously-prepared rationale (Exhibit 11), I emphasized that science is on a *diverging* — not *converging* — path of knowledge.

As Monod has said regarding investigations of the structure and functioning of the human brain:[291]

About the only convergence they show at present is in the difficulty of the problems they all raise.

There were two separate indications that I had made an impact by my impassioned discourse — both of which had long-lasting effect.

First, Virla Krotz, the State Board member assigned liaison duties with our Commission, came to me afterward and strongly suggested that the State Board of Education also be allowed to hear my reasoning on the origins issue. To that end, she asked me to make a long-distance call to Newton Steward, Board President, and request a place on the agenda for their July meeting (the June agenda was already finalized).

I did exactly as Mrs. Krotz instructed me – and later regretted it. President Steward was less than receptive to Mrs. Krotz's suggestion. I even sensed reluctance on his part when he weakly suggested, "You might send me any written materials that you have prepared on the subject." Because of his half-hearted response, I delayed sending anything to him for a couple weeks. More about that later...

The second indication that my message had struck home came from the publishing community. The moment that the Commission adjourned, I was surrounded by publishers' representatives (known in the trade as bookmen). Their praise was quite embarrassing. Having spent years in industry, I was well aware that I was cast in the role of a *customer*, with them as *vendors* eager to sell to me.

But this reaction was different. For one thing, I seemed to be not just one of sixteen Commissioners — I was virtually the only one they were trying to greet. Then what they said to me was far more than a standard "sales pitch." They spoke of how they had been *awakened* to a new slant on the subject — how I had *changed* their minds — why they now considered origins from a *new* perspective. This was not ordinary "glad-handing" to which I had long ago developed a thick skin.

I departed the Sutter Room of the El Rancho Motor Hotel in Sacramento on 19 May 1972 not only with a fistful of publishers' business cards...I also had a strong conviction that I had gotten the Science Textbook Struggle back on track!

Chance Versus Design

Maybe it was *frustration*. Or it could have been *disgust*. Even *impatience* might have provided the impetus that first produced an idea in my head: why not substitute *chance* for evolution and *design* for creation?

As recounted earlier, I first proposed this switch to Glenn Leslie during a visit we had in Fresno, California in June 1971. The substitution of terms might have died right there had I not included it in my draft on publisher compliance for John Ford that he issued on 8 July 1971 (see Exhibit 14). His effective delivery of this statement immortalized the terms *chance* and *design*. From then on, we were *stuck* with them — whether we liked it or not.

Many times after that, I tried to reason with myself on why I had chosen these two terms. In all honesty, they turn out to be "apples and oranges" under certain conditions. They certainly are not universally comparable. For example, *chance* is sometimes presumed to be a cause, a source, a producer of life. Note how Monod refers to chance:[292]

> Natural selection operates *upon* the products of chance and can feed nowhere else...It is clear (after the fact) that only such a source as chance could be rich enough to supply the organism...

In contrast, *design* really turns out to be a *description* of something rather than its *source*. Since the issue of interest is *origins*, it follows that we are more interested in originating causes or sources than we are in descriptions of a process. Therefore, *chance* turns out to be a more appropriate term than *design*.

While it may not be obvious (and it certainly wasn't initially obvious to me), both the Godly and Darwinians subscribe to design in nature. The Darwinians believe that the design is the result of chance while the Godly believe that the design testifies to an intelligent Designer.

So the Darwinians would probably call theirs "chancy" design (without assignable cause) while the Godly would refer to theirs as "derived" design (assigned to a Designer).

The primary reason for using *chance* and *design* was to avoid two inflammatory and emotional terms — *evolution* and *creation*. While this objective was partially achieved, it failed for a number of reasons. Some of the more consequential reasons were:

1. The Darwinians tended to view the exchange of terms as prejudicial against them.

2. The Godly tended to view the interchange of terms as enhancing their position by enabling them to exclusively claim *design* as theirs alone.

3. The terms *chance* and *design* had both popular and technical meanings that, if not properly defined, could lead to confusion.

4. Many people failed to realize that "evolution versus creation" was meant to be the *equivalent* of "chance versus design."

Each of these four reasons, in addition to being expressions of truth, are valid and logical. People have good reason to believe any one of the four.

First, the exchange of terms — though unintentional on my part — *was* prejudicial against Darwinians. Without doubt, the average layman is *offended* when asked to believe — especially in the name of objective science, that everything he sees in nature was an "accident."

Everyone realizes that any product they purchase had a *designer* — whether it's a house, car, lawnmower, bed, or typewriter. Even a novice recognizes that any product of nature is far more wonderful and marvelous than products made by man. This profound gap then between the complexity of nature and man-made products only intensifies the unbelief that the universe and all it contains could have happened by accident — without a plan, purpose or intelligence.

George Bernard Shaw's observation is certainly valid here: "A thing that nobody believes (such as *origins by chance*) cannot be proved too often."

It is not that Darwinians wish to deny the "accidental" connotation of the Big Bang origin of the universe, the "lucky strike" origin of life, or the unbelievable transformation of an ape-like creature into man. They seem to simply want to avoid having it exposed and focused for public consumption. Therefore, to translate *evolution* into a term called *chance*, is, in a sense, "rubbing salt in the wound."

But that is not the only reason the exchange of terms was prejudiced against Darwinism. The exchange also seemed to exclude the possibility that Darwinians could observe the obvious design within nature. That was unduly prejudicial because such design is widely observed and discussed by Darwinians. A chief Darwinian spokesman, George Gaylord Simpson, has even devoted an entire chapter in a book to the subject — "The Problem of Plan and Purpose in Nature.[293]

The second reason the exchange of terms failed to totally avoid emotional response — that is, the Godly claiming design exclusively for themselves — likewise had basis in fact.

The reasons the Godly viewed the exchange of terms as *favorable* are almost the inverse of why the Darwinians saw the exchange as *prejudicial*. All along the Godly had claimed that there was a Designer for the universe and all that it contained, so the "plan and purpose in nature" that troubles Simpson was obviously a product of their Designer. However, whether due to chauvinism or ignorance, the Godly position was a bit unfair because it did not acknowledge that Darwinians also recognize design.

The third reason for failure of the word interchange — the *popular* versus *technical* meaning of terms — was also sound. The wider the audience to be reached, the more an expert is forced to abandon *precision* of terminology for *universality* of understanding. It could be easy to criticize writers who must digest highly technical scientific writings and then summarize them in popular language. They perform a vital function in society. However, consider this contrast between the *popular* and *technical* versions of laboratory findings:

> (The abstract of an article in a scientific journal[294]) These findings suggest that gene fusion is a possible mechanism for the generation of complex proteins in the course of evolution.

> (A newspaper article summarizing the scientific journal article[295]) The experiment, a unique scientific feat, is important in showing how evolutionary changes may occur in nature on the most fundamental level of molecular activity.

This popular-versus-technical problem is compounded when well-known people use technical terms to communicate with the general public. Because NASA's top goal has always been a "search for life beyond earth,"[296] great importance has been focused on statements by astronauts concerning origins.

Therefore, when Apollo 17 returned from the Moon, its commander, Eugene A. Cernan, said that he was convinced that the universe "didn't happen by accident." In poetic language intended to reach a wide, non-technical audience, he went on to say:[297]

> The earth looks big and beautiful and blue and white. You can see from the Antarctic to the North Pole and the continental shores. The earth looks so perfect...you think of the infinity of space and the infinity of time. You feel...like you are looking back at earth as God must be looking now and as He must have when He created it. I am convinced of God by the order out in space.

Likewise, Astronaut Edgar D. Mitchell of Apollo 14, during private conversation we had shortly after he returned from the Moon, emphasized that as far as he was concerned it was a foregone conclusion that the universe is an ordered, structured, directed place and that its happening by *chance* is ridiculous. He elaborated some of our conversation in a later letter (see Exhibit 17):

> The contemporary interpretation of molecular mechanics and the electromagnetic nature of molecular fields leaves only one philosophic possibility for origins. They must be "chance" occurrences. Yet, there is a persistent and subconscious hope, if not a strong conviction, that in such an ordered universe as we observe scientifically man is more than simply a "chance" synthesis of matter.

In connection also with NASA's pursuit of life in outer space, one scientific observer referred to chance in noting:[298]

> Evolution of chemical molecules into amino acids and amino acids into proteins and proteins into rudimentary life forms, such as viruses, is a long and chancy process.

Even in a dictionary, chance has a variety of meanings that could confuse the average layman:[299]

> (Chance is) an opportunity; a possibility, fate, risk, hazard or gamble; a fortuitous, and usually unfortunate event, mishap or accident; (chance means) not expected or planned; accidental; casual.

To contrast with the wide spectrum of definitions in the dictionary, I intended by using the term *chance* a more technical description similar to the one Monod uses in his famous book, *Chance & Necessity*:[300]

> The initial elementary events which open the way to evolution in the intensely conservative systems called living beings are microscopic, fortuitous, and utterly without relation to whatever may be their effects upon teleonomic functioning.

Importantly, *design* appears in scientific literature in a more limited, and therefore more universally understood, sense. The word *design* may not always be used, but the idea of "a plan with a purpose, aim or intention which accounts for arrangement and order of form and function" is generally implied.

To illustrate, Joseph Kraut, professor of chemistry at the University of California in San Diego, conducted some enzyme studies that produced unique examples of nature's having come up twice with the same solution to a specific problem of molecular engineering. From these studies, Kraut concluded:[301]

> Given a particular function to perform, nature will devise the same machinery to perform it every time. The method nature chooses is presumably the most efficient. We can suppose that nature does these things the same way because there is only one way to do them most efficiently. We can then hypothesize that there is only one way that living things can be assembled, here or anywhere else in the universe.

Kraut concluded from his studies that there is nothing haphazard about the way in which the so-called "building blocks" of life are put together. Thus, we could say that he is describing *design* without using the word specifically.

A similar example by George Wald, Higgins professor of biology at Harvard University, indicates design without using the word. Wald, recipient of the 1967 Nobel Prize in Medicine and Physiology (for his research into the mechanisms of visual perception) said during his delivery of the 1971 Mike Hogg Lecture at the University of Texas:[302]

> The human eye is a magnificent thing. It is easy to see why some people think the eye is the best evidence of the existence of God.

Both *chance* and *design* became misunderstood during the Science Textbook Struggle. I was ultimately to regret that I had substituted them for evolution and creation. Yet the net score in trying to clarify the issue by different terminology was probably a plus.

The most significant point that finally emerged from the replacement of terms was the discovery that design was *common* to both evolution and creation. Thereby, design was a pivotal word or concept. It was the central commonality to two viewpoints which otherwise had apparently nothing in common.

Even more significantly, that *design* which is universally observable by scientists is truly a *scientific* matter. Whereas, the point around which this agreement has swirled (the *cause* of the observed design) is *outside of science!*

In 1962 while I was program manager at Litton Industries for an Air Force space research contract, I was involved in a lively discussion that perfectly illustrates this point.

I had engaged a biophysicist, Iben Browning, as a consultant for this project. He and I, along with four or five other specialists from various scientific disciplines, were engaged in a brainstorming session one day. The main subject had to do with how technology could best duplicate the ways of nature.

To set the stage, Browning offered three interesting observations of natural phenomena:

> *Observation No. 1*: The frequency of light to which the human eye is most sensitive (or sees most efficiently) is 5550 Angstroms. (Normal human vision ranges between 4000 and 7500A, with threshold values of 3600 and 14,000A.) Coincidentally, the Earth's atmosphere acts as a bandpass filter for the Sun's energy that is highly tuned for 5550A. In other words, the atmosphere allows the same frequency of light to pass through to the Earth's surface that the human eye sees best.

> *Observation No. 2*; The human ear is most sensitive (or hears most efficiently) at that frequency of sound which is best transmitted (with least loss) in air.

> *Observation No. 3*: The porpoise sonar frequency in water is 80,000 cycles per second that, coincidentally, is the optimum frequency for energy transmission in water.

"That proves it!" Joel Greenberg, an electronics engineer, exclaimed as he jumped to his feet.

"Proves *what*?" I asked.

"Proves *evolution*!" He was really excited.

"That's funny," I replied. "Those observations cause me to believe that there has to be a Designer behind them all. Further, the idea that all these rather remarkable coincidences could have happened entirely by *chance* takes more faith than I can muster."

Greenberg looked puzzled. In no way could he see what made me think like that. On the other hand, I could see how he easily fitted those observations into the idea that, over billions of years, man and other animals had *adapted* to their environment so that they (both the animal and environment) are perfectly aligned with each other.

Gradually, by separating observable facts from personal beliefs, Greenberg began to see that I, with equal *scientific* right, could fit those observations into the idea that, from the very beginning, man and animals were *designed* to be perfectly aligned with their environment. (Remember Figure 3 in Chapter 1?)

Chance *versus* design — it was a poor choice to set in opposition to one another. And eventually I would not only see the error but acknowledge it publicly. Still, it became a nerve-jangling buzzer that stimulated both the Godly and Darwinians as they awoke with alarm from their deep sleep of 1970-71.

Preposterous Panic?

Firemen have always intrigued me.

As a young boy, I often rode my bike on hot summer days to Fire Station No. 16 at Northwest Boulevard and Euclid Avenue in my hometown of Spokane, Washington. I'd spend hours admiring the shiny red fire truck and talking with the firemen about their work.

Of course, admiring fire trucks and visiting firemen wasn't my *real* reason for being at the fire station. My highest hopes were that the huge fire gong would begin to clang so that I could watch the firemen drop everything, run for their coats and helmets, board the fire truck and roar away — with red lights and screaming siren — to a fire. I would have especially loved to have been there about 3 AM to see the sleeping firemen jump up from a deep sleep at the sound of the bell, step into their well-positioned trousers and boots and then complete the sequence I had often observed in the daytime.

Many times since those days, I have wondered about the adverse effects on human physiology of the sudden shift from deep sleep to immediate physical exertion that not only firemen but physicians, military jet pilots on alert, and other "on call" personnel experience. Undoubtedly, it does not promote good health.

One of the obvious effects of sudden and unexpected awakening is a *temporary disorientation.* Where am I? What's happening? Myriads of like questions race through your mind as your heart begins pounding to respond to the anticipated need for physical reaction. Under certain conditions, false panic may be generated.

When the Godly and Darwinians "awoke with alarm" in 1972, panic did occur! However, it was *one-sided* panic. Only the Darwinians exhibited it. Why did they panic? There could be several reasons. Most likely, however, is their frightening realization that their monopoly concerning origins (that they presumed, as they went to sleep in 1970, they had successfully rescued from attack) seemed once again to be endangered.

Was the Darwinian panic of 1972 warranted or based on a false premise? To answer that question, perhaps we should establish some criteria. If panic was a *proper* response, then Darwinians should be shown as facing "clear and present danger." If panic was an *illogical* reaction, they should be shown as "jousting at windmills." The key seems to be in the concept of *threat* — real or imagined.

So let's examine what might threaten Darwinians.

Imagine that some politically potent pressure group decided to resurrect the "flat earth" theory. (There is a small faction in England that still holds to the concept of a flat earth.) This group would begin to demand that "flat earth" theory be taught in public school science classes comparatively with currently-taught theory. What kind of reaction would this produce from the scientific community?

Once they believed that these folk were serious, scientists would handily "shoot the theory down in flames" by documented evidence from both space exploration and experiments like Echo I that were visible to everyone on Earth. Further, the general public would rise up *en masse* to support the scientific community in this "put-down" effort because they are overwhelmingly convinced that Earth is a spheroid.

Perhaps the "flat earth" idea is too easily disproven. Consider another possible demand.

Suppose a powerful group insisted that a new theory for the Moon's origin be taught alongside the currently prevalent theory — that the Moon broke away in one big chunk from Earth about 4.5 billion years ago. (As mentioned in Chapter 1, this current view has a wide endorsement in the scientific community, including that of Nobel laureate Harold Urey.[303])

We need not even speculate on this proposal — because it already has happened! Another Nobel laureate in physics, Hannes Alfven, proposed in 1972 that the Moon originated quite differently than the prevalent theory indicates. He says that the Moon is actually an amalgamated collection of five to ten smaller moons that formerly revolved around Earth.[304]

What reaction did the scientific community exhibit when this happened? Did they say that only one theory should be taught at a time? Was there a cry that *multiple theories* would confuse students? Not at all. There are even several additional theories for the Moon's origin that are allowed to be taught side-by-side.

Therefore, we have shown two interesting characteristics of science teaching. First, scientists are not bashful about producing massive evidence that disproves obviously ridiculous theories. Second, alternative theories are permitted to be taught in comparative style in science classrooms.

These two points then narrow our search as to whether Darwinians panicked properly or illogically in 1972, when an alternative to the *chance* hypothesis to origins was proposed.

Following this line of reasoning, we could expect (1) that if the *design* concept was as ridiculous as the "flat earth" idea, scientists would have no difficulty in rapidly eliminating it with the blessing of the general public, and (2) if they couldn't thus negate the concept, they would have no qualms about letting it be taught side-by-side with the currently prevalent *chance* hypothesis. But was that the reaction of Darwinians in 1972? By no means.

This anomaly in Darwinian behavior then leads us to ask additional questions. Was there something unique about the *design* concept that caused Darwinians to deny it side-by-side status even though no evidence against it had been presented? If so, what was this uniqueness? If not, why did they not either present evidence that would eliminate the *design* concept or else allow it to be taught comparatively with the *chance* hypothesis?

Regarding this latter question, Nobel laureate Arthur Kornberg at Stanford University is most emphatic in his conviction that existing data *do* eliminate the *design* concept (see Exhibit 52):

> Conditional statements are appropriate when multiple theories have been proposed and none of these can be eliminated by the existing scientific evidence. However, this is not the case in the present argument. The

> "creation theory" of man does not stand as an alternative to the theory of evolution in this scientific sense. *It is eliminated by existing data.* (Emphasis added)

Where are these "existing data" that eliminate the "creation theory of man?" Why are they being withheld? Surely they would receive wide recognition if released — perhaps even warrant a Nobel Prize for whoever published them.

The Commission on Science Education of the American Association for the Advancement of Science was not quite as dogmatic as Kornberg, but they alluded very strongly to a conclusive body of knowledge that had eliminated the *design* concept of life's origin in a resolution they sent to all fifty states (see Exhibit 40):

> Biologists have intensively studied the origin...of living organisms...(These) scientists ... have built up the body of knowledge known as the biological theory of the origin and evolution of life. There is no currently acceptable alternative scientific theory to explain the phenomena.

If Kornberg and the AAAS hadn't assured everyone that "existing" data" are available that disprove the *design* concept, some people might be tempted to believe that a smoke screen of scientific snobbery was being thrown around a pet theory to protect it from comparison. But because such experts are unbiased and objective, we can trust their word — even when they decide to withhold such data. Or can we?

Speaking of experts and their trustworthiness, perhaps the Darwinians' panic was due, in part, to their *dependence on experts.* Don't misunderstand — expertise is critical to good science. However, when experts become *personally impressed* with their own expertise, they are no longer dependable. And the Darwinian foundation often seemed to be laid on such *experts* rather than on *scientific evidence.*

Consider this definition of a *scientific expert* proposed by a member of the National Academy of Sciences (see Exhibit 26):

> A scientist is a person who does experiments, gathers original observations, reads first hand papers in which experiments and observations are described, and goes to scientific meetings or conferences to discuss problems with his colleagues. A person who merely writes about science by quoting second hand, third hand or old, obsolete statements can by no means be regarded as a scientist. Furthermore, scientific opinion can be regarded as informed *only when the scientists in question have acquired first hand information about the subject being discussed.* (Emphasis added)

This definition is very restrictive — and undoubtedly it should be. As medical physicist Thomas H. Jukes says (see Exhibit 51):

> "Science is hierarchical and elitist. In other words, science is ruled by only a select, distinguished group of experts. So, it is important that *nuclear* physicists discuss only *nuclear* physics — not optical, medical, acoustical, or astrophysics. Otherwise, they fail to qualify as a scientific expert."

The reason I said that Darwinian panic could have been attributed to their dependence on experts is that, when they had to find experts to bolster their position, *there were none available*!

Who ever heard of a PhD in Evolution? What university has a School of Origins? There aren't even any science *courses* in "First Causes," "Natural Selection," or "Environmental Adaptation." So what could Darwinians do? They were forced to borrow experts from other disciplines and "christen" them as *evolution experts*!

Using proxy experts is probably expedient until your expertise is challenged. Then, it could easily lead to panic — especially if you have been insisting that "only bona fide experts and elitists should be allowed to speak in science." Would *you* want to be such a *hypocrite?*

Even more disturbing is the identical parallel between the Darwinian position as they awoke in 1972, and that of medieval church authorities as they were forced to evaluate the theories of Copernicus during their confrontation with Galileo.

Here is how Caltech physicist Albert R. Hibbs describes the church's dilemma at that time:[305]

> Although the theories of Copernicus did indeed seem to be against official church doctrine, the highest authorities had studiously avoided making any official pronouncement on the subject. They wanted to keep their options open. Suppose further studies should tend to show that Copernicus was right? The best move was to take no stand at all, but now the hand was forced.
>
> The action was an official examination of the theories of Copernicus by a group of scholars and philosophers called "qualifiers." The result of the examination was that the Copernican theory on the solar system was held to be absurd, false, against theology, and in part heretical. *No one was to hold or to teach this theory.* Galileo was notified of that decision. (Emphasis added)

What were the Darwinians going to have to do when they awoke in 1972? They were forced to either disprove the *design* concept or allow it to be taught side-by-side with the *chance* hypothesis. Neither alternative was attractive to them. And for the very same reason — they either lacked or did not want to expose the *scientific evidence* for their action.

This dilemma, that undoubtedly generated panic in the Darwinians, forced them down an age-old path. They became *hypocrites*, demanding of their opposition what they could or would not do themselves.

As will be described later, they organized a parade of "elitists" from every conceivable branch of science who were instructed to parrot the same "company line" used by the medieval church — that the *design* concept is "absurd, false, against science, and in part heretical. No one is to hold or teach this theory." This heterogeneous herd of heroes thus became the 1972 version of the "qualifiers" used centuries earlier by the church against Galileo.

"But *these* qualifiers of 1972 are much different than the qualifiers of 1614," you might protest.

"After all, they are all PhDs in science, and many hold Nobel Prizes for their unique scientific contributions. Further, they all do experiments, gather original observations, read first-hand papers in which experiments and observations are described, and go to scientific meetings or conferences to discuss problems with their colleagues."

Very true. But alas — you didn't read the entire definition of a *scientific expert*. You stopped short of the last sentence:

> Furthermore, scientific opinion can be regarded as informed *only when the scientists in question have acquired first hand information about the subject being discussed.*

On what single *scientific* subject could the following list of experts have all acquired "first-hand information" in order to speak as *authorities* on it? Note the diversity of their interests:

Paleontologist G. G. Simpson (Exhibit 21)

Zoologists Alfred S. Romer (Exhibit 22) and William Z. Lidicker, Jr. (Exhibit 31)

Bird Curator Ned K. Johnson (Exhibit 27)

Physiologist Donald O. Walter (Exhibit 28)

Biologist Harlan Lewis (Exhibit 29)

Chemist Harold C. Urey (Exhibit 32)

Biochemists Richard M. Lemmon (Exhibit 50) and Arthur Kornberg (Exhibit 52)

Medical Physicist Thomas H. Jukes (Exhibit 51)

Certainly these men form an awesome assemblage — but on what basis could they have *first-hand information* about the *chance* occurrence of the universe, life, and man? On the basis of their diverse specialties, it might be easier to list those scientific fields that do *not* produce first-hand information on how the universe, life, and man conclusively occurred by chance.

So much for the obvious hypocrisy of Darwinians in using "proxy experts" to support their case for *chance* origins. Without doubt, this hypocrisy could generate panic — and may well have.

On the other hand, it appears that Darwinian panic *caused* the hypocrisy rather than vice versa.

What else, then, could have caused Darwinians to suddenly realize that their monopoly on origins was once again endangered? I propose that if Darwinians had only had extensive contact and discussion with the non-scientific public about origins, they would have easily sensed at least four major difficulties with the *chance* hypothesis that the public cannot seem to swallow as credible.

Lest Darwinians attribute these hindrances to *religious indoctrination* that stubbornly refuses to be erased, these problems have nothing to do with religious preconception. In fact, these four obstacles have legitimate *scientific* overtones. Let's refer to these apparent absurdities as *Public Puzzles.*

Consider **Public Puzzle No. 1**: *The persistent failure to acknowledge that belief rather than observable data forms the basis of abiogenesis (life originating from lifeless matter).* The average layman is virtually mesmerized by what he has been sold about "the scientific method." Yet he is puzzled about how such an apparently overpowering intellectual activity could produce thoughts that even he can identify as mental rubbish. The public is ready to accept mind-boggling revelations from science. That is not the problem. But *mind-boggling* does not mean *absurd.*

How do scientists slip into making absurd pronouncements? It may be that, being involved in an intellectual activity, they thereby lose touch with common sense and the average layman. As Herman Kahn has said, intellectuals deal exclusively with "second-hand information." Thereby, they fail to realize that their own personal belief overrides the facts that they are supposed to be observing in nature.

Sir Peter Brian Medawar, eminent biologist of the British National Institute for Medical Research, goes one step further. He says that scientists even deceive themselves about what they believe:[306]

> Unfortunately, a scientist's account of his own intellectual procedures is often untrustworthy. "If you want to find out anything from the theoretical physicists about the methods they use," said Albert Einstein, "I advise you to stick closely to one principle. Don't listen to their words,

> fix your attention on their deeds." Darwin's case is notorious. In his autobiographical sketch, contemporary with the sixth edition of *The Origin of Species*, he said of himself that he "worked on true Baconian principles, and without any theory collected facts on a wholesale scale" (p. 83); but later in the same work (p. 103) he said that he could not resist forming a hypothesis on every subject...Darwin's self-deception is one that nearly all scientists practice.

Public Puzzle No. 1 is an indication that the scientific community has failed to communicate with the lay public. The average person readily realizes that abiogenesis is *belief* not *science.* To illustrate this perception by the public, note this letter to a newspaper editor:[307]

> The creationists and the evolutionists are again in one of the world's most futile arguments. However, today it is the evolutionists who regard their theory as undisputable truth.
>
> Stripped of its mumbo-jumbo, godless evolution requires us to believe that at some remote time and place the just-right combination of elements came together and without intelligent intervention became, by pure accident, a living cell. Instead of immediately dying, this great god Unicell acquired continuous life, divided itself into two cells, thence multiple cells (what they fed on is not clear), thence groups of cells that by some rough magic became multicelled plants and animals. By some mysterious development of intelligence these multicelled organisms realized they could not reproduce themselves by simple cellular division, so following the rules laid down for them by great god Unicell they devised the stamen-and-pistil, sperm-and-ovum processes of reproduction. With no difficulty at all, the offspring of Unicell then proceeded to develop the complex laws of embryology, bacteriology, botany, entymology, zoology and their law-giver biochemistry, and all of these sciences were put under the direction of the great guide Evolution. Quite a god that Unicell — that accidental, finite daub of mud that developed our complex life systems with their finely adjusted interactions and laws without intelligent help.
>
> Make sense? Not to me...An accidental "in the beginning" without intelligent planning, and the consequent enthroning of the great god Unicell, to the exclusion of Elohim, seems to me more assinine than the 144-hour creation of the fundamentalists.

If Darwinians are going to regain credibility with the public, the solution seems to lie in their differentiating of Evolution I (science) from Evolution II (belief). The determined effort by Darwinians to overlook or deny the difference will only further alienate the non-scientific. Occasionally, scientific journals like *BioScience* call attention to the philosophic (Evolution II) implications of the *chance* idea:[308]

> Due to its historical and cosmic orientation and the human questions it raises, evolution is by its very nature philosophical!...It is unfortunate that many evolutionary texts do not give sufficient weight to the philosophical issues that are inevitably raised in class discussions...Facts are stressed while concepts and implications may be neglected.
>
> Contrary to public opinion, scientific research and judgments are not value-free or free from assumptive reasoning. Students may not be aware of the meaningful interrelationship between the sciences and philosophy in the history of evolutionary thought.

Should Darwinians heed this suggestion to recognize Evolution II as a *philosophic doctrine* — completely separate from Evolution I, the *scientific theory*, the public may choose to evaluate Evolution II's merits against other cosmological alternatives.

However, the required distinction between Evolution I and II will not be easily made because most of the spokesmen for Evolution II are also renowned scientists. Though not necessarily by intention, Evolution II pronouncements and beliefs become disguised as *scientific* statements — so attributed because of the scientific eminence of the spokesmen.

Here are a few examples of Evolution II declarations made by distinguished scientists without a *shred of scientific validity.* Unfortunately, they are not identified as *private personal beliefs.* So they pass as scientific statements. First, Sir Julian Huxley:[309]

> Our present knowledge indeed forces us to the view that the whole of reality is evolution — a single process of self-transformation.

Next, the late Theodosius Dobzhansky. Note carefully how he even attributes *personality* to Evolution II:[310]

> In giving rise to man, the evolutionary process has, apparently for the first and only time in the history of the Cosmos, become conscious of itself...Evolution comprises all the stages of the development of the universe: the cosmic, biological, and human or cultural developments.

Finally, Rene Dubos enthrones Evolution II by conferring on it the zenith honor — the origin of all things, from heavenly bodies to humans. And only the unenlightened would dare disagree with his exultation:[311]

> Most enlightened persons now accept as a fact that everything in the cosmos — from heavenly bodies to human beings — has developed and continues to develop through evolutionary processes.

It should not be surprising, then, that lesser scientists than these luminaries should be swept up in rapturous — if not ridiculous — projections of Evolution II for the future. Picking up the torch from their illustrious heroes, they see no limit to what Evolution II can or should become.

Consider the vision of this Belgian scientist:[312]

> Science itself offers a ray of hope. It stems from the discoveries of Darwin, the implications of which are only now, one hundred years later, being fully grasped. The view of the world as a world of evolution enables us to conclude that the aim of being is becoming...Contrary to the former view that the aim of being is being, it is not self-destructive. What is perhaps not so easily appreciated are the very important practical consequences of such a philosophy for everyday life. For it follows immediately that every individual is responsible not only for his existence, but for all future existence...It means the end of politics, business, economics, and trade unions as we know them.

Or ponder this revolutionary mission projected for Evolution II:[313]

> The doctrine of evolution can answer crucial ontological, epistemological and ethical questions...The purpose of teaching evolution is not merely to interpret natural and social phenomena quantitatively, but to change the human condition qualitatively through responsible, collective action...It is hoped that evolutionists will be increasingly concerned with the synthesis of conceptual frameworks as they will be with the synthesis, engineering, and preservation of Life.

Is it any wonder that the public is disenchanted with such spurious spooferies swaggering as science? *Public Puzzle No. 1* demands a responsible explanation from Darwinians. Will it happen?

Another possible explanation for Darwinian panic in 1972 involves a second enigma for the non-scientific public — **Public Puzzle No. 2 —** *The obvious struggle to "invent" data and explanations to verify a non-existent "amoeba-to-man" chain.*

Darwinians are neither subtle nor clever enough to hide their missionary zeal to confirm the "amoeba-to-man" *belief* as a scientific *fact.* The general public may be unschooled in science, but they are not easily duped. Even children, in some cases, can see through the desperate fight waged by Darwinians to prove, in the name of science, the apparently unprovable. Due to this overselling, unspoken but real *tension* exists between the public and Darwinians. If Darwinians became aware of such tension in 1972, they could have well panicked.

Would you like an example of the absurdity that has generated *Public Puzzle No. 2*? Consider this explanation of evolution by geneticist Richard B. Goldschmidt:[314]

> The evolution of the organic world, from the synthesis of the first complex molecules endowed with the faculty of reproducing their kind to the most advanced type of life, must have taken place roughly within the last two billion years on our planet. All the facts of biology, geology, paleontology, biochemistry, and radiology not only agree with this

> statement but actually prove it. Evolution of the animal and plant world is considered by all those entitled to judgment to be a fact for which no further proof is needed.

He has stated that the "amoeba-to-man" chain (Evolution II) is a *proven fact.* No further proof is needed because "all those *entitled to judgment*" have declared it to be a fact. One wonders whether anyone who disagrees with this "fact" is *ipso facto* not "entitled to judgment." Rather frightening, indeed!

Pause just a second and reflect on the position of the medieval church regarding the earth as the center of the universe. Could not their official decree be stated as: "Geocentrism is considered by all those entitled to judgment to be a fact for which no further proof is needed"? Have we reentered the Dark Ages of unquestionable authoritarianism — only with different tyrants demanding obeisance?

Goldschmidt's obvious indiscretion is turned into an even more bizarre conclusion by continuing his explanation of evolution:[315]

> But in spite of nearly a century of work and discussion there is still no unanimity in regard to the details of the means of evolution. This statement should not be misunderstood; all biologists who have mastered the available facts are agreed upon the main points, the big outlines of the explanation of evolution...The differences in opinion mentioned above, refer only to the technical details within this framework.

What Goldschmidt is saying is that the "amoeba-to-man" *belief* is a "fact" but the *scientific* basis for it is unknown. Backwards, isn't it?

There is no scientific explanation for the "amoeba-to-man" chain of life — yet we are told that it has been decreed by "all those entitled to judgment" to be a fact. That kind of logic is what boggles the public's mind, producing *Public Puzzle No. 2.*

Goldschmidt affirms that Neo-Darwinians (including the majority of geneticists) are convinced that living things have become increasingly complex through the ages by the accumulation of small genetic mutations and subsequent adaptation to new environments.

Supposedly, this is how plants became fish, fish became reptiles, reptiles became birds, and all the other illogical transformations in the "amoeba-to-man" chain described in Exhibit 66 occurred. In other words, sub-species became species, that became genera, that became orders, that became classes, that ultimately became phyla — a building up process from specificity to generality.

Yet he is candid in admitting that the historical facts furnished by the fossil record show *exactly the opposite* from what the Neo-Darwinians believe. Phyla are the *oldest*, with species being the *youngest*. Even a more serious difficulty in making the facts match the preconceived Darwinian belief is described by Goldschmidt this way:[316]

> As far back as early Darwinian days, critics of the theory of selection of small mutants as an explanation of the origin of adaptations had expressed doubt as to whether a complicated adaptation could be accomplished by such means, because selection cannot work before the adaptive change has reached a working level. The wing shape, type of flight, and correlated structure of the lungs of a hummingbird, together with honey-sucking bill and tongue, are a beautiful adaptation to a definite ecological niche. But could selection favor such a combination before it had reached a working capacity?

The answer to Goldschmidt's question is obviously "No!" So what does he propose to bolster the belief-now-decreed-fact?

Something equally untestable and unobservable in nature — a mutation that runs contrary to all known rules. He calls it a "macromutation." What caused him to propose it? Anything he *observed* in nature? No. Some *experiment* he conducted? No. What was it then? A desperate need to "invent" an explanation for what he already believed but could not prove scientifically!

By his own words, "Biologists and paleontologists who realized these difficulties (of Neo-Darwinism) began to think in another direction." With good reason!

Grasping wildly for some imaginary, fanciful justification for their belief in the "amoeba-to-man" chain, here's how macromutation is explained:[317]

> As a rule, one may *assume* that mutants affecting developmental processes at earlier stages are *likely* to disturb the fabric of interwoven and closely attuned individual features of development in an adverse way, making orderly development impossible. But it is *imaginable* — and there are examples of this — that in *rare* cases such an early-acting mutation is not lethal but permits development to continue. The result is a large deviation from normal, which *might* affect in one step major features of development, producing a more or less divergent pattern of the entire organization. Such a macro-mutation *might* accomplish in a single step major morphological or physiological changes at the level of the higher categories from the species to the phylum. *Thus the conclusion is reached*...that the higher categories are built up not by slow accumulation of micromutations...but by large initial macromutations. (emphasis added)

Is that scientifically convincing to *you*? Does it not resemble a "cover-up story" more than a rationally-developed and inevitable conclusion proceeding from unbiased observation of nature? Can Darwinians *really* expect the public to be gullible enough to accept that fabrication as *scientific truth*? It does seem to stretch credibility a bit too far...

In a rather backward fashion, Goldschmidt *concludes* his technical paper with observable facts rather than *starting* with them. Here are the facts that have forced him and other biologists to "invent" a story that will convert these facts into an otherwise unobservable "amoeba-to-man" chain:[318]

> A large body of facts is available, but there is no unanimity in its interpretation. The facts of greatest general importance are the following. When a new phylum, class, or order appears, there follows a *quick, explosive* (in terms of geological time) diversification so that practically all orders or families known *appear suddenly and without any apparent transitions*...Down to the generic level *abrupt major steps without transitions* occur...The decisive steps are *abrupt, without transition*...
>
> New types *never* appear to be formed from specialized families, genera, or species but rather from simple, generalized forms. Specialization seems to lead into a blind alley...In spite of the immense amount of the paleontological material and the existence of long series of intact stratigraphic sequences with perfect records for the lower categories, *transitions between the higher categories are missing.* (emphasis added)

The *facts* then *deny* the existence of an unbroken chain of living things. Higher categories of life (including man himself) are totally disconnected from one another — unless bridged by some invented, unverifiable myth.

As Porter M. Kier, Director of the Museum of Natural History of the Smithsonian Institution in Washington, D.C. once said to me, "It would be a lot easier to believe that every single species was independently created, as far as the record goes." Why then do Darwinians fight the record? Why do they not present the facts to schoolchildren? What impels them to foist on the public a myth disguised as science. A *belief* or *faith*. Why should only *one* belief or faith be granted scientific respectability?

If Darwinians are to gain public confidence, they should address *Public Puzzle No. 2* head-on. Clearly separate *science* from *fiction*. Will they do so?

A third possible source of panic in 1972 for Darwinians, attributable to their rejection by the public, could be summarized as **Public Puzzle No. 3** — *The obscurity, uncertainty and conflict in dating methods that force constant and confusing revision of prehistoric dates.*

The public gets dizzy trying to keep up with revised dates for the origin of the universe, life, and man. In the first place, no one can imagine a *million*, let alone a *billion*, of anything! Numbers of that magnitude are meaningless to the average person. Down deep, we realize that such numbers are exceedingly large, so that realization is what perplexes most people about the wild revisions of important prehistoric dates that occur so frequently.

It is well recognized that Darwinians are super-sensitive about dates. They need *time* like living things need *oxygen*. The longer the time between prehistoric events, the better they like it. This is because they treat *time* as though it were a *causative agent* — like it could make something happen. And they need all such "causing" they can muster to accomplish the required miracles to link the "amoeba-to-man" chain together.

Medical physicist Thomas H. Jukes illustrates the importance of time for Darwinians in the question of the origin of life by stipulating *slow* evolution (see Exhibit 51). Likewise, biochemist Richard M. Lemmon in Exhibit 50 states:

> "Spontaneous generation" is scientifically recognized as a definite *probability* for the appearance of life on our planet, given that the "spontaneity" may have required one billion years.

Why are one billion years *required*? What *precludes* the "spontaneity" from occurring in 10 seconds or even a day? An inherent belief in the scientist's mind that *time makes things happen*.

Here is another example of this blind belief in time as a *causative agent* in the origin of life:[319]

> Much to their frustration, however, biologists can still only speculate on how these simple organic molecules emerged through the eons as proteins and genes.
>
> More baffling still is how these proteins and genes got together in the first self-replicating cell. The odds against the right molecules being in the right place at the right time are staggering. Yet, as science measures it, so is the time scale on which nature works. *Indeed, what seems an impossible occurrence at any one moment would, given untold eons, become a certainty.* (emphasis added)

More than likely, what appears to be Darwinian belief in incredibly long periods of time as the means of converting a staggering impossibility into a certainty is actually a slightly different *belief* — but a belief nonetheless.

They probably believe that the chance synthesis of life from non-living material was due to a very large number of random collisions of materials. In other words, if there were enough time (and thereby collisions), one of those collisions might have produced the first living cell. Mathematically, this is known as the "concept of randomness."

What do mathematicians say about the use of the "concept of randomness" to explain the origination of the first living cell? Most of them say that it is an unwarranted and unsound assumption. In a symposium held exclusively to address this subject. Murray Eden, Professor of Electrical Engineering at the Massachusetts Institute of Technology, said:[320]

> It is our contention that if "random" is given a serious and crucial interpretation from a probabilistic point of view, the randomness postulate is highly implausible and *that an adequate scientific theory of evolution must await the discovery and elucidation of new natural laws — physical, physiochemical and biological.*

As for specific dating methods used to set prehistoric dates, they — like all other methods — are updated as new information becomes available. However, much of the obscurity, uncertainty and conflict of dating will remain inherent in the premise that dates can be set for distant events in the past. In other words, revising and refining *dating methodology* will not remove its basic difficulty. There are at least four facets to the problem of establishing prehistoric dates:

1. The *selection* of a measurement technique from among alternates.
2. The measurement *process* employed on test samples.
3. The *extrapolation* of a measurement backward in time.
4. The role and nature of *time.*

Let's examine briefly each of these facets. First, many dating techniques are available. Some are simpler to use than others. Some, like Carbon 14, cannot be used on objects much more than 10-15,000 years of age. Selection is primarily dependent on radioactive elements being present in the test samples, and the radioactivity of C_{14} is too weak to measure much beyond this time scale.[321]

For skeleton fragments, potassium-argon dating has been the most common method used. It measures the amount of argon gas that has accumulated, through the slow decay of radioactive potassium, since the molten rock surrounding these fragments solidified. Yet, another method of measurement was used to more than double the estimated time span that man has apparently lived on the North American continent.[322] It is known as "Aspartic Acid Racemization" and measures the ratio of one type of amino acid to another. Because the proportion of D— to L— amino acids in a fossil steadily increases with time, it is assumed that the age of a fossil can be measured by determining the extent of racemization.[323]

A variety of methods have been used on Moon rocks recovered in the Apollo program. Caltech geologist Gerald J. Wasserburg used the rubidium 87—strontium 87 method to date Apollo 11 materials.[324] The uranium-lead technique, on the other hand, was selected for crystalline Moon rocks like basalts and anorthosites returned on later Apollo missions.[325]

The so-called "Genesis rock" recovered by Apollo 15 astronauts was dated by the argon 39-40 method.[326] So there's always the opportunity for error as well as disagreement on dates when selection must be made from a variety of measurement techniques. The selection process is an *art* rather than a *science*, and frequently scientists disagree among themselves on which method should be used.

Regardless of which measurement technique is ultimately selected to date material, it is also open to questions of accuracy. Although reasons for selecting a particular technique will probably be based on the types of radioactive elements thought to be in the test sample, quite often the radioactivity may be so weak that accurate measurement is not possible. Also, many techniques are *inferential* rather than *direct*. They measure not the bones or tools themselves but the material from which the samples were dug.

To illustrate, it was the *volcanic ashes* from the Olduvai Gorge in East Africa that were dated by potassium-argon rather than the *Olduvai Skull* itself to arrive at the skull's age.[327]

Further, the techniques are often quite different from one another. Radioactive decay is measured in a variety of ways. The decay of thorium and uranium in the Olduvai volcanic ash deposits was measured by fission-track analysis that counted the number of microscopic marks left in the rocks as a result of the decay.[328] The potassium-argon method produces a gas that is analyzed while the rubidium-strontium and the uranium-lead do not.

When one technique is used to confirm a date set by another technique, they seldom agree. For example when dating Apollo 16 anorthosite rocks, potassium-argon tests yielded ages as high as 3.9 billion years. U. S. Geological Survey scientists, on the other hand, used the uranium-lead method and got only 3.6 billion years — a difference of 300,000,000 years! One report of this discrepancy explained:[329]

> Scientists do not fully understand the reason for the differences between the potassium-argon and uranium-lead ages of these rocks. However, the differences may be caused by experimental uncertainties.

So methods and techniques for prehistoric dating are far less precise and reliable than those, for example, used to measure distance, velocity or time.

The third problematic facet of radioactive dating — the *extrapolation* or *projection* of today's measurement backward for millions and billions of years — introduces another large set of uncertainties. Anyone acquainted with surveying knows the importance of a transit theodolite — a telescope used by surveyors to measure horizontal angles. This optical instrument was developed to reduce angular measurement error because the tiniest mistake is multiplied by the distance one travels at that angle. Likewise, the miss-distance of a rifle bullet depends, given the same aiming error, on the distance of the rifleman from the target. The greater the distance, the farther the bullet misses the target.

The key to understanding how we can measure something today and stretch that measurement backward billions of years lies in the *assumptions* that are made.

One basic assumption is that everything in the universe, including the geological strata, has been developing in a uniform or predictable manner without catastrophes of any type ever since its Big Bang origin. This assumption is called "uniformitarianism." A companion assumption to it is a belief that the laws of nature have always been the same as they are today. Thus the *present* state of nature is the explanation of its *past* state and of its *future* state as well.

The average man on the street, like the small child in Hans Christian Andersen's *The Emperor's New Clothes*, wonders about the Big Bang itself — "Wasn't it a *catastrophe*?"' he reasons.

"It could not possibly have been a *development* based on past existence! Further, it apparently was a one-time-only type of catastrophe since nothing like it has occurred since."

There are some obvious advantages to these assumptions. For one thing, they keep calculations *simple*. Second, they eliminate the need to know anything about the past — whether, for example, there actually *were* any catastrophes in prehistoric history — because you simply *ignore* them. Third, they permit relativistic *inference* — from one measurement many conclusions can be drawn about objects that cannot or have not actually been measured themselves.

Finally, it allows one to be *dogmatic* like plant chemist Junji Kumamoto, my fellow Commissioner on the Curriculum Development Commission, when he made this unbelievable statement in a public hearing on 26 July 1972:

> One of the basic laws of science is the Law of Conservation of Matter and Energy in which we state that matter can neither be created nor destroyed. And so speaking in terms of creation, to presume that a Creator can create matter in any form is simply contrary to that Law.

Not only dogmatic but a bit *arrogant*, isn't it?

Another assumption relating to dating geological formations is that the oldest formations contain *only* the simplest forms of life. As Wilder Smith says:[330]

> The more complex forms of life were, according to theory, not yet developed at the time the older formations were laid down. If, therefore, a formation contains trilobites, for example, from this one fact alone it is deduced that the formation belongs to the Paleozoic age. Certain types of fossils are thus indicative of certain ages, so that wherever these fossils appear, according to theory, the age of the formation may be diagnosed with certainty. Thus the trilobites *prove* the formation to be Paleozoic.

The circular logic used here — Darwinism being used to prove Darwinism — is obvious. By assuming that the oldest formations contain only the most primitive and least complex organisms, one is adopting one of the basic assumptions of Darwinism. Then, *ipso facto*, fossils of simple organisms are used, *a priori*, to date formations.

G. A. Kerkut at the University of Southampton comments on the index-fossil method of dating as it relates to establishing the "amoeba-to-man" continuum:[331]

> We have as yet to obtain a satisfactory objective method of dating the fossils. The dating is of the utmost importance, for until we find a reliable method of dating the fossils we shall not be able to tell if the first amphibians arose after the first choanichthian or whether the first reptile arose from the first amphibian. The evidence that we have at present is insufficient to allow us to decide the answer to these problems.

There are myriads of unanswered questions about these assumptions that are required to extrapolate a current measurement back for billions of years. These interesting questions really should be included in public school science textbooks to stimulate the inquiry process of children. Instead, they are absent.

What about the sudden extermination of large numbers of mammoth in northeastern Siberia — quick-frozen with fresh food in their mouth? Does this not bespeak some rapid, catastrophic climate change, unaccountable when assuming uniformitarianism? Or what about the clear evidence that the force of melting ice on an ancient South Pole left its record in sandstone that is now located in the middle of the Sahara Desert?[332] How do such clues that the crust of the earth has shifted radically jibe with uniformitarianism? Or what about the Hanks-Anderson inhomogeneous accretion hypothesis that describes the earth being formed in a "geological instant" of perhaps 10,000 years instead of the millions of years apparently required by uniformitarianism?[333] Was this hypothesis "necessary" to overcome the otherwise inexplicable (under uniformitarianism) solid inner core and molten outer core of the earth?[334]

The fourth aspect of radioactive dating is one of large magnitude — and certainly beyond the scope of this book. It has to do with the *nature of time*. The rotational velocity of the earth — our primary basis for measuring time — is not constant but gradually slowing down. So right away, we know that time is *relative* – not *absolute*.

Einstein showed, to illustrate relativity even more dramatically, that while a person traveled in space at the speed of light for 60 years, the earth would age about 5 million years.[335]

Sir James Jeans, as quoted in Chapter 6, makes the point that there is no obvious reason to believe that time has progressed at the same rate (or even in the same direction!) in the past. The puzzle of the nature of time, he says, "brings our thoughts to a standstill."

What if time were not the steadily onflowing phenomenon we have considered it to be? What if it had stood still or even moved backwards, as Jeans mentioned, in the past? Then Carl Sagan's "cosmic calendar" would be totally meaningless. Sagan proposes that, if we arrange time (as an assumed steadily ongoing measure) into a one-year calendar, the first humans would not have appeared until late on 31 December![336] Not only does this treatment of the past assume something that cannot be verified (the rectilinearity of time) but it contributes to a credibility gap between the public and Darwinians.

Sagan, in explaining his calendar, says that even though we know little about the past, "We are able to date *events* in the remote past." How? By geological stratification and radioactive dating, both of which involve unverifiable and fanciful assumptions.

Further, Sagan reveals his personal bias in favor of "deparochializing" the present while admitting that the evidence for his calendar is rather shaky. Amazingly and while still down-playing human consequence in cosmic history, Sagan concludes:[337]

> Despite the insignificance of the instant we have so far occupied in cosmic time, it is clear that what happens on or near the earth at the beginning of the second cosmic year will depend on the scientific wisdom and the sensitivity of mankind.

What absurdity! How could something happening in the very last second of an entire year lead anyone to reach Sagan's grandiose conclusion? He has reduced human contributions in the universe to total insignificance — not only in terms of time *scale* but in the dimension of any *leverage* that we could exercise in an overwhelmingly preset construct of events. Yet he concludes that what happens on earth will depend on *human sensitivity!* Can he be serious? Does he expect the average human to understand him?

Or take another tack on the role of time — work backwards from the answer. A rather amusing proposal caused me to reevaluate the antiquity of man. We already know the population of the earth today. So it was proposed that if we were to assume (1) that Noah and his wife lived about 2300 years B.C., (2) that they and their offspring survived a universal deluge, (3) that the average lifespan was 43 years for everyone, and (4) that each person reproduced 1.25 children for the 100 generations from Noah to now, we would arrive at the present population of the earth.[338]

To compare that calculation with one that assumes man to be the age projected by Darwinians, we would get an answer of 10^{2700} enough people to not only cover the earth solid but perhaps "fill our entire universe out as far as we are now able to observe."[339] Such a comparison involves many assumptions that need validation, but it does raise practical questions about the astronomic dates projected by Darwinians.

The stupendous magnitude of dates now projected for prehistoric events probably masks the large margins of error that are tolerated by Darwinians. The date for the Big Bang origin of the universe varies tremendously — anywhere from 7 to 20 billion years. For example, the Sandage-Tammann estimate is 10 billion years.[340] On the other hand, Sagan uses 15 billion years for his "cosmic calendar."[341] That's a 50% difference! What if the velocity of light was that vague and sloppy? Or the atomic weights of elements, tensile strengths of metals, or coefficients of thermal expansion? We would be in trouble.

The toleration of huge error is not limited to the origin of the universe. The same slovenly approach to dating error exists regarding the origin of *life*. F. Clark Howell pegs that important milestone at 2 billion years.[342]

Yet, the illustrious committee that prepared the *Science Framework* for California public schools set it at 3 billion years.[343] Again, another 50% difference. And the blasé attitude that accompanied the 1975 announcement of a 1-billion-year revision in the date for the origin of life further boggles the mind of a layman.[344]

The wild range of dates for the origin of *man* far exceeds the speculation of dates for the origin of *universe* and *life*. One thing that permits this guessing game to perpetuate itself is that no one can yet define *what* or *who* man is! By keeping that definition sufficiently vague, there is no compulsion to find the accurate date for Human One.

The reader should be aware that this criticism of dating errors is not directed at the work of individual scientists who are proposing dates. Instead, it is leveled at Juke's "hierarchical elite" (see Exhibit 51) who alone rule science and protect it from the incompetent and the spurious.

Why is this group not concerned about the reckless use of these dates by textbook writers and the popular press? Don't they realize that such dates not only *mislead* the public about what science knows but, at the same time, *alienate* these folk?

Maybe, on the other hand, Darwinians did finally realize that they had misled and alienated the public on *Public Puzzle No. 3* — and the panic of 1972 resulted!

In case *Public Puzzles No. 1, 2* and *3* failed to induce Darwinian panic in 1972, maybe the source for panic was in **Public Puzzle No. 4** — *The construction of whole people, cultures, and even civilizations out of one or two chips of bone.*

We have already examined at length in Chapter 1 the implausibility of the anthropologists' explanation of man's origins. Their willingness to invent myths based on a few bone fragments and then attempt to brand these myths "scientific" should be suspect. But it apparently isn't.

Even when an authority like F. Clark Howell, quoted extensively in Chapter 1, openly admits:[345]

> We don't know the answers to almost any question you might have about these earlier phases of human evolution.

In explaining that statement, Howell said that each new discovery in anthropology raises *ten fresh questions* for every old one it solves.

One week previous to Howell's admission, Richard E. F. Leakey had unearthed a human-like skull near Lake Rudolph in East Africa. This skull, according to Leakey, was more than 2.6 million years old. His discovery pushed man's origin back more than one million years!

Only one year earlier in November 1971, headlines heralded the discovery of a 200,000-year-old skull in the Pyrennes Mountains as "ranking among the two or three most historic finds in the history of the study of human evolution."[346] It was the first entire skull that had been found to fill the gap between Java Man (400,000 years old) and Neanderthal Man (90,000 years old).

The non-scientific public has to be completely confused by both the frequency and diversity of reports concerning human fossils. It's like anthropologists are playing, "Can You Top This?" Consider a few examples that illustrate this contest.

Richard E. F. Leakey, son of the late Louis S. B. and Mary D. Leakey (both well-known anthropologists), is often at the center of controversy on human bone fragments. Interestingly, Leakey has *no formal education beyond high school!* Yet, he is granted top billing by leading scientific organizations such as the American Association for the Advancement of Science (AAAS).[347] Since these same groups *deny* recognition to bona fide scientists, with doctorates from top-ranked institutions, who do not endorse the "chance" idea for origins, it could appear that *formal education* is not nearly as important in certain scientific circles as subscribing to the correct *belief.*

Leakey, addressing the 140th annual meeting of AAAS in 1974, proposed that man lived in East Africa three million years ago *alongside* Australopithecus — that "missing link between ape and man" whose dramatic photo-painting appears in the *Time-Life* series called "The Emergence of Man."[348] (The mystery of how anyone could produce photo-paintings of something that has never been found — a "missing link" — was discussed in Chapter 1.)

How did Leakey reach this revolutionary conclusion of coexistence? On the basis of a few supposed Australopithecus knee, shin and thigh fragments none of which were from the same individual! From these chips, Leakey said *it would seem* Australopithecus was a sedentary, highly-specialized creature that fitted comfortably into a well-defined vegetarian niche in the grassy savanna of Africa.[349] Quite revealing bone fragments, weren't they?

Only months after Leakey's stupendous supposition, he was upstaged. A French-American-Ethopian team found one complete jaw, a half upper jaw, and a half mandible that they dated as 4 million years old. Using "conservative" terms like *unparalleled* and *major revolution in all previous thinking,* these three men held a press conference (only eight days after uncovering the scraps of bone) in which they released this prepared statement:[350]

> We have in a matter of merely two days extended our knowledge of the genus homo by nearly 1.5 million years. All previous theories of origin of the lineage which leads to modern man must now be totally revised.

How did they set the age of these bits of bone? Quite easily — and very scientifically. Here's the method:[351]

> They were in a stratographic level 150 feet below a volcanic basalt recently dated by potassium-argon technique to 3.00 to 3.25 million years, thus apparently making the specimens far older than previous finds...By comparing the ones (they had found) to others discovered at different sites in the East African Rift Valley system and already well dated, the team has determined that some of the animal fossils are 4 million years old. Since some were located in the same stratographic level

as the human fossils, this has led them to the conclusion their specimens must also be from the same time period.

Not only was the *dating* accomplished fast and without difficulty. Apparently many other details of *life style* also jumped right out of these fossil scraps:[352]

> The small size of the teeth has led Johanson (one of the finders) to the hypothesis that the genus homo "was walking, eating meat and probably using tools — perhaps bones — to kill animals" 3 to 4 million years ago. "It also means there was probably some kind of social cooperation and some sort of communication system," he said.

To throw a bit of salt in Leakey's wound of dethronement, the team fired a parting shot at him by saying that they believed the experimental team approach they had used in their expedition had shown its superiority over the "usual individualistic one" used in the search for fossils.

But Leakey wasn't long in returning the fire. While he didn't yet have another cache of even older fossil chips to top the team's claims, he attacked the team's conclusions that their find was really human. Pointing out that some prehumans had very human teeth, Leakey said:[353]

> You do not define man by teeth, but by the size of his cranium. Until we find it, it is impossible to tell if Afar man is genus homo.

That credibility setback for the team lasted long enough for Leakey to be rescued by none other than his mother, anthropologist Mary D. Leakey. She resorted to rather questionable logic, however, to put down the Afar fossils discovered by the team. Accusing them of having to estimate dates by *comparison*, she herself *extrapolated* from rocks to fossils to get dates for some new specimens she uncovered late in 1975. Here is an account of the Leakey snatching of the prized "oldest date for man" title back from the team:[354]

> Hominid jaws and teeth thought to be as old as (Mrs. Leakey's) new Laetolil find were discovered last fall by D. Carl Johanson and co-workers near the Hadar River in north-central Ethiopia. The Johanson team was not able to establish a firm radiometric date, however, and was forced to arrive at the estimated date of three million to four million years by comparison with associated fauna. The Leakey find, potassium-argon dated to 3.35 to 3.75 million years, thus supersedes the Johanson find as the oldest hominid remains yet reported.
>
> The teeth and mandibles (lower jaws) of 11 distinct individuals were found scattered over an area of volcanic ash beds, some specimens as far as five miles apart. They were found ... in the same types of rock beds and at approximately the same levels, "This," she said, "makes the

> individuals geologic contemporaries." Even so, they spanned 400,000 years. Some were found near the top, in the beds dated 3.35 million years and others were near the deeper 3.75 million-year-old beds."
>
> Four of the fossils were found in place in the rocks, making it possible to establish the firm dates. Garniss H. Curtis of the University of California at Berkeley carried out potassium-argon dating tests on biotite crystals associated with the fossil jaws, and gave them "watertight dates." Even though seven of the fossils were not found embedded in the volcanic ash that originally covered them, Leakey says, "they are identical (to the others) and therefore we could extrapolate from the rocks safely."

And so we can expect this fanciful fabrication of fossil fantasy to forever fulminate with frequent fanfare. If it's not fossils of human beings, it will be fossils of something — like the teeth of a cuttlefish (1/8 inch long) which recently pushed back the whole evolutionary time scale by 100 million years![355] Don't you wonder what will be next?

Public Puzzle No. 4 is real to the average layman. Even more, it is justifiably *absurd* to him. Darwinians would be well advised to educate the non-scientific public on how they arrive at their conclusions. Not many people realize that:[356]

> The facts are observable, but their *organization* is not. Man invents an *organization* which seems to fit the facts he observes. (emphasis added)

It's time for Darwinians to openly announce to the world that the "organization" of facts is what the public would call a hunch or educated guess. Nothing more. Further, it is a *belief*. Since all scientists are as human as anyone else, they will stick to their beliefs as tenaciously as the average person. When beliefs about fossils overpower the fossil facts, the public is wholly justified in their suspicion of absurdity.

Let's compare what Charles Darwin *believed* about the fossil record with the *data* (facts) he could assemble to support his belief. In his famous book, *The Origin of Species*, Darwin clearly states that he *believed* that all existing species are descended from a single progenitor. In truly scientific manner, Darwin listed and candidly discussed many possible reasons why his belief may not be true. These reasons were obviously based on observable *facts*. One chapter in his book is entitled, "On the Imperfection of the Geological Record." So we know that Darwin was aware that the fossil record did not support his belief.

One of the most serious difficulties with fossils that Darwin acknowledged was the "sudden appearance of groups of allied species in the lowest-known fossiliferous strata." He devoted several pages to this problem. Here are some direct quotations from Darwin's discussion:[357]

> To the question why we do not find rich fossiliferous deposits belonging to these assumed earliest periods prior to the Cambrian system, I can give no satisfactory answer...The difficulty of assigning any good reason for the absence of vast piles of strata rich in fossils beneath the Cambrian system is very great ... The case at present must remain inexplicable; and may be truly urged as a valid argument against the views here entertained (Darwin's belief).

Despite the scarcity of facts to support Darwin's belief, many were persuaded by his proposal, One hundred years (and thousands of man-years of scientific searching) later, *the belief was still alive — and the supporting facts were still missing!* Norman D. Newell, celebrating the 100th anniversary of Darwin's announced belief, said:[358]

> A century of intensive search for fossils in the Precambrian rocks has thrown very little light on this problem. Early theories that these rocks were dominantly nonmarine or that once-contained fossils have been destroyed by heat and pressure have been abandoned because the Precambrian rocks of many districts physically are very similar to younger rocks in all respects except that they rarely contain any records whatsoever of past life.

It is true that a few fossils were found in the intervening 100 years that had the appearance of being possible connections or links to fill in the gaps. However, Newell's comments on these meager findings were not encouraging:[359]

> These isolated discoveries, of course, stimulate hope that more complete records will be found and other gaps closed. These finds are, however, rare; and experience shows that the gaps which separate the highest categories may never be bridged in the fossil record. Many of the discontinuities tend to be more and more emphasized with increased collecting.

This persistent, dogged search for something unfound raises the question of the century – "*What drives Darwinians to continue their fruitless pursuit?*"

The answer does not lie in scientific *facts*. It can only be due to their *blind belief* in the "amoeba-to-man" myth proposed by Darwin. And, believe it or not, I endorse the use of such myths — provided we recognize them as such.

The Science Textbook Struggle in California provoked an English professor at California State University in Fullerton to prepare a major article for the *Los Angeles Times* on the usefulness and propriety of myths in education. Rita D. Oleyar wrote:[360]

> Let us return to a more mythological approach to education — no, let us simply admit that we all live by myth...Since the onset of the scientific

> revolution, at least, we have fooled ourselves into thinking that we live without myth — which in itself is the greatest myth of all...It is a myth that any part of the academic community has answers which preclude other answers. And, finally, it is a myth that doubt endangers the security of the young. Or that a variety of answers will confuse him...Placing the biblical account of man's creation by a master artist side-by-side with the equally mythical evolutionary story might help our children in a small way to live more rationally with themselves and their pieced-up environment.

Of course, Professor Oleyar risks the wrath of science's hierarchical elite by her suggestion. But before they are permitted to burn her at the stake for her heresy, could it be irrevocably dangerous to momentarily consider what she is saying?

Just for sake of discussion, what if public school science textbooks contained a side-by-side exposure of *two* myths about origins — the "amoeba-to-man" continuum and one that described a series of instantaneous or discontinuous origins? The latter would not propose an Originator nor would it be an appeal for scientific recognition of supernaturalism as an explanation for the discontinuities. Such appeals to explain difficult problems leads to a "god of the gaps," and such a god dies just as soon as the gaps are filled with natural explanations.

One major advantage to this comparative approach would be the elimination of the currently-enforced *indoctrination* of only one belief about origins. But beyond that, imagine the mental stimulation that public school children would receive by having *two* myths to compare with the known data or facts of science.

After enjoying a monopoly for their myth for a considerable time now, Darwinians would be expected to be not only *reluctant* to share their unique pinnacle, but also *frightened* by the prospect. Maybe their myth would not survive such comparison with the facts. Or perhaps they would be forced to explain *why* they believe in a continuum that cannot be supported by fact.

But setting possible Darwinian embarrassment aside for a moment, let's ask a few questions about what would happen to *science* if such a comparison of myths were to occur. First, would there actually be any *retrograde* in scientific research and discovery? Secondly, would there be any *inhibition* or *misdirection* of scientific thought if this comparison were to be made? The answer to both of these questions would seem to be "No" — particularly if we believe in the classical liberal position:[361]

> Liberalism is a tolerance of views differing from one's own. It favors reform or progress which tends toward democracy and freedom for the individual.

Science has historically flourished in a climate of free thought. Only when indoctrination (based not only on enforcement of a *single* view but also on enforced *denial* of other views) has been imposed has science suffered — as in the case of Lysenkoism in the Soviet Union. So it would seem both logical and scientific to follow Professor Oleyar's suggestion of comparison.

But probe further for hazards. For all the cries of anguish that arise upon the suggestion of possible vulnerability of the Darwinian myth, what are the *specific dangers* that might emerge from comparison with another myth?

Or flip the coin over. What scientific *achievements* can be traced directly to an unshakeable faith in the Darwinian myth? What scientific blessings have come *directly* from this concept (and thereby would have to be denied or rejected as a result of having it compared with another myth)? What house would fall? What scientific *advancement* would have to be purged from scientific textbooks (like the removal of all references to Stalin by the Khrushchev regime)? In other words, is the Darwinian myth essential to *science* or only to the *elimination of God* as an explanation for origins?

R. E. F. Leakey found over 150 fragments of a skull, KNM-ER 1470, in 1972. After assembling these bits together, he announced to the world, "Either we toss out this skull or we toss out our theories of early man. It simply fits no previous models of human beginnings."[362] Leakey went on to say that the skull's surprisingly large braincase "leaves in ruins the notion that all early fossils can be arranged in orderly sequence of evolutionary change."[363]

When a 27-year-old high school graduate with no formal scientific education can make such sweeping statements and have the scientific community pay attention to them, the Darwinian myth must be considered as yet quite tentative. Therefore, it might be wise to have alternative myths available.

So we can conclude that *Public Puzzle No. 4*, as well as three other puzzles, may have generated the Darwinian panic of 1972 — but only if the Darwinians knew of the public's puzzlement. The panic *was* preposterous — whether or not the Darwinians were aware of the public's viewpoint. If there was Darwinian awareness of layman bafflement, the panic was traceable to the preposterous claims of Darwinians. If not, the panic was preposterous because it occurred without explanation.

PARADE OF PERSISTENT PUBLISHERS

My telephone was ringing off the hook.

Hour by hour, day by day, it rang. Day and night, weekends as well as weekdays, it never quit jangling.

It started as soon as I arrived home on 19 May 1972 from my first meeting as a Commissioner. Every publisher in America, it seemed, wanted to meet privately with me — probably to tell me the reasons I should accept their textbooks as well as to learn more details of my concern about origins.

While honestly empathizing with the bookmen and their need to communicate directly with each Commissioner, I was totally unprepared for such an onslaught of salesmanship. In addition, I hadn't imagined or foreseen a heavy demand on my already overcrowded schedule.

Since most of the other Commissioners were either classroom teachers or school administrators, they had relatively fixed work schedules. They were on fixed salaries. Their superiors had all concurred in their appointment as Commissioners, recognizing that their work schedules would have to accommodate the duties associated with their new assignment.

Another factor cramped my ability to meet with publishers. I was the only Commissioner who traveled extensively. Most of my income-producing services are performed outside of California — many of them in Washington, D. C.

I was scheduled to teach at The George Washington University in Washington on 5-16 June 1972 two of the four weeks immediately following the May Commission meeting. Due to my travels, the time available when I am in my California office is even more limited.

Perhaps the most critical restraint on my ability to allocate time with publishers was an economic one. Adolph Moskovitz, a Sacramento attorney, and I were the only Commissioners from the business community. Neither of us were independently wealthy, and every moment we spent as Commissioners was not only *uncompensated* — a gift of public service. It *reduced* our income potential because we were paid only when we performed services in our respective professions.

The other Commissioners felt no impact on their incomes, regardless of how much time they allocated to Commission activities. (Significantly, Moskovitz resigned his commission within four months due to the amount of time demanded by Commission business.)

While I kept turning down one publisher after another, I weighed various alternatives. As a matter of principle, I wanted to treat each publisher equally. I had no way to decide which one to see first. Further, I had very strong feelings about accepting any form of gratuity, thereby becoming beholden or compromised to a publisher.

As an aerospace executive, I had managed large organizations that controlled the selection and approval of millions of dollars worth of electronic and electromechanical equipment. In that capacity, I had imposed stringent standards and controls on the relationship between subcontractor marketing personnel and those under my direction. And I meant to maintain those same high standards in my dealings with publishers — even to the acceptance of a cup of coffee.

At one point, I felt as though I might totally resist talking to *any* publisher. Just read and evaluate science textbooks on my own. Yet I was aware of my ignorance of the entire process of determining the values of teaching materials.

At least in California, textbook selection is a complicated process.[364] I knew absolutely nothing about it. Very soon, I realized that the publishers were (1) a vital part of the system of textbook acquisition, and (2) a valuable source of education about that acquisition process. I could not really ignore them.

But how could I possibly grant each of dozens of publishers an hour or two of my time? It would be a full time occupation!

I can't honestly recall how the thought of combining publisher interviews into one session occurred to me, but it did. The more I pondered it, the more I felt it was the only practical means I could employ. After discussing the idea with several of my fellow commissioners, I decided to go ahead…

Collective Coffee

"This is not an official meeting in the sense that it has been sanctioned by the State. I talked to other Commissioners to make sure I wasn't doing something illegal. But to me, this is no different than going to lunch with 30 or 40 of you, one at a time. And in that sense, it's a whole lot cheaper on the whole publishing industry to do it this way. Secondly, it doubles up on our time, and all of us are busy people," I explained over the roar of the air-conditioner in a crowded suite at the Holiday Inn in Woodland Hills, California.

It was a hot Tuesday evening, 30 May 1972. Representatives — science editors, textbook authors, corporate executives, regional and state managers as well as local bookmen — from 17 different publishers had responded to my invitation to meet collectively, rather than individually, with them.

Some had come from Chicago, New York and other Eastern cities on very short notice. Realizing that I had undoubtedly failed to notify *every* publisher who was interested in the science textbook adoption, I felt obliged to inform them of my reasons for the invitation.

"I intended to exclude no publisher from this meeting. However, as you realize, I am new to this publishing world and know none of you personally. The first thing I did was contact every person whose business card I received in Sacramento at the Commission meeting week before last. When I called Dean Hurd of Holt, Rinehart and Winston, he graciously offered to send out an announcement to all members of the American Association of Publishers, since he is the current chairman of AAP for this area — "

"We sent that announcement out to every company office a week ago," Hurd boomed as he momentarily interrupted me.

(I later learned, with regret, that AAP did not include all the interested publishers. Therefore, some publishers that belong to other trade organizations, like the California Bookmen's Association, could have missed notification of my meeting.)

"Thank you, Dean, and I deeply appreciate your effort," I responded before returning to my explanation.

"I've been in the business world, like all of you are — wined and dined, offered fishing trips, free tickets to everything and all other kinds of payola. Frankly, I wanted to be as honest and aboveboard as I could with every one of you, without showing one bit of favoritism. As you know, I've gone to lunch with *nobody*. I can honestly say that. I felt that this meeting tonight was the only way I could be fair with all of you in the compressed time scale we all face in the science adoption."

At least those present seemed to agree with my rationale. But I wanted to clarify the nature of the meeting one more time — just so they wouldn't later misconstrue it.

"Now I am sure that what I've done might appear wrong in some people's eyes. If so, I apologize. Many of you have tape recorders here, and if you want to share your tapes of the meeting with other publishers, you know that it's perfectly all right with me. But I emphasize once again — this cannot be a State-sanctioned meeting. I didn't clear it with our Chairman, Chuck Terrell, or with Wilson Riles. However, I don't clear luncheon dates with them either. So this is a *glorified lunch without food!*"

The place broke apart with howls of laughter. I sensed that a spirit of open camaraderie and informality now prevailed. And such a spirit was absolutely required — because we were all physically uncomfortable.

I had selected this particular Holiday Inn because it was close to my home. One of the publishers sent four or five of their key personnel to the meeting, including their executive science editor from the East Coast. Since they had reserved a suite in the hotel, they suggested that I hold the meeting in their suite. Even though the meeting didn't start until 7 PM, the suite that was located on an upper floor was still unbearably hot.

Even before 7 o'clock, the room was crowded. The hotel furnished chairs, but they couldn't get enough of them into the room to accommodate everyone. Approximately 40 people, representing 17 different publishing firms, were crammed into the suite — seated, standing all around three walls, and overflowing out the doorway. All that body heat was more than the noisy air-conditioner in the room could handle — but it roared on, making it very difficult to talk above its racket.

Going out for a cup of coffee together is a traditional American rite performed to mix relaxation with business. This "collective coffee" I was simultaneously having with 40 or more publishers was not intended to relax anyone, however. It was *all* business...

Resolute Review

"The purpose of tonight's meeting is *clarification* — not defense. I intend to clarify what has turned out to be a particularly emotional issue here in California. It has been vastly misunderstood. Therefore, I wish to start by reviewing its background and development."

I wasted no time in setting a firm direction for our get-together. "With your indulgence, I would like to make a rather formal statement for about 30 minutes, after which we will throw the meeting open to questions from anyone. By all means, we want to have a wide-open session with lots of questions, answers, rebuttals, defenses, or whatever you want to say. We are here to learn from each other."

After that introduction, I began to explain how I had gotten into the issue. For an opener, I said that my daughter had brought home a textbook from her sixth grade class that contained a ridiculous myth. I proceeded to read this fairy tale (see Exhibit 66) to the publishers. It wasn't even necessary to finish reading it. I had made my point of its total absurdity.

"As we can all see, this figment of imagination is *pure mythology*," I remonstrated. "But it wasn't labeled that way. It carried the impressive title, '*The Beginning of Life on Earth*.' More critically, it was the *only* explanation in the book! That disturbed me."

By being specific with this indoctrinating fantasy, I was able to justify my reaction to the *Los Angeles Times* editorial that had drawn me into a maelstrom of controversy on origins. I used an overhead projector and screen to rapidly summarize the 4 major points in my address before the State Board in November 1969 (see Exhibit 4). Likewise, the three curriculum criteria that I had written for the State Board concerning origins (see Exhibit 13) were read.

"I am trying to get rid of this religious fervor that frankly exists on both sides of the question. It's not just church-going creationists who are fanatic in their beliefs. I've never seen emotion like I did once I got involved in this issue and ran into biologists. They believe in their myth more than anyone I have ever met believes in God. I had hoped to pull the stinger out of this fight, but as it turned out, I apparently 'loaded it for bear.' It turned out to be 180 degrees from what I intended."

The publishers seemed to be enjoying it. I scanned their faces to detect any negative reaction, but they were hanging right with me.

At times, my comments undoubtedly appeared quite tutorial — especially for those science authors and editors present. However, I was determined to drive home my key points. Regarding the nature of science, I continued, "The scientific method is a continuing asking of 'why?' and 'how do we know?' as well as a clear identification of gaps in our knowledge. And there are *plenty* of gaps. In the kindergarten through eighth grade textbooks, we ought to be really exposing these questions for the kids, so that they can see that there is still room for them — that they too can make a contribution to science someday. We certainly fall short of our goal for children if we simply elaborate what science has decided to be factual."

The two paragraphs that the State Board of Education had unanimously adopted in 1969 were an obvious topic for the evening. So I de-emphasized their intrinsic importance right away.

"I wasn't happy with the two paragraphs that were adopted," I acknowledged. "And the news media have skillfully and deliberately managed a miscarriage of their intent. Therefore, I openly say here and now that under no conditions did I ever desire that the Book of Genesis or the Genesis account would appear in the textbooks. I never mentioned, for inclusion, the Bible nor God. In fact, I disavowed both of them. And if you read carefully the two paragraphs that were adopted, you will see that's true."

The next point I made, with heavy emphasis, was that we can and should pay attention to the fact that objectivity about origins *demands* a case for design as well as a case for chance. "We have two large umbrellas of philosophy — *randomness or chance* on one side and *design or purpose* on the other — under which we can group all beliefs concerning origins. That's what I am really interested in putting into the textbooks. Beneath these two umbrellas, you can put all the specific theories that you want. For example, it's my understanding that people who believe the Genesis account have as many as 10 schools of thought on how to interpret that account. So I am not after an interpretation of Genesis in the textbooks. I am just simply saying that the scientific evidence *demands* that we allow these two large umbrellas or schools of thought to be heard. That's what we're after!"

From there, I simply but systematically reinforced my message. I had brought a suitcase full of books, scientific journals and periodicals. Brief excerpts from them were read or noted to support specific questions about the randomness concept, the entropy conflict, abiogenesis, and uniformitarianism.

In addition, I brought several science textbooks from a variety of publishers who were present that evening. These books were up for adoption in California, and I had marked certain passages in them that illustrated my concern. Naturally, I was careful not to identify the publishers. Among those that illustrated my contention of scientific inadequacy were those that:

1. Depicted man as a part of an unbroken chain of living things.
2. Told teachers to "emphasize and dramatize the *fact* that the development of living things goes back more than a billion years."
3. Stated without qualification that "from just a few bones of early living things, scientists can tell what they looked like."
4. Exclusively used "adaptation" as the explanation for differences among living things.
5. Used only "development" to explain the origin of physical characteristics.
6. Offered several "chance" hypotheses for origins — like the nebular, collision, scientismal, or dust cloud hypothesis for the origin of the earth — but none that suggested that the origin of the earth was other than an accident.
7. Dogmatically spoke of life originating in a soup of amino acid-like molecules three billion years ago.

At this juncture, I felt it appropriate to introduce a good friend of mine, Roy E. Cameron, a Caltech scientist whose specialty is desert microflora. He has done extensive botanical research on every continent. Much of his work has involved numerous extended trips to the Antarctic. He was due to leave for Moscow in only four days to deliver a research paper as a guest of the Soviet Union.

"I have asked Dr. Cameron to attend this evening's meeting just in case anybody disagreed with me. As you know, I am claiming no scientific expertise myself. So, Roy, if I say — or have said — anything you object to scientifically, blow the whistle on me!"

My resolute review was virtually complete. However, before resting my case and throwing the meeting open to questions, I read a brief statement I had prepared on the interrelationship of science and religion — since so many had raised the religious specter:

> Science and religion are not mutual exclusives. Further, science is not anti-religious. It is simply a-religious. Therefore, if evidence from scientific observation favors the case for DESIGN, science has no choice but to be objective and state its findings, even if those findings raise questions which can only be answered outside of science; e.g., "WHO is the Designer?"
>
> Since none of us were there to observe and record the origin of the universe, there is the *possibility* that there is a God (in Whom we claim to trust on our money, and Whom we acknowledge when we pledge allegiance to our flag) Who *actually did* create the universe with purpose and design! I am personally unaware of any scientific evidence to the contrary.
>
> If this creation actually was the manner by which the universe, life and man originated, I am quite certain that the true scientist would not want to be quoted as saying:
>
> 1. That science would deliberately ignore, discard or destroy data and evidence that revealed this fact.
> 2. That science would be totally unable to discover such a fact.
> 3. That science currently has data and evidence that irrevocably eliminates the possibility of this fact.

I was later to change my mind about the second of those three quotations. However, my case was now established. "I now throw the meeting — or even the *windows* — open. Let's have a good time exchanging ideas..."

Responsible Reaction

Ideas didn't exactly start popping like popcorn. Many in the room seemed to relax noticeably. I was almost feeling like I'd have to prime the pump when the first question came.

It was a cautious probe to see if I had urged or invited any certain publishers to submit revisions to their textbooks to comply with my views. I hadn't. But I guessed aloud that several publishers might have approached John Ford on the State Board of Education along that line.

That reply evidently struck a raw nerve or two because before we could satisfy everyone that no one had a "head start," many had chimed in with explanations and comments on how the textbook selection process works. The merits of modifying one's own books versus waiting for the State Board to do it for you were warmly debated.

The next question was a request for me to provide them with some biographical information — who I was, what my professional responsibilities were, and what background I had in education. We all had a good time at that moment, laughing together about my diverse interests and activities — none of which seemed to qualify me as a Commissioner. (Several attendees later told me privately that they believed my interdisciplinary orientation was critical to overcoming what had become intellectual incest in the previous Curriculum Commission.)

"Any other questions?" I asked.

"Yes, I'd like to make a *statement* rather than a *question.* I think your presentation was, indeed, very scholarly. I felt as though I were sitting in a graduate course in physics at your institution, Doctor. However, after listening to it, I can't help but think that perhaps there's some feeling that — among the publishers, the editors and authors who are close together — there has been a tremendous lag in knowledge from people who have the background that perhaps you have. Most of the books, as we look at them, are indeed very *prosaic* — they lack something. And I'm amazed that in the publishing industry we haven't gone to people who have this kind of understanding. The items that you've picked out of those books were *specifics.* You're not recommending *massive overhaul* of the textbooks — but rather specific items."

He went on to explain how textbook changes to bring them into line with my thinking could be made. He was joined by many others who also were apparently impressed to respond positively to what I had said. The meeting continued in that constructive mode for at least another hour.

As we drew near the end of the meeting, one more question was raised about the Biblical account of creation — evidently to seek further assurance that I was solely interested in a non-religious approach.

"Is it not possible that the method of presentation of these two schools of thought could be hazardous — specifically with respect to the Biblical version of creation? Are we not getting ourselves into the same predicament by limiting ourselves to only these two positions?"

"They are as general as you can get," I responded. "If you become more specific, you get into trouble..."

"But specifically as far as the *Biblical account...*" he persisted.

"You haven't heard *me* mention the Bible, have you? I am *not* in favor of putting a Biblical account in the textbooks. In fact, I'm not even interested, as some of the books have done, in saying that religious people think one way but scientists think another way. That's a joke! Ralph Gerard is one of the most religious persons I ever knew, and he's an *atheist!* But he's certainly religious about biology. What I'm saying is that we ought to open up to the fact that, first of all, we *don't know* how origins occurred. Then there should be a statement that there are two general schools of thought — one that says it just *happened*, with no reason…"

Roy Cameron broke in at this point — "Let me make a comment here. Being a biologist, I was brought up to believe that evolution is not a theory — it is a *fact*. And I was almost ashamed about that last slide you put up on the screen about the truly honest scientist. I am sorry to say that scientists are emotional and biased, just like anybody else. I can easily write a paper, which I have — 70 of them — that supports my beliefs. On the other hand, I can name the things that I don't agree with. Having been in the Antarctic, I think it is especially interesting that some of the seals walking around down there are 3,000 years old — based on C-14 dating techniques."

An outburst of laughter temporarily broke his observation. When they quieted down, he continued, "It's important to point out that we don't know where 'time zero' is and that we only *believe* that things continue along in a steady state. These are the kinds of things that can be brought out with reference to these two schools of thought. While they are both philosophic, you can still bring out the influence they have on scientific findings."

With Cameron's succinct confirmation, the meeting came to a rapid close. It had lasted two hours. There had been no rebuttal of what I proposed, and the collective reaction of all present seemed to be one of serious and responsible intent to right some longstanding wrongs.

As I was packing my materials to leave the suite, the Executive Editor for Science for one of the nation's largest publishers, who had flown out from the East specifically to attend this meeting, told me that he did not have one single difference of opinion with me on any of the points that I had made that evening. And there were others who said the same thing.

Within the next few days, I began to receive mail that further confirmed my conclusion that the publishing community had listened to my arguments, *per se*, rather than kowtowing to my status as a Commissioner.

"I felt the meeting was very worthwhile. It gave me a better idea of the background and depth that you have gone into in developing your concerns in the areas of science…In each case that the State Board of Education has indicated desired changes in our books, we have honored those requests. We have found them reasonable and have made the requested changes. We see no reason why we would not be able to honor similar requests by the State Board of Education in the science adoption," one of the publishers who attended the meeting wrote.

Another editor wrote:

> An editor finds his job satisfying because he occasionally has an opportunity to meet someone who can speak intelligently and authoritatively on a subject of universal interest. I enjoyed very much having the privilege of sitting in on your discussion in Los Angeles last week. Thank you for setting up the meeting in behalf of publishers and their representatives.
>
> You can be sure that I returned (to the East) with the feeling that some firm guidelines had been established for our future handling of textbooks. Your points were well stated and backed up with convincing evidence. You did, indeed, clarify a subject in which those of us who edit science textbooks have long been interested.

The president of another large publishing firm addressed a letter to the Curriculum Development Commission that reflected the impact of our meeting on 30 May 1972. In part, this letter read:

> Since there has been discussion concerning possible changes that may be necessary in all the series of science books offered in this year's adoption, (our company) wants to assure you that we will be happy to ... make changes in (our science program).
>
> Possible changes to meet the California requirements might be necessary in chapters of our series dealing with the origin of the universe and the formation of the earth. In these chapters, greater emphasis could be used to show that man has always speculated about the origins of the universe and the solar system. Many times these speculations resulted in teaching theory of design. The theory of design can be tied in with the current "scientific" theories (Big Bang and Steady-State for the universe, and nebular, collision, tidal, etc., for the solar system).
>
> Other changes might concern radioactive dating as a way of determining the age of an object. It might be pointed out more specifically that these methods are not absolutely perfect and the ages obtained are not absolutely accurate but are working data for scientists.
>
> In addition, content involving the origin and development of various forms of life could be modified so that the students are exposed to the idea that various forms of life might exist by design, rather than strictly as the result of adaptations that occur by chance.

Three days after the meeting, I decided to respond to the lukewarm request for some written materials on my viewpoint that State Board President Newton Steward had made during our telephone conversation two weeks earlier.

I sent him my November 1969 speech (Exhibit 4), the *Los Angeles*

Times editorial that initiated my involvement (Exhibit 1), the background statement on origins that I prepared for Ralph Gerard (Exhibit 11) and a number of other reference materials. In a cover letter to Steward, I enthusiastically told him about the 30 May 1972 meeting in Woodland Hills — about the excellent reception and lack of disagreement among publishers.

I closed my letter by repeating the same request that Virla Krotz had urged me to make — and that I had made of him by telephone on 19 May from Sacramento — that is, to appear before the State Board of Education for 15 minutes during their July meeting to expose my viewpoint on the origins issue.

Steward didn't bother to answer me for three weeks. When he did, he soundly thrashed me. First — and deliberately overlooking the fact that I had contacted him at the direction of a State Board member, Virla Krotz — he condescendingly instructed me that all Commissions must work together as *units*.

Therefore, any "official requests" to appear before the Board would have to come from our Commission Chairman. "I therefore cannot grant your request for a personal presentation before the State Board at its July meeting."

Obviously, Virla Krotz had felt that the State Board needed my viewpoint. Just as obviously, my personal viewpoint — or any other single person's — would never be presented as the viewpoint of an entire Commission. Therefore, by a typical bureaucratic maneuver, the State Board was denied the opportunity that Virla Krotz, one of their own members, had felt was critical to their understanding.

But that wasn't the end of Steward's castigation of me. He had "great concern" about the propriety of the 30 May 1972 meeting in Woodland Hills. He warned that individual Commissioners should not call meetings that might be considered, due to the Commissioner's status, as a command appearance by publishers.

I think this is a valid criticism — until you learn the circumstances. And he obviously had not sought the details. No publisher was threatened or punished for not attending the meeting. Ample disclaimers of any official status for the meeting were made. One wondered whether Steward viewed meetings between Commissioners and individual publishers, on a one-to-one basis, as also unethical. If anyone was forced into a "command appearance," it was *me* — not the publishers!

NATIONAL ALERT

The infamous Watergate break-in was about to occur.

It was late on Sunday evening, 4 June 1972, when I checked into the Howard Johnson Motor Lodge that faces the Watergate complex across Virginia Avenue in Washington, D. C.

I was to teach another two-week accelerated course in the School of Engineering and Applied Science at The George Washington University. Since 1969, I have stayed at Howard Johnson's an average of 12 weeks a year. They probably have no more frequent customer.

"How about Room 621?" Al Ayers, the desk manager, asked as he scanned the reservations list.

"I've tried them all, Al, and you know every room is the same," I joked.

Unknown to both Al and me, only about 50 feet away from Room 621, Alfred C. Baldwin III was very busy in Room 723.

Starting on 30 May 1972, (the same evening I met with the publishers), Baldwin began to monitor the telephones in the Democratic National Committee (DNC) headquarters across the street in Watergate after a two-day delay in picking up wire-tap radio bands set up during an earlier break-in. Baldwin made transcripts of the conversations and gave daily logs to James W. McCord, Jr., who had hired him.

During the two-week period that I was registered in Room 621, Baldwin monitored and documented approximately 200 conversations in the DNC headquarters. Without question, I must have ridden up and down the hotel elevator with Baldwin, McCord, and the other four Watergate burglars.

Some months later, I became a close friend of Charles W. Colson, President Nixon's "hatchet man," We have since worked together in his program of Federal prison reform.

One evening as Chuck and his wife Patty were riding with Phyllis and me in our car, I told him of my immediate proximity to the Watergate burglary. His reaction? "You probably knew about it before I did!"

Meanwhile on Monday morning, 12 June 1972, the phone in Room 621 at Howard Johnson's rang while I was rushing to get ready to leave for the University.

"Vern, I've been discussing your textbook controversy in California with some people in my office. Would you be willing to meet and review it with some of the top officials of various scientific organizations before you return to Los Angeles?"

It was Glenn S. Pfau on the line. Pfau, Director of Project LIFE at the National Education Association (NEA) headquarters in Washington, was a consulting client of mine as well as a personal friend. During my frequent trips to Washington, he and I had had a chance to exchange a few ideas on the struggle I was having out there.

Pfau received his doctorate at Ohio State University in speech pathology, but he had an avid avocational interest in the scientific question of origins. As he discussed the issue around the office with scientists and educators in NEA, he discovered a surprising amount of inquisitive reaction.

"Oh, I suppose that I could spend a little time in informal discussion of the issue," I said. "However, I brought no materials with me. I would have to simply explain in generalities what I have in mind."

"That's all right. I think that it's important that these people hear directly from you about your ideas. Let me see what I can do to set up a meeting. Is there any time that you have free from your teaching at the University?"

"Let me see...yes, I could possibly get over for about an hour at 11 o'clock on Wednesday morning."

"I'll get back to you as soon as I find out everyone's schedule."

Pfau was not long in arranging a meeting. We met in his conference room on the ninth floor of NEA headquarters in the Coyne Building in downtown Washington on 14 June 1972. Frankly, I was surprised that he was able to assemble such an impressive group of people on such short notice.

Seated at a long conference table, the following people introduced themselves-

Arthur H. Livermore	Deputy Director of Education, American Association for the Advancement of Science (AAAS)
C. Charles Peterson	AAAS Staff Associate
Jerry P. Lightner	Executive Director National Association of Biology Teachers
Robert Carleton	Executive Secretary, National Science Teachers Association (NSTA)
Mary Hawkins	Editor, NSTA *Science Teacher*
Dorothy K. Culbert	Director, NSTA Division of Field Services
Richard B. Glazer	Assistant Director, American Institute of Biological Sciences
Glenn S. Pfau	Director of Project LIFE, National Education Association

Although there had been nationwide exposure of the Science

Textbook Struggle through the news media as well as within the publishing community, I had not had any discussion with science educators at the national level. Via this meeting, the issue was escalated to the top officialdom of science education.

Suspicious Sparring

The meeting was anything but impressive — at least at the start. Glenn Pfau opened by recounting our personal and professional friendship. Trying to describe a payoff for all attendees at the meeting, Pfau explained that (1) it was important for the scientists and educators present to be apprised first-hand about the California issue, rather than rely exclusively on news reports, and (2) it was critical that I hear the response of these executives at the national level.

With that inauspicious start, I carefully explained that my role had been that of a private citizen, father of six children in the public schools, and not as an expert scientist. Since my appointment as a Curriculum Commissioner was only two months old, the attendees at this meeting were not yet aware of that appointment. Further, they had not previously recalled my name so that they would not have connected my appointment with my initiating role in the issue anyway.

"Politeness" best describes the early part of the meeting. Everyone sat quietly and listened. Though I was trying to stimulate some dialogue, they were not quick to respond. I felt as though I was shadow-boxing. The first spark of life in the meeting did not occur until I mentioned my recent appointment as a Commissioner. This fact, of course, gave some political clout to what I was saying — whether or not they felt there was any validity to my ideas.

There is no doubt in my mind that most of those attending the meeting had anticipated some sort of religious argument similar to those raised by William Jennings Bryan in the Scopes trial of 1925. After all, that is how the news media had portrayed the issue.

Jerry Lightner led off the first discussion. "Are you demanding that the Genesis account from the Bible be included in science textbooks?" It was a question I had obviously anticipated.

"Absolutely not! From the very beginning of my involvement, I have denied that we were talking about God, religion, the Bible, or Genesis," I responded with emphasis.

Lightner did not look impressed. In fact, he looked like he didn't believe a word I'd said. Did you ever talk to someone who already had their mind made up about what you were going to say? I felt a little bit that way myself.

I couldn't really blame any of them for their attitude because they had been totally captive to press reports — and the news media never had told the truth. Yet, I was still hoping that they might have open minds on the subject. So I continued on.

"It is my contention — and that of the State Board of Education in California as well — that origins lie wholly outside the domain of science. No one was present at the origin of the universe, life or man. There are no actual data concerning those events available for anyone to observe. Therefore, while scientists may propose mechanisms of how these events *might* have occurred, they will be based on cosmological *beliefs* rather than on repeatable, verifiable scientific *data*."

The mood of the meeting began to warm with that statement. Several around the table gave me eye contact, whereas they had been simply looking at the table until this time. Evidently Lightner's first question had encouraged others to get involved.

"Are you saying that science textbooks at the present time discuss origins?" Mary Hawkins asked.

"Yes, I am. They discuss them both directly and by implication. And furthermore, they are committed to only one philosophic predisposition when there are other viable ones as well."

"I don't think I understand what you're driving at," she responded.

I then proceeded to expound on the points I had made only two weeks earlier with the publishers. Though speaking without notes, I was able to recall almost all my arguments. Periodically, those around the table would interrupt me to test whether I wasn't really promoting religion disguised as science. Although the meeting lasted well into the lunch hour, I was never convinced that I had successfully persuaded all of them to my viewpoint.

You can win a debate without converting your opponent to your position. I had no doubt in my own mind that I had successfully answered all their questions in a scientific context. The key, however, lay in sticking with the subject of *origins* rather than the *diversity of life*.

They continually wanted to discuss the genetic basis for diversity of life, for example. And of course, it was not a debate after all. They were there basically to learn of my position rather than vice versa.

The tenor of the meeting at its conclusion was one of guarded suspicion of my motives and yet an apparent inability to crack my argument that origins lay outside of science, that current science textbooks were biased in favor of one cosmological belief, and that the California State Board of Education was serious about changing the situation.

The only objective evidence I ever received of the NEA meeting was a letter from John R. Mayor, AAAS Director of Education. Mayor was invited to our get-together but had to attend the annual AAAS Committee on Publication meeting instead. So he had sent in his place his deputy, Arthur Livermore, and Charles Peterson from his staff. According to Mayor, both of these men made "a very favorable report" of the meeting. Saying that "it was a great disappointment to me to miss your presentation," he expressed hope that we could meet soon in Washington.

Jerry Lightner, though unable (or unwilling) to refute any of my arguments, was later to play a flamboyant role as a "savior of scientific sanctity" in reaction to what we discussed that June day in Washington.

As a matter of record, I checked out of Howard Johnson's on 16 June 1972. In the wee hours of the next morning, Alfred Baldwin looked on helplessly from Room 723 as McCord and four others were captured red-handed by Washington police in the Watergate offices of the DNC.

I escaped just in the nick of time!

Man Battle Stations!

However subdued the meeting in Glenn Pfau's office had been, there was no question in my mind later that it must have been analogous to a nerve-jangling alarm clock going off for Darwinians nationwide. Within weeks and even days, prominent science educators and members of professional scientific organizations were writing letters madly, professing alarm at what appeared to be an "irrational throwback to the Dark Ages."

One such desperate "call to arms" was forwarded to me by Ellsworth Chunn, Chief of the Bureau of Textbooks in Sacramento, shortly after the June meeting in Washington. Chunn and I had become warm personal friends shortly after my appointment as a Commissioner. He was on a long list of key people — including those from the State Senate and State Assembly, as well as Wilson Riles — who received official copies of a fear-stricken letter by Gordon B. Oakeshott, Past President of the National Association of Geology Teachers (NAGT).

The letter, on NAGT letterhead, was addressed to the incoming NAGT President, William D. Romey. It was a classic. Evidently written by a man who had spent many years defending Darwinian concepts, it dripped with exposed emotion. Because he wrote with his heart instead of his head, he committed foolish errors like saying that the State Board of Education had required "teaching of the *creationist theory of evolution*." Without doubt, he had done no research to ascertain what the Board had done. Press reports make very poor sources of fact. Clearly, they were his only source.

Because Oakeshott made such a point of his 19 years of teaching geology and an additional 24 years with the State Division of Mines and Geology, my realization that *the degree of emotion shown by Darwinians varies directly with their age* was reinforced one more time. The *older* they are, the more *emotional*. In fact, I have concluded that Darwinism is an "old man's belief." (Theologians and philosophers could have a field day with that thought.)

Although Chunn forwarded Oakeshott's letter to me without comment, I decided to respond to Oakeshott — hopefully to allay some of his fearfulness. Two specific statements in his letter required a rebuttal, I felt. To justify my comments on these points, I said that it appeared that he had either (a) stepped off from some baseline of preconception (even prejudice), (b) believed, without taking time to verify sources, the news media versions of the State Board action, or (c) overstated the facts to an indefensible limit (perhaps knowingly for emphasis sake).

His first statement, "The creationists' viewpoint is wholly unscientific and completely unsupported by scientific investigation," caused me to comment:

> This statement seems to ring with emotion and an exclusive certainty, out of character with the calm, rational and objective statements generally made by men of your experience. Further it implies that you or others may have some very conclusive evidence as the result of careful investigation. Do you personally have such evidence? You could render a valuable service to the scientific community if you could provide scientific data and/or evidence that completely rules out "the case for design."

His second statement that drew my rebuttal was this: "Creationism, is a religious *belief* — and certainly any man has a right to subscribe to it; however, it has no place in the science classroom of the public schools." This statement prompted this reply:

> I am perplexed as to how the postulate that the universe was *designed*, instead of happening by *chance*, automatically becomes a "religious belief." In the field of space technology, I never felt that I raised a *theological* question by examining the *design* of a space vehicle. In fact, I would have been thought most unscientific by my colleagues if I had "believed" that the Apollo space vehicle had happened by chance (without design, purpose or rationale)!

Oakeshott never acknowledged my 4-page letter. There could be many reasons why he didn't. I have mentioned his letter here simply as a typical example of many that were written in consternation, dismay or terror as Darwinians awoke in 1972. However, it was most rare for any of the alarmists to answer letters I wrote to them.

Writing to them was like "sinking to your armpits in a bag of feathers." They neither denied, rebutted nor debunked. They just ignored me.

One could easily take such snubs *personally*. But I suspect that these discourtesies depict, more accurately, a *general* haughtiness and disdain by the scientific elite for all those outside their exclusive set. And I would suggest that such scientists would do well to reconsider their attitude. I am not alone in this suggestion. Here is another voice of warnings:[365]

> The life sciences have come to crisis, but neither the political leaders of the nation nor the scientists seem to sense how serious it is. So much science in the past has produced unpleasant results...that people are frightened of the next breakthrough...These fears, deep and widespread, are often greeted with a paternal shrug by the scientific establishment...Scientific research has simply become too important, too effective and too conspicuous to remain private or secret any longer. What is needed is not less research but more politicization of science.
>
> It would seem appropriate for government to establish an independent public body to publicize all aspects of research, discovery and the known implications of scientific knowledge. The decision to use the available knowledge is properly a public decision, and articulating that knowledge — including the best criticism — is the proper function of government.

Since the State Board of Education is constantly changing membership due to new appointments by the Governor, the collective viewpoint of the Board on any subject likewise changes. I had seen the reaction in my meeting with science educators at NEA headquarters in Washington. I assumed that the flood of letters of protest from various people in the scientific community had resulted from that meeting. Therefore, I felt impelled to bolster whatever opinion in favor of objectivity about origins might exist in the State Board of Education.

When Donn Moomaw, the first Board member I met in October 1969, completed his term of appointment, Governor Reagan named David A. Hubbard to replace him.

On 11 July 1972, Hubbard and I met for lunch at the Pierpoint Inn in Ventura, California. For more than an hour, I briefed him on the state of the origins issue.

Because Hubbard is President of Fuller Theological Seminary and a well-known theologian, many might think that he accepted my viewpoints with enthusiasm. Such was not the case. His reaction was guarded, reflective and even a bit testy. As will be revealed later, he played a critical role in the science textbook revisions.

Though the Godly and Darwinians had been lulled into sleep, they had awakened with alarm in 1972.

The war cry on both sides of the Science Textbook Struggle was "man battle stations!"

Chapter 8

MOBILIZE EN MASSE

"Which has been indulged to excess almost always
produces a violent reaction." — *Plato*

Battles require people — lots of them. Skirmishes may involve just a few. But forces engaged in battle are generally massive.

As the Godly and Darwinians awoke with alarm in 1972 and issued orders to "man battle stations," the response was a bust. There was no one *available* — on either side! Neither faction had a standing army. There weren't even any reserves or "weekend warriors" to call up! Recruitment had to start at the grass roots.

How do you recruit for an ideological battle? Who makes a good soldier in such a fight? What characteristics should he have? True, ideas are *weapons*. But not everyone can use an idea effectively. Not even *scientists*.

Mobilization is generally keyed to anticipated conflict. The bigger the war expected, the more massive the mobilization. Because the Darwinians evidently foresaw a major conflagration, they far outmobilized the Godly. In fact, the mobilization was so lopsided that it could be described as a totally *Darwinian* mobilization.

The magnitude of the Darwinian "call to arms" was so disproportionate to the threat that it became amusing. Their mismatch of force resembled using a sledge hammer to drive thumbtacks or a fire hose to get a drink.

Never in the history of modern science has so much scientific power been assembled to resist a supposed threat to science's chastity. The Wilberforce-Huxley debates were peanuts along this display of power. The Scopes trial of 1925 was played by bush leaguers when compared with the major league all-stars who suited up for the Science Textbook Struggle.

What were the earmarks of this Darwinian mobilization? *Emotion* was the primary trait or characteristic. Those mobilized, with few exceptions, also exhibited almost *total ignorance* of the issue. In other words, they blindly reacted to a few catchwords generated by news media instead of examining the facts of the matter.

Of course, the combination of *emotion* and *ignorance* produces other undesirable products — and it did here. *Chauvinism*, *bitterness*, *irrationality* and *revenge* all surfaced — embarrassingly out of tune with the public's perception of scientists as being calm, cool and collected.

The Darwinian mobilization was aimed down three main boulevards — a letter campaign by scientific illuminati, testimony in public meetings, and exposure in the news media. Each of these avenues had its own peculiarities.

MOBILIZED MAIL

Politicians are the usual postal victims. The "bum's rush by mail" is most often dumped on them. And they are virtually immune to such onslaughts. Perhaps immunity is *inevitable* because the clues of a letter campaign are so blatantly obvious.

The letters all arrive during a short span of time. They all use virtually identical wording. They all profess alarm about the same issue. They all allege impending doom lies immediately ahead. They all demand the same action from the recipient. They all threaten the addressee with some bludgeoning if he fails to heed their arm-twisting. And on and on...

Yet crusades of postal paper are perpetually perpetrated by passionate partisans pressing for particular predominance. The larger the number of letters, the less likely any single letter will be read. At best, politicians either *count* or *weigh* the letters as they impatiently wait for the campaign to run out of gas.

The mobilized mail in the Science Textbook Struggle was not identical with that mobilized for politicians on a pet issue, however. First of all, there were far fewer letters. Further, a large percentage of the letters were written by distinguished scientists. But there was another anomaly — the more *distinguished* the author, the more probably he was a *parrot* of an emotional but erroneous issue.

Why would notable men of science fall into a trap and "tilt at windmills?" Why would they react to second-hand information and fail to carefully go back to primary sources as would be expected of scientists? We can only speculate that they implicitly trusted the organizers of the letter campaign. Or they could have yielded to peer pressure to join the campaign.

But we will never know what prompted their response to the mobilization call...

Disgust, Denunciation and Denial

"I am deeply concerned about this issue. I feel strongly enough about it that if equal time for creationism in science classes becomes a fact, I will insist that my children not be subject to this fiction in science classes at their public school. As a college Biology teacher, I would refuse to teach it other than in a sarcastic fashion."

Isn't that an objective, rational response? It is in one of the first letters that I received as a Curriculum Commissioner. The author, a biologist at one of the California State Colleges, had somehow heard that "creationism" (whatever he imagined the term to mean) would be given equal time in science classes. Neither the State Board of Education nor the Curriculum Commission had ever proposed such action.

Firing blindly at this phantom, he listed five reasons for his objections:

1. It violated the First Amendment to the United States Constitution.
2. It required teachers to become theologians.
3. It could not be scientific because it had no testable hypothesis.
4. It threatened the very basis of religion.
5. There should never be outside dictation of what should be taught in a classroom.

It is conceivable that one could conjure up some straw man to which all these objections might apply, and maybe he had. But, as far as the *facts* were concerned, this misguided professor had emptied his weapon at a purely mythical target. The pitiful thing is that his letter was *typical* rather than exceptional.

I answered his letter immediately — outlining what was *really* being proposed and offering further dialogue if he continued to have uncertainties. My letter concluded with, "I would be very interested in your version of the 'creationist viewpoint' so that I could better understand the five points of your letter."

He never answered my letter.

That was typical. With the exception of Ralph W. Gerard and Norman D. Newell of the American Museum of Natural History, I never received answers to the many explanatory letters I wrote in response to mail from alarmed scientists.

Were they too busy? Did they not believe me? Was my response unworthy of an answer? Maybe their mother never taught them good manners...

The majority of letters that were mobilized were addressed to the State Board of Education or the Curriculum Development Commission. However, I also received private letters like the one just described.

Another personal letter manifests the *attitude* of scientists that prompted the writing of this book. The author of this letter, a physics professor at Georgia Tech, refused to let me reveal his identity when I asked for his permission to do so. His letter, handwritten on Georgia Tech letterhead, read:

> Your remarks quoted in C&EN (Chemical and Engineering News) Dec 11, 1972 have come to my attention. The *ignorance* upon which they are based is noteworthy. It is far better to remain humble in the face of Nature's Laws, known or unknown, than to presume knowledge and act upon it, when one has *no* knowledge. Your remarks about entropy suggest a need for more education. The enclosed manuscript is intended to help you get started.

The manuscript that this professor enclosed with his letter was one written by none other than *himself!* That's an effective way to get someone to read what you've written, isn't it?

First, you accuse them of "noteworthy ignorance" (without bothering to find out whether they were accurately quoted). Then, you advise them to be humble because they possess "*no* knowledge." By that time, since you have reduced them to absolute zero, you propose that they need education — *starting with your own publication!*

What a unique approach to marketing one's writing. Perhaps all authors who have difficulty getting someone to read their literary work should employ this technique.

Some months later when I wrote asking that he permit his authorship of the letter to be acknowledged, he said, "Because you failed to answer my letter then, I have liquidated your file."

What was there to *answer*? Did he want me to thank him for his profound conclusion that my ignorance was noteworthy? Did he wish me to swear to remain humble until I could assimilate his masterful writing? He berated me further by accusing me of lacking sufficient interest to even look up the C&EN article and failing to "take to heart the advice offered."

But I *had* read the article! And what evidence, pray tell, did he have that I *failed* to take his advice to "remain humble" and to get educated by reading his masterpiece? His parting words to me were touching and consonant with the humility he had consistently shown:

> It has been said that "ignorance is bliss", and it has also been said that "the man who knows not that he knows not is a fool, shun him."

This recounting of scientific arrogance is meant to alert the non-scientific public to the fact that scientists suffer the same human frailties and weaknesses as anyone else — *and it spills over into their work in science!* This physics professor is not exceptional for his pomposity, impudence, and inordinate haughtiness. I was to run into literally dozens of others just like him in the Science Textbook Struggle.

For those who wish to verify first-hand the emotion, prejudice, and irrationality of prominent scientists, a representative group of their letters are reprinted *in toto* and included as exhibits in this book. Permission to reprint each letter has been granted in writing by the author in those cases where the author had not already released the letter publicly. Thus, the letters evidently represent viewpoints that these scientists are willing to defend as valid.

Perhaps the most exasperating and irritating topic in the Science Textbook Struggle has been a phrase that I used in one of the two paragraphs adopted unanimously into the *Science Framework* by the State Board of Education in 1969. As recounted earlier, those two paragraphs were never intended to stand alone and were arbitrarily selected from two different portions of my writing. The critical phrase occurs in the following sentence:[366]

> Some of the scientific data (e.g., the regular absence of transitional forms) may be best explained by a creation theory, while other data (e.g., transmutation of species) substantiate a process of evolution.

"The regular absence of transitional forms." Six innocent words. However, to Darwinians, those six words arouse the passion otherwise reserved exclusively for defending your mother's virtue.

Somehow, with total innocence, I had apparently selected the "singularly sensitive six." I could have just as well used as an example, "the high mathematical improbability of abiogenesis" or "the occurrence of catastrophes in uniformitarianism." The most vehement — as well as absurd — letters in the mobilized mail of 1972 were written in response to this six-word phrase.

Let's examine a few of these responses.

G. Ledyard Stebbins, renowned geneticist at the University of California at Davis and a member of the National Academy of Sciences, was one of the first to attack this innocuous but annoying phrase. He said that "the regular absence of transitional forms" is "an error of statement that is commonly made by those who advocate so-called "creation theory" (see Exhibit 18).

But he didn't stop there. He sent out a clarion call for colleagues to confirm his contention. Among those who immediately responded was Alfred S. Romer, former president of the American Association for the Advancement of Science and Director of the Museum of Comparative Zoology at Harvard University from 1946 to 1961. Romer said (see Exhibit 22):

> I was astounded to read in this (*Science Framework*) the phrase "the regular absence of transitional forms." Such a statement is directly opposed to the known facts of paleontology...For all higher groups transitional forms are definitely known...The statement that there is a "regular absence of transitional forms" is a direct untruth which should not be stated in any document put forth by any responsible public agency.

One can hardly get more emphatic than that! Without question, it would seem that I had not only made a serious error but had gone even further — I had sought to *deliberately portray an error.* (Unfortunately, Romer choked on a mouthful of food a short time later, fell into a two-week coma and passed away.[367] I would have been delighted to explore the phrase further with him.)

Stebbins, in his desperation to mobilize mail, evidently used the "snowball technique" to exert pressure on others to join the letter writing campaign. For example, William Z. Lidicker, Jr., Professor of Zoology at the University of California in Berkeley, in Exhibit 31 refers not to *facts* that he had personally verified but to *materials* that Stebbins had sent him (and that were incorrectly represented).

Stebbins also sent Lidicker copies of his own letters and those of others he had received in support of his crusade. Naturally, Lidicker fell into line and called "the presumed lack of transitional forms" an error.

Without specifically naming "the regular absence of transitional forms," Stebbins succeeded in mobilizing the Curator of Birds at the University of California in Berkeley, Ned K. Johnson (Exhibit 27), UCLA physiologist Donald O. Walter (Exhibit 28), the Dean of Life Sciences at UCLA, Harland Lewis (Exhibit 29) and even Nobel laureate Harold Urey (Exhibit 32) to all parrot concern about mortal danger to science if my words were allowed to influence textbooks.

Stebbins' prize collaborator in letter writing had to be the dean of paleontologists, George Gaylord Simpson. Without doubt, Simpson enjoys top billing among scientists for his research and writing about fossils. President Lyndon B. Johnson awarded him the National Medal of Science for 1965 "for penetrating studies of vertebrate evolution through geologic time, and for scholarly synthesis of a new understanding of organic evolution based upon genetics and paleontology."[368]

If anyone in the world could speak with authority on the subject of transitional forms in the fossil record, it would be Simpson. Stebbins apparently sought Simpson's support before anyone else's because Simpson's letter (Exhibit 21) is dated the earliest. What does Simpson say about my choice of words — "the regular absence of transitional forms?"

> I have paid special attention to Professor Stebbins' discussion of transitional forms between major groups of organisms, as this is a subject that I have studied for many years and on which I have published extensively. Professor Stebbins' statement agrees with my own views and with those of virtually all paleontologists...*Anyone who cites me or my work in opposition to Professor Stebbins' statement is either woefully ignorant or willfully misrepresenting the facts.* (emphasis added)

That unequivocal statement by such a scientific luminary should slam the door, once and for all time, on the case. Those six sensitive words that provoke such inflammatory reaction have become the prime proof of my scientific ignorance. The fingers of ridicule and derision all point to my stupidity in imagining such heresy.

But I must make a confession — those six words were *plagiarized.* I did not dream them up. I borrowed them. From whom? *George Gaylord Simpson*!

Of course, it is always possible to quote someone out of context — especially when you use only six words. So all the disgust, denunciation and denial may still be valid.

In what sense did Simpson coin the words? *In discussing gaps in the fossil record!* Figure 9 is taken from one of Simpson's most famous books.[369] His extensive discussion of this picture of the fossil record is reprinted in Exhibit 48. How did *he* use the phrase "the regular absence of transitional forms?"

> This regular absence of transitional forms (in Figure 9) is not confined to mammals, but is an almost universal phenomenon, as has long been noted by paleontologists. It is true of almost all orders of all classes of animals, both vertebrate and invertebrate. A fortiori, it is also true of the classes, themselves, and of the major animal phyla, and it is apparently also true of analogous categories of plants. Among genera and species some apparent regularity of absence of transitional types is clearly a taxonomic artifact: artificial divisions between taxonomic units are for practical reasons established where random gaps exist.

Some may protest that this quotation is not from one of Simpson's latest books, and that is a proper observation. However, this book was reprinted in 1965 (21 years after original publication) with no disclaimers or updating comment. And lest someone suggests that a major revision of Simpson's thinking simply failed to get incorporated in his reprinted book, Norman D. Newell, another noted paleontologist from the American Museum of Natural History, confirms Simpson's original conclusion in a studied treatise entitled, "The Nature of the Fossil Record":[370]

> Experience shows that the gaps which separate the highest categories may never be bridged in the fossil record. Many of the discontinuities tend to be more and more emphasized with increased collecting (of fossils).

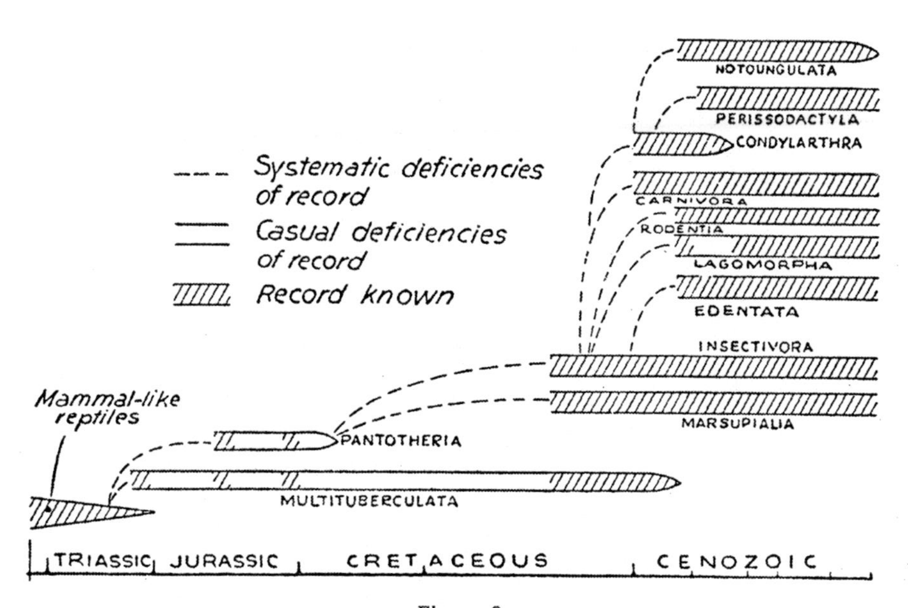

Figure 9

Systematic Deficiencies of Record of Mammalian Orders

Known spans of some of the orders of mammals with relatively good records and inferred phylogenetic and temporal relations of their systematically lacking origin sequences. No ordinate scale. Abscissal scale arithmetic, approximately proportional to absolute time. *(reprinted by permission)*

The mail mobilized by Darwinians in 1972 fell far short of the lofty criteria Stebbins established in Exhibit 26 for "informed scientific opinion." Almost without exception, even those who personally responded to Stebbins' call to arms violated his criteria — especially his stipulation that "scientific opinion can be regarded as informed only when the scientists in question have acquired first-hand information about the subject being discussed."

For example, Donald O. Walter, a *brain physiologist*, posed as an expert on the conditions existing on the early earth, using examples from the *physics of gases*, *rock chemistry* and *geophysics* (see Exhibit 28). Are we to suppose that brain physiologists are daily gaining first-hand information on the chemistry of rocks? Maybe that popular saying, "He has rocks in his head," has a scientific basis…

Stebbins, in establishing his criteria, seemed particularly incensed that I had mentioned in a July 1972 Curriculum Commission meeting that Wernher von Braun, like Stebbins, might also be classified as a scientist and entitled to his opinion about the origin of the universe — even if it was counter to Stebbins'.

Misquoting me (in Exhibit 26) as saying that von Braun was an "authority on the origin of *life*," Stebbins then made his firm statement about the necessity for first-hand information. Yet, who could have *first-hand information* on the origin of life?

Apparently thinking that he knew such an expert, Stebbins urged Harold G. Urey, Nobel laureate in chemistry, to join the fray. Exhibit 32 is Urey's response. Note his sweeping statement about the unanimous *belief* in evolution that every single scientist of his acquaintance subscribes to (as though evolution was being threatened!). But what has *that* belief got to do with the origin of life? Does *belief* establish *truth*?

Further, Urey's previous statements on the subject hardly sounded authoritative or conclusive:

> All of us who study the origin of life find that the more we look into it, the more we feel that it is too complex to have evolved anywhere. We all believe as an article of faith that life evolved from dead matter on this planet. It is just that its complexity is so great, it is hard for as to imagine that it did.[371]

and even more recently...

> I think those who have made the most extensive study of the very complicated chemistry of living organisms are those who are most amazed that life could have evolved at all.[372]

There is no question that Stebbins fully intended for Urey's letter to shut off all further debate on the origin of life. A Nobel laureate is an automatic trump card, isn't he? *So what* that Urey says, "the problem of the origin of life is being investigated vigorously at the present time?" Isn't that analogous to saying that no other theories for curing cancer will be considered because "the problem of curing cancer is being investigated vigorously at the present time"?

As mentioned earlier, *emotion* was probably the prime denominator of all the Darwinian's mobilized mail. And it was difficult to differentiate the emotion of the *scientific elite* from the emotion of the *uneducated layman*. The scientist's emotion was as unsophisticated and uncouth as anyone else's.

Compare zoologist Ned K. Johnson's irrational and irresponsible discussion in Exhibit 27 with that of Lorne Albert Lenaghan, a self-taught layman with no formal education (see Exhibit 23). Johnson uses inflammatory words like *woeful ignorance*, *insult*, and *betrayal* — words which evoke passion rather than logical thought.

Actually, with the exception of the reference to "fundamentalist rot," Lenaghan's letter is far more credible than Johnson's. Zoologist William

Z. Lidicker, Jr., whose letter (Exhibit 31) was mentioned earlier, was as guilty as Johnson when he spoke of *unfortunate intrusion* or *now notorious* in referring to my two paragraphs that had been adopted into the *Science Framework*.

A telegram from two professional research biologists is included too as Exhibit 33 because it is a classic of absurd reaction. I still have no idea whether they were serious or attempting a parody.

Harlan Lewis, UCLA's Dean of Life Sciences, was properly restrained in the use of emotional *words* but unnecessarily raised an emotional *non sequitur* — that California school children would fail to receive the same exposure to scientific understanding of evolution as do children in the rest of this country if textbooks included an alternative view. In his eyes, evolution is apparently a fragile concept that, without *exclusive* exposure, cannot be understood scientifically.

In addition to Stebbins' straightforward appeal for a barrage of mail, there was another attempt to enlist letters that could only be described as "sneaky." Junji Kumamoto, my colleague on the Curriculum Development Commission, nearly pulled off the attack with success. I was tipped off — almost too late — that he intended to publicly "stab me in the back." The tipoff came quite innocently.

"Hey, Vern, I got a letter the other day from a guy named Kumamoto at UC in Riverside that made me curious."

It was Mark Biedebach calling in late August 1972. Biedebach, it will be recalled, had been closely involved with me in the Science Textbook Struggle from the beginning. He had even been present in the auditorium when the two paragraphs were adopted in November 1969.

"What did it *say*?"

"For one thing, it looks like some sort of entrapment — like the old 'Have you quit beating your wife yet?' sort of thing," he replied.

"Why don't you read it to me?" I suggested, beginning to suspect some sort of foul play.

Biedebach then read Kumamoto's short letter that started by saying, "You have been cited by Mr. Vernon Grose as one who not only endorses but also teaches the concept of the principle of creation."

I interrupted Mark at that point. "It sounds like he is referring to my speech before the State Board of Education in November 1969 — you know, when I mentioned you and several others in academia who held the position that one could view scientific data regarding origins as supporting the creation hypothesis as well as one of chance."

"Oh, I had forgotten about that...You're probably right. But that isn't the disturbing part of the letter. Let me read on..."

The next sentence was the clue that Kumamoto intended to secretly discredit me (and thereby the issue of origins) by first misrepresenting what I had said so that it looked like I had put words in the mouths of others and then by securing letters of denial of my misrepresented words that he could suddenly spring on me, without warning, at the September meeting of our Commission.

Mark continued reading, "It is implied (by Grose) that creation by Divine design is a viable alternative hypothesis to the teaching of evolutionary theory in the subject area of science."

"You know good and well that I really referred to you and the others I mentioned as someone who teaches a creation concept as a *complimentary* rather than an *opposing* idea to evolution," I protested as I dug furiously through my desk file for a copy of my November 1969 speech (see Exhibit 4). Finding it, I quickly located my reference to Biedebach and the others.

"Listen to what I said. 'Creation, on the other hand, does provide rationale for "first causes," and here the complimentary feature of the two concepts is illustrated.' My whole speech was devoted to *eliminating* the polarization or pitting of evolution against creation."

I was upset.

"Well, it is evident that Kumamoto recognizes that he may be misquoting you because he goes on to say, 'If there has been an error in the citation or a misunderstanding of what is meant by creation, I would appreciate a reply before September 7, 1972.' So, you have to give him that much credit."

"Sure, but if he *really* wanted to avoid error, why didn't he ask me to secure a letter from you that would have circumvented both his error in introducing 'Divine' and the possibility of you reacting to the 'strawman' he created? If he still wanted to exclude *me*, he could have sent *you* a copy of my speech so you could see the context of citation...I have a hunch that he may have sent identical letters to all the others I mentioned, too — Knoblock, Hearn, Anderson and Bullock. If so, I have even a worse problem because they weren't there in person, like you were, in November 1969."

"Oh, I haven't any doubt that he probably has! Maybe you'd better get in touch with them right away. But it's probably too late because I got my letter over a week ago."

My fears were confirmed. Long-distance calls were immediately made to Wilbur Bullock in New Hampshire, Walt Hearn in Iowa, Elving Anderson in Minnesota, and Irving Knoblock in Michigan. It *was* too late. Their answers to Kumamoto's misleading implication had already been sent — and Kumamoto had apparently scored a coup.

Fortunately, although I had used their names in November 1969 without notifying them and further they had no idea of the context in which I had cited them, they had independently seen through the ruse that Kumamoto was attempting. I was relieved when I learned the contents of their replies.

V. Elving Anderson, Assistant Director of the Dight Institute for Human Genetics at the University of Minnesota as well as former President of the Minnesota Academy of Science, was forthright with Kumamoto while chiding him for attempting to put words in his mouth:

> Yes, I do accept and endorse the concept of creation. I find it difficult, however, to describe what that concept means to me by agreement to a simple brief statement written by someone else.
>
> I would be very happy to expand on my views if you could give me an idea as to what kind of statement would be most helpful to you at the present time. Specifically, are you interested in views of evolution as science or as world-view or both?

There was no question as to Anderson's scientific qualifications. At the time of his letter, he was the Secretary of the Behavior Genetics Association (of which Theodosius G. Dobzhansky was the President) and on the National Executive Board of Sigma Xi, the honorary society for natural sciences. He had clearly differentiated between Evolution I (science) and Evolution II (world-view) — something that Kumamoto was never to understand throughout the Science Textbook Struggle.

In similar manner, Wilbur L. Bullock was a spoiler of Kumamoto's planned attack. Professor of Zoology at the University of New Hampshire, Bullock was also Chairman of the Subcommittee on Acanthocephala for the Committee on Systematics and Nomenclature of the American Society of Parasitologists at the time he wrote his reply. Like Anderson, he faulted Kumamoto for his sophomoric attempt to oversimplify:

> The only misunderstanding implied in your letter is that creation is an *alternative* hypothesis to evolution. The issue is not that simple and I don't believe Mr. Grose is so simplifying it. The issue is that the teaching of evolutionary theory in high school and university is dominated by a materialistic and often antichristian philosophical outlook with emphasis on "chance" to the exclusion of "design." Such one-sided presentations have no place in a free society and they are not objective positions in a science classroom...
>
> It seems to me that Mr. Grose's position is quite reasonable. The fact that some people suspect an "anti-evolutionary plot" is further evidence

some of the people involved need to be a bit more open and listen to what others are saying.

The other responses were equally lucid in unraveling what Kumamoto undoubtedly thought would be a brilliant maneuver. He never introduced or mentioned those letters, nor did I ever acknowledge that I knew of their existence. *That* mobilized mail backfired on him in more than one way:

1. I had cited these five scientists correctly, so he couldn't discredit me personally.
2. All five were bona fide scientists who fit Stebbins' criteria perfectly.
3. Every one of the scientists confirmed my contention that creation or design deserves consideration when teaching about origins.
4. They all had made my points better than I could.

In retrospect, I have wondered whether Kumamoto personally conceived of this attempt to discredit me or whether he was only the "front man." Many times, I pitied him for the role he was being forced to play by frustrated scientific illuminati.

A slight, obscure plant chemist at the University of California in Riverside, he had no academic rank. He was simply a lecturer. He was neither articulate nor organized in his public speaking. Yet he had been vaulted into prominence as the focal point of all those who felt evolution needed defense. His mind seemed to be cluttered with minutiae that prevented him from focusing on major issues. His main contribution during Commission meetings, for example, frequently involved technicalities regarding Roberts' Rules of Order.

Another one of Kumamoto's quirks was his apparent obsession over racial or ethnic minorities. His fervor was understandable, on one hand, because of his own Japanese ancestry. More than likely, he may have even been appointed to the Commission for his racial status rather than his scientific background. But, given those factors, he still carried his passion for minorities to an inordinate extreme.

One irony of his fanaticism in defending minorities was his obvious *hypocrisy* when it came to *scientific* opinion. He was totally insensitive to a potentially minority viewpoint among scientists about a *non-scientific issue* — origins!

At one point, Kumamoto had spelled out his strategy. "It is my feeling that if the scientific community can present a united stand, it should be possible to resolve the issue of the mandatory issue of divine creation as a scientific theory."

He then set about to show that the *vast majority* of scientists believe in evolution as the sole explanation for origins. *Ipso facto*, any scientist who thinks otherwise is either unqualified as a scientist or so unaware of scientific opinion as to be unworthy to be heard.

Where was his passion for an oppressed minority? It may well be that the size of the minority of scientists who do not believe that Evolution II is the only possible belief about origins may far exceed the size of the minority of Americans with Japanese ancestry. And even if they don't, are they entitled to be heard — or only stomped into oblivion? What if the non-believers in Evolution II had all been of *Japanese ancestry?*

I honestly feel that the burden of representing the scientific elite in defending the status quo was beyond Kumamoto. Further, the task was thrust upon him without warning — much like the sudden release of an avalanche.

At least my involvement in the Science Textbook Struggle had *gradually* grown. Undoubtedly, too much was expected of him. Nobel laureates and members of the National Academy of Sciences all expected him to be their spokesman — on a subject with unsurpassed *emotional* commitment. Kumamoto's pattern of thinking typified the tenacious nit-picking that a research chemist must have — quite the opposite from that needed in a philosophic issue like origins.

Because of these inherent limitations, Kumamoto avoided personal confrontation with me whenever possible — even to the point of *dishonesty*.

For example, in the September 1972 meeting of our Commission, he personally handed copies to each Commissioner of a report that he titled "Report of the Science Committee." Although I was a member of the Science Committee, I had never seen the report. And it maligned several things I had written. Of course, I challenged him on this blatant attempt to pass his own ideas off as those of our Committee.

The other members of the Committee agreed that they too had not seen it. So he was forced to withdraw it.

I also tried numerous times to get him to respond in writing to written questions I had asked him, but he refused. Therefore, my wonder continues as to whether he *conceived* or only *executed* the "sneak attempt" to mobilize mail that would have had my own friends testifying against me.

Less than a week after I blocked his railroading venture to trick the Commission into accepting his personal view as that of our Science Committee, Kumamoto openly mobilized a *mail* campaign. It was a desperation move. Quite obviously, he felt the pressure mounting as he repeatedly failed to convince the Curriculum Development Commission that dogmatism was essential for preserving the purity of science.

Even though he could wave letters in Commissioner's faces from such prominent spokesmen for science as Thomas H. Jukes of the University of California at Berkeley, there was still open-mindedness among us all. Probably, it was the chauvinism of people like Jukes that turned out to be most counterproductive to Kumamoto's strategy of disallowing alternatives.

Consider, for example, this statement that Jukes sent to Kumamoto for our consumption:[373]

> The time record of evolution is accepted by all knowledgeable scientists as being shown by the fossil record, by isotope dating and other well established scientific procedures. The scientific community, with its experts in specific fields, has examined such findings and has accepted them.
>
> It would be most unfortunate to confuse our school children by representing that there is a division of informed scientific opinion on these matters.

Clearly an attempt to persuade laymen to believe that scientists are *unanimous* in their acceptance of evolution, isn't it? In addition to inciting Jukes, Stebbins had also succeeded in provoking Nobel laureate Melvin Calvin to overshoot in a letter to Kumamoto. Calvin complained bitterly that while a few revisions proposed by one publisher (but by neither the Commission nor the State Board) to open up the dogmatism concerning origins were technically truthful, "the overall impression which is conveyed in reading them is seriously misleading." Only because an alternative to the chance origin of the universe was offered, Calvin became an incensed reactionary:[374]

> In every case there is the general impression created that *scientists* are about equally divided on the nature of the origin of the earth and the life upon it. The division is not among scientists but between scientists and nonscientists who profess a belief in special creation.

With such heavy artillery as Jukes and Calvin, why would Kumamoto still feel pressured to organize a barrage of mail? Maybe because of the five surprising (and disturbing) letters he had received as a result of his aborted "sneak attack."

Why didn't he ever wave the letters from Anderson and Bullock while waving those of Jukes and Calvin? When Jukes and Calvin made their unwarranted pretext that scientists were unanimous, why was Kumamoto silent when he had proof positive in his hands that they were *not* unanimous? Where was his sense of *fairness* and *objectivity?* He had none. All's fair in love and war ... and this was *war!*

So Kumamoto sent out his urgent call for letters to 16 prominent scientists. He summarized his three-page appeal thus:[375]

> The informed scientific community is united in the position that creationism is not appropriate for inclusion in science. We would urge that California not knowingly introduce non-science into the science curriculum!
>
> May I solicit your comments and support for this opinion statement before the end of October because the public hearing on science textbooks is scheduled for November 9th. Could I further impose upon you to have your colleagues read the opinion statement with a request for comments or support in a letter to me?

And he got just what he asked for...emotional tirades laced with phrases like *no scientists of any standing, overwhelming percentage of qualified scientists*, and *no currently acceptable alternative theory.*

With no exceptions, every response ignored the real issue — *presuppositions* or *belief systems* that underlie "scientific" conclusions about origins. The one ability that all his respondents unanimously displayed was to "tilt at the same windmill." Like puppets on a single string or a row of tin soldiers, they unthinkingly kowtowed in perfect unity to a straw man — spewing forth disgust, denunciation and denial...

Supplication, Sustenance and Support

I'll confess. It was lonely.

The organized letter campaigns by Stebbins and Kumamoto took their toll. There was no way I could match the impact of those onslaughts — either in terms of scientific notoriety or number of letters.

As recounted earlier, Astronaut Ed Mitchell and I had a personal acquaintance. We had spent three hours in private conversation shortly after he returned from the Moon on Apollo 14. One topic that we had discussed in considerable depth was the origin of the universe, life, and man. We continued to correspond with one another by mail after that conversation.

I mentioned my discussion with Mitchell at the publisher's briefing on 30 May 1972. After all, it seemed to me that Mitchell's credentials in science were at least equal to high school graduate Richard E. S. Leakey. Mitchell had earned his doctorate in astronautics at Massachusetts Institute of Technology and had been the sixth man in history to walk on the Moon. His first-hand observation of natural phenomenon (i.e., the Moon) had to be at least as significant as Leakey's.

During my next exchange of letters with Mitchell following the publishers' meeting, I suggested that it would be helpful to me if he would put in writing some of his thoughts about the possibility that the universe, life, and man had occurred entirely by chance. This he kindly did (see Exhibit 17).

In contrast to the letters mobilized by Stebbins and Kumamoto, Mitchell's letter to me was strictly private — and it was so treated throughout the Science Textbook Struggle. It simply corroborated what he and I had been privately discussing for many months.

In mid-July 1972, I saw a news article that mentioned that Wernher von Braun had delivered a series of lectures at Taylor University on the subject of "Science and Religion." Because of our common NASA ties, I wrote to him and asked for a copy of his two lectures so I could study them. He immediately replied, saying that he gave the lecture series "without the benefit of a prepared script." However, he enclosed a copy of a talk that he had given along similar lines at Belmont College in North Carolina subsequent to his Taylor University series.

I was impressed by the rational and penetrating line of thought that von Braun had expressed in his Belmont College address. It described precisely the position that I believed the State Board of Education was desirous of having presented in the science classroom.

Therefore, I wrote once more to von Braun, asking if he would be agreeable to condensing some of his thoughts in letter form so that I might share them with the California State Board of Education. He graciously did so, and Exhibit 24 is a copy of that letter.

His letter became widely distributed. It was read on radio and television and was reprinted in a number of periodicals. Undoubtedly, his letter will live on as one of the more beautiful and ageless expressions of wisdom that emerged from the Science Textbook Struggle.

One example of how von Braun's letter influenced the news media appeared in a newspaper editorial. This editorial chided scientific arrogance while quoting extensively from von Braun's letter:[376]

> For years, mankind has wrestled with the problem of whether he was the creation of God, as assured us by the Bible, or was the inspiring result of evolutionary process that started with a bubble in the mud.
>
> The Supreme Court, which sometimes takes its title much too literally, has ruled that the children of atheists cannot be force-fed the Biblical story of creation, but the Court has looked casually aside while children of God-fearing families have been force-fed the theory of evolution…
>
> There must be a certain variety of special arrogance reserved for the person who feels that he, a human being, is the ultimate, the final

> triumph of an evolutionary process. There must be a special kind of satisfaction in the makeup of a person who feels that there is nothing higher than he, no Creator, no God, no great Designer...
>
> Admittedly, the actual incorporation of the Theory of Evolution into education came at the end of a long and relentless process, but strangely, the elimination of God was quick and ruthless. Strangely, some of those educators who demand the right to inspect all aspects of all questions under the cloak of academic freedom, renounced God with a cavalier shrug of the shoulder...
>
> We owe it to our young people to bring God back into their lives, if only as an alternative to those scientific adventures that only isolate questions they do not answer. Just because He is no more visible than the electron, is no assurance to the open mind that He does not indeed exist. A loving God will smile, and understand.

Much of this editorial was based around von Braun's penetrating question and observation from his letter (Exhibit 24):

> What strange rationale makes some physicists accept the inconceivable electron as real while refusing to accept the reality of a Designer on the ground that they cannot conceive Him? I am afraid that, although they do not really understand the electron either, they are ready to accept it because they managed to produce a rather clumsy mechanical model of it borrowed from rather limited experience in other fields, but they would not know how to begin building a model of God.

Von Braun's stature in the world apparently sent shivers of terror into those who were committed to scientific dogmatism on origins. The widespread positive response to his letter caused many who thought they had an airtight case against any alternatives to Evolution II to worry about the extent to which von Braun might become involved in the struggle.

One such frightened individual was William V. Mayer, Director of the Biological Sciences Curriculum Study at the University of Colorado. On 9 October 1972, Mayer wrote to von Braun inquiring about rumors that von Braun was to appear in California in support of "the inclusion of fundamentalist religion in biology textbooks." (Try, if you can, to extract any "fundamentalist religion" from von Braun's letter in Exhibit 24.)

In responding to Mayer, von Braun confessed that Mayer's rumor was a complete puzzle to him. Von Braun then explained that he had written a letter to me. He had later learned that the letter was read at the September meeting of the California State Board of Education and had been favorably received by educators and the news media. He also confessed that his letter to me seemed to have attracted more attention than he ever expected.

To set Mayer back on track as far as what his original letter had said or implied, von Braun made two observations:

1. If fundamentalistic religion means belief that the Book of Genesis gives a correct scientific account of how the world came into being; that 4004 B. C. is the date of the origin of the earth, and that all living species were "created" in their final form rather than developed through evolutionary, "survival-of-the-fittest" processes, then I am most emphatically not a believer in fundamental religion.
2. If, however, the question is whether behind the many random processes which are operating in nature, there is a "divine intent", my answer is an equally emphatic "yes." With this position I am only sharing and accepting the views expressed by giants of science such as Newton, Kepler, Faraday, Pascal and Einstein.

In case Mayer did not recognize it, he had been exposed to a kind but firm lesson on "whose mouth not to put words into."

I also received support for my position from the Curriculum Development Commission itself. Curriculum Commissioner Carroll (Bud) Creighton had served on the antecedent Curriculum Commission. He was among those carryover commissioners who felt strongly that continuity on the issue of origins was important, even though there had been a re-organization.

I found him to be a wise and sustaining counsel throughout the struggle — not only because he frequently supported my position but also because he had extensive and important involvement in the educational process. I lacked any experience in the public school system. Bud was recognized as both an educator and an administrator of public education by virtue of his position as Assistant Superintendent of the Orange County School System.

Bud kept the issue of origins before the Orange County Board of Education because he saw its importance to the minds of school children. As the battle began to heat up, Bud urged the Board to support the 1969 position of the State Board of Education by resolution. They heeded his counsel. Exhibit 25 is the *unanimous* resolution passed by the Orange County Board of Education on 17 August 1972.

This was a particularly significant action. While other resolutions were later sent to the Curriculum Development Commission and the State Board of Education, they did not represent public school interests but, rather, elitist professional scientific organizations who were far from the real world of children in a classroom. And I always felt that the child deserved our greatest attention.

Because of my strong conviction that there was danger in being swayed by scientific notables away from the public school classroom, I valued most highly those commissioners whose experience and advice focused on the teaching of children.

Commissioner Zelda E. Dawson was one such person. She was a classroom teacher. In a letter dated 1 September 1972, she expressed her belief in the orderliness of the universe that she said had been reinforced by Wernher von Braun's letter (Exhibit 24). Her chief concern about children's minds — was summarized this way:

> Teachers have such varied beliefs themselves that textbook material is reinterpreted anyway. My chief concern for the classroom is to teach children that theories are tentative, and that the degree of certainty with which one can accept things as fact, not belief, is based on tremendous amounts of "testable" data. My only WHEREAS would be that all books adopted as basic instructional systems be edited to ensure that there is a clear and obvious distinction made between what is fact and what is belief.

While her letter was not the result of supplication, it was a source of support that I really appreciated. Such unsolicited encouragement was indeed rare.

I admit to inviting one more letter. Astronaut Jim Irwin, who on the Apollo 15 mission became the eighth human to walk on the Moon, and I were conversing by long-distance telephone in early August 1972. Since a major confrontation was anticipated to occur during the September meeting in San Diego of our Commission and the State Board of Education, I asked Jim whether he might be free to be with us and offer his personal observations on the "chance-versus-design" controversy.

He was uncertain about his schedule at that time, so I asked him to drop me a letter on his views if he was unable to come out to California. His schedule did not permit him to make the trip, so Exhibit 30 is a transcription of his hand-printed letter.

Correspondence also was provoked over a very inflammatory error: the controversial and widely-publicized claim that the State Board of Education had mandated "equal time" for the Biblical account of man's creation alongside the Darwinian version.

As this book's chronology of the issue should amply verify, the State Board of Education never intended nor proposed such action. And it should be even more evident that I personally never was a party to the "equal time" idea.

The apparent history of this controversy started in 1963. In that year, a small group of scientists formed an organization in Michigan known as the Creation Research Society. By 1971, it had grown to about 350 members. Also in 1963, two Southern California housewives, Jean E. Sumrall and Nell J. Segraves, initiated action to seek legal relief from the exclusive teaching of "atheistic evolution" in Mrs. Segraves' son's science classroom.

The latter effort produced a legal opinion by the California Attorney General to the Superintendent of Public Instruction that included the following statement concerning religious doctrine:

> Those constitutional and statutory provisions that provide "no sectarian or denominational doctrine" shall be "taught or instruction thereon be permitted directly or indirectly in any of the common schools of this state" apply equally to all forms of religious belief irrespective of whether they embody a belief in the existence of God. Thus the "teaching of" atheism or agnosticism in the public schools is prohibited if by the words "teaching of" it is meant the teaching of doctrine with a view toward obtaining an acceptance as to the truth of that doctrine...

The Attorney General further pointed out that the State Education Code contains penalties that are applicable to:

> The making of statements, in such schools and colleges, which advocate, tend to advocate, or implant in pupils' minds a preference for, atheism or agnosticism or which reflect unfavorably upon any particular religion, upon all religions, or upon any religious creed.

Armed with that ruling, Mesdames Sumrall and Segraves mounted a campaign for "equal time" for Biblical creation and atheistic evolution by reading that premise into that Attorney General's opinion statement. (It would seem far more logical that the Attorney General was more specifically *disallowing* the teaching of Evolution II than *insisting* that all philosophies of origins be taught with equal emphasis.)

However, nothing significant in these women's battle occurred until November 1969, when the State Board of Education unanimously adopted my words into the *Science Framework*. By that adoption, the women *assumed* that the adopted paragraphs mandated "equal time". Of course, that was an error of assumption — one that continually plagued me.

Shortly after the November 1969 State Board of Education decision, the Scott Memorial Baptist Church in San Diego was persuaded to organize a core or team of people who might prepare "creationist" science textbooks and teaching materials for submittal to the Curriculum Commission for adoption in California public schools. It looked to this church as though a brand new market had opened up, and they would be the first to capitalize on the supposed "equal time" ruling.

In the fall of 1970, the church formed this team at the time they also established Christian Heritage College. The team was named the Creation Science Research Center (CSRC) and was designated an affiliate of Christian Heritage College. Henry M. Morris, who had been Professor of Hydraulic Engineering and Chairman of the Department of Civil Engineering at the Virginia Polytechnic Institute since 1957, resigned his position to accept the directorship of CSRC in 1970. Mrs. Segraves' son, Kelly, was named Assistant Director of CSRC.

Within two years, Morris and those with scientific qualifications withdrew from CSRC and formed another organization, the Institute for Creation Research. Morris was simultaneously serving as President of the Creation Research Society. So by 1972, there were three organizations devoted to "creationism":

1. The Creation Research Society (CRS), a national society of scientists "dedicated to research and publication in support of creation versus evolution as the most likely explanation of origins."

2. The Creation Science Research Center (CSRC), a non-scientific layman's organization whose primary function is marketing creationist materials and information and directed by Kelly L. Segraves.

3. The Institute for Creation Research (ICR), a division of Christian Heritage College and directed by Henry M. Morris.

Throughout the Science Textbook Struggle, the news media repeatedly tried to link me with one or all of these groups. I must emphasize that I am not, and at no time in the past have been, a member of any of these organizations. As recounted earlier, I became acquainted after November 1969 with a number of individuals in all three associations and consider several of them to be personal friends.

A very disturbing article from the *Oakland Tribune* was mailed to me by a fellow Commissioner in August 1972. It carried bold headlines, EQUAL TIME FOR BIBLE, DARWIN IN CLASS ASKED, and was an Associated Press release carrying a Sacramento dateline. The lengthy article included such statements as:

> (Nell Segraves of the Creation Science Research Center says) the state Board of Education adopted a policy several years ago that the Darwinian theory and the Biblical account of man's creation get equal time...
>
> "We could easily end up with 50 per cent of the tax dollar used for a religious, Bible-based philosophy in public education," Mrs. Segraves

said. "At present, 100 per cent is used for education which is basically agnostic or atheistic."

"Based on what has already been done to protect the religious beliefs of the atheist child from offense in the classroom, they would have to give equal time to the Christian child," she added.

That kind of thinking, if Mrs. Segraves was correctly quoted, was totally divisive to what I was trying to accomplish. I had learned only a few days earlier that ICR had spun out of CSRC as a separate entity, so I decided to find the ICR position on the "equal time" idea.

I minced no words with Henry Morris, Director of ICR.

"I must ask you directly whether you and the Institute endorse the position as reported in the (*Oakland Tribune*) article. If so, I feel that we are on a divisive and destructive collision course," I warned him. My letter closed with, "Please advise me of your Institute's official position on this reported interview as soon as possible."

Morris immediately responded. His answer was *unequivocal* — ICR was *not* advocating "equal time":

> We were as distressed as you apparently were when we saw the newspaper account of Nell Segraves' views on religion in the schools...We most certainly do not concur in any way with what she was quoted as saying. We do *not* favor teaching religion in the public schools...Neither our own organization nor the Creation Research Society endorses them in any way. All of us specifically reject and oppose the opinions she expressed in the news release.

That assurance was solace indeed! Seldom had one single letter provided such optimism for me that this issue was still on track. What little mail I had mobilized had provided me with sustenance and support...

Collective Contempt

Not all the mobilized mail was *individualized.* Much of it came from powerful *coalitions.* Prestigious bodies of scientific illuminati, distinguished fraternities of academicians, and influential orders of the scientific elite all jumped into the fray.

Never in the history of modern science has such a massive reaction been provoked! And, unfortunately, never has there been such an irresponsible reaction.

Why was this organized revulsion so far beneath the dignity of those who got trapped in it? For at least three reasons:

1. *They "tilted at windmills."* Without exception, they fired salvos at the same phantoms that their colleagues, as individuals, had.

(Of course, the collective contempt was organized in most cases by scientists who had already protested as well by personal letter.) They attacked *divine intent*, *religious accounts*, *special creation*, and *religious philosophy* as though those ghosts were real.

2. *They mistakenly thought evolution was under attack.* Every one of their resolutions of protest mentioned *evolution.* Why? In no way was the teaching of evolution threatened, maligned, excised, or modified by the State Board of Education's action.

3. *They were hypocrites.* They loudly complained that a public body like the State Board of Education had no right to establish what constitutes the content of science. Yet, all their resolutions, addressed to the State Board of Education, begged this public body to establish *their* version of the content of science!

The public generally assumes that scientists, of all people, are scrupulous in seeking *data* before jumping to conclusions. They expect scientists to ferret out *facts* before forming an opinion. Scientists are the last people in the world that a layman would suspect of "going off half-cocked."

However, this is exactly what happened! In a 3-week period from mid-October to early November 1972, there was a furious but foolish fusillade of fulminatory fireworks focused on fallacies.

Why, why, why? Maybe the key lies in a letter written at the very outset of the Science Textbook Struggle and signed by 15 Caltech biologists, including Max Delbruck, 1969 Nobel laureate. Note the unqualified leadoff statement in their letter (Exhibit 3): *"Biologists consider evolution to be a fact, not a theory."* When something is a fact, it no longer is debated or reconsidered. As Nobel laureate Harold Urey says in Exhibit 32, "The evidence in regard to evolution as a fact impresses me as strongly as the evidence for the theory of universal gravitation." So their minds are slammed shut on that subject.

But which "evolution" are they talking about? *Evolution I* (science) or *Evolution II* (belief)? Probably both — although there obviously can be no "scientific evidence" for Evolution II. Of course, this failure to differentiate between Evolution I and II is most unfortunate. Personally, I have no quarrel whatsoever with their letter if they mean exclusively Evolution I. However, since they were challenging a statement attributed to the State Board of Education concerning *origins*, they obviously meant Evolution II as well as Evolution I.

It probably did not occur to them that they were less than scientific in saying, "Biologists consider..." instead of, "*Many* biologists..." or "*Most* biologists..." Despite closed minds at Caltech, there *are* qualified biologists who *accept* Evolution I and *reject* Evolution II. In fact, such biologists might rewrite the Caltech letter, if origins (not variation) were their concern, by substituting "creation" (meaning Creation II) where "evolution" formerly appeared:

> Biologists consider creation to be a fact, not a theory...When scientists speak of the "theory of creation" they are not questioning the reality of creation, but are saying that we do not yet understand in detail the mechanisms by which creation comes about. There are theories about these mechanisms, and these constitute the "theory of creation." But there is no doubt that creation as such has occurred and is occurring. This is a fact that can be demonstrated in nature and in the laboratory.

It may be helpful to turn back to Figure 5 and reexamine the definitions of Evolution I and II, as well as Creation I and II. Because those scientific and academic organizations that launched the three-week tirade of collective contempt in October 1972 failed to understand these definitions, they all misfired.

Let's look at individual examples of their misfiring. Consider first the resolution by the AAAS Committee on Science Education (Exhibit 40). Here are their specific errors:

1. The California Board of Education was not requiring "that religious accounts of creation be taught in science classes."
2. The "evolution of life" was not an issue in California and therefore is a *non sequitur.*
3. There is no such concept as "*the* biological theory of the origin of life." As discussed in detail in Chapter 1, biology is rampant with widely varying ideas (hardly theories!) of how life might have started here on earth.
4. The AAAS failed to realize that statements about origins can *never* be scientific statements "because such statements are not subject to study or verification by the procedures of science." To illustrate, the last two paragraphs of the AAAS resolution read equally well when "Evolution II" is substituted for "creation."

Without doubt, the biggest gun fired in the barrage of resolutions belonged to the National Academy of Sciences.

Established by Act of Congress on 3 March 1863, the National Academy of Sciences, according to its act of incorporation, "shall, whenever called upon by any department of the (Federal) Government, investigate, examine, experiment and report upon any subject of science or art." It is clearly a *national* body, although by 1974 it also had elected into its membership 138 *foreign* associates. Hardly anyone would doubt that it is the most important and prestigious scientific body in the *world.*

On 17 October 1972, the National Academy of Sciences took an unprecedented action. *For the first time in its 110-year history, it involved itself in a state issue.*[377] The Academy's ostensible reason for interfering in California's affairs was that there were "national implications" that 35 of the 1065 members who were present to vote that day "felt very strongly could affect the study of science for a generation."[378]

In fairness, it must be pointed out that this resolution provoked debate among the 35 members who were present. The resolution was apparently the work of only one Academy member, Preston E. Cloud, a biogeology professor at the University of California at Santa Barbara and a member of the Executive Committee of the Council of the Academy.

James G. Horsfall objected to Cloud's resolution on the proper grounds that the Academy would be overstepping its charter to interfere in a matter that was solely of concern to the people of California.

The vote for Cloud's resolution was *not* unanimous, even among those 35 members who represented only 3% of the Academy membership! Nonetheless, due to the pressure that Cloud exercised to "not delay transmission of the resolution," the full weight of the National Academy of Sciences was thrown into the California issue *with 97% of the membership being totally unaware of this violation of charter.*[379]

There is a note of irony about this resolution. Immediately upon its approval, William B. Shockley, whose affliction with Type B scientific heresy was noted in Chapter 2, attempted to get the Academy to approve a resolution on the "Eighty Percent Geneticity Estimate for Caucasian IQ." He had been presenting resolutions on this topic since 1966 without success, and this time was no exception.[380]

If there ever was an issue in which scientific prejudice was rampant, this had to be it. In the 1973 annual meeting of the National Academy of Sciences, Shockley tried once again — this time basing his cause on the *Academy's interference in California concerning creationism*![381] But it didn't help — he failed again.

Unfortunately, for its first foray into forbidden territory, the National Academy of Sciences made the following errors (see Exhibit 41)

1. *It based its action on a fallacy.* The California State Board of Education was not "considering a requirement." All requirements had already been established on 1 March 1971 when science textbook criteria were adopted. *None of those requirements* mentioned "parallel treatment to the theory of evolution and to belief in special creation."

2. *It raised "appeal to supernatural causes" as an issue.* There is no evidence whatsoever (apart from erroneous news media reports) for this false issue. Repeatedly, this issue had been denied by the State Board. Exhibit 34, that was widely disseminated, is one example of such denial by Vice President Ford.

3. *It declared religion and science to be "separate and mutually exclusive realms of human thought."* This foolish and unbelievable declaration ignored, perhaps under a precipitous impulse of high emotion, the thoughtful musings of the noted bacteriologist Rene J. Dubos, "As pointed out by Whitehead, religion and science have similar origins and are evolving towards similar goals."[382] Likewise, it is not likely that Albert Einstein would have signed the Academy's resolution because he said:[383]

> The most beautiful and profound emotion we can experience is the sensation of the mystical. It is the power of true science.

Without question, the tiny minority of Academy members who voted for this unprofessional resolution were pressured by vocal colleagues like Preston Cloud to "rush to judgment." Undoubtedly, because they implicitly trusted their illustrious colleague to have the facts (rather than independently verify such facts themselves), they were stampeded into a wholly unwarranted position.

Such hasty and reckless reaction, by those who should be most trusted, forces the layman to be cautious in accepting any future conclusions of this illustrious body — especially if they choose to enter once more an arena outside their expertise, as they did this time.

What about the other resolutions by scientific and academic illuminati? Were they just as error-ridden? Did they violate the facts as flagrantly as those already analyzed? Judge for yourself...

Exhibit 42 is another AAAS resolution — apparently more significant than Exhibit 40 because it is signed by top AAAS officialdom, the *Board of Directors*!

Here are some of its errors:

1. It states that teaching alternative theories on origins "represents a constraint upon the freedom of the science teacher in the classroom." Is not the requirement that the teacher must teach *only one theory* (for something which cannot be verified) much greater constraint on teacher freedom? What about the Soviet insistence on the *exclusive* teaching of Lysenkoism?

2. It decries "dictation by a lay body of what shall be considered within the corpus of a science." Yet the same resolution "strongly urges" the California State Board of Education (a lay body) to dictate that the AAAS viewpoint be imposed on California schoolchildren. Is that hypocrisy?

3. It urges, *on 22 October 1972,* that the California State Board of Education adopt the version of the *Science Framework for California Public Schools* prepared by the California State Advisory Committee on Science Education. However, the Directors of AAAS are apparently unaware that their urging is almost exactly *three years late* — the *Science Framework* was adopted on 13 November 1969!

Or look at Exhibit 43, a resolution by the University of California Academic Senate, theoretically representing 7,000 professors. Note its errors:

1. It repeatedly speaks of a religiously-based "special creation" as being specified in the *Science Framework*. Maybe the only way to expose this flagrant error is to offer a $100,000 reward to anyone in the University of California Academic Senate who can find such a term in the *Framework!*

2. It speaks of "our ideas of biological evolution." Who believes that these chairmen of the nine UC campus senates, the chairman and vice chairman of the academic assembly, and the chairmen of the three state-wide faculty committees possess "*biological* ideas?" What happened to the scientific concern about "dictation by a lay body of what shall be considered within the corpus of a science?" Surely the Academic Senate is not comprised entirely of *biologists!* Could it be that Nobel laureate Calvin may have provoked his non-biologist colleagues to adopt *his* ideas of biological evolution as *theirs?*

3. It acknowledges that the Academic Senate does not care about the *truth* of how origins may have occurred — only that theories of origins come from a restricted pattern of thought known as

"scientific investigation." Yet it was conclusively shown in Chapter 6 that such famous scientists as Mendel, Planck and Einstein derived their theories on entirely different bases than what could be called "scientific investigation." In fact, under the Academic Senate's rule, the work of Kepler, Newton, Faraday, Maxwell and most other heroes of science would have to be discarded as unscientific because its origin was "in philosophical thought and religious beliefs."

4. It concludes by urging that the State Board of Education "reject inclusion of an account of special creation in State-approved science textbooks." Ironically, the State Board of Education never considered taking any such action. Therefore, the resolution was written entirely in vain!

Exhibit 44 provides additional irony of scientists being misled into putting their reputations on the line for an absurd cause. It is particularly ironic because this resolution by the American Association of Physics Teachers claims that its signers have "carefully reviewed" the situation for which it feels compelled to write a resolution. How careful do *you* think they were? Their resolution contains at least four errors:

1. It lists as "of extremely great importance" the fact that California is adopting "the study of a particular religious philosophy of creation as part of science programs." That is a *patently false premise* as shown throughout this book.

2. It recommends that the State Board of Education apply *criteria* specified in the *Science Framework* (sans my two paragraphs that were adopted into the *Framework*). Any careful study would reveal that the *Science Framework does not provide criteria.* Textbook criteria are published as a separate document.

3. It mentions "various county review committees." *None such exist.*

4. It recommends specific actions for textbook decisions by the California State Department of Education. The State Department of Education *does not make such decisions.* They are made by the State Board of Education.

Although there were many other resolutions and indicators of indignation marshaled — all of them based on errors — in that 3-week campaign of October-November 1972, only one more will be specifically noted.

Exhibit 46 is the official position on the textbook issue of the

American Chemical Society (ACS), one of the largest scientific organizations in the world. It seems incredible that this responsible body could be provoked into a Pavlovian response by completely false charges. Why did they not seek *facts* before putting the respected reputation of ACS behind the strawman? Would they react so irresponsibly regarding the banning of pesticides or taking a position on nuclear testing?

Strictly on the basis of (1) an anonymous article in *Scientific American* (that erroneously stated that "the stage is being set for the mandatory teaching of divine creation as a scientific theory on the same footing as evolution"[384]) and a letter from Junji Kumamoto (that erroneously proposed that California was mandating "special creation performed by an ultimate or supernatural intelligence"), the ACS Board of Directors issued their irrelevant suggestion "that California science textbooks should not be distorted by inclusion of non-scientific material by legislative fiat."

ACS even introduced the *book of Genesis* that had not been mentioned in the two references that triggered their reaction! What causes eminent scientists to irrationally jump to false conclusions on the subject of *origins* while they are fastidiously cautious about conclusions in other areas of their expertise? Is there a *special* danger about origins that doesn't exist elsewhere?

The Postal Service certainly carried a heavier load in late 1972 due to the massive mobilization of mail. But mail wasn't the only mobilization target...

MOBILIZED MEETINGS

Most of the mobilized mail either resulted from or was intended to influence public meetings of the Curriculum Development Commission or the State Board of Education. These meetings were mobilization points themselves — in two ways.

First, such meetings provided a forum for publicly introducing the letters and resolutions that had been mobilized. Second, they supplied a platform for the public appearance of influential personalities. The second aspect had some advantages over the first.

People will *watch* and *listen* when they will not *read.* Therefore, a personal appearance can be much more influential than a letter.

Another advantage of a personal appearance over a written message is that public speaking is *dynamic* while writing is *static.* One can take advantage of the direction and flow of thinking that spontaneously arises in a meeting, whereas something written can easily be made irrelevant by changes of opinion occurring in a public group.

My emphatic statement on the viability of the origins issue during my first meeting with the Curriculum Development Commission in May 1972 may have stimulated Darwinians to consider mobilizing key witnesses for future meetings. With 20/20 hindsight, it probably would have been wiser to have kept a lower profile, thereby not provoking Darwinians to mobilize eminent scientists to offset or refute my voice on the Commission.

The Commission's Science Committee met alone on 18 July 1972 at the University of California in Riverside. Chairman Kumamoto supplied the agenda for the meeting.

The first business item, following the introduction of Commission members, was "correspondence." This was his strategic point to introduce the mail he had mobilized. The next agenda item was not *origins* (as it should have been) but "creationism."

As chairman, Kumamoto maintained control of the meeting. Building on the one-sided correspondence he had amassed, he immediately announced "creationism" as the next business item and launched into reading a position paper he had prepared against "divine creation."

Droning on in high-pitched voice, he finally finished his prepared statement. To that point in the meeting, no one but Kumamoto had said anything. He had made numerous errors, not only of fact but also of assumption. One of the most absurd statements he read was one that he later also made frequently during Commission meetings:

> Science is limited by universal principles. One of these, the *law* of conservation, states that matter or energy can neither be created nor destroyed. There can be no exceptions to this, and any notion to the contrary cannot be a part of science.

When Kumamoto finished, I asked to be recognized. I had no prepared materials, but I extemporaneously attempted to correct a number of errors he had made — particularly related to his continued emphasis on *divine* creation. I also gently chided him for his compulsion to defend evolution when it was not under attack.

Most of the publishers had representatives present at this meeting. My conversation with them afterward convinced me that they at least understood my points. Kumamoto, in attempting to demonstrate scientific opinion that supported his position, distributed a statement by G. Ledyard Stebbins (Exhibit 18). He announced that this statement would be one of many similar ones to be presented at the upcoming July meeting of the Curriculum Development Commission in San Francisco.

That was my first tipoff that Kumamoto was mobilizing for the meeting.

There was no way that I could mobilize counterforces on such short notice. Further, I had agreed with Chuck Terrell, our Commission Chairman, that the origins issue should be kept as low-key as possible.

For me to match Kurnamoto's mobilization would have done just the opposite. In a letter to Terrell on 24 July 1972, I urged:

1. That the Commission should not only limit the *time* for each speaker who appeared before us but also the *number of speakers* for any given argument or position.

2. That the Commission should attempt to locate spokesmen for opposing viewpoints on *all* major issues — not just the origins issue. Thereby, we could all *learn* from such an adversary process.

3. That all speakers on the origins issue should be reminded that our Commission is simply an *advisory* body to the State Board of Education and must carry out requirements unanimously passed by the Board regarding origins.

4. That since no one would be present to rebut the arguments of those speakers mobilized by Kumamoto, the "silence in rebuttal" should not be construed as *consent* to their arguments. Therefore, I offered to read a brief comment I had prepared in response to Stebbins' written statement (the only one available to me prior to the meeting) if Terrell thought it appropriate.

Kumamoto's decision to distribute Stebbins' statement eight days before its delivery in San Francisco, proved to be a "blessing in disguise." At least to me, it allowed me to study it carefully — something I don't think *Stebbins* did himself!

Stebbins perfectly demonstrated my observation in Chapter 7 that Darwinism is an "old man's belief." Remember the adage: "The degree of emotion shown by Darwinians varies directly with their age"?

Conscientious study of Stebbins' statement in Exhibit 18 forces one to conclude that he wrote *emotionally* — not *rationally*. Because of his long list of scientific achievements and recognition, I expected formidable arguments. I found just the opposite. He actually made it easy to locate his errors.

Two days before the scheduled July meeting of the Curriculum Development Commission in San Francisco, I prepared a rebuttal to Exhibit 18. I reproduced enough copies for each Commissioner and carried them with me to the meeting. I had no idea how the rebuttal might be introduced — if it was at all.

There was no way I could have anticipated what really happened in the first of the mobilized meetings...

What a Warm-up!

The word had gotten out. You could sense it as you walked into the meeting room at the Jack Tar Hotel in San Francisco on the morning of 26 July 1972.

Confrontation...conflict...anticipation. The atmosphere was electric with excitement. Publishers milled about in small groups or buttonholed individual Commissioners as they entered the room. People's eyes continuously swept around the large room — even as they carried on whispered conversation.

It was more difficult than usual to get the Commissioners to take their seats around the large U-shaped table. Tension seemed to prevail as Chairman Terrell called us to order and led the salute to the flag.

Evidently rumors had been flying about the *mobilized scientific assault.* I had heard that the number of scientific giants might be as high as ten. And I also knew that it would be a one-sided attack. If there were to be any counter-arguments, it would be up to me alone.

Once more, a feeling of insufficiency and loneliness crept over me...

Somehow, I had been unable to anticipate this scene. Perhaps two hundred spectators — publishers, interested citizens, the press — all settling down to watch a head-on battle.

While the opening business items — approval of the minutes, adoption of the agenda, and announcements — were being accomplished, my head was awhirl with tactical planning. Should I rebut each scientific speaker one-on-one? If I don't, maybe they'll build an impregnable case! Yet if I interrupt too frequently, I will lose my effectiveness with my fellow Commissioners ... Maybe I should wait until all the arguments have been presented...I wonder what points they will raise?

So it went — alternative options ricocheting and rebounding in my head.

I finally decided to become an *actor.* My instructions to myself were harsh. I would have to struggle to carry them out. "Your posture must be totally relaxed, as though you haven't a care in the world. Your facial expression (that least subtle part of you) must continually appear smiling, interested, and involved — with constant eye contact on each speaker. Most of all, you must remain silent until someone calls on you to speak. Do not *volunteer* — be *drafted* — don't *react* — play the *underdog* all the way!"

With my fighting Irish temperament, those were tough orders. And they got no easier as the meeting proceeded. In fact, I tensed up as the moment for the public hearing approached.

The U-shaped table around which the Commission was seated opened out toward the audience. The spectator seating was arranged with a center aisle separating two large sections of individual chairs arranged in parallel rows. In the open side of our U-shaped table, a small table equipped with a microphone and chair was placed facing Chairman Terrell. I was seated about half-way down the left leg of the U (as viewed by Terrell).

I soon spotted the star scientific witnesses. They were lined up on the front row of the audience — on the opposite side of the room from me. It was a beautiful setting because I could keep my eyes on the guest speaker seated at the small table right in front of me while also looking beyond him to the whole row of his colleagues.

The tension of anticipated combat crested as the first speaker was introduced. Instead of leading off with a Nobel laureate, they unbelievably chose to start with a *non-scientist*!

David H. Ost was only an associate professor of education at California State College in Bakersfield. I could hardly believe it. His statement was an illogical tirade about *usurping* academic responsibility, *denying* academic freedom, and *prostituting* the *Science Framework* — all based on the false assumption that the science classroom was being transformed into a Sunday School class teaching the book of Genesis!

Ost's objective was to get our Commission to disregard that portion of the *Science Framework* that contained my two paragraphs. Of course, that was a legal impossibility, so his testimony went for naught. I breathed a little easier...

Each speaker was limited to five minutes. That gave me a couple of advantages. First, unless they had highly organized their thoughts, they couldn't score too many points. Second, I could carry out my play-acting of smiling unconcern without too much difficulty for that short period of time.

The second mobilized spokesman — in contrast to Ost — was indeed a scientist. In fact, he was a Darwinian hero who was to play a leading role in the Science Textbook Struggle. This was his "baptism of fire."

Mentioned earlier for his correspondence with Wernher von Braun, William V. Mayer was about to deliver his first salvo. Like his predecessor, he aimed his gun at my two paragraphs on page 106 of the *Science Framework*.

Ranting and raving in an emotionally-choked voice, Mayer laced his short sermon with terms like *religious dogma*, *theological statements*, *Biblical accounts*, *religious beliefs*, *theological instruction*, and *fundamentalistic religion*. His summary point? "Let us not smuggle a minority religious belief into classrooms on the mistaken notion that we are dealing with a scientific position."

Mayer might just as well have urged our Commission to not vote for Woodrow Wilson — something else we had no intention of doing.

Mayer had missed the mark of relevancy so far that my tension eased back a few more notches — even while he was still speaking.

As I relaxed a bit, my eyes drifted beyond Mayer to a man seated in the row of scientific notables. Like a sweeping radar gunsight seeking and then locking onto a target, my gaze became riveted on an unbelievable scene.

It was such a startling sight that I was immediately confused as to whether this man was having some sort of seizure, was relieving some excessive physical discomfort, or was losing control of himself due to rage.

He was seated on the far end of the row, so I could view his whole body. He had his legs crossed. In one hand, he held a paper that he appeared to be reading intensely. His other hand was wildly flailing the air. At the same time, his leg that was crossed over the other was in violent motion — kicking outward and upward in a cadence out of phase with his swinging arm. His face grimaced in frightening contortions as his head jerked spasmodically from side to side. His mouth worked furiously.

Gradually, as I continued to watch him, I began to conclude that this person (and I had no idea who he was) was rehearsing his five-minute speech. Like a pitcher in the bullpen he was warming up...and it was some warm-up!

Meanwhile, the third speaker — another non-scientist, Richard J. Merrill — had been introduced and was delivering his views to the Commission. Merrill was a consultant in the small town of Concord, California on curriculum for *high school*, and our Commission was considering science curriculum requirements for *kindergarten through only the eighth grade.* So he was speaking outside his area of expertise in education as well as not being a qualified scientist.

Merrill's speech was far more relevant than those of his two predecessors. He attacked the idea of having to present more than one theory for origins. Overlooking two of the three criteria that had been unanimously adopted by the State Board of Education on origins (see Chapter 5), Merrill accused Vice President Ford in his statement of July 1971 (Exhibit 14) of attempting: "to read into both the text adoption criteria and the framework upon which the criteria are based a requirement that has no real basis in either." He concluded by saying, "Dr. Ford's statement seems to me to be an untimely, unfortunate, and totally unwarranted extrapolation of the framework, and seems to have no relationship at all to the criteria."

Although I later became personally acquainted with Merrill and came to respect him, he disappointed me greatly in this speech — not only by his failure to read the criteria carefully, but also because he had knowledge of the Board's intent from the very beginning. You may recall from Chapter 3 his correspondence with State Board member Moomaw on the issue of origins that even preceded my involvement.

So there were three up and three down — with no telling or meaningful points as yet. While I could hardly afford to relax, my play-acting was beginning to become natural. Meanwhile, I continued to be intrigued by the elderly witness who was still vociferously rehearsing his speech...

The fourth speaker was the second bona fide scientist to appear. Herman T. Speith was a zoology professor and colleague of Stebbins at the University of California at Davis. His brief talk was much less emotional. He even included a little humor. Unfortunately, he made an error that assumed evolution was going to be omitted in the teaching of biology. "As an actively-practicing Christian and as a scientist," Speith said the inclusion of creationism in biology courses is "morally irresponsible."

The mystery of the agitated expert's identity was removed as the fifth speaker was called to the microphone by Chairman Terrell. It was none other than *G. Ledyard Stebbins*!

I was flabbergasted! As Stebbins approached the speaker's table, I pondered why this 66-year-old star witness had felt it necessary to "warm-up" prior to delivering his pitch to us.

Warming up certainly paid off for Stebbins. He didn't use the gestures and body language that I had been observing, but he delivered his speech (Exhibit 18) faultlessly. His emphasis, though exaggerated, was correctly placed. Since we all had copies of his talk in writing, we simply followed his reading of it.

Stebbins was the first speaker to be questioned. Commissioner Zelda E. Dawson asked him whether scientists didn't have varying degrees of certainty about events in the past.

"It doesn't matter whether we have varying degrees of certainty. The important difference is whether a person can get facts or whether he is told *not* to get facts," Stebbins responded.

"But there is a point to which even science comes where the certainty is so low that you are really in a realm of *belief*...not in the realm of anything else. In other words, there are hypotheses which are so untestable, about which there is so little information, that your degree of certainty is very small. And you are really in the realm of *belief* rather than in the realm of *scientific documentation*."

Mrs. Dawson wasn't going to be cowed by Stebbins' expertise.

But Stebbins didn't listen to her. The vital point she had made went right over his head. He continued to make his own point...

"No matter *what* the degree of uncertainty, a scientist is able from all the hypotheses he accepts, to look for more facts — to *seek* more facts. With respect to the idea that the earth was created — that animals were created by a Supreme Being, no one can look for facts. No one is *expected* to look for facts. This, I think, is the basic difference between science and religion — whether we are able to continue and look for knowledge or whether we must accept on faith without question."

With that unresponsive response, the star of the show returned to his seat. Since I had already written a rebuttal to Stebbins' talk and the first four speakers had not made any points that I could not refute, I was still on top of the mobilized attack.

There were only two more guest speakers to appear on the issue. Both of them were biology professors — Lawrence R. Cory from St. Mary's College in Moraga, California and George Johnson from California State University in Hayward. Both were much more open and objective about presenting alternative concepts about origins, although both rejected the idea of "demanding the false juxtaposition of creationism and evolutionism as antagonistic scientific alternatives." Of course, I was in agreement with them in their concern.

Johnson was quizzed briefly by Commissioners Adolph Moskovitz and Shirley S. Myers about his views on including a theory of special creation in the science curriculum. As soon as he answered their questions, a recess was declared. It was a needed respite from the intensity of the public hearing.

I stood, stretched, and walked over toward the group of science witnesses who were huddled together in conversation. As I approached, they scattered. Singling out Stebbins, I extended my hand to introduce myself. Clumsily and probably only because he could not gracefully avoid shaking hands, he took my hand.

Trying to overcome the obvious awkwardness of the moment, I suggested to Stebbins that we were both interested in protecting the purity of science. Recoiling, he glowered at me. "What you are doing is worse than what Lysenko did in the Soviet Union! Believe me, we will settle this issue in court."

With that, he abruptly turned on his heel and walked away, leaving me standing there shocked.

Chairman Terrell shortly reconvened our Commission, saying:

"I think it is appropriate at this point that we engage in conversation relative to the public hearing. As you know, you may call back an individual to clarify a point. I'd rather not see a person come back to make a whole new statement — but simply to clarify a point, if that would be appropriate. Perhaps the objective of this part of the session would be to comment on the hearing, the messages brought to us by very dedicated eminent educators and people who are concerned about the education of children and the materials that are going into our schools."

Commissioner Bob Rangel broke in at this point. "I have a question. At what point are we going to hear from Dr. Grose? After the reaction by the others?"

"That's up to him. As soon as he raises his hand, I'll call on him."

"I would certainly like to hear Dr. Grose *now!* Rather than have a time lag here..." Commissioner Bud Creighton boomed.

Looking toward me, Terrell said with a smile and a twinkle in his eye, "Vern, no one has accused you of not being heard, I am sure."

There was a roar of laughter not only among the Commissioners but in the audience as well. My game plan was working perfectly. I was being *drafted!* Taking the microphone in front of me, I asked, "How much time do I have?"

Still in a joking mood, Terrell banged his gavel down and declared, "That will do, thank you!"

Even louder laughter burst forth. While the laughing was still out of control, Commissioner Shirley Myers, who was seated next to me, leaned over and jokingly said, "You can have *my* three minutes!"

As the chuckling died down, I joined in the merriment. "I now have *Shirley's* three minutes," I cried in mock protest. Terrell reassured me that there was ample time for me to be heard.

Somehow, those few moments of mirth seemed to break the tenseness that had prevailed during the hearing. In fact, the joking really set the stage for *informality* — a quality my remarks would require since they were not in writing. Taking advantage of this relaxed atmosphere, I mentally outlined several key points and began to present them...

"I attempted at the last meeting to try and clarify this issue. The news media have certainly made a different issue out of it than the State Board or I, by my personal involvement, had in mind. There are several factors that should be clarified," I explained.

The following points were then elucidated in conversational voice:

1. The Curriculum Development Commission is simply an *advisory* body to the State Board of Education and is powerless to change or ignore the *Science Framework*.

2. The State Board of Education unanimously adopted in March 1970 three criteria to carry out the intent of the controversial two paragraphs on page 106 of the *Framework*.
3. Evolution is not threatened by any wording in the *Science Framework*, and there is no reason why it cannot be taught exactly as it has been for many years.
4. Evolution has *two* meanings — one dealing with *variation* within living forms (Evolution I) and the other involving belief about how *origins* have occurred (Evolution II).
5. The issue before the Commission is exclusively about *origins* — that of the universe, life, and man.
6. The current textbooks in the classroom present *myths* about these origins which are *not scientific* — they are *beliefs* or *philosophies.*

The points just seemed to flow in a sequential pattern, without any notes or prior preparation. The sixth point led logically to a summary statement...

"This singular *belief* or *philosophy* about origins — and this alone — is what concerns the State Board of Education. They want to open the door a little bit, instead of teaching only one belief. The Attorney General in California ruled in 1963 that we can teach evolution, providing we don't *indoctrinate.* If you have only *one* theory or one story, I submit that you are not even within the law because we should provide an alternate view. This is what the State Board is seeking. Now to cast that effort off as a fundamentalist religious activity is far from the facts. It is also far from the desire of anybody I am familiar with in the issue. The words that were adopted in the *Science Framework* on page 106 *specifically disavow* the Bible, God, Genesis, and anything else that's religious. For that reason, I think that it's time that we quit talking about putting things in juxtaposition — namely, religion and Darwin."

At that point, I turned to the Chairman and mentioned that I had reproduced copies of my rebuttal to Stebbins' statement that I could give each Commissioner. On the other hand, I offered to read the rebuttal if that was more appropriate.

"We probably spent more time on those two paragraphs in the *Science Framework* than we have on the entire remainder. They will not change. We therefore ought to concentrate on complying with the *Framework* and criteria. If your rebuttal is within that context, I think it would be perfectly proper. In fact, it should be *necessary* that you read it."

Terrell passed the ball back to me.

"I'm not sure it fits exactly in that context. However, I think that Dr. Stebbins has very well articulated a premise that perhaps summarizes the position of all the guest speakers today. Therefore, I think that at least another position should be presented in contrast."

Without introduction, I began to read my written rebuttal (see Exhibit 20).

When I finished reading, I pointed out that biology was only one subdivision of science and that all the speakers had exclusively represented that branch. The general theory of evolution (Evolution II) presents some problems in other branches of science that were not represented. Examples of such problems include abiogenesis, the concept of mathematical randomness applied to natural selection, and the apparent conflict of evolution's increasing complexity with the Second Law of Thermodynamics.

"There may be some scientists who have a belief, a religion, a dogma to which they subscribe that would rule out the possibility of a Creator. They are entitled to that belief, but it cannot be called *scientific*. On the other hand, it just may be that there is a God who just happened to have created it all. If *one* belief about origins is permitted, we ought to leave the door open a little bit for the possibility that a Creator does exist."

I felt that the case was adequately clarified. However, Commissioner Creighton asked me to share some statements by Wernher von Braun relative to the issue. I knew that Stebbins et al would consider von Braun to be a *technologist* instead of a *scientist*, and I said so. Yet, I opined that I would personally consider von Braun to be a legitimate scientist.

With that endorsement, I recounted that he is "clearly on record both in writing and in conversation that there is no question to him that a chance hypothesis for the origin of the universe is ridiculous." (Later, Stebbins falsely accused me in Exhibit 26 of saying that von Braun is a scientific expert on the origin of *life*.)

Except for an innocuous comment by Commissioner Kumamoto about the potential creation of life in the laboratory (that I quickly pointed out would support the thesis of an intelligence behind it — rather than "chance"), the mobilized assault had been repulsed.

Not one of the scientific illuminati stood to challenge or rebut me. On the other hand, I am sure that they didn't really listen to my arguments either.

About a month later, Stebbins acknowledged that my rebuttal had been effective. What acknowledgement did he send me? A rebuttal of my rebuttal! (See Exhibit 26)

As a footnote, there was apparently another scientific notable who was mobilized but failed to appear at this meeting. The President of the National Association of Biology Teachers, Claude A. Welch, prepared and submitted a noteworthy statement to the Commission (see Exhibit 19). It was more rational and logical than any of the statements that were read in person to the Commission. Had Welch not stumbled over the strawman of "religious doctrine" challenging "the biological theory of evolution," his statement might have become the focal point for resolving the whole controversy.

For example, Welch asked four proper questions, the answer to which could lead to understanding of the issue. They are certainly not profound questions, but, if asked in sincerity, could lead to more meaningful dialogue. (All four questions are answered in this book.) The unfortunate aspect of his statement is that he prefaced his questions with a conclusion that was false.

September Sequel

While the mobilizing of *mail* occurred on a continuous basis, mobilized *meetings* could only occur when scheduled. And scheduling was not in the hands of the Darwinian mobilizers.

Following the organized confrontation in San Francisco in July 1972, there was no meeting scheduled until a joint meeting of the Curriculum Development Commission and the State Board of Education in San Diego in September. Whereas the purpose of the July meeting had been for the Commission to *select* the science textbooks to be recommended to the State Board, the joint meeting in September was for *presenting* and *discussing* the Commission's selection to the Board for their approval and ultimate adoption.

These intervening seven weeks between meetings allowed "capacitors to be recharged" on both sides of the issue.

Aside from a small and unsuccessful effort by the Creation Science Research Center (CSRC) to get their text materials reinstated on the list approved by our Commission (their materials had been removed by the previous Commission), I am unaware of any effort by the Godly to mobilize between the July and September meetings. On the other hand, the Darwinians were wasting no time rallying in hysteria over "mandated divine Genesis creationism."

Stebbins probably took over the management of mobilization from Kumamoto during this period. He had connections among the elite that Kumamoto would have no means of even recognizing. To trust such a consequential campaign to a plant chemist with no academic stature was undoubtedly more than Stebbins could bear.

Apparently within hours after I had rebutted him in San Francisco, Stebbins launched a crusade to crush the crazy creationists. To a number of scientific notables, he sent a copy of his 26 July 1972 statement (Exhibit 18), page 106 of the *Science Framework*, and selected pages from textbooks by American Book Company and Leswing Communications, Inc. (that compared their national version with one that they had modified to hopefully comply with page 106) together with a letter appealing for these experts' support of his position.

Of course, Stebbins got just what he asked for — unqualified support. It didn't seem to bother all these "qualified" scientists that Stebbins had erroneously fallen for the God-versus-Darwin strawman. They fell right along with him. Nor did these eminent scientists investigate the bait Stebbins put on the hook to see if it was valid.

For example, Leswing textbooks had been *eliminated* by the old Curriculum Commission and were *not being considered* for adoption. Further, the modifications by American Book Company of their national version to comply with page 106 had neither the endorsement of the Curriculum Development Commission nor the State Board of Education.

Exhibit 49 will prove that *I rejected* the American Book Company version that Stebbins sent out to his colleagues! The same exhibit confirms that I also rejected similar proposed modifications by Holt, Rinehart and Winston (that Stebbins didn't send to his colleagues) for the very same reason.

Stebbins evidently asked that the endorsements of his position be sent to Kumamoto as well as to himself. Thereby, Kumamoto received a foolish letter from Thomas H. Jukes of the University of California at Berkeley calling for rejection of the Leswing series (that had already been eliminated months earlier). Nobel laureate Melvin Calvin likewise sent an unnecessary three-page letter to Kumamoto decrying the American Book Company modifications.

Why weren't these giants of science *careful*, *objective* and *thorough* before jumping? What made them act just like any other mob whipped into a frenzy? Are *all* their letters of support as irrational as these?

Since the September meeting was not a public hearing, Stebbins' mobilization effort simply provided Kumamoto with a portfolio or base of support from the scientific community for fighting the phony strawman. Thus charged with such illustrious backup, Kumamoto tried the maneuver described earlier in this chapter of issuing to the Curriculum Development Commission a personal position paper in the name of the Commission's Science Committee. Of course, that trick backfired on him, but it's hard to handle the heavy honor of hustling for heroes...

It had been more than a year since I last saw John R. Ford, Vice-President of the State Board of Education. During that interval he had sponsored my appointment as a Commissioner, and the new Curriculum Development Commission had met three times.

Therefore, I took advantage of a joint meeting scheduled in September in San Diego between the Curriculum Development Commission and the State Board to suggest to Ford that we have dinner together so that I could apprise him of the status of the origins issue. Because I had missed the first of the three Commission meetings, I asked Commissioner Creighton whether he would like to join my dinner engagement with Ford.

Creighton had been a member of the antecedent Curriculum Commission and was present in the April 1972 meeting that I had missed. Therefore, he could provide additional perspective of the issue.

Creighton agreed to the dinner date and suggested further that State Board member Clay Mitchell might also be invited. Mitchell had been appointed to the State Board after the adoption of the *Science Framework*. Since Mitchell and Creighton were personal friends, we agreed that he might profit from our discussion.

The four of us met at 5 PM for dinner in San Diego on Wednesday evening, 13 September 1972. I was shocked by Ford's appearance when he picked us up in his car to go to dinner. He was still on crutches and bore evidence of injuries sustained in his tragic automobile accident on 14 May.

As we drove to the restaurant, he recounted the horrifying details of how, at dusk on a remote highway, a car swerved over the centerline and hit his car head-on — instantly killing his wife seated beside him, as well as the driver of the other automobile. Miraculously, though he was critically injured with broken ribs and leg, he was on the road to recovery.

Trying to brief someone about as complex an issue as the Science Textbook Struggle had become to that point in time — especially while trying to eat a meal in a lovely restaurant — can be quite frustrating. During this period of rather intense communication, punctuated by probing questions, it became evident that the issue needed further public clarification by the State Board. Both the news media and enraged scientists refused to believe that it was not a *religious* issue.

"It looks like my statement of July 1971 (Exhibit 14) hasn't done a complete job of clarification," Ford said with puzzlement. "Maybe I should make another attempt to alleviate the fears that this is a religious rather than a scientific effort."

Reluctantly, the three of us nodded our heads in agreement. In a unique way, Ford was the only person on the State Board who could be convincing to those who believed otherwise. Only he had played a consistently central role in the issue from its very inception.

"We'd better go back to my hotel room and put our heads together on some definitive statement," Ford observed as he grabbed his crutches and started to hobble out of the restaurant.

Driving back to Ford's room at the Royal Inn, we all discussed how the issue could be reduced to simplest terms — free from all possible ambiguity. We admitted among ourselves that it was not necessarily to the news media's best interest to prepare such a clean explanation. They had profited from making the issue ambiguous — and even *erroneous*!

The view of the San Diego harbor at night from Ford's veranda in the Royal Inn was spectacular. It made it difficult for us to concentrate on a simplifying statement about origins.

"Your July 1971 statement spoke of a requirement for science textbooks to discuss more than one theory for origins and stipulated that the 'case for design' or a general theory of creation should be one of those," I recalled turning to Ford. "The hangup or misunderstanding seems to center on *what is desired* when discussing the theory of creation. Maybe the statement you make tomorrow should present *two* angles on creation — both what you are *not* proposing and what you *do* want in the books."

Everyone seemed to agree with this suggestion. True to form for all those who volunteer ideas, I was asked to take a stab at drafting such a statement. While Mitchell, Ford and Creighton discussed other matters, I sat at a small desk and wrote out some thoughts on Royal Inn stationery. When finished, I handed them to Ford who made a few minor revisions before sharing them with the other two men. Exhibit 34 is the final version of this statement.

Before leaving Ford's room that evening, we also discussed how Kumamoto should be quizzed the next day by the Board regarding the origins requirement. He was slated to present the Commission's textbook selections to the Board. Creighton and I knew that the Commission had failed to carry out the direction of the Board to reject all books that, when discussing origins, did not present alternative theories. We warned Ford and Mitchell of this deliberate failure so that they could elicit such an admission from Kumamoto.

As the State Board of Education convened on 14 September 1972, there was a last-minute rush in the Curriculum Development Commission to get all the science textbooks and supplementary materials displayed on a matrix of 15 science material options for each of 9 grades (kindergarten through grade 8) on a huge blackboard. The State Board could then rivet their attention on this single point of reference.

The science adoption, of course, was the primary topic of the Board meeting. As Board President Steward came to that point in the agenda, Commission Chairman Terrell introduced Kumamoto who, as the Science Committee chairman, was to make the presentation to the Board.

Kumamoto began to explain the selection of materials displayed within the large 15x9 matrix on the blackboard. It was a routine, colorless explanation. Board members stifled yawns and gazed around the room. As Kumamoto completed his reading, Board Vice President Ford was ready for him...

"Mr. Chairman, I have a number of questions that I would like to ask Dr. Kumamoto. But before I ask the questions, I think it would be quite apropos at this time for me to make a statement since I was the member of the Board in 1969 who raised some objections in regard to the *Science Framework* that was being presented to us. I was the one who was called an *idiot*, *stupid* and *non-scientist*, so I would like to reiterate some things which are very important prior to asking my questions."

The audience present in the large meeting room suddenly became focused on Ford. His voice — almost spellbinding in its resonance and sharp articulation — commanded the attention of everyone. Clearly and with deliberate emphasis, Ford read the statement we had prepared the previous evening (Exhibit 34). He was eloquent.

As he finished reading the statement, Ford paused. The pause was just long enough to be dramatic. Then, he slowly picked up a copy of the letter Wernher von Braun had written to me (Exhibit 24) that I had given him earlier.

"I think Dr. Wernher von Braun, whom all of us know so well, can express my feelings in this area. I would like to read a letter that he wrote." Ford then read the entire letter as TV cameras rolled.

The stage was now set for Ford to quiz Kumamoto. Eloquence and class versus dull routine. From Kumamoto's point of view, Ford's was a hard act to follow. It was evident that Ford was going to bear down on *specifics*...

"With this letter in mind, Dr. Kumamoto, I would like to ask these questions. They are simple questions. They are not meant to be of any embarrassment, but really in order to complete what the State Board has requested in the *Science Framework* and criteria." Ford, recognizing Kumamoto's speaking limitations, tried to allay any of his fears.

"First, was the Science Committee aware that the necessity of presenting alternative hypotheses for the origin of the universe, life, and man was not only a *unanimous* conviction stipulated by the State Board, but was *reinforced* since then? Were you aware of this position?"

Kumamoto blinked, hesitated, and then replied, "I was not aware of the reinforcing. We were aware of your previous statement (Exhibit 14) which was read to us at our first Commission meeting."

What kind of an answer was *that*? Was Kumamoto aware of the *requirement*, reinforced or not? Was he saying that Ford's July 1971 statement was *not* a reinforcement of that requirement? At best, his answer was gobbledygook — at worst, an acknowledgement of the requirement.

"All right. Now did the Science Committee consider this requirement in its recommendation of science materials?" Ford was fast closing the door on Kumamoto's escape from accountability.

"Yes. We did not direct ourselves directly to it, in the sense that where origins is (sic) discussed is an alternative aspect given to it."

Is that an *answer*? It sounded as though Kumamoto was admitting that, although the Science Committee knew about the requirement and should have used it as a criterion of selection, the requirement was *ignored.* But don't be too sure because Kumamoto was trying to slip through a tightening noose.

Ford, intent on getting a straight answer, kept the heat on Kumamoto. "Do *any* of the books then include an alternative idea about origins? Or do most of them *ignore* an alternative?"

"Most of them do not contain origins in either sense."

If Kumamoto had not previously lied to Ford, he certainly did by that answer. I wanted in the worst way to rise and protest his answer, but he was designated as the sole voice for our Commission. To my knowledge, there was not a textbook series submitted to the Commission for approval that did *not* discuss origins.

"What did the Science Committee recommend to the publishers to fulfill the origins requirement?" Ford continued his pursuit.

At this point, Kumamoto took another tack to escape the web of logic that Ford was weaving around him. "Creationism, as such, was referred to only in the appendix of the *Science Framework* and did not appear to us in the form of a mandatory requirement there or in the criteria which is (sic) more explicit."

Although earlier *acknowledging* a requirement, Kumamoto was now denying there *was* one!

This is all Ford really needed. From that moment on, Ford was free to play with Kumamoto like a cat toys with an exhausted mouse. Kumamoto appeared to be rescued, short of total disgrace, by Commission Chairman Terrell who rose, walked to the microphone and intervened. Terrell's words, however, turned out to be an indictment of Kumamoto for claiming that he didn't believe a requirement existed.

"Dr. Ford, when this matter came up in the Commission, I stated that as far as I was concerned the Commission was supposed to go by what was actually written into the *Framework* and the criteria. I want to make very clear that I had made that public statement to Dr. Kumamoto and the full Commission," Terrell emphasized.

"It is my understanding then that this has *not* been carried out, and that is what concerns me this morning," Ford responded to Terrell.

Kumamoto was thoroughly discredited. By attempting to push his personal views through the Board, he had hit a stone wall. He sat down, and Terrell remained to respond to a variety of questions posed by Board members as they discussed the situation among themselves.

During the ensuing discussion, Ford proposed that, if there were textbooks available that did discuss alternative theories for origins, the Board could add them to the recommended ones in the matrix.

This proposal raised some procedural questions — particularly about the legal implications of the Board making selections that had not previously been recommended by their subordinate Curriculum Development Commission. Legal counsel from the Department of Education was present and consulted on this matter. But that only muddied the waters...

The on-the-spot verdict was that the Board was totally captive to its own creation — the Curriculum Development Commission — and could not adopt a textbook that the Commission had not recommended. Of course, it was a ridiculous ruling and was reversed before the next Board meeting.

However, the *Los Angeles Times* article that reported the reversed opinion carried the fallacious headlines:[385] BIBLE CREATION THEORY IN SCHOOL TEXTS OKD.

Was there any truth to those headlines? Absolutely none. All that had happened was a confirmation of the law under which the State Board of Education had operated for many decades; i.e., the Board had the legal right to edit, excise, or modify any textbook to comply with its directives. The State Board therefore is not limited to only those materials that have been approved by its own advisory body, the Curriculum Development Commission.

To prove that the *Los Angeles Times* deliberately chose to inflame the issue, compare how the *Sacramento Bee* reported the same news:

> LOS ANGELES TIMES:[386] "The State Board of Education got legal clearance Thursday to insist that the religious version of the origin of life be placed in proposed new California science textbooks."

SACRAMENTO BEE:[387] "Board members have the final say on textbook selection...Ford said he was not advocating that textbooks teach the Biblical account of creation, but he added that the theory of evolution should be taught as a theory, not as unchallenged fact."

Returning to the discussion in the September Board meeting, Board member David A. Hubbard, President of Fuller Theological Seminary, offered a keen observation that incisively drove to the heart of the origins problem:

I have a couple of concerns along this line which are not unlike those which Dr. Ford has expressed. One is that we make as clear as possible in our science texts the differences between — to use oversimplified language — *fact* and *theory*...I know that we cannot protect our young people completely from the distorted attitudes or ideas that they will develop when they crystallize subjunctives into indicatives — that is, theories into facts. But I would hope that in our review of textbooks, there would be the maximum protection against that tendency ...

Whenever we deal with origins, particularly, we are dealing more with philosophy and theology than we are with science because we are not dealing with that which is repeatable and verifiable within a laboratory situation. We are dealing with something that happened back there, and it has to be part of our pedagogy that we make clear that any speculations that we make about origins are *speculations*. They are really more in the realm of *philosophy* than they are in *theology*. Therefore, my concern is twofold. First, protection of the young person against this tendency to override the subjunctive. Second, a clear distinction in the disciplines between *philosophy*, on the one hand, which deals with theories of origins — how it happened, why it happened, and *science*, on the other hand, which deals with the observable data of a repeatable experiment.

Here was yet another confirmation of the difference between Evolution I (science) and Evolution II (belief or philosophy). But once again, neither the news media nor the scientific elite listened.

The *Los Angeles Times* reported the September Board meeting with the same distortion of truth that they had employed since 1969. Here is part of their account that should be compared with the facts that have just been described:[388]

Proposed new state science textbooks ran into stiff opposition Thursday from members of the state Board of Education, who want to see the *religious* version of the origin of the universe given equal treatment with the *scientific* theory of evolution in the schoolbooks...

Several members also charged that a state commission which selected the proposed series of schoolbooks had failed to follow the board's instructions to insert the *Biblical account*. "It is my understanding that this has not been carried out," Ford said. (emphasis added)

Go back and read Ford's exact statement in Exhibit 34. Did he not say, "I do not wish to see the Bible, God, or the Genesis account of creation mentioned in science materials used in the public classroom?"

Figure 10. Let's Be Objective

What can a person do about *direct lies* in a newspaper? Is it any wonder that some alarmed citizens, realizing the power of the press to influence people's minds, question that interpretation of the First Amendment of the United States Constitution that allows *deliberate falsehood* to pass as "freedom of the press?" How does this "manipulation of the public" differ in principle from that employed by Goebbels of the Nazi regime? At what point does *freedom* become *tyranny*?

Another Los Angeles newspaper, *The Valley News*, printed a cartoon on 24 September 1972 that depicted the State Board's reprimand of

Kumamoto (for ignoring the *Science Framework* and criteria) as applying to all of us on the Curriculum Development Commission (see Figure 10).

Depicting Kumamoto's interjection of his own personal decision to ignore Board direction as representing the entire Commission was correct since Kumamoto was our sole spokesman. However, note that, despite the cartoonist's proper insight about the need for alternatives regarding origins, the State Board's position in Figure 10 is mistakenly shown as endorsing "the Biblical theory" of origins as one of those alternatives.

The battle for objectivity regarding origins was heating up as autumn arrived in 1972. The arguments were sharpened, the objectives were clarified, and the opponents were identified in sharp relief as the result of the September sequel to the July warmup...

There was one more major force in the battle that needed mobilization — the publishing community.

Publisher Parties

Publishers were mobilized for combat differently than the scientific illuminati. In several ways.

First, they weren't mobilized by the enraged elite of science. The Curriculum Development Commission, a neutral party in the conflict, *officially* mobilized them.

Second, the publisher's mobilized meetings were *private* rather than *public*.

Third, publishers weren't asked to take sides in the battle.

Throughout the Science Textbook Struggle, the function of textbook publishers was underplayed. The State Board of Education and the scientific elite seemed to be the titans locked in mortal combat — with the news media playing an agitating role. Yet, the textbooks were the *grist for hostilities*.

Even more importantly, textbooks ultimately had to be the mechanism for resolving the issue and determining who won!

One of the primary results of the September meeting of the State Board of Education was the setting of a date for a public hearing on the recommended science materials. This is a normal procedure. Once the Curriculum Development Commission has made their recommendations to the State Board of Education, the Board, as they did in September 1972, *tentatively* adopts recommended materials. A period of 45-60 days is established for review of these tentatively adopted materials by the general public, school administrators, parents, teachers, and any other interested groups.

Then, a public hearing is set for the Board to learn the reaction of those reviewers. It is this meeting that generally produces the information that the Board uses in editing, excising, rejecting, or approving materials.

The public hearing on the science textbooks and supplementary materials was scheduled by the State Board of Education for 9 November 1972 in Sacramento. However, this public hearing was going to be more than just an *average* hearing.

It was to be the public culmination of the Science Textbook Struggle — the last chance for the scientific colossi to impress laymen with their righteous cause — the final opportunity for religionists to beg for relief from atheistic indoctrination under the guise of science.

Wisely, Chuck Terrell, Chairman of the Curriculum Development Commission, perceived the unusual character of the upcoming public hearing. Within a few days after the hearing was scheduled by the State Board in their September meeting, Terrell decided to reduce the force of impact in November by introducing the publishing community to the specific arguments in the origins issue earlier than they normally become involved.

In fact, he decided not only to *expose* publishers to detailed points of contention, but to allow publishers an *independent response* to this exposure prior to the public hearing. Thereby, the publishers were afforded a meaningful and determinative role rather than simply responding to the fallout from the inevitable and grinding crunch of collision between two irreconcilable forces.

On 22 September 1972, Terrell wrote separate but identical letters to every publisher of a basic science textbook series. His letter described a task that Terrell had assigned me — to propose possible changes that would align textbooks with the *Science Framework* and criteria concerning the origins issue. In part, Terrell said:

> This letter is to verify and support the task of Mr. Vernon Grose, of the Curriculum Development and Supplemental Materials Commission, as he meets with you and/or your representative to suggest possible revision of your basic science materials in relation to several sections of the Framework and Criteria.
>
> Mr. Grose will be one of several members of the Committee on Science from the Commission preparing materials for presentation to that committee and the Commission for the Public Hearing before the State Board of Education scheduled for November.
>
> It is understood that the suggestions for possible editorial change will not be in final form since a number of other factors are involved in this process after the Board adopts the material. However, if we can find an area of agreement before the Public Hearing, the issues may be better defined.

As soon as they received Terrell's letter, publishers immediately contacted me for an appointment. They all seemed excited about Terrell's obvious desire to promote them from *passive observers* to *active participants* in the process of resolving textbook discussion of origins.

Of course, with the ball clearly in my court, I had an almost overwhelming responsibility. If I was going to meet with publishing executives and furnish *specific details* in their textbooks that might be revised, *I had to actually read their books!* Rhetoric and generalities would not suffice.

If you haven't yet read Exhibit 15, you may be unaware that we Commissioners "serve without compensation." Since there were 7 major publishers with whom I was to meet and suggest possible revisions, and each publisher's series consisted of 9 books each, I would have to find free time to read and edit at least *63 books!* And it all had to be done in the month of October 1972.

Meanwhile, I had accepted a full-time, 7-month appointment by Governor Reagan to the Governor's Select Committee on Law Enforcement Problems in August (See Figure 11). Since 3 of the 5 members of this Committee resided in Sacramento, the Committee work was centered there. Likewise, our personal meetings with the Governor had to occur there. So I was flying once or sometimes twice a week to Sacramento.

The Governor had high expectation from our Committee, and the pressure on us was correspondingly great. I had managed to get permission from the Governor's office to teach an already-scheduled system management course at The George Washington University in Washington, D.C. during the last two weeks of September. Already burning the candle at both ends, how was I going to accomplish this impossible task of editing 63 science books?

To compound my scheduling dilemma, I was haunted by my own scientific inadequacies. By no means did I want to pose as a qualified scientist because I knew my qualifications were totally unacceptable to the scientific nobility. What I needed was a group of scientists whose credentials met Stebbins' criteria in Exhibit 26 to work with me — people with doctorates from recognized institutions in a broad distribution of specialization in science who were actually engaged in research or academia.

In desperation and on short notice, I contacted the ten scientists listed in Exhibit 37 and secured their agreement to work with me in my editing efforts.

Figure 11. Governor's Select Committee on Law Enforcement Problems Century Plaza Hotel, Los Angeles – 1 August 1973 (V. L. Grose, right)

Believe it or not, the impossible was somehow accomplished! Aided by a personal friend, Richard J. Schultz, a Rocketdyne engineer with whom I share many interdisciplinary interests, all 63 books were read and edited. Individual meetings with all 7 publishers were held during evening sessions at the Holiday Inn in Woodland Hills, California.

A typical meeting started by having dinner at 5 PM in a nearby restaurant with publishing personnel. Dinner conversation centered around the objectives of the editing process, specific difficulties the publisher might experience in accomplishing the changes, the publisher's approach to origins, and what format our Commission would require for any revisions.

I also presented the publisher two items in writing that he could take back to his office as aids in making revisions. The first was entitled, "Review Criteria for Science Textbook Compliance With 'Chance Versus Design' Requirement" (see Exhibit 38), and the other was a short elaboration of what I was calling, "The Case for Design." (see Exhibit 39)

By the time dinner was finished and we had returned to the Holiday Inn, we were ready to immediately start the critique of textbooks. This intensive page-by-page review generally lasted about 3 or 4 hours. With few exceptions, publishers sent not only sales personnel but those charged with editorial management responsibilities to these meetings.

In each meeting, one of the scientists whom I had enlisted to verify scientific accuracy was present to confirm or modify anything I said. Publishers were also urged to contact any of the others in Exhibit 37 for additional assistance. Here is the schedule and attendees for the publisher review meetings:

9 October 1972 — HARCOURT, BRACE, JAVANOVICH, INC.

Donald B. Chapin	Sales Manager, Western Region
Lance G. Day	Manager, Western Region
Roy E. Cameron	Science Advisor for Commission
Richard J. Schultz	Editorial Assistant for Commission
Vernon L. Grose	Commissioner

12 October 1972 — AMERICAN BOOK COMPANY

Ralph H. Shawhan	National Marketing Coordinator
Robert B. Fischer	Science Advisor for Commission
Vernon L. Grose	Commissioner

16 October 1972 — CHARLES E. MERRILL PUBLISHING COMPANY

Richard L. Brown	Sales Representative
Robert L. Dees	California Area Supervisor
Mark C. Biedebach	Science Advisor for Commission

Richard J. Schultz	Editorial Assistant for Commission
Vernon L. Grose	Commissioner

18 October 1972 — LAIDLAW BROTHERS

Peyton Hurst	Regional Manager
Richard Mohr	Executive Vice President
Robert B. Fischer	Science Advisor for Commission
Vernon L. Grose	Commissioner

23 October 1972 — HOLT, RINEHART & WINSTON, INC.

Carl J, Pendleton	Manager, Pacific Division
Roy E. Cameron	Science Advisor for Commission
Richard J. Schultz	Editorial Assistant for Commission
Vernon L. Grose	Commissioner

25 October 1972 — SILVER BURDETT COMPANY

James W. Gregg	Sales Representative
Joseph W. Foraker	Executive Editor, Science
Mark C. Biedebach	Science Advisor for Commission
Richard J. Schultz	Editorial Assistant for Commission
Vernon L. Grose	Commissioner

26 October 1972 — HARPER & ROW PUBLISHERS, INC.

Robert T. Hay	Executive Editor, Science
Glen A. Hoggan	District Manager
Lorne Holtmeier	Sales Representative
Robert B. Fischer	Science Advisor for Commission
Richard J. Schultz	Editorial Assistant for Commission
Vernon L. Grose	Commissioner

The response to these meetings was remarkable. There was an openness that went well beyond the typical sales reaction of "anything the customer wants, the customer will get." Follow-up letters from executives of each firm confirmed that we had made sense. Not once did anyone suggest that something religious was being proposed. The proof of our efforts was many specific revisions that were prepared by publishers prior to the November meeting.

During this series of publisher meetings, I suffered the greatest personal blow in the entire Science Textbook Struggle. It was a shattering experience! Just recalling it makes me shudder.

However, I want to share it because it produced an abrupt redirection in my thinking. Only by revealing my errors, faults and mistakes, can *progress* in the struggle for objectivity regarding origins be understood.

At 5 o'clock sharp on Wednesday, 18 October 1972, I walked into the lobby of the Woodland Hills Holiday Inn. Peyton Hurst of Laidlaw

Brothers walked up to me. "Vern, meet Richard Mohr, our Executive Vice President," he said with obvious pride.

"I am very pleased to meet you," I replied. Looking around the lobby, I spotted Bob Fischer as he walked in the door. Hurst, Mohr and I walked over to greet Fischer, and almost immediately the four of us left to get in the car and drive to a restaurant for dinner. This was the fourth such meeting with publishers, and despite the panic of having to read and edit nearly a dozen books for each meeting, I was enjoying the camaraderie.

Bob Fischer, Dean of the School of Natural Sciences and Mathematics at California State University in Dominguez Hills, had already taken time from his busy schedule to meet with me to edit the American Book Company textbooks only 6 days earlier. I was deeply indebted to a man of his stature who would be a watchdog on the scientific accuracy of revisions I was proposing to the publishers. I had not previously known Fischer, but after our first meeting, I was delighted that he had been recommended and had accepted my invitation.

Just as Peyton Hurst was proud to be able to introduce Mohr as a top executive of his firm, I was equally proud to introduce Fischer as a notable scientist and educator. His recent book, *Science, Man and Society*,[389] had amply demonstrated his insight into the *philosophy* of science as well.

As the four of as sat down to dinner, the confidence that I was making a positive contribution to objectivity in science — a confidence that had been cumulatively building with each meeting — seemed to reach a new high.

With restrained exuberance, I summarized our 3 previous meetings with publishers — recounting with enthusiasm how receptive everyone so far had been to our revision proposals. Publishing officials, particularly those from outside California, had all been initially skeptical. Mohr, who had flown out from Illinois, was no exception. I could see it in his face. But I was confident that he too would change just as the others had.

As I handed Mohr, Hurst and Fischer a copy of "The Case For Design" (Exhibit 39) and started to explain it, Fischer broke in — with a scowl on his face. "Vern, I don't think I can agree with you," he said.

I couldn't believe my ears! My face flushed immediately as my pulse jumped into high gear.

"Why not?" I weakly asked, still hoping that I hadn't heard him correctly.

"Well, what I think you have written here is a case for a *Designer* — not a case for *design*! I read it carefully after you gave me a copy last week, and that's what I'm forced to conclude."

If there was anything I felt sensitive about, it was that I was promoting "religion veiled as science." And my supposed strong ally had turned on me, accusing me of doing so!

Maddening thoughts raced through my mind. Why hadn't he called me *earlier*? Why did he have to *expose me* right in front of those I was trying to convince? Did he mean to *deliberately* discredit me? Why *now*? Why couldn't he have *waited* until we were back in the hotel?

I wanted the floor to open up and swallow me. Mohr and Hurst stared at me, glanced at Fischer, then riveted their eyes back on me. This is probably what they *wanted* to hear — that I was a religious phony. And now they had the evidence — from the mouth of my own star witness! What could I do?

The shock was so great that it has wiped out most of my memory of the dinner. I mentally staggered, floundered, and grasped wildly for some temporary stability until I could recover my verbal equilibrium. Slowly — and it must have been after an embarrassingly long period of silence — I threw enough thoughts together to attempt conversation.

"Well, I appreciate your honesty, Bob — I guess." Pain from his devastating blow to my ego had eased very little. "Maybe you could explain your reasoning."

Fischer then made the point that *every* scientist — regardless of his presuppositions about origins — is convinced of design in nature. Therefore, there is no reason to propose a case or a defense for *design*. The key question is whether or not the design occurred *with* or *without* intelligent assistance.

He explained that those who subscribe to a wholly mechanistic origin for the universe, life, and man can rightly be said to endorse a "case for chance." However, those who are convinced that some external intelligence planned and participated in the origin of the universe, life, and man should be described as endorsing a "case for a Designer."

Fischer was quick to point out, much to my relief, that the arguments in favor of a Designer do not necessarily constitute a *religious* document — at least not until a particular concept of that Designer emerges. Therefore, his main disagreement with me was over the *title* of Exhibit 39 rather than its *content*.

Despite Fischer's partially-supportive explanation, I remained near-numb for the rest of that Wednesday evening. My buoyant confidence was gone. I am sure that Hurst and Mohr sensed it. Worries about the impact of having to revise my strategy from "chance versus design" to "chance versus Designer" plagued my mind. I could hardly wait for Hurst and Mohr to say goodnight so that Fischer and I could be alone.

As the two of us stood alone in the Holiday Inn parking lot — rehashing for another hour what had taken place that evening, I perceived more and more clearly the error I had committed.

Fischer was a wise but firm teacher. Although it hurt deeply to admit that I had been wrong, I could acknowledge to myself that it was good to be corrected — even at this late stage in the struggle. Fischer reassured me that my arguments were not demolished — they only needed a slight redirection. Nevertheless, I did not sleep well that Wednesday night…

Although only the last 4 publishers profited from my newly-learned lesson, the textbook revisions that all 7 companies submitted were unaffected by the redirection of my thinking. As I was to say in a speech before the State Board of Education (see Exhibit 47) only two weeks after the last meeting with publishers:

> I had hoped by substituting the words, *chance* and *design*, for *evolution* and *creation* that I might thereby reduce the emotional content of discussion and point to what I thought at the time was the key issue. I now confess that I have made an error. That substitution turned out to be a "straw man." The key changes in the curricula are essentially the same, but the public issue is quite different … I now realize that it was not fair to suggest, by placing *chance* and *design* in juxtaposition, that all those who endorse *chance* automatically reject *design* in nature. It would have been more accurate had I said that the issue is whether or not a Designer was at all involved because, as everyone knows, all scientists must believe in some universal order in nature.

Bob Fischer had made a lasting impact on my thinking — and to him, I shall always be grateful.

Likewise, the foresight of Chuck Terrell to mobilize the seven "publisher parties" set the stage for my mid-course correction, thereby placing me in debt to him as well.

MOBILIZED MEDIA

That the news media became mobilized in the fall of 1972, no one would deny. Whether their mobilization was self-induced or provoked by Darwinians, however, remains debatable.

As described in earlier chapters, California newspapers periodically stoked the fires of controversy from the outset of the Science Textbook Struggle in 1969.

The new mobilization dimension, as the battle completed its third year, was the rallying of *national* and *international* news coverage. *Scientific American*, a leading national organ of science since 1845, was among the first to give extensive coverage of the straw man issue — "the mandatory teaching of divine creation as a scientific theory."[390]

That *Scientific American* account contains numerous errors (in addition to the quoted straw man) such as (a) making the California *Science Framework* applicable through Grade 12, when it only applies through Grade 8, (b) saying that Grose "favors the teaching of evolution in the schools but feels that divine creation also provides a scientific rationale of first causes," (c) claiming that the Creation Research Society has lobbied strongly for adoption of books (that they have never done), and (d) misnaming State Board Vice President John R. Ford as *James* Ford.

Perhaps such errors are due only to careless reporting rather than deliberation, but laymen expect higher accuracy from science periodicals. The unfortunate aspect of this article is that it was widely re-quoted — particularly that erroneous phrase "the mandatory teaching of divine creation as a scientific theory." Due to its stature among science magazines, *Scientific American* thus triggered false alarms all over the nation.

National and international news coverage of the struggle wasn't limited to *science* periodicals. An extensive summary of the situation, "God and Darwin in California," appeared in *Christianity Today*, a respected and scholarly journal among Christians. It too was loaded with errors that had the potential of misleading its readers:[391]

1. It credited the revision of the *Science Framework* to a petition by Nell Segraves.

2. It said Governor Reagan appointed me to the Curriculum Development Commission.

3. It asserted that the Commission (instead of the State Board of Education) held the November public hearing on science textbooks.

4. It stated that the Commission voted to include creationism along with Darwinism in science textbooks.

5. It alleged that the problem of determining which textbooks were to be selected was yet unsolved in December 1972.

6. It misnamed the Creation Science Research Center as the Creation Research Center.

7. It declared that publishers (instead of the State Board of Education) edit textbooks for compliance.

So error in the news media was not one-sided. Without doubt, people on all sides of the issue were unduly provoked to fallacious reaction by occasional articles they read in periodicals.

Coordinated Campaign?

In contrast to randomly scheduled, occasional articles by the news media on the Science Textbook Struggle, there was evidence in a few cases of a well-orchestrated and systematic release of news that apparently involved *collaboration*.

Not unlike some of the skullduggery that my friend Chuck Colson was accused of masterminding in the White House, these articles bore threads of *continuity* and intent to *cumulatively* build a case.

As the date for the public hearing on science textbooks approached in 1972, the *Los Angeles Times* revived the crusade in defense of evolution that it had abandoned in 1970 — when it apparently believed that it had been successful.

The campaign opened on 15 September 1972, with the large headlines: EVOLUTION ISSUE STIRS FURORE ON STATE TEXTBOOKS.[392] The irony of this headline is that the lengthy article that accompanied it didn't discuss *evolution*. Instead, it falsely claimed that the State Board of Education wanted to see "the religious version of the origin of the universe" introduced in science textbooks. The use of the inflammatory catchword, *evolution*, was obviously to sell newspapers since evolution was not an issue. Evolution wasn't threatened, debated, or impugned by the State Board.

On 13 October 1972, even bigger headlines screamed: BIBLE CREATION THEORY IN SCHOOL TEXTS OKD.[393] Again, although the headlines kept the momentum of the *Los Angeles Times* crusade rolling ahead, they were *totally false!* The captive public whose only source of State Board activity was a fallacious newspaper story was being stampeded into irrationality. The headlines incensed the scientific community while elating some religionists who had no scientific understanding.

Only five days later, another alarming headline screeched: SCIENCE ACADEMY HITS TEXTS' CREATION THEORY.[394]

The sub-headline read: URGES THAT THE RELIGIOUS CONCEPT OF LIFE'S ORIGIN BE KEPT OUT OF NEW SCHOOL BOOKS.

This was an account of the National Academy of Sciences' resolution (Exhibit 41) that was passed by less than 3% of the Academy's members. The article quoted the Academy's public information officer as saying that the resolution, "unprecedented in terms of the Academy involving itself in a state issue," was urged by some members who "felt very strongly that this (insertion of religious views) could affect the study of science for a generation." It certainly might have done so, if it had only been what the State Board was proposing.

But it wasn't...

News Report, an official publication of the National Academy of Sciences, chose much more accurate and subdued headlines in reporting the Academy's resolution: NAS EYES CALIFORNIA CONTROVERSY, URGES THAT SCIENCE TEXTS DEAL ONLY WITH SCIENCE.[395]

Two weeks later, the *Los Angeles Times* hammered home another straw man headline in a coordinated campaign of contradictory charges:

SPECIAL CREATION THEORY IN SCHOOL BOOKS OPPOSED.[396] Reporting the action of the University of California Academic Council in passing their resolution (Exhibit 43), the article emphasized that the Council was the third major scholarly organization to oppose the State Board's attempt to supposedly insert "a religious creation theory in public school science books."

Thus, brick upon brick, layer upon layer, the public was progressively set up to attack the God-versus-Darwin straw man from two sides at the State Board's public hearing on 9 November 1972 in Sacramento. By collusion or coincidence, there had been a coordinated campaign.

Editorial Enforcement

There can always be debate about which scientific periodicals are most prestigious. Journals published by specialized scientific societies are obviously more erudite than those that report on scientific activities in general. To illustrate, *Physical Review — Particles and Fields*, a journal of the American Physical Society, more likely contains papers at the cutting edge of the physics state-of-the-art than does *Scientific American.* The language in the former is much more technical than that in the latter. Thus for a physicist, it may be a higher honor to publish in *Physical Review — Particles and Fields* than in *Scientific American.*

On the other hand, the circulation of *Physical Review — Particles and Fields* is 4500, while *Scientific American* publishes 600,000 copies of each issue. So if *exposure* of one's findings is important, it could be a greater honor to publish in *Scientific American.*

Another dimension to prestige of publication might be its geographic distribution. If so, *international* periodicals would be considered more prestigious than *national* ones. The international journal *Nature*, published in Great Britain with the circulation of 21,000, might be a more respected platform for exposure of one's ideas than *Science*, the journal of the American Association for the Advancement of Science, which prints about 150,000 copies of each volume.

The Science Textbook Struggle surfaced in a number of prestigious scientific journals — *Scientific American*, *Nature*, *Science*, *Daedelus* (Journal of the American Academy of Arts and Sciences) and *Chemical and Engineer News* (organ of the 120,000 member American Chemical Society) to name a few. Not only was the origins issue given exposure in extensive articles in these journals. It also drew *editorial* fire.

Since only a small percentage of scientific subjects ever warrant editorial comment in scientific publications, the significance of the Science Textbook Struggle is established by the attention that editors gave it. Again, it is unfortunate that even editors of respected scientific organs failed to get the facts and got trapped by a phony issue.

Two editorials — one in *Science* and another in *Nature* — adequately illustrate both the *apparent importance* of the issue to the scientific community and the *poor scholarship* of the respective editors.

"Two Cooks for the Same Kitchen?" That was the title of the editorial in the 29 September 1972 issue of *Science* (see Exhibit 35). Written by *Science* publisher William Bevan, it not only pits Creation II against Evolution I as though they were meant to be complementary, but also totally ignores Evolution II!

In addition, it raises unfortunate irrelevancies like the Wilberforce-Huxley debates, the Russian biologists who defended Lysenko, and the need for a Fund for Freedom in Science Teaching.

Bevan shortly thereafter became Executive Officer of the American Association for the Advancement of Science that is comprised of 290 affiliated scientific societies. Because of the high regard and trust he obviously enjoys with the scientific elite, Bevan's inability to differentiate the real issue from the straw man invented by the lay press was both a shock and a disappointment.

The *Nature* editorial — "Creation in California" — appeared on 20 October 1972. Calling the origins issue "absurd" and "especially foolish," this anonymous editorial contains at least a half dozen errors. I have always admired the British for their ability to be succinct, pungent and accurate in the use of the English language. So to see so much error in an editorial in their renowned scientific journal was indeed disconcerting. But my greatest disillusionment with the *Nature* editorial resulted from a cheap shot they opted to take — the type one only expects from crass, crude crimps:[397]

> This (origins issue) is not a conflict in which a brave and embattled minority can reasonably be expected to restore Creation to the status of a scientific theory. To pretend otherwise is to misunderstand not merely the status of Darwinism but the nature of science. And who in any case is the minority? The State Board of Education in California has been

> indecently reticent on that subject. Who are the others? In the hope of helping to clarify the board of education's mind, *Nature* is prepared to send a free subscription to *Nature* to the first ten scientists working or teaching in a field of science bearing on the evolutionary question who are prepared to affirm that present observations are in their opinion inconsistent with the now commonly accepted views of Earth and species evolution. Applications must be received before October 30 and must give the present occupation of the applicants, who must be actively employed in the science department of a university. A list of names (if any) will be published.

Since *Nature* is published in Great Britain and third class mail deliveries from Europe do not arrive overnight in the United States, it is quite possible that American subscribers had not received their copy of this rather childish and narrow-minded offer before the deadline had expired for response. If *Nature* truly desired open-minded debate and was honestly seeking a measure of the minority, the editor would have done well to have read his editorial over once to check for possible bias prior to setting it in print.

So the long list of otherwise notable scientists who fell for a ruse was lengthened by the addition of editors of respected science journals — editors who enforced and reinforced the God-versus-Darwin phantom.

Rousting Retort

Neither the *Science* nor the *Nature* editorials escaped readership wrath. Due to the much wider circulation of *Science*, the reaction to Bevan's editorial was greater in magnitude.

My personal disappointment with "Two Cooks for the Same Kitchen?" caused me to write to Bevan. He graciously answered me. Exhibit 36 contains both of our letters.

He said that my letter was the kindest that he received in response to his editorial, so evidently he got considerable mail from those who were upset with his position.

Bevan returned to academic life as the William Preston Few Professor of Psychology at Duke University. Since Bevan's expertise is neither in the physical nor life sciences, Stebbins would declare with emphasis that "Bevan cannot be regarded as an informed scientist, qualified to make an authoritative statement about origins" (see Exhibit 26).

It may be that Bevan's incomprehensible answer to me, "At best, I come off as close-minded and rigid," should be attributed to his lack of scientific qualifications rather than to his personal prejudice. At least, one hopes that is the case.

Other letters that Bevan received in response to his editorial were printed later in *Science*. E. C. Lucas of the Dyson Perrins Laboratory in Oxford, England was one of the fairest and most objective respondents. He made several important points:[398]

> Though not a convinced creationist, I have acquainted myself with modern creationist thinking and consider that Bevan's editorial is rather unfair. Bevan seems unaware of the changes in creationist thinking since the early 1900's.
>
> He objects to creationism being taught in schools because it is a "theory of primordial history," but surely evolutionism is also a theory of primordial history, as it is used to explain the origin of species. Any theory of origins is of necessity scientifically unprovable in that it is not capable of being verified experimentally. The best we can do is demonstrate what could have occurred, but not what did occur. However, it is wrong to say that creationism is not subject to any form of testing; like evolutionism, creationism makes postulations about the past history of life which can be checked against the evidence. Thus the evidence can be used to rule out theories, while not being able to prove them.

Lucas went on to point out that all creationists postulate a polyphyletic origin of life with subsequent speciation. Regarding the fossil record, Lucas said:[399]

> Creationists would predict that the fossil record will show the absence of transitional forms between each of the older, independently created forms. So far creationism fully agrees with the evidence, whereas evolutionists have to have faith in the original existence of the missing transitional forms.

Summarizing his views on the idea of having alternatives for origins, Lucas clearly showed the relationship between Evolution II and Creation II:[400]

> I consider that those creationist views that accept the antiquity of life and its gradual appearance are just as acceptable in the light of the evidence as is neo-Darwinism, and so should be taught alongside it.
>
> The modern antipathy toward all forms of creationism is partly due to a widespread philosophical prejudice that arises from the *a priori* nonscientific assumption that there can be no divine intervention into the workings of the universe and that the scientific method is the only way to truth. However, there are other approaches to truth, and application of the scientific method can only determine whether or not divine intervention might have occurred in a given case, and not whether it can or cannot happen in general. I hope that the "freedom fund" will be used to help free science from this naturalistic influence as well as from undue religious influence.

Another letter from a physicist, D.E. Iviartz, attacked the dogmatism of biologists:[401]

> It is interesting to note that some biologists are taking an obstructionist position toward the California State Board of Education ruling that the origin of life by natural processes must be taught as a model rather than as dogma. The creationist model may not be scientific, but as a physical scientist I cannot believe that many of the models presented by biologists of the origin of life and the evolution of species are less speculative. It is in the highest tradition of science to allow opposing viewpoints to be heard rather than insisting that only one side be taught in a school textbook. This was the fatal mistake of fundamentalists in the 1925 Scopes trial, and it should not be repeated by evolutionists in the 1970's.

Even a letter from an avowed creationist was reprinted in *Science*. William J. Tinkle, a founder of the Creation Research Society, responded in a thoughtful manner to Bevan's editorial:[402]

> I am convinced that discussion of beginnings, although interesting to many people, is not science but philosophy. Science consists of facts that must be organized and interpreted; without a body of observed facts you can have no science. It is evident that there was no one to observe the beginning of the world, hence there can be no *science* of beginnings.
>
> We can study present life processes and make inferences about past developments. We creationists do not object to teaching the inferences of the evolutionists, but as careful scientists we ask to be excused from teaching these ideas as scientific truth.

What about the response to the *Nature* editorial, "Creation in California"? Was anyone able to respond within the 10-day period from publication of the offer in *Nature* to having their letter arrive in England? Apparently, two readers were quick enough to meet the almost impossible deadline.

The 3 November 1972 issue of *Nature* acknowledged that Garret Vanderkooi, a professor in the Institute for Enzyme Research at the University of Wisconsin, and Harold Van Kley, a chemistry professor at St. Louis University were willing to be listed among the minority of qualified scientists who find Evolution II wanting in the light of scientific evidence. One wonders how many would have responded if there had been 30 days to accept *Nature's* offer. Maybe the enticement of a free subscription to *Nature* was too much…

In retrospect, the *Nature* editorial seemed to be 180 degrees out of phase with the scientific spirit. It *taunted* those it should have been blessing. It *ostracized* when it should have encouraged. As E. T. Jaynes, Professor of Physics at Washington University and under whom I completed some graduate studies in statistical mechanics, once said:[403] "Physics goes forward on the shoulders of doubters, not believers."

The editor of *Nature* obviously thinks otherwise about evolution.

Paranoid Paradox

Think about it for a minute. Reflect on the entire history of the Science Textbook Struggle. Was evolution ever *attacked* by the State Board of Education? Was it *maligned*, *desecrated*, *revised* or *banned*? Was the freedom to teach evolution ever *reduced* or *threatened*? Why then was there such a massive uprising in defense of *evolution* and *Darwin* during the Science Textbook Struggle?

Paranoia has been defined as "a mental disorder characterized by systematic delusion of persecution." People suffering from paranoia honestly believe a lie — that they are being persecuted. Adolf Hitler, on 29 April 1945 — the day before he committed suicide in a Berlin bunker, dictated his political testament. It contains a clear example of Hitler's paranoia:[404]

> It is untrue that I or anybody else in Germany wanted war in 1939. It was desired and instigated exclusively by those international statesmen who were either of Jewish origin or working for Jewish interests. I have made so many offers for the reduction and limitation of armaments, which posterity cannot explain away for all eternity, that the responsibility for the outbreak of this war cannot rest on me...
>
> As late as three days before the outbreak of the German-Polish war, I proposed to the British Ambassador in Berlin a solution of the German-Polish problem — similar to the problem of the Saar area, under international control. This offer cannot be explained away, either. It was only rejected because the responsible circles in English politics wanted the war, partly in the expectation of business advantages, partly driven by propaganda promoted by international Jewry.

Poor Hitler believed that a vast and wealthy conspiracy had forced the war upon him! Ironically, only *he* was prepared for war. Only *he* took the initiative of invading another country. Only *he* screamed berserk accusations against others. Only *he* refused to listen to reason.

History fails to confirm Hitler's observations. He was *not* being persecuted — quite the *opposite!* As uncomfortable as it may be, there is a close parallel between Hitler's paranoia and the reaction of Darwinians in the fall of 1972.

Just as Hitler saw an apparition of conspiracy, so the Darwinians imagined a powerful coalition was about to "do them in." One of the most hysterical Darwinists was William V. Mayer, director of the

Biological Sciences Curriculum Study at the University of Colorado. His writing was built around phrases like "*assault on the theory of evolution...fundamentalist attacks...bastardization of terminology...smuggling religion into classrooms under the guise of science...insidious attempts...smokescreens...fear of reprisals...infringement of freedom...mercenary motives ... a drive to co-opt science...most flagrant disregard for logic and reason...and diatribes against the theory of evolution.*"[405]

Mayer, whom you may recall appeared in person before the Curriculum Development Commission in July 1972, spoke of "a small but well-financed group operating throughout the United States"[406] in the same tone that Hitler used to describe "the universal poisoners of all people, international Jewry" in his political testament.[407]

While Mayer had a different cast of characters in mind, his Messianic language style was not unlike Hitler's paranoid ravings:[408]

> The attack on biology teaching uses evolution simply as a vehicle for an entering wedge of antiscientism. Evolution is not the issue — the issue is the integrity of science and the ability to present its dimensions in the classroom without outside interference. That such attacks cause us to defend academic freedom and force us to reconsider our definitions of science is revivifying. If they cause us to examine our stance on science teaching and to state a position on the necessity for freeing it from the unreasoned pressures of anti-scientism, attacks such as these could have the salutary effect of freeing science once and for all from the chains of the nineteenth century.

That stirring call to arms should "warm the cockles of the heart" of all red-blooded scientists. Just point them in the right direction, and they will attack — enemy or not.

Riding high in the spirit of Mayer's whipped-up frenzy, the National Association of Biology Teachers (NABT) issued a call for arms. NABT had earlier formed an ad hoc committee, allocated NABT funds for legal action and engaged legal counsel because "it became evident to the Ad Hoc Committee that implementation of the controversial sections (the two Grose paragraphs) of the *Science Framework* posed a threat to the academic freedom of California biology teachers, and indirectly, to science teachers throughout the nation."[409]

One of the 3 members of the NABT Ad Hoc Committee was Jerry P. Lightner, who as recounted earlier attended the meeting at NEA headquarters in Washington in June 1972. The following month, in July 1972, the NABT Board of Directors established a Fund for Freedom in Science Teaching. Claiming that "academic freedom of science teaching has been endangered," NABT warned:[410]

> If appropriate action is not taken to offset this trend, the decade of the seventies may witness a continuous erosion of sound classroom and laboratory teaching techniques, delayed or decreased development and implementation of new and scientifically sound curriculum materials, and a potentially disastrous limitation on the freedom of secondary school and college science faculty to teach in a professional manner. Continual attacks are being made on the teaching of the theory of evolution...There is every indication that expenses to counteract these attacks to infringe freedom in science teaching will rise, possibly into tens of thousands of dollars, in the future.

Sounding amazingly evangelistic in fervor, the NABT rhetoric had the ring of 19th century oratory:[411]

> It is not enough to pass resolutions to be inscribed in our minutes, published in our journals, and stored in our archives. History will not care what we thought unless we are also willing to act according to our beliefs. We must be willing to assist in the defense of our principles in courts of law and before governmental bodies, as well as to inform the public, through the communications media, of the rationality of those principles. None of this is possible without the intellectual and financial support of all of us who are concerned with protecting academic freedom.
>
> If you are interested in supporting the NABT *Fund for Freedom in Science Teaching*, please send a check...Not a penny accrues to the organizational or operational expenses of NABT; on the contrary, the *Fund* will be used exclusively for the furtherance of academic freedom in the teaching of science. Your contribution is tax deductible...and urgently needed.

Even though the National Science Teachers Association joined hands with NABT in soliciting money for their *Fund*,[412] it was another *false alarm*. Legal action was never required.

How humiliating it must have been for them to learn that they had cried "Wolf!" when a small ant innocently walked by...More consequently, the lay public should seriously wonder what it is about evolution that produces hysterical overreaction. Only a deep-seated *belief* would provoke such paradoxical paranoia. What *is* that belief?

Despite the ruckus that the fund-raising generated in the news media, there was even an occasional ray of objectivity in the press.

On 27 October 1972, Cunnington produced another tongue-in-cheek cartoon (Figure 12) that depicted the State Board of Education's role quite differently from that of an oppressor of academic freedom.

There were also other lighter moments that offset the impassioned paranoia of biologists. Right in the middle of the "kill-em-by-resolution" campaign conducted by many professional scientific societies during

November 1972, a humorous event occurred on Johnny Carson's "Tonight Show" on NBC-TV.

Johnny's first guest one evening was William F. Buckley, Jr., the noted columnist and intellectual. Buckley had just finished a new book entitled, *Inveighing We Will Go.*

"You've just written another new book," Johnny said. "Inveighing…*inveighing*…why in the world did you use *that* word? I frankly didn't even know what it meant myself until I looked it up in the dictionary!"

AND THE BOARD SAID, "LET THERE BE LIGHT!" AND THEY SAW THAT IT WAS GOOD, AND IT WAS THE FOURTH DAY.

Figure 12

With mock surprise and wrinkled forehead, Buckley responded, "Oh, you've never *heard* of that word?"

"No, and I'll betcha I'm no different than anyone else...All right audience, all those here tonight who know what 'inveighing' i-n-v-e-i-g-h-i-n-g means, please clap your hands," Johnny taunted.

Two or three weakly clapped. "Now, how many out there had never heard of the word until now?" The place went wild with applause. Turning to Buckley with a sweep of his hand, Johnny gloated, "See, *nobody* uses that word."

With amusement written all over his face and moving his hairline in that inimitable manner, Buckley softly retorted, "The meaning of words is a most interesting subject itself."

He then gave a brief definition of "inveighing" and why he had chosen it for the title of his book. Without condescension but with subtle clout, he decried the imprecision of the English language when compared with some of the classical languages in the world.

"For example," Buckley said, "there is a word in the New Testament which, when it was translated from the original Greek into English in the King James version of the Bible, was used interchangeably to replace 23 different Greek words with similar but slightly different meanings. It set theology back for 300 years!"

Since Johnny Carson is neither a semanticist nor a Bible scholar, he was obviously a bit nonplussed by this radical turn in the conversation. Reaching desperately for something that might keep continuity in the dialogue, he rebounded by saying:

"Out in California right now (the 'Tonight Show' was being produced that evening in New York instead of Burbank, California), we've got a rather interesting issue. There is a controversy concerning how schoolchildren should be taught about origins of the universe, life, and man in science classrooms. There are some apparently sincere and competent scientists on both sides of the question of whether the Bible's version of creation should be taught alongside that of Darwin's theory. What do you think about *that*?"

Buckley, without a smidgen of hesitation, leaned back in his chair and said, "I'll tell you what I think. Science changes its mind so often. There are those who are so quick to defend Galileo and recall the persecution that the church forced him to undergo. Yet much of Galileo's thinking has been replaced by Einsteinian physics. While I would have some arguments about the manner in which the Genesis account would be discussed in a science textbook — for example, they'd certainly have to take into consideration the current state of science, I'd turn much more willingly myself, for an explanation of how everything began, to the Book of Genesis than to the Institute of Advanced Studies at Princeton!"

Rather stunned and with no ready quip to offer, Johnny gulped and said, "Let's go to a commercial."

The massive mobilization of mail, meetings and media in 1972 certainly alerted the world of the conflict over science textbooks.

However twisted and distorted the issue of origins had become, the public was aware of an impending confrontation...

Chapter 9

DEFEAT OF DOGMATISM

"No one is exempt from talking nonsense; the misfortune is to do it solemnly." — *Montaigne*

The unavoidable finally happened...the head-on collision of the Godly and the Darwinians.

Produced and carefully orchestrated by the news media for more than 3 years, the phony issue, "Religion Masquerading as Science," finally provoked open combat.

This culmination of conflict was a lot like Armageddon. To the participants, it had overtones of *good* and *evil* locked into mortal battle. To the on-looking world at large, it had both *finality* and *inevitability*. *Everyone* would have prophesied it.

The full impact of the mobilized mail, meetings and media were to be felt in this meeting. The unleashing of pent-up forces awaited only the naming of the date and place.

Yet when the *date* — 9 November 1972 — and the *place* — the State Resources Building in Sacramento — were announced, they were likely as innocuous as the Valley of Megiddo will be on that fateful day in the future when all the armies of the world are gathered together to fight the Battle of Armageddon.[413]

I'll admit that 9 November 1972 did not sneak up on me. I foresaw it as a day of perhaps greater influence in science teaching than the day the verdict was rendered in the Scopes Trial of 1925.

Because I foresaw it as a special day, I did some special preparation for it — preparation that would hopefully *combine* the excessive energy being generated on both sides and *channel* it into a common cause.

But like my previous efforts of reconciliation, this one was also doomed to failure...

REFUSE TO LISTEN

My strategy was both sweeping in scope and magnanimous in motive. Once again, it was based on the presumption that the opposing parties were noble in their objectives. They, I reasoned, had only been hornswoggled by the news media into false combat. If I could expose the straw man, everyone could join hands to seek better science textbooks.

The seed thoughts for my strategy of reconciliation between the Godly and the Darwinians came from my shattering encounter with Bob Fischer during the publisher review meetings in October.

He had pointed out an error in my thinking that could be turned to the good of all — *if I would only confess it!* My strong ego naturally rebelled against publicly exposing any of my weaknesses. But my conscience prevailed.

However, if I was to publicly confess, I would need the right audience. Otherwise, my confession would be counterproductive to the objectivity that I was seeking for origins.

The only appropriate audience for my acknowledgement of error would have to include spokesmen from both camps — in addition to the Curriculum Development Commission and the State Board of Education. *The one possible time that audience would be assembled would be during the upcoming public hearing before the State Board of Education on 9 November 1972.*

But there was no way that I could address such a broad audience. Though it was to be a public hearing, Commissioners traditionally do not request an opportunity to appear on the agenda.

And even if I did break tradition and request to speak, I would have only five minutes — the limit imposed on all speakers. There was no way that I could effectively present both my confession of error and my proposal for unity within five minutes.

There seemed to be no possibility of implementing my strategy of "unity-via-confession." Yet an inner conviction persisted that I would get an opportunity...

Less than two weeks before the public hearing, I was invited to address a dinner meeting of over 200 educators, attorneys, scientists, and physicians at a yacht club in Seattle. Local chapters of four professional societies had scheduled this joint meeting to learn more about the Science Textbook Struggle. Interviewed after the meeting by Seattle's two major newspapers, I assured them that we expected success on 9 November in the battle to remove religion from science.

The next day, the *Seattle Post-Intelligencer*, in a lengthy article, quoted me as saying, "Science may come clean on Nov. 9 for the first time in 60 years and get out of the religion business...We're not talking about the Bible's Book of Genesis, we're talking about anti-theism."[414]

An even longer *Seattle Times* article carried my prediction that November 9th would be a historic day. "For the first time in the history of science, we are going to admit we don't know anything about the origins of man — as scientists," I was reported as saying.[415] There was no doubt that I was convinced of my strategy.

Early in November, I began to draft a detailed statement that would form the core of the strategy. It contained an admission that I, like others I had accused, had also raised a straw man that I now wished to withdraw in a spirit of compromise. Included also was a prophecy for both sides — a summary of the arguments each would be offering in the public hearing. (This prophecy came true!) Even more significantly, I proposed a new statement of the problem that faced all of us.

As I had done three years earlier, I asked Ralph Winter to criticize my tentative strategy. He had valuable thoughts to contribute, as usual. We both agreed that my decision to publicly recant as a step toward reconciliation of the polarized debate should be kept as a *surprise*. Neither side should have any inkling of my intent. That was the fairest approach.

I finalized this statement — dubbing it "Science And Origins." (see Exhibit 47) In blind faith, I called it a report to the California State Board of Education and dated it 9 November 1972. At worst, I felt that I might have to hand a copy to each of the ten Board members. At best, I could deliver it orally to what I previously described as "the only appropriate audience."

Even though it turned out to be 18 pages long, I reproduced 100 copies of "Science And Origins" — again with the unwarranted conviction that I would be permitted an opportunity to present it. And that opportunity did come about...in a most unlikely manner.

The Curriculum Development Commission was scheduled to meet on Wednesday, 8 November — the day before the momentous confrontation. We had to finalize the Commission's position on the science textbook adoption so it could be presented to the State Board. The approved books, as recommended to the Board in September, were either acceptable or they needed editing on the subject of origins.

In order to be in Sacramento early on Wednesday morning for our meeting, I had to fly from Los Angeles the night before — Election Night 1972. It was the night of the Nixon landslide. My interest in politics — though strictly amateur — is sufficiently intense that I delayed my flight to Sacramento until the very last minute to learn the election outcome.

Pacific Southwest Airlines (PSA) has a "Midnight Flyer" that leaves Los Angeles shortly after midnight and arrives, after an intermediate stop in San Francisco, at 1:50 AM in Sacramento. No reservations are accepted for the "Midnight Flyer." For this quirk, PSA knocks $5 off the airfare.

Of course, you can never be sure you'll make this flight, so you are forced to get there more than an hour early — and take your chances. Hippies, backpackers, students and the wildest collection of air travelers in the world are all amalgamated into line on a first-come, first-serve basis.

My duties with the Governor's Select Committee required me to be in Sacramento on Monday. I flew home to vote in the election in Los Angeles on Tuesday, and here I was back at the airport that same night. I felt like a yo-yo!

Loaded down with dozens of copies of "Science and Origins" as well as all the other material needed for the Commission meeting, I staggered wearily into the LAX terminal. As I entered the large waiting room, the line was already long — winding all around the room like a serpentine.

No sooner had I taken my place at the end of the line when up walked Chuck Terrell, Commission Chairman.

"I see you're a glutton for punishment, too!" Terrell chuckled as he set his luggage beside mine.

"I hope our punishment doesn't include missing this flight," I responded as I apprehensively surveyed the very long line ahead of us. "I wonder how many will get on..."

I had high regard for Terrell. He possessed an almost uncanny ability to bring harmony out of discord — and it wasn't by being a wet noodle, either. He could be strong and assertive, while never losing his sense of humor. Above all, he was fair and objective. I am sure no Commissioner felt that they had him in their camp on any subject.

My respect for Terrell's neutrality caused me to be most reticent in sharing with him my inmost thoughts about the origins issue. Though I was totally preoccupied with the desire to communicate the message in "Science and Origins," Terrell could rightfully decline any lobbying I might do during the flight.

On one hand, I had a golden opportunity to "sandbag" him while I had him captive in an airplane. But on the other hand, I knew I could destroy rapport by taking advantage of the opportunity. Besides, hadn't I agreed with Ralph Winter that my position was to be a *surprise*?

We did manage to get aboard the Midnight Flyer. It was crowded — every seat taken. Terrell and I sat side-by-side on the aisle. As we sipped hot beef bouillon together, we discussed the need to reach a common position in the Commission the next morning about editing the science textbooks. He knew that I favored editing — and that Kumamoto didn't.

I could see that he dreaded a clash between the two of us. As Terrell openly mused about how such a clash might be averted, I began to waver in my resolve to keep my strategy secret.

Because of Terrell's open and fair nature, I feared that if I shared my intent to confess, he might feel obliged to disclose it to others. Yet *that* fear was counterbalanced by *another* fear — that, in his ignorance of the real issue, Terrell might take an approach that could totally preclude me from ultimately exposing my strategy. I wavered between these two fears as Terrell continued to show consternation.

Above 30,000 feet somewhere between Los Angeles and San Francisco, I decided to confide in Terrell. Opening my briefcase, I pulled out a copy of "Science and Origins." Terrell glanced over to see what I had in my hand.

"Before I show you this, Chuck, I want to admit that I have been presumptuous," I explained. "I have prepared a report to the State Board even though I had no request for it from them. Further, I fully intended to keep it secret until I could disclose it publicly in the joint presence of the Commission and the Board — and hopefully antagonists on both sides of the issue as well. In showing it to you now, I respectfully ask you to keep it confidential until I can read it in public because it contains a *bombshell!*"

"Really? Well, you can trust me to keep it under my hat." Terrell sounded reassuring.

As he read hurriedly through the report, we discussed key points. He appeared shocked that I would so boldly confess my own error. Yet he acknowledged that a confrontation could be prevented only when both parties gave some ground.

Gently, I explained how important it was that I read my report *publicly* — so that it could set a scene of reconciliation. Terrell only listened.

The next day was a disaster — as far as reaching any consensus in the Commission regarding science textbook revisions. I had asked each of the 7 publishers with whom we had met in October to have some typical changes ready for the November meeting as well as a letter from their top management that expressed willingness to incorporate such changes.

Publishers, of course, watched the wavering among individual Commissioners all day long. One by one, publishers would quietly approach me, handing me management letters and a few changes. However, they cautioned me to treat them as tentative because it was very uncertain which way the Commission would go on revisions.

As late as 5 PM, nothing approaching unanimity had developed. Debate had been acrimonious most of the day. Kumamoto and I both had supporters, but neither had a clear majority. About 5:30, a hotel official came in our meeting room and told Terrell that we would have to clear the room by 6 PM, as there was another meeting scheduled at that time.

Kumamoto was still reading aloud resolution after resolution from prestigious scientific organizations and letter after letter from scientific elite decrying the introduction of religion into science textbooks.

Ostensibly, he was rehearsing — before the entire Commission — the Science Committee report he would be presenting to the State Board the next day.

At 5:45, as Kumamoto fumbled through his sheaf of resolutions to find the next one, I addressed the chair.

"Mr. Chairman, I would like to request a 5-minute recess for the purpose of meeting privately with you and Dr. Kumamoto. It appears that we will be expelled from this room within a few minutes without the slightest hope of a Commission-wide position on this subject."

I sensed a sigh of relief all around the table. Evidently Terrell did too.

"If there is no objection, I declare a 5-minute recess. Please stand by and do not leave the room. Junji, please join Vern and me in private conversation."

The three of us adjourned quickly in a far corner of the room. I opened the conversation. "There is no way that our Commission can be considered responsive to the clear direction that the State Board has given us over the past months by reading a bunch of resolutions and letters!"

"I agree completely," Terrell replied. Kumamoto was mute.

Looking directly at Terrell, I continued, "I would like to propose that I be permitted to draft in the next 5 minutes a resolution that I think the entire Commission can support and —"

Terrell broke in, "— do you *really* think you can do so?"

"Yes, I know it sounds impossible, but I have an idea that should work. However, such a resolution, if it passes, could mislead the Board into believing that we are in total harmony. Therefore, I think it's imperative that Junji and I both address the Board tomorrow from our two positions — after presenting the resolution. That way, the Board will be granted the benefit of the breadth of thought that has occurred in our Commission."

Kumamoto was still silent. He looked at Terrell. Slowly, Terrell began to nod his head in agreement. "You know, I think you're right. Why don't you go ahead and try writing a brief resolution?"

He turned to Kumamoto. "Meanwhile, Junji, maybe you can wind up the rest of your resolutions while Vern puts his thoughts down on paper."

"Let's get the Commission back in session — we've only got about 10 minutes before they kick us out of here!"

Terrell immediately called the Commission to order, gave the floor back to Kumamoto who then resumed his high-pitched recitation of another resolution from the scientific illuminati.

Meanwhile, I grabbed a pad of white, lined paper and began to furiously scratch out a motion with black felt pen. It was rather difficult to concentrate while Kumamoto was droning on with his harangue to "leave the science textbooks alone." The Commission appeared confused because nothing had seemed to change as a result of our brief recess and caucus.

In less than 3 minutes, as Kumamoto completed another resolution, I finished my draft. I glanced up at Terrell. He had his eyes riveted on me.

"Mr. Chairman, I have a resolution to offer."

"Junji, with your permission to yield, I would give the floor to Vern." Terrell had urgency in his voice. We had only 5 or 6 minutes to go. Kumamoto nodded agreement. Terrell then instructed the Commission to listen very carefully to what I was about to say.

"Mr. Chairman, it is moved that, on the subject of discussing origins in the science textbooks, the following editing be done prior to final adoption:

1. That dogmatism be changed to conditional statements where speculation is offered as explanation for origins.
2. That science be shown as *not* involved in determining ultimate causes for origins.
3. That questions yet unresolved in science be presented to the science student to stimulate interest and inquiry processes."

While it momentarily stunned the Commission, there began an almost immediate murmur of relief and endorsement as the motion was seconded.

Terrell called for any discussion, explaining that the motion was the result of our brief recess only minutes earlier. Commissioner Robert A. Bennett then raised a point about the phrase "prior to final adoption." He suggested that it be changed to "prior to execution of a contract." I agreed.

Terrell then suggested that Kumamoto, who was seated on the opposite side of our large U-shaped table, walk around to my place at the table and make sure he agreed with the wording.

Upon doing so, Kumamoto suggested a change to the second editing criterion so that it read: "That science discusses 'how' and not ultimate causes for origins." I agreed with his amendment.

As Kumamoto returned to his chair in agreement, Terrell asked if there was any last-minute discussion. There was none. He called for the question. The vote was unanimous!

It was 2 minutes until 6 PM, and we were done. Terrell explained that Kumamoto need not read any more resolutions, since he would be making a presentation to the State Board the next morning as an *individual* — not on behalf of the Commission. *He also announced that I would be making a contrasting statement to the Board.*

My gamble had paid off! I would be allowed to deliver my surprise strategy to "the only appropriate audience" after all!

Even more critically, a tremendous victory in the Science Textbook Struggle had been won in the closing minutes of the meeting. *The science books would now be edited* — something that Kumamoto and his supporters had fought desperately to avoid.

As Terrell banged his gavel to adjourn the meeting, bedlam broke loose. Publishers in the audience rushed forward and began pounding me on the back. They couldn't believe how quickly the scene had shifted.

"You've just won your case!" one of them yelled. He was right. If *one single point* on the subject of origins in one single book required editing, my point had been made. And ultimately there would be *hundreds* of such points...

Confession, Clarification and Challenge

The great day of confrontation — 9 November 1972 — got off to an insipid start. The early routine business on the agenda for the State Board of Education took less time than expected. The science textbook hearing, scheduled for 10:30 AM, was announced by Board President Steward at 9:55.

Because most of the Commission's presentation to the Board had to be typed *after* its adjournment the night before, there was still material to be reproduced for the Board. Terrell was thereby forced to ask the Board to recess until 10:30 — the scheduled time.

This 35-minute recess seemed to heighten the anticipation of conflict — in the audience, among Commissioners, and certainly amid the many television crews who were waiting to record the battle.

At 10:30 sharp, Steward recognized Terrell, who set the stage by reading 4 criteria — the 3 that I had written in 1970 plus No. 2 in the Instructional Program:[416]

> The program shall reveal science not as a set of complete, final, and unalterable truths but as an ongoing search whose inferences, hypotheses, theories, and conclusions are subject to continuous evaluation and change.

These 4 criteria, Terrell said, formed the limits for the public hearing. He explained that, prior to opening the public hearing to those who had asked to speak, the Science Committee would make a 2-part presentation to the Board regarding the science books.

The first part would be *formal*, consisting of brief reports on various aspects of conformity to criteria. The second would be *informal* — as Terrell said, "...statements by Junji Kumamoto and Vernon Grose on the issue of origins that has apparently been the central concern of the people who are speaking here today."

The Science Committee's formal presentation took only a few minutes. Commissioner Cora Mary Jackson reported on *ethnic and racial balance* in the textbooks, Commissioner Loren D. (Bud) Good on *ecological considerations*, Commissioner Alpha Quincy on the *role of women*, and Kumamoto on *technical accuracy* of the books. The only aspect left was the subject of *origins*, that was assigned to me.

"Commissioner Grose will present a report on origins," Kumamoto announced to the Board when he finished his report.

There was a flurry of activity among the TV cameramen as they moved to their cameras. This was apparently the first shot to be fired in the gigantic struggle. Cameras began to whir.

"President Steward, Dr. Riles and distinguished members of the Board, I am pleased to present to you a unanimous position of the Curriculum Commission reached yesterday on the subject of discussing origins in the science textbooks..."

A hush blanketed the overflow audience in the auditorium of the State Resources Building in Sacramento as I began. It took less than 30 seconds to read the unanimous resolution we had passed in the dying seconds of our meeting the previous evening.

"Thank you. That's the end of our resolution," I explained.

The shock was too much. Even though Terrell had explained that we would divide the Commission discussion into formal and informal sections, everyone had apparently failed to make that distinction. They thought that the battle was underway.

Yet, I was all done! They couldn't believe it.

After asking if I had a copy of the resolution for the Board, Steward asked incredulously, "Is that the end of your statement at this time? Any questions by any members of the Board?" There were none. They appeared stunned.

Although it was brief and caught everyone off guard, that resolution was the most significant item the Board was to hear all day on the subject of origins. Unless the Board overruled their own Commission, *science textbooks would be edited and revised concerning origins.*

As I took my seat, Terrell returned to the microphone.

"Now Mr. Steward, that concludes this part of the Commission's presentation of reports that were approved by the Commission. We are asking now that you give an opportunity to Commissioners Kumamoto and Grose to make further statements in an attempt to clarify what we feel is the central issue or what some people are concerned about in the textbooks. Our desire is to show as eloquently as we can — and these gentlemen will make this point — that there are many more areas of thinking *together* than there are *differences*. I think that this would be a help to the Board and to the people here today in the hearing. We ask your permission for these positions to be heard — understanding that they are not papers that have been presented to the Commission. They represent the feelings of the Commissioner as an individual."

My presumptuous (and almost impossible) hope of simultaneously addressing the Commission, State Board and antagonists on both sides of the issue was about to be realized, thanks to Chuck Terrell. I could scarcely believe how miraculously it had all happened!

"I want to explain that I have talked with Dr. Terrell about this," Steward announced, "and I feel that this would be important information to help the Board. The official presentation by the Commission has ended, and we are now having supplementary information which may be helpful to the Board and others to understand this particular matter."

So Terrell had even made a private overture in my behalf to Steward ahead of time! I had given Terrell a dozen copies of "Science and Origins" the night before, so he had probably shown a copy to Steward as well as distributed individual copies to each Board member.

Regardless of what he had done, the stage was beautifully set for me to make my public confession exactly as I had hoped. I had only one nagging fear..."Science and Origins" is *long*.

As G. Ledyard Stebbins does, I also believe in rehearsing a speech. I had not only rehearsed "Science And Origins," but also *timed* it. It took me *29 minutes* to read it!

I was well aware that there were 50 other people in the audience who had secured written permission to speak in the public hearing. Each of them — including some Nobel laureates — would be granted only 5 minutes. I was going to take 6 times as long!

I feared the possible embarrassment of being stopped midway through my speech — for the obvious reason of wasting the Board's time. The one thing that could save me would be the length of time Kumamoto took.

Fortunately, he preceded me.

As Kumamoto walked to the microphone and began his presentation, I noted the time. He began to read a report that he had originally prepared to be a Science Committee document. However, he had not secured the endorsement of it from the Committee nor the Commission, so he read it as his personal opinion.

Kurnamoto's report was not a cohesive single document. He had a sheaf of loose papers in his hand — resolutions, letters, science articles and other references. Due to the pressure he evidently felt as the spokesman for the scientific elite as well as the tension generated by a dozen TV cameras focused on him, Kumamoto was to be pitied.

Disorganized and halting frequently while he searched for individual references to quote, he even dropped pieces of paper on the floor. During the long periods of silence that punctuated his talk, I continued to count the minutes he was taking.

The longer he took — whether speaking or trying to organize his bundle of papers, the better I felt. I was amazed as he approached 20 — then 25 minutes. As he passed the magic 29-minute mark, I knew I was home free. He lasted about 32 minutes.

When Kumamoto completed his presentation, there was no applause. In fact, many on the Board seemed irritated that he had taken so much time. Board member Tony N. Sierra asked Steward to clarify why, if Kumamoto was speaking as an *individual*, he was not held to the 5-minute limit that was announced for all individual speakers in the public hearing.

"Dr. Kumamoto and Dr. Grose are both members of the Curriculum Commission," Steward responded. "This issue that is here before us — I feel that they have the two differing viewpoints, so we could use their expertise on it as members appointed by the Board to this Commission. Dr. Kumamoto has made the first presentation. Dr. Grose will take the other viewpoint, which he will present now. In this way, hopefully, the Board will have the benefit of this contrast. Then we move to the public, which will be limited to 5 minutes each because we have 50 speakers. Now to Dr. Grose."

My moment had arrived. Kumamoto had done nothing to reconcile or resolve the dilemma facing the Board. He had tilted at windmills to the very end — crying out against religious mythology in science textbooks.

My strategy of reconciliation via confession of error would be a total surprise, as intended. If the 50 speakers waiting in the wings would listen to me, their pent-up emotion — on both sides — could be released toward a *common* goal. Their stingers would be pulled before being plunged into an imaginary adversary.

I felt at complete ease as I read "Science And Origins" (Exhibit 47). To this day, I believe it to be my clearest expression of the issue. My desire to succeed in "heading the army of speakers off at the pass — before open war could erupt" was sincere. And my presentation drew considerable comment in the press — all of it *positive!*

The *California Journal* editorial for December 1972 reported me as "a serious man with a crewcut and a sense of humor" who was "standing at the podium before the intently listening audience and facing the expression-less board members above him."[417] Regarding my delivery, the editorial continued, "It was an eloquent presentation, and drew warm applause and a few amens."

That editorial was not the only press coverage of my hair style (that then was just starting to grow out). The *San Francisco Chronicle*, in a feature article (see Exhibit 62), noted that Grose was "wearing his blond hair in a crewcut." Perhaps my dark brown hair appeared blond because it is sufficiently thin that plenty of scalp was exposed to the bright TV lights! Though misreading my hair color, this article correctly understood my position.

> Grose was starting out with two strikes against him. But as he began, in tones of quiet confidence, to address the Board, it soon became clear that this was no wild-eyed religious fanatic. No, far from launching into an uncompromising, Bible-quoting harangue, Grose, his tall frame draped in a smartly tailored navy blue suit, led off his statement with a series of admissions and concessions.

When I finished, the "warm applause" mentioned in the press still continued as I returned to my seat in the auditorium.

But the real interesting part was yet to come. President Steward, when the applause finally died down, asked that Kumamoto and I stand by for any questioning by Board members.

I had foreseen this possibility of being questioned as an *additional opportunity* to clarify much of the misunderstanding I attributed to the news media. Betting that questions might follow my report (were I allowed to read it), I had prepared, before leaving for Sacramento, 5 key questions that Board members might ask me.

I ran into John Ford, Board Vice President, just before the Board convened. Handing him a copy of these questions, I suggested that he get several Board members to ask them of me.

As soon as Steward asked if there were any questions by Board members, Clay Mitchell responded, "Yes, I have a question for Dr. Grose. Could you give us a specific example of what you consider to be the type of philosophy that is currently in textbooks and which is *not* scientific?" I recognized it as the first of the 5 questions I had prepared and handed to Dr. Ford. And I was obviously ready to answer it.

"Yes, I have an example from a textbook my son in the sixth grade brought home. It is a story entitled, 'The Beginning of Life On Earth' and is accompanied by pictures in full color. I would like to read a portion of that story."

I then began to read a ridiculous myth (Exhibit 66).

In no time, snickers — then chuckles were heard throughout the auditorium. The chuckles soon turned to loud giggling — then guffaws, and finally howls of uncontrollable laughter. By the time I got to the place in the story where plants "*reluctantly* left the water," the audience was obviously out of control. I decided to stop reading the story. I had made my point. And it wasn't a *partisan* viewpoint either...

Even though I stopped reading the story, the packed auditorium continued their hilarity. Finally, President Steward began to rap his gavel loudly to call the meeting back to order.

Vice President Ford, as soon as the auditorium had quieted down, fired the next question to me. It was the second question in my list of five.

"If I am not mistaken, it sounds as though you are saying that the science books are *already* 'religious' and that you want that 'religion' removed rather than the introduction of additional religion as 'creationism,' Is that correct?"

"If it is, the press will have an interesting time today," I joked, taking a swipe at news media bias. "But my personal commitment to the position expressed by Dr. Kumamoto, the National Academy of Sciences, the American Association for the Advancement of Science, and other distinguished bodies about keeping religious beliefs out of the science classroom is *without qualification!* Although I articulated (along with many others in and out of science) the 'case for design' as an alternate for the 'case for chance,' it was only as a balancing suggestion for anti-theism. Any teaching that expresses, or even implies, that all the order we see in nature *does not have* and *cannot have* involvement of a Supreme Being is, by definition, *anti-theistic.* I am pleased by the unanimous position of the Curriculum Development Commission which excludes *all* philosophic, religious and metaphysical beliefs regarding *ultimate cause* and which further says that science is limited to only trying to explain the 'how' of nature, rather than establishing 'who' or 'why'."

My third prepared question was a natural follow-up to the second. It was critical for destroying the news media's persistent straw man about sneaking Genesis into the science textbooks.

Ford was the logical person to ask the question...and he did. "Well, you're asking that the Genesis account also be *removed* from the textbooks. Is that correct?"

"Yes, and if you'd like a specific example, I'm prepared to say that in several books I have already asked to have Genesis removed." I then read several typical samples of wording that I had requested publishers to delete concerning Genesis.

If those in the audience who were waiting to decry the introduction of Genesis were listening, my answer should have taken the wind right out of their sails.

"Dr. Grose, you stated that you have already met with seven major publishers in this regard." It was Board member Sierra asking the next question in my list. "What evidence do you have, after spending many hours with these 7 primary publishers affected by this requirement, that the changes you have suggested are *scientific* rather than religious?"

I had no difficulty fielding that question. In fact, I had plenty of specific revisions that various publishers had prepared and handed to me the day before. I offered to let anyone present review them, if they suspected anything religious in them.

The first 4 questions I had prepared were apparently sufficient. The fifth one I had furnished was a *technical* one — dealing with the term "adaptation." Quite appropriately, it was never asked because it was too detailed and narrow in scope.

Just as Kumamoto received no *applause* for his presentation, he also received no *questioning* from the Board concerning it. On the other hand, prior preparation of leading questions for the Board had paid off for my strategy.

By deliberately using the question period, I had involved the Board as well as the audience in my attempt to defuse the time bomb they had all gathered to watch explode. It seemed impossible to me, after the Board's questions had been answered, that anyone present could ignore the challenge to face what I had called "the primary issue:"

> THE *TEACHING* OF SCIENCE HAS OVERSTEPPED THE RECOGNIZED LIMITS OF SCIENCE BY ENTERING THE AREA OF ULTIMATE CAUSE AND HAS COMPOUNDED THIS TRANSGRESSION BY BECOMING DOGMATIC AND DOCTRINAIRE IN EXCLUDING ANY ALTER-NATIVE CONCEPTS WHERE THE DATA WOULD CLEARLY ALLOW ALTERNATES.

As I sat down following the question period, the lunch hour was approaching. Rather than barely start the public hearing and then quit,

Steward elected to adjourn for lunch a few minutes early. His rap of the gavel was a signal for the news media to converge on me.

As I stood up at my seat, I was immediately surrounded by reporters.

Microphones were thrust at me from almost every side. Portable TV cameras were aimed and began to grind. Like machine-gun fire, questions pelted me. I did my best to answer them carefully, one at a time.

In the midst of this intense cross-examination, a tall, thin, distinguished-looking gentleman rudely interrupted.

"Do you have any *degrees*?" he screamed.

"Yes," I replied quizzically, turning to look at this obviously disturbed man.

"Are any of them in *science*?" he hollered.

"Yes."

"Have you ever published any *papers*?" His agitation was increasing.

"Yes — dozens of them. "

"On what *subject*?"

"On a *variety* of subjects."

Not only was I getting a bit irritated by this impromptu inquisition, but all the newsmen around me began to buzz with exasperation.

"How *recently* have you published?" he persisted.

I'd had enough. Deciding to reverse roles from *askee* to *asker* without losing my Irish temper, I gently asked him, "What's all this interrogation about?"

"I'm trying to determine your *right* to speak on science," he said with disdain.

Oh ho! I had run up against one of Galileo's medieval "qualifiers" who had been reincarnated! Even the onlooking reporters laughed at this pompous one who had come to either open or close my mouth on the basis of his own prejudice.

Instead of paying attention to *what I said*, he would start by deciding whether I had a *right to say it.* Still trying to keep my cool and with a smile on my face, I suggested softly that he would do well to read Hans Christian Andersen's, *The Emperor's New Clothes.*

"Even a child can speak the truth, sir. Perhaps you didn't understand my definition of the issue facing the State Board. If so, I'd be glad to explain it one more time —"

With no more than an arrogant snort, he turned his back and pushed his way through the ring of reporters. Later, I was to learn that he was Richard M. Lemmon, Associate Director, Laboratory of Chemical Biodynamics at the University of California's Lawrence Radiation Laboratory in Berkeley. Although he was not even a member of the

National Academy of Sciences, Lemmon evidently felt called to judge me. Perhaps he believed that his close association with Nobel laureate Melvin Calvin entitled him to occupy the senatorian seat of science sentencing.

I apparently flunked Lemmon's test as a person who should be allowed to speak because he wrote a lengthy, scathing letter (see Exhibit 50) to the State Department of Education — the wrong party, of course, since I was appointed by and reported to the State Board of Education — for allowing me to serve on the Commission's Science Committee.

Here's how I scored with Lemmon (see Exhibit 50):

> I submit that Mr. Grose has thoroughly disqualified himself as an expert on what should, or should not be, in science textbooks. I have no doubt that this opinion would find concurrence in the representative bodies of all of America's recognized scientific societies.

That wasn't the last broadside that Lemmon would fire at me either. Later incidents will reveal even greater desperation on his part.

Meanwhile, after Lemmon departed in a huff, the press conference continued for some time. In fact, it made me late for lunch in the cafeteria on the eighth floor of the State Resources Building. Hurriedly finishing lunch, I returned my tray of empty dishes. State Board member Hubbard spotted me from the opposite side of the cafeteria and came over to greet me.

"Vern, that was the finest of presentations," he said warmly. "I couldn't have stated it better myself." From Hubbard, that was a real compliment — especially when I recalled my first meeting with him months earlier in Ventura.

Even with Hubbard's glowing approbation, however, I had only set the stage for the victory I sought — turning a destructive head-on collision into a common effort to upgrade science textbooks. I had *confessed*, *clarified* and *challenged*. Had the combatants *listened?* Would they *change* anything they had come to Sacramento to say or not?

The public hearing, due to start immediately following lunch, would mark *success* or *failure* of my strategy...

Prepared, Predisposed Predicament

It was a *public hearing*, so 50 people came to be *heard publicly*. They had all visualized a life-and-death struggle at stake that needed their assistance.

They had all written *letters* to Sacramento requesting the opportunity to speak. They had all spent time *thinking* about something they could say in the public hearing. They had all *documented* their thoughts in a 5-minute speech. They had all taken *time* from their busy schedules. They had all *traveled* to the State Resources Building in Sacramento.

On the other hand, with few possible exceptions, none of them had *seen* — let alone *read* — the science textbooks that were the subject of the public hearing. None of them had seen the *changes* that were being proposed by publishers as a result of the October meetings.

None of them had come to Sacramento prepared to change their minds by anything they might hear in the public meeting prior to delivering their primed recitations. None of them heard my radical compromise because apparently *none of them listened to what I said!*

The State Department of Education in Sacramento published a list of those speakers who had been given the right to appear before the State Board of Education at the public hearing. The list of 50 names read like an *intellectual smorgasbord* — Nobel laureates, housewives, members of the National Academy of Sciences, eminent theologians, engineers, fundamentalist pastors, Buddhists, Catholics, Jews, Protestants, professors, publishers, consultants, educators, attorneys, physicians, and even a representative for the feminist movement!

With that variety, what *single subject* could be discussed? Numerous people discussed earlier in this book were scheduled to speak — including Nobel laureate Arthur Kornberg, Mark C. Biedebach, Jerry P. Lightner, Richard J. Merrill, Robert B. Fischer, Nell Segraves and her son Kelly, Herman T. Spieth and Roy E. Cameron.

It was going to be a formidable onslaught of speakers — at least from the State Board of Education's viewpoint. Probably in great hope but little expectation, President Steward made a prefacing statement to all listed speakers:

> If you find that your statement is a duplication of several statements that have been made before, and you feel that you can just step up and indicate that you have this opinion or that opinion and that you are filing your paper (if you have a paper), this would be very helpful to the Board in cutting down the time required for the public hearing.

No one evidently heard him. Everyone had come there to *speak* — not to mutter "me too" as they laid their speech on a pile of others.

The kickoff spot was won by C. Julian Bartlett, Dean of San Francisco's Grace Cathedral. And indeed an eloquent speaker he was. Noting that he had received a chemical engineering degree prior to turning to theology, Bartlett felt qualified to deliver a dissertation on how modern science had "rendered Biblical religion an inestimable service" by destroying its scientific credibility.

Evidently disappointed that the State Board had failed to heed his scientific insight in 1969 when they adopted my 2 paragraphs into the *Science Framework*, Bartlett closed his pontifical discourse with a scornful "non-benediction":

> Without any intention whatsoever to offer personal offense to any member of this honorable Board, I state again what I stated publicly in 1969 when the news of the amendments were first publicized, I would find your final approval of those amendments incredible, appalling and preposterous. Thank you.

Bartlett apparently and erroneously presumed that the *Science Framework* was a tentative document still to be finally approved. It had been approved and published in 1970! But he was not alone in railing at the 2 paragraphs I had written. Someone must have spread the word among the scientific elite that the *Science Framework* itself was being adopted by the Board — instead of *science textbooks!*

"I urge the Board to adopt the *Science Framework* without the inclusion of the opinion by Grose," Thomas H. Jukes of the University of California at Berkeley pleaded. Likewise, Max E. Crittenden, Jr., a research geologist with the U. S. Geological Survey, called on the Board to "rescind the present wording" in the *Science Framework* that I had authored.

Crittenden had snapped at Simpson's "singularly sensitive six" and even misquoted them:

> The *Science Framework* contains an outright error. The phrase, "the characteristic absence of transitional forms," is both misleading and erroneous. Many paleontologists who deal with ancient life and many biologists who deal with modern forms devote much of their careers to the problems of identifying and classifying and separating transitional forms. And the paleontologists who are dealing with the fossil record are doing precisely the same thing! The evolution of new forms, the gradual differentiation of these forms and their actual geographic distribution over the earth can be traced in the fossil record for literally hundreds of kinds of animal life. Moreover and more importantly, transitions between major kinds of animals, between fish and amphibia, amphibia and reptiles, reptiles and birds — which have been denied today — all of these transitions are well known in scientific references.

There were quite a few others that day beside Crittenden who were revulsed over "the regular absence of transitional forms" that I had borrowed from George Gaylord Simpson. Is it not mystifying why — if, as Jukes protested in his speech — "transitional forms *are not absent*," someone doesn't make a fortune by publishing a book loaded with photographs of transitional forms? Or why doesn't someone update Simpson's work that has been cited in Chapter 8?

The forms either exist or they do not. If they do, they are made of stone and can easily be photographed. Why only *talk* when one can *show?*

Though he was No. 18 on the list of speakers, the senior protagonist, Ralph W. Gerard, was moved up to 4th place. I had not seen him since our breakfast meeting in San Francisco on 12 February 1970.

Despite our irreconcilable disagreement over my 2 paragraphs, I had a very warm spot in my heart for Gerard. The nearly 3 years since we had been together had taken their toll. He was noticeably slower and more frail.

I listened intently as Gerard repeated the same arguments he had used from the beginning. As always, he used a clever little anecdote to illustrate the point that my two paragraphs were most inappropriate in the *Science Framework*:

> The essential violence being done to the *Framework* and science texts by the injection of a religious matter is well shown by an inverse situation.
>
> As a curate of a large church walked with a visitor toward the exit, the visitor said, "I'll give you $50,000."
>
> "Sorry," replied the minister, "I can't do it."
>
> "Well, $75,000."
>
> "The answer is still no."
>
> "$100,000 — my final offer."
>
> "Really quite impossible. I'm sorry."
>
> And the man left. The Assistant Minister who overheard this then said, "$100,000 would repair the church roof and it needs this badly. Are you *sure* you couldn't do it?"
>
> "Judge for yourself. He wanted me to say at the end of the Lord's Prayer — instead of 'amen' — 'Pepsi Cola'!"

Gerard tickled everybody's funnybone with that one. It was good to have a little humor to break the tension of the long, boring hearing.

However, Gerard did not stay with humor too long. He closed his short address with a rather plaintive plea not for the Board to *remove* my 2 paragraphs but just to *ignore* them:

> I and many others strongly urge the Board on the basis of science, religion, decency, reason and the best education of California's children to list as acceptable those textbooks which follow the original statement of the *Science Framework* guidelines.

Gerard turned toward my side of the auditorium as he left the podium. I left my seat and went to meet him as he slowly moved through the crowded auditorium to an exit. I put my arms around him and told him how good it was to see him again. He took my hand, smiled warmly, and thanked me.

It was a touching moment for me because I could sense that most of the fight had left him. It was to be his last involvement in the Science Textbook Struggle. A little over a year later, on 17 February 1974, Ralph Waldo Gerard passed away.

Since Gerard's death, I have often pondered Gerard's "test for God" that he shared with me in the faculty dining room the first day we met. Earlier, in noting the correlation between age and emotion among Darwinians, I proposed that Darwinism is an "old man's belief." In a profound way, Gerard's vociferous demands in the Science Textbook Struggle represented his final and clinching argument against the existence of God the Creator.

Obviously, there either is or is not a Creator. Gerard's "test" came out negative. But *that* "test" was on *Gerard's* criteria. If there *is* a Creator, *His* test may be different...

And Gerard was not alone in having his lifelong crusade against a Creator culminate coincidentally with the final days of the Science Textbook Struggle. Within a few months before and after Gerard's passing, so many of the giants of Darwinism — the "old men" and pillars of strength — departed.

Just a month before the public hearing, on 1 October 1972, *Louis S. B. Leakey* died at 69 years of age. *Alfred S. Romer*, at 78 years, followed on 5 November 1973, only three months before Gerard died at 74 years of age. Then came *Sir Julian Huxley*, at 87, on 14 February 1975. At 75 years of age, *Theodosius Dobzhansky* was next on 18 December 1975. Six months later, *Jacques Monod*, at 66, died on 31 May 1976.

Maybe it was all coincidental and happened by chance. Yet there was the persistent and intriguing correlation. All of these intellects who individually and collectively denied that the subjects of their research had a Creator...have they changed their minds? Or do they any longer have a mind to change?

Richard M. Lemmon, who had given me the "scientific litmus test" just before lunch, was the 8th speaker. He exuded confidence that life had originated *inevitably* — it couldn't help but happen.

One, of course, wondered whether the origin of the universe was also inevitable, but Lemmon ignored that one. Lacing his brief talk at several points with reference to "scholarly, professional, scientific societies," he explained how all options on the origin of life were reduced to only one:

> Research clearly indicates that appearance of life on our planet was not a function of chance. Rather, it was the result of the properties of matter itself...Given the inherent properties of elements such as their electrical properties, their bond-forming properties, and given the conditions of the pre-biological planet, *life was sure to appear.* It is unnecessary to invoke any unexplained chance. This broad outline of the present scientific theory of the appearance of life would be agreed to by the overwhelming majority of members of our scholarly, professional, scientific societies. *There just are no other scientific theories* about the appearance and proliferation of life other than those of chemical and biological evolution.

So far, it was evident that *no one had listened to my challenge.* Instead, they had stopped their ears and rushed into the fray — like lemmings charging pell-mell down a hill to drown in the sea. They were going to prove a point, whether or not it was relevant.

The 20th speaker, Robert Bulkley, Pastor of the Portalhurst Presbyterian Church in San Francisco, who was speaking as Protestant co-Chairman of the San Francisco Conference on Religion, Race and Social Concern, was typical of those who heard what they wanted to hear — even if it was in error. There was no excuse for Bulkley's statement about Genesis:

> We understand proposals are being made — *and I have learned today that our understanding is correct* — that (science) textbooks include as an alternative to the theory of evolution, the special creation theory which seems to be implied as in the early chapters of Genesis.

In Chapter 4, it was noted that Bulkley had written a resolution in 1969, representing 44,000 United Presbyterians in San Francisco, against introducing Genesis into science textbooks. Even religious journals like *Christianity Today* had clearly shown that he was in error at that time — that he had fallen for press reports rather than facts.[418]

Yet, here Bulkley was — 3 years later — still singing the same tune, and claiming that "I have learned *today*" that Genesis is being included as an alternate to evolution. How much more clearly could I have said *that very morning* that Genesis was not desired? Had I not listed, in response to Vice President Ford's question, specific instances of *removing* Genesis from textbooks?

The refusal to listen to my challenge was not limited, by any means, to those who feared religious intrusion into science. There were many speakers who urged that Genesis be *included* as a model for origins.

I was just as upset with this group's failure to listen to my challenge as

I was with the Darwinians. While I might agree that Evolution II is no more valid scientifically than Genesis, my goal was the removal of *all* non-science. Only if Evolution II could not be removed would I favor introducing a theistic alternative to counterbalance it. Even then, Genesis, per se, would be an inappropriate account in a pluralistic society.

There was one truly unique presentation at the public hearing — the only one that was accompanied by visual aids. Ronald S. Remmel, who received his PhD in elementary particle physics from Princeton was working in the Department of Physiology at the University of California at Berkeley on a National Institute of Health post-doctoral fellowship. His research in neurophysiology at the time involved implanting stereotaxic micro-electrodes in cats to study potentials in single neurons related to eye movement.

Remmel, armed with large charts that depicted details of quantum mechanics, squeezed into his 5-minute time frame a remarkable message. While there probably weren't over a half dozen people among the hundreds present that day who followed every step of Remmel's technical jargon, he was able to summarize the idea that "quantum mechanics says that there is a basic *unknowableness* about the world." Mendel's laws and mutations that are so critical in the evolutionary concept are based on the randomness of quantum mechanics. Remmel's brief paper was later published.[419] His conclusion in the public hearing is worthy of attention:

> As a scientist I feel that textbooks which declare that present-day events and the origin of life are the result of mindless and meaningless chance are expressing an *assumption*, not a scientific fact. Science teachers in texts should stress not only the limitations and uncertainties of present data and theories, but also the basic quantum-mechanical *unknowableness* and its implications for unpredictability in evolution and history. Humility in the face of this unknowableness is certainly in order.

I was most disappointed that anyone viewed the issue as religious. Yet it was apparently inevitable that a couple fundamentalistic ministers would want to be heard — since the press had set up the "God-versus-Darwin" straw man. These preachers represented enough "non-scientific impurity" to get all those asking for an alternative to Darwinian dogma branded as either "religionists" or "creationists" — derisive terms to the scientific elite.

Yet an ironic thing happened at the public hearing. *There were twice as many religionists on the Darwinian side!* A Buddhist priest, Catholic nun, Presbyterian pastor, Jewish rabbi, Catholic priest, and the Dean of an Episcopalian cathedral all testified in defense of Darwin.

In no way could the 9 November 1972 hearing be truthfully described as *scientists* fighting *religionists.*

An even more amazing statistic emerged. When all the speakers had been heard, the number of qualified scientists (by Stebbins' criteria) calling for an alternative or open-mindedness concerning origins outnumbered 3 to 2 those scientists insisting on enforcement of Darwinian dogma. Likewise, since the two sides were roughly equal in number, there were twice as many *unqualified defenders of Darwinism* as there were so-called *religionists fighting for creation*!

Long before the final speaker had spoken at the public hearing, I realized that my strategy of offering a compromise that both sides could endorse — of combining the polarized forces into a single force for cleansing science textbooks of *beliefs* — was a failure.

Why did it *fail*? Why would none of the speakers *listen*? Why would they continue to attack positions that had clearly been shown as *non-existent*?

Seriously searching for answers to such questions, I was forced to conclude that Darwinians — even if they heard mc — believed either that I had a hidden ulterior motive, that I was a "front man" for a religious cause, that I was a scientific idiot unworthy of consideration, or that I was deliberately lying.

On the other hand, those who wanted Genesis quoted directly in science textbooks evidently thought either that I had abandoned their noble cause, that I had been duped by Darwinians, or that I was absolutely wrong in deleting Genesis.

Net results for my efforts? Zero. All the speakers had come *prepared.* All the speakers had come *predisposed.* My compromise had only presented them with a *predicament* — they would have to give up their party line! This they could not do...

Monotonously Muddled Media

Within minutes after the final speaker finished in the longest public hearing I'd ever witnessed, I was on my way to the Sacramento airport to catch a plane back to Los Angeles. I was both pleased and disappointed — *pleased* because I had been permitted to present my strategy to "the only appropriate audience" but *disappointed* that it had failed to reconcile two inflamed poles of thought.

The Sacramento airport terminal was crowded. It was at the height of the early evening rush hour. I had just confirmed my flight when ABC's Los Angeles correspondent, Dick Shoemaker, walked up to me and introduced himself. He had attended the public hearing, heard my speech, and wanted an in-depth interview with me for the ABC *Evening News* for the next day.

My flight was so imminent that I suggested that I could be interviewed in Los Angeles in the morning, handed him my card, and ran to catch my plane. The flight back to Los Angeles was loaded with publishers, people who had spoken in the hearing, and Los Angelenos who had flown up to Sacramento just to observe the hearing. The spirit aboard the plane was jovial — as everyone, now relaxed, recounted specific incidents and anecdotes of the day.

Early on Friday morning, 10 November 1972, ABC called me at home. "We have two men in a car ready to leave our Hollywood studios if you will just give us directions to your home," Dick Shoemaker said.

Before giving directions, I began to discuss with Shoemaker what story he intended to develop.

"We have miles of footage of video tape from the hearing which needs editing before we can even build a story," he said.

A different idea popped into my head. "Why don't I come down to your studios instead of you people coming out here? I'll bring all the necessary materials with me..."

"I'll tell you one thing we want for sure," Shoemaker broke in.

"What's that?"

"We need your son's textbook that you quoted yesterday. The one with that ridiculous story about how life began. Do you still have it?"

"Sure, I'll bring it. I should be there in less than an hour."

When I arrived at ABC in Hollywood, Shoemaker was going at a dizzy pace. He was running in and out of a dark closed-circuit video viewing room, studying intently a few minutes of film, then barking orders to several technicians. Cameramen grabbed my son's textbook and set it up for filming.

There is an early afternoon deadline for all Pacific Coast news to be shown on the network's evening newscast. Not only is there a coast-to-coast 3-hour time difference, but news from all over the world must be integrated into a single program to be released at 6 Eastern time. I stayed with Shoemaker all morning and began to understand how it can take several hours to produce a 2 or 3-minute story for a news program.

Our family watched Howard K. Smith on the ABC *Evening News* only a few hours later. Smith listed the key issues around the world for that evening — a Haig-Thieu peace plan, ecology problems in India, British Tommies in Belfast, and *EVOLUTION!*

Dick Shoemaker's story lasted about 3 minutes — including random footage of various speakers on both sides of the controversy in the public hearing. The greatest excitement occurred as our children — almost in unison — cried out, "There's our book!" (All six of them eventually used that same book.)

Sure enough, there was the same textbook from which I had read in Sacramento the silly myth, "The Beginning Of Life On Earth" (Exhibit 66) with its full-color pictures, being seen by millions across America.

Later that evening, KABC Radio, the ABC affiliate in Los Angeles, called and asked me to appear on the "Marv Gray Show" the next night, Saturday, 11 November.

I had already accepted an invitation to speak at a dinner meeting in Seattle and was flying there early on Saturday. However, Marv Gray himself was so insistent about having me as a guest that he asked if I would agree to be "patched in" by telephone from Seattle on his show!

Having agreed, arrangements were then made for me to participate by telephone — that was also piped over loudspeakers to the more than 100 dinner guests in Seattle, since I was delayed from speaking to them until after the radio show.

Marv Gray was a well-known radio personality in Southern California. His talk show was probably the most popular one in the area. Such shows capitalize on controversy, and Gray intended to do just that with me.

I disappointed him. He'd feed me bait — and I wouldn't bite. One typical example:

> GRAY: Mr. Grose, it's coincidental that, just when this argument with the State Board of Education occurs over whether or not to include the creation theories beside those of evolution in the scientific textbooks and has become such a heated discussion, that at that very moment, Dr. Leakey in Africa announces that he has discovered a skull from 2.5 million years ago that might now cast doubt on the Darwinian theory. (Gray was referring to Richard E. F. Leakey's announcement on 9 November 1972 that he had found two mandibles that he suspected might represent some early form of the genus *Homo*, to which modern man belongs, and which he dated at 2.5 million years.)
>
> GROSE: I thought that was rather interesting from a number of points of view. First of all, I don't think that the person who made that announcement is himself a doctor — he is the son of the famous Dr. Leakey.
>
> GRAY: Yes, he's a son, that's right.
>
> GROSE: He's unschooled as I understand it...a 28-year-old son of the famous Dr. Leakey now deceased.

GRAY: You know, it's funny that you'd say that, because I thought that you'd be delighted to hear that his findings raised a question about the Darwinian theory of evolution.

GROSE: It really doesn't delight me as much as some might suppose, because I am not an anti-evolutionist in spite of all the media stories to this point in time. I really deplore the controversy that has developed over this because it seems to be a deliberate desire to drive a wedge between science and religion or between Biblical creation and evolution. From the very beginning, 3 years ago last Monday when my words were adopted unanimously by the State Board of Education, I denied at that tune that I had an interest in introducing religion into the science classroom.

This exchange was a mystery to Gray. It naturally led him to ask how the controversy developed — since I was apparently against religion being put in science textbooks. That gave me an excellent opportunity to expose the role of the news media in generating a controversial straw man. Whatever *Gray's* purpose had been in having me on his show, *my* purpose was to destroy the straw man.

Of course, there is nothing quite so counterproductive to a talk show built around controversy as to learn that the controversy is a *phony*.

So Gray was obviously disgruntled. I wouldn't tangle with him. Instead, I read the resolution I had written for the Curriculum Development Commission — which they passed *unanimously* — calling for science textbooks to be rid of dogmatism. In doing so, I drove my point home:

You will note in that resolution that there is no mention whatsoever of religion, creation, God, Bible or anything else non-scientific that the press has been reporting.

By this time, Gray was out of gas. He had been so sure that he had a classic and universal rhubarb that ought to light up his switchboard like a Christmas tree. Instead, it was a *bust* — like steel turning into warm butter!

He had introduced me to the radio audience as "the chief advocate for the creation theory on the Curriculum Development Commission" — apparently because the *Los Angeles Times* had called me that the day before.[420]

If I was worthy of that title, how could I be urging "no mention whatsoever of religion, creation, God, Bible or anything else non-scientific?"

"We've got some people on the line who I am sure will want to ask you questions about this, rather than my pursuing it personally any further at this particular time," Gray said, trying to maintain a running dialogue. Quite evidently, he had given up trying to pump life into a phantom. Instead, he would turn me over to the radio audience.

"We are talking with Vernon Grose. He's the chief advocate for the creation theory on the Curriculum Commission. Of course, we're discussing an age-old thing — the conflict between scientific evolution and religious creation, although Mr. Grose says he's not opposed to the theory of evolution nor does he feel that the conflict between evolution and religion should get into the textbooks. Rather, he says that we should get a broad-minded look at everything to stimulate the interest of the students. Now let's hear what some members of our audience have to say. KABC Marv Gray, you're on talk radio with Mr. Grose. Go ahead, please."

Before describing the first caller's question, I have another confession to make. Just as I had reaped benefits from preparing questions for the State Board of Education to ask me two days earlier in Sacramento, I decided to enlist additional help for the "Marv Gray Show."

Prior to flying to Seattle, I called Ralph Winter, who had been with me in this struggle as long as anyone, and suggested that he listen to the show. Further, I proposed that the two of us do a "straight man-funny man" act. By calling in and asking the right questions, Winter could set me up to shoot down the "God versus Darwin" straw man better than I could do single-handedly.

Despite our prior planning, I have to confess that I was shocked when I heard the first caller's voice. It was *Winter's!*

"I'd like to ask why it is that Mr. Grose has been called the 'chief advocate of the creation theory' and yet this resolution of the Curriculum Development Commission which he read makes no reference to a creation theory. It does not apparently seem to propose such a thing."

Barely able to hide my pleasure and surprise at our success, I brought Gray into a three-way conversation with Winter.

"I think Marv Gray is probably responding to a *Los Angeles Times* article which, incidentally, I am not happy about because it calls me a 'chief advocate.' I was going to correct you later, Marv, on that subject because I am not advocating a creation theory. What I *am* saying is that we ought to cleanse science from *all* religion — that's been my first and foremost objective from the very beginning."

Thanks to Winter, I went ahead and explained that religion need not be *theistic* — it can also be *anti-theistic*, like Madelyn Murray O'Hair's version. Therefore, if I was an advocate of anything, it was of *objectivity* in the name of science.

"Well, that seems very different from the way it's been in the newspapers — where it seems to be the 'Monkey Trial' continued. I think that clarification somewhere needs to be made. I am quite mystified by the whole thing."

Winter was doing a superb job! Because Winter called for clarification of the issue, Gray finally caught on and also asked me to clarify my position.

"Well now, there's one aspect of this that I think we *ought* to get clarified. You say that you want to remove...you want to *cleanse* science from all religions, right?"

Gray was beginning to listen to me. "All right, now when you say that, at the same time my understanding is that you want to *include* in scientific textbooks various theories other than those of evolution, including that of creation. Is that right?"

Winter and I had brought Gray around to asking a meaningful question. I was able to answer Gray's inquiry with a double-edged reply.

My first choice would be to have *absolutely no reference* to origins in science textbooks. That would clearly *cleanse* them from religion. However, since scientists are human beings who like to speculate about origins, we may not be successful in preventing them from expressing their beliefs on how the universe, life, and man originated.

If that happens, I explained that I definitely would demand that an anti-theistic belief, such as evolution, be *balanced* by a theistic position so as to maintain the neutrality of science about religious beliefs.

The tone for the rest of the hour-long "Marv Gray Show" was set. Winter and I had succeeded in exposing the straw man. All other callers were forced to address the real issue.

That broadcast was my last contact with Marv Gray. He was a nervous person with a heart condition for which he was taking medication. Less than a year after the broadcast, on 5 November 1973, Gray attended a party and had a drink. He was not considered drunk, however, upon leaving the party to go to work. On the way to work, he was driving erratically and was stopped by police. They took him into the police station "drunk tank," and he died there tragically at 56 years of age.

My encounters with the *electronic* media — television and radio — following the public hearing had been meaningful. The truth of what had happened in Sacramento had been expressed. The struggle was correctly described as involving *objectivity* in science.

But the *written* news media, for the most part, continued the same religious harangue they had invented 3 years earlier.

The *Los Angeles Times*, that has been heavily criticized in this book thus far in the struggle, published its most objective report thus far regarding the public hearing.[421] Jack McCurdy, *Times* Education Writer, who had consistently built the "God-versus-Darwin" straw man, wrote an excellent and accurate review — until he quoted me:[422]

> Grose, a Canoga Park engineer, said the teaching of science "has overstepped the recognized limits of science by entering the area of ultimate cause" without sufficient evidence.
>
> Prof. Junji Kumamoto, head of the curriculum commission's science committee and a UC Riverside scientist, said evidence for evolution is abundant...

Evidently, McCurdy felt obliged to pit what I had called "the rudimentary issue at stake" against Kumamoto's impressive defense of Darwinism. This was simply a *non sequitur*.

Compare what I *said* with what McCurdy *reported*:

> The teaching of science has overstepped the recognized limits of science by entering the area of ultimate cause and has compounded this transgression by becoming dogmatic and doctrinaire in excluding any alternative concepts where the data would clearly allow alternates.

Did I suggest any "shortage of evidence?" Particularly, was the issue of the "evidence for evolution" raised? Why then did McCurdy twist my words? Either because he didn't understand what I said or he needed to keep the straw man alive.

McCurdy's reason for muddling, I am convinced, was to sustain the straw man that he was largely responsible for creating 3 years earlier. What basis is there for that conclusion? *The very first sentence of McCurdy's report*:[423]

> The age-old conflict between scientific evolution and religious creation was revived anew and passionately debated by scientists, theologians and educators before an overflow audience at a state Board of Education hearing Thursday.

There is no denying that a vast majority of the 50 speakers on 9 November 1972 may have agreed with McCurdy's lead-in. However, I was not part of that conflict.

Look once more at Figure 6 in Chapter 4. My objective at the public hearing is unmistakably depicted by Option III. Only if Darwinians insisted on keeping Evolution II (Box D) in the textbooks would I insist on Creation II (Box C) being introduced for balance. That, of course, would mean selecting Option II.

Note McCurdy's first sentence again. It is a classic description of Option IV in Figure 6 — "*scientific* evolution versus *religious* creation." He totally ignored *religious* evolution (Box D) and *scientific* creation (Box A).

After 9 November 1972, that kind of reporting must be connoted as "muddled" — *monotonously* so. McCurdy hadn't changed in 3 years!

To contrast McCurdy's implacable sustenance of a straw man, a front-page report in the *San Francisco Chronicle* on the Sacramento public hearing, MOVE OVER, DARWIN, took a completely different twist. Whereas southern Californians would be led by the *Los Angeles Times* to believe that *an age-old battle was revived* on 9 November 1972, northern Californians would be persuaded by the *San Francisco Chronicle* that *a 3-year fight had ended* in capitulation:[424]

> The State Curriculum Commission gave up its three-year fight yesterday to keep Eden from getting equal billing with Darwin in the State's science textbooks.
>
> The State Board of Education adopted this policy three years ago, but the commission stood fast until its surprise announcement here yesterday...The new criteria adopted by the commission, which recommends material in textbooks, would allow use of both divine creation and evolution as opposing theories in the new science textbooks...

This report had to be a much greater distortion of truth than McCurdy's. Ron Moskowitz, Education Correspondent for the *Chronicle*, had no basis in fact for what he wrote.

There is no evidence whatsoever that the Curriculum Development Commission ever took a position in opposition to the State Board of Education — publicly or privately. Likewise, the resolution that was unanimously adopted by the Commission, by no stretch of the imagination "would allow use of both divine creation and evolution as opposing theories."

In fact, Moskowitz later reports *in the same article*:[425]

> The new guidelines would allow books to discuss the "how" of man's origin but would prohibit them from mentioning "ultimate causes" of life. They would have to avoid, then, naming either God or the random chemical reactions of several billion years ago that evolutionists say led to living cells.

Which of Moskowitz's statements is *true* — that selection guidelines would (1) *allow* use of both divine creation and evolution as opposing theories or (2) *prohibit* naming God and the randomness of evolution as alternate ultimate causes of life? You can't have both — in spite of Moskowitz.

Maybe Graysmith saw *compromise* as the answer to Moskowitz's confusion because his cartoon (Figure 13) appeared 3 days later in the *Chronicle.*

The editor of the *California Journal*, among many others, was as muddled as anyone could be. He reported, in error, that (1) none of the current science textbooks discussed origins, (2) the Curriculum Development Commission announced in the public hearing that none of the proposed science textbooks discussed origins, (3) the footnote regarding my two paragraphs in the Science Framework was written by the Curriculum Development Commission, (4) no publisher had produced textbooks responding to a creation theory requirement, and (5) the State Board of Education had "forced" my two paragraphs into the *Science Framework* over objections of the Curriculum Commission.

Figure 13.

Human beings make errors — but that is not the point. All 5 of these errors are on *one side* of a straw man issue! They all reinforce the concept that science was being prostituted by idiots. Is it fair to ask whether this editor was biased?

Don Speich, Education Writer for the *Sacramento Bee*, gave a fair and balanced account of testimony during the 9 November 1972 public hearing.[426] Enigmatically however, Speich invented a version of the unanimous Curriculum Development Commission resolution on editing that was passed on 8 November (Exhibit 45) that does not even resemble the resolution:[427]

> Uncertainty can be seen in a resolution adopted by the Commission Wednesday night which ends up by first posing a question then answering it:
>
> "Would alternatives to (English Naturalist Charles) Darwin's theory (of evolution published in his book 'Origin of the Species' in 1859) be allowed under these guidelines?
>
> "That would be up to the board to decide."

Speich likewise reported me as summing up best the arguments for including creation in science textbooks. Quite obviously, that was *not* my position. Instead of demanding equal time for creation theories, as Speich reported, I had proposed *two alternatives* for publishers whose books explained, for example, how plants "became" animals (see Exhibit 47):

> The first and preferable alternative, by far, was to remove *all* explanations for this supposed phenomenon, since there is absolutely no record of such an event occurring — there is simply a break in the fossil record at that point and *any* explanation would be speculative rather than testable. The second alternative was to provide an optional possibility that animals may never have come from plants at all but that they may have been designed from their inception as a unique class of living things.

Since Speich had a written copy of my speech, one wonders how such errors occur. Errors like that are emotional in character and provoke irrational cartoons like Figure 14 that appeared in the *Sacramento Bee* two days after Speich's article. What more can be done to preclude errors of that nature?

One reason that someone should be hesitant in accepting one-sided error as unavoidable is that there *was* some accurate reporting and analysis of the 9 November 1972 public hearing.

Some newspapers correctly recounted not only individual testimony in the public hearing but also the issue of evolution having a *religious* (Evolution II) as well as a *scientific* (Evolution I) connotation. The *Los Angeles Herald-Examiner* was one such example.[428] The *Santa Ana Register* likewise devoted an entire section of their Sunday newspaper to a series of 9 articles on all sides of the issue.[429]

The *Seattle Post- Intelligencer* also carried a lengthy article that correctly explained my objective of either *cleansing* all science textbooks of origins or else *balancing* the anti-theistic Evolution II with a pro-theistic Creation II.[430] So the message was comprehensible for those who were open-minded.

Figure 14.

National periodicals too covered the public hearing. *Newsweek* chose to muddle the truth by reporting Board Vice President Ford as believing that the Judeo-Christian theory of the Creation deserves a fair shake in the textbooks.[431] Again, biased error occurred. Why?

The longest recap of the public hearing appeared in *Science*, the journal of the American Association for the Advancement of Science.[432] Nicholas Wade, author of this article, had interviewed me by long-distance telephone. His in-depth search for accurate information is revealed in his report.

At several points, he does mix truth and error as in this passage:[433]

> Creationists, although they personally do not believe that evolution occurred, are not asking that Darwin be evicted from the classroom. Nor, as they are sometimes accused of doing, are they trying to put Genesis into the biology books.

Creationists *do* believe that evolution occurred. They *do not* believe that evolution explains *origins!* However, Wade's account is a worthy source of truth, for the most part. Anyone desiring an objective review would be well advised to consult it.

The great confrontation of 9 November 1972 had become history. But what *kind* of history? If one depended only on news media for that history, it could be most difficult to know the truth. With few exceptions, the written news media were monotonously muddled.

RESOLVE WITH POMPOSITY

The Titanic collision on 9 November 1972 in Sacramento — that unleashing of forces that had been inevitable ever since the news media invented "God-versus-Darwin" — was no more than that. Only a *collision.* It produced no conclusions.

In a way, it was like the Battle of Gettysburg in the War Between the States. Union and Confederate forces had done nothing more than maneuver for position on 1-2 July 1863. However, on 3 July, Robert E. Lee, commanding the Confederate Army, decided to directly attack the center of the Union line under command of General George G. Meade. Advancing across an open field and up the slopes of Cemetery Ridge into murderous Union fire, General George E. Pickett's troops advanced in a now famous charge. They reached the crest of the ridge but were driven back.

Both sides suffered roughly equivalent losses. Instead of pursuing battle to a point of victory or defeat, both Lee and Meade stopped to "lick their wounds." Lee was able, before Meade could assess the situation, to retreat and regroup for two more years of war. Rather than producing a decisive victory for one side or the other that would have ended the struggle, the Battle of Gettysburg only dissipated the two armies.

The public hearing before the State Board of Education in November 1972 was never intended to be a decision-making meeting. The Board had announced that it wanted to delay any decisions until their December meeting. This meant that the 34-day period between discharge of all forces in the battle and toting up the score could be critical as to which side actually *won*!

In athletic contests, courtroom trials, or political elections, we all like to know the final outcome as soon after the two sides have ceased their competition as possible. Can you imagine waiting 34 days to learn the outcome of a championship heavyweight boxing match or the winner of a presidential election? The longer we had to wait, the more we would suspect the results.

Yet, the State Board of Education deliberately refused to announce the winner of the public hearing until 14 December 1972. What was to take place in this intervening period? Should the combatants attempt to further influence the State Board as each member allowed the testimony of 50 witnesses to "simmer on the back burner?" Or should the combatants "leave well enough alone" and withdraw quietly to await the reasoned outcome without interference (as is expected in jury deliberations)?

The Godly chose the non-interference course. But the Darwinians decided to use the 34-day deliberation period for additional combat.

There may be several reasons for the Darwinian decision to continue the fight beyond the public hearing. First, they may have felt that their case was not well stated during the hearing. Second, they may have feared that the Godly were also lobbying during this time and should be counteracted. Third, they might have had some additional ammunition that was inappropriate to use in a public hearing but that could have impact in pressuring the Board to go along with their view.

It was probably the latter of these reasons that caused Darwinians to keep the heat on the State Board in November and December 1972.

For one thing, there is a limit to how blunt you can be to a person's face — like in a public hearing. Even if State Board members were considered "stupid" by Darwinians (and they were), Darwinians probably thought it counterproductive to tell them so *in person.* It would be more effective to bring the *impersonal* force of public opinion against them for their stupidity.

Darwinians had also mounted the very visible and massive campaign of resolutions by influential scientific groups just prior to the public hearing. It would be natural for them to just let this campaign continue forward with inertia until the December Board meeting.

The most important Darwinian resolution, of course, had been the unprecedented one by the National Academy of Sciences (Exhibit 41). The day after the public hearing, the *Los Angeles Times* published a lengthy and thoughtful criticism of the National Academy of Sciences resolution by a philosophy professor at Pepperdine University, Arlie J. Hoover.[434] Hoover's treatise was later published in the international journal, *Christianity Today*.[435] It undoubtedly was a stinging blow for the Darwinian's prize resolution, and may have been a factor in their decision to continue their fight beyond the public hearing.

Whatever provoked Darwinians to pursue a campaign between the end of the *public hearing* on 9 November 1972 and the *decision date* of 14 December 1972, the basic approach Darwinians used was condescending. Perhaps feeling that they could not win their case with the non-scientific public in a point-by-point debate with their opposition, they resorted to a highhanded and pompous attack on the State Board — probably to *intimidate* rather than *convince* the Board.

Ridicule was a prime weapon in this attack.

Ricocheting Ridicule

The public hearing ended on Thursday night, 9 November 1972. I spent the following weekend in Seattle with two speaking engagements, flying back to Sacramento on Sunday evening in order to meet with the Governor's Select Committee on Law Enforcement Problems on Monday.

A long-distance call from Chuck Terrell, Curriculum Development Commission Chairman, interrupted our 5-man Select Committee on Monday morning.

"Vern, I hate to bother you, but Clarence Hall (Wilson Riles' Assistant Superintendent of Public Instruction) has called a meeting in his office on Tuesday, 28 November, to prepare some typical changes to the science textbooks. Could you make that date?"

"It looks tight, Chuck. I am to address the annual Alumni Banquet of Southern California College in Costa Mesa on Monday evening and must deliver a paper in San Diego at the Naval Safety Conference on Wednesday morning. If I can juggle airline schedules, I'll make it."

"It's critical that you be there to present 5 to 7 specific changes that deal with origins," Terrell urged. "You are the only Commissioner who has both reviewed the textbooks for revisions on origins and met with publishers on the subject. Other Commissioners will be present in Hall's meeting to have examples of editing for ecology, scientific accuracy, the role of women, and minority representation in the tentatively-adopted materials."

The increasing pressures of my full-time assignment for Governor Reagan plus my other professional responsibilities were really squeezing me. Terrell wanted my editing suggestions on origins to be in *writing* — illustrated by showing the original version in the textbook on one side of the page and a suggested revised edition on the other side. Thereby, a comparison could be made by simply looking across the page.

Of course, all this work was to be done gratis, without any financial compensation. By this time, I had been in the Science Textbook Struggle for more than 3 years and had never received a dime for my investment of thousands of hours. That may explain why I was not overly enthusiastic to receive Terrell's phone call.

Within 4 days however, I had selected 7 typical editing proposals — one for each of the 7 publishers with whom I had met privately in October. I not only showed the "before" and "after" versions, but also tied each change directly into specific curriculum criteria (see Exhibit 49). Two of these 7 changes called for the *removal* of all references to Genesis!

These proposed changes in the textbooks obviously drew the fire of Darwinians. But I expected their wrath. Any change at all would be an acknowledgement that they had erred — and no one enjoys admitting error.

The meeting called by Clarence Hall in Sacramento in 28 November 1972 was ostensibly to collect typical changes in science textbooks so that the State Board could get an idea of the type and magnitude of editing that would be required before they could finally adopt the books, As Terrell had mentioned to me, changes were being proposed not only for *origins*. Other Commissioners on the Science Committee came prepared with editing revisions regarding *ecology*, *scientific accuracy*, *the role of women*, and *minority representation* in the materials that had only been tentatively adopted by the State Board.

Yet when I arrived for the meeting, the only changes that drew any criticism were mine. Kumamoto had prepared a resolution that said: "Be it resolved that the Commission recommends to the State Board of Education that no editing on origins is necessary except for the deletion of references to Genesis and related theological material." It was never adopted.

No one present at the meeting, beside Kumamoto, had any science education or experience. Commissioner Alpha Quincy had fought as a member of the antecedent Curriculum Commission against the introduction of alternatives on origins. It was no accident that she got herself appointed vice-chairman of the Science Committee in the new Curriculum Development Commission — a position that I challenged unsuccessfully but that was abolished as a result of my challenge.

Supposedly, Mrs. Quincy's contribution was to propose changes about the role of women in the textbooks. While she did get all the aprons removed from pictures in the books because they were "sex stereotypes," she obviously came to the meeting to block any changes concerning origins. Her vehement but unknowledgeable support of Kumamoto was never greater than during this meeting.

As I presented my proposed revisions (Exhibit 49), I found myself in an awkward position. Whereas all the other presentations had been accepted without any significant discussion, Kumamoto, joined by Mrs. Quincy, immediately challenged every change I proposed — except the deletions of Genesis in the Holt, Rinehart & Winston and American Book Company materials. While I had given far greater justification for my changes based on the curriculum criteria than anyone else, my proposals were the only ones being attacked.

I do not shirk a good fight. However, my opponents were a diminutive Asian (member of a racial minority) who was the committee chairman, and a vociferous woman who knew absolutely nothing about science. To fight either was to fight a member of an oppressed minority — both victims of stereotyped prejudice. Clarence Hall, an educator himself, was in no position to adjudicate. It was a frustrating scene, to say the least.

Our meeting, that was attended by only the 5 members of the Science Committee — not the full Commission, was for gathering information. It was not for *approving* changes. We were not *voting* for specific changes.

They were to be only *examples* of editing that the State Board would order to be done throughout all materials. Yet, they should represent a consensus of the Commission and not just the private view of one Commissioner. As Kumamoto and Mrs. Quincy hammered away, I finally pointed out that, unless they could specifically find either technical error or criteria non-compliance, the changes should be presented to the State Board just as the other proposed revisions would be.

I won my point. Exhibit 49 survived and got to the State Board.

Since our meeting was not a public one, news media were not present. Evidently, someone who had been present gave their version to the press because the following account resulted:[436]

> On Nov. 28, members of the curriculum commission met in a secret, closed session in Sacramento to hammer out a compromise based on the November resolution (see Exhibit 45).
>
> The meeting, called by Dr. Clarence Hall, assistant state education superintendent, was essentially a bargaining session. Grose presented some 20 pages of suggested changes.

> The commission made compromises but Grose became angry and quit negotiating when it balked at many of his ideas. What came out is language that clearly softens evolution's credibility.
>
> Kumamoto is still claiming a victory of sorts. The changes, he says, do not reduce Darwin's concept of evolution to simple speculation that man may have evolved from lower forms of life.

Though Kumamoto may have claimed "a victory of sorts," he immediately called for reinforcements. He forwarded my proposed revisions (Exhibit 49) to Richard M. Lemmon, Associate Director of the Laboratory of Chemical Biodynamics at the University of California in Berkeley.

Lemmon, you will recall, confronted me as the self-appointed "qualifier" of my credentials during the public hearing. His analysis of my proposed changes is included as Exhibit 50. Lemmon's fairness in notifying me that I was "unqualified to act as an advisor on the content of science textbooks" is indeed overwhelming.

Lemmon, in turn, enlisted Thomas H. Jukes, a medical physics professor in the Space Sciences Laboratory at the University of California in Berkeley, to join him in his vituperative vitriol against me. In calm, cool, objective language, Jukes cries (see Exhibit 51):

> It is an intellectual affront that Mr. Grose should be attempting to enforce opinions on biological matters that would result in an undergraduate student receiving a non-passing grade...I object to the presence of Mr. Grose on the State Curriculum Development and Supplemental Materials Commission for just this reason. Other examples of Mr. Grose's scientific incompetence are listed by Dr. Lemmon.

Before reaching conclusions on Jukes' outcry, be sure to read carefully (in Exhibit 49) what I said about Louis Pasteur that so upset Jukes.

The *personal* attacks by Lemmon and Jukes (as contrasted with an *impersonal* technical refutation of error that is normal in scientific work) were one of the *private* forms of ridicule that the Darwinians employed against the State Board of Education.

Public forms appeared in both scientific literature and the general news media.

My editing examples of Exhibit 49 were reviewed critically in the journal of the American Chemical Society, *Chemical and Engineering News*.[437] Although I felt it was an extremely biased review, I had no idea that the ridicule it contained would ricochet the way it did. Beyond reiterating the tiresome but persistent errors commonly made from the beginning of the struggle and thoroughly documented thus far in the book, this article pitted *creationists* against *scientists* in a most unfair manner.

A real ruckus resulted in reviling reaction that repulsed the ridicule. There is no way to determine the number of letters to an editor on a given subject, their majority opinion, or the ratio of one view over its opposition.

However, the editor of *Chemical and Engineering News* published a group of 21 letters in response to this article — 15 of which denounced its non-scientific bias![438] If this reaction was at all representative of either the letters received or the 120,000 members of the American Chemical Society, the anonymous author of the article should have recognized that his ridicule had ricocheted.

Newspapers, of course, had a field day during the 34-day period of waiting for the State Board decision. Arthur Hoppe, a columnist for the *San Francisco Chronicle* composed a parody called, "A Simply Divine Theory of Evolution."[439] It was a supposed textbook account that would qualify for the "stupid" State Board of Education requirements. It ended with this mockery of the State Board:

> So now you understand the world you live in. If you have any questions, please don't ask your teacher. Ask the California State Board of Education. They obviously have all the answers.

The State Convention of 260,000 Southern Baptists in California during this 34-day waiting period also added fuel to a fire that had been generated only by the news media. The fire? The State Board had supposedly made a "decision to add the Biblical account of creation to science textbooks." At least that was what the news media had persistently reported.

So, the Southern Baptists unanimously passed a resolution urging the State Board to abide by its decision.[440] This resolution, of course, provided the Darwinians with exactly what they needed to sustain their campaign of coercion —- a *cause celebre.*

In their typically biased manner, the *Los Angeles Times* also ran two letters to the Editor responding to Arlie J. Hoover's earlier-mentioned critique of the National Academy of Sciences' resolution (Exhibit 41). The first letter, that *opposed* Hoover, was thoughtfully and carefully written. The second letter, in *favor* of Hoover's position, was a facetious caricature only half the length of the first.[441] The conclusion one was expected to draw from this mismatch is obvious.

About the same time, the *Los Angeles Times*, along with 150 other newspapers across the country, carried Conrad's explicit cartoon (see Figure 15) calling for the banishment of Adam and Eve from science textbooks. The prime target for that editorial message? The California State Board of Education.

Figure 15. Adam and Eve Banished

A news analysis in the *San Francisco Chronicle* only 2 weeks before the State Board's "day of decision" spiced the issue up a bit by inventing a *revolt*:[442]

> Textbook publishers have been notorious in the past for doctoring up history if they thought it would help sell their books.
>
> With many millions of dollars at stake in the huge California adoption of scientific textbooks, the publishers will be under great pressure on December 14-15 when the state Board of Education meets to take final action on the science books...
>
> The *Chronicle* has learned that three major publishers whose books are recommended have privately told commission members they will not doctor up their books to get the contract.

> If other publishers join in this revolt, the state Board of Education might well be forced to back down on its demand that evolution be balanced with Genesis.

Is there any *truth* in that fabrication? Judge for yourself. The same issue of the *Chronicle* carried an additional article based on a different falsehood. It described the Leswing Communications series of science textbooks.

Leswing had proposed a California version of its national edition in which they replaced a brief biography of L.S.B. Leakey with Michelangelo's painting of the Creation — evidently to comply with the requirement to have a creation alternative. *This series of books was never considered for adoption by the Curriculum Development Commission!* It had been *eliminated* — not for its creationist modification but for its scientific inaccuracy — by the antecedent Curriculum Commission.

Despite this fact, what did this fallacious article say?[443]

> Why should there be a special California edition of a science textbook? And why should it include Genesis?
>
> Because the State Board of Education has told publishers it wants the theory of evolution balanced with the Bible's theory...
>
> Trading Michelangelo's art for Leakey's findings is only one of the remarkable differences between the California and national edition of this book, *which will be considered for adoption when the Board meets in Sacramento next month.* (emphasis added)

Ron Moskowitz, Education Correspondent for the *San Francisco Chronicle*, seemed determined to outdo Jack McCurdy, his *Los Angeles Times* counterpart, in using falsification and deception to keep the straw man alive. He kept up his sniping, jibing campaign throughout the 34-day waiting period. Only 6 days before Decision Day, Moskowitz commented on the suggested changes on origins (Exhibit 49):[444]

> (Grose's proposed editing) probably will have scientists all over the nation giggling...
>
> The scientists have wanted only evolution discussed in the science texts, on grounds that it is the sole scientific explanation for man's origin.
>
> The commission's compromise proposal, drawn up at a meeting last week, would leave all mention of God and Genesis out.
>
> But, through some suggested literary alchemy, it would change Darwin's Theory of Evalution (sic) from scientific dogma to simple speculation that man might have evolved — not did evolve — from lower forms of life.

A THEORY ON THE EVOLUTION OF AN ATTITUDE

Figure 16.

The *Oakland Tribune* reported that Thomas H. Jukes, the Berkeley medical physicist, was suffering a severe case of paranoia about Genesis. Responding to direct quotes by Board Vice President Ford as "not interested in a Biblical account of creation" and Grose as ordering "the publishers to remove Genesis," Jukes was still cited as saying:[445] "This could just be the first step in getting creation theory back into science textbooks."

Jukes' reasoning seemed analogous to worrying that young men might be joining the Marine Corps as a means of overthrowing the United States government! Yet, it was also a form of ridicule — accusing responsible public servants of doing exactly the opposite of what they *say* they are doing *before they do it.*

The plight of the State Board of Education in searching for a rational and objective answer in the face of the ridicule that was ricocheting about their heads during the 34-day decisionary period was portrayed by another Cunnington cartoon (Figure 16) that appeared in Los Angeles' *Valley Times* on 30 November 1972.

As the campaign of harassment and ridicule of the State Board of Education approached the countdown date of 14 December 1972, Darwinians got ready to fire a surprise weapon...

Trump Card

Two dates — 6 August 1945 and 10 December 1972 — have great commonality. On both dates:

- A unique and *unprecedented weapon* was employed in the dying days of a war that had lasted for years.
- The weapon was intended to deliver a *final* knockout punch.
- The onlooking world was *shocked* that such a weapon could be manufactured. The aspect of *fear* played a bigger role in the weapon's impact than the actual damage wrought.

What were these unimaginable weapons? The first atomic bomb dropped on Hiroshima on 6 August 1945, and a joint statement of mortification by 19 Nobel laureates in science dropped on the California State Board of Education on 10 December 1972.

Nobel laureates had been involved in the Science Textbook Struggle prior to 10 December 1972, but always as *individuals.* Statements by Harold Urey (Exhibit 32) and Melvin Calvin — both of them Nobel laureates in chemistry — had received wide distribution in the press.[446] In addition, Arthur Kornberg, 1959 Nobel laureate in medicine and a Stanford biochemistry professor, had requested and been granted permission to be one of the 50 speakers in the 9 November 1972 public hearing. He did not appear, however.

Kornberg's failure to deliver his scheduled 5-minute speech before the State Board in November disappointed many present that day. But it should not mislead anyone into believing that he had lost interest in the origins issue. Quite the contrary!

Kornberg, instead, had bigger plans. He conceived what he must have imagined to be the ultimate weapon to be unleashed in the Science

Textbook Struggle — mustering 18 other Californians who had also won a Nobel prize in science to join him in bludgeoning the State Board into submission.

This brilliant piece of strategy was without parallel in the history of science. Such an august assembly appeared to absolutely assure abandonment of asinine and abominable aberrations of academic authenticity. Who could possibly *stand* against them? Who would dare *ignore* them? Would not a "peril worse than death" await anyone who failed to heed their derisive denunciation? Surely Kornberg thought so.

It is probably difficult for all of as, who have never achieved the pinnacle of success that a Nobel Prize represents to imagine the sense of importance it brings to its recipient. Nobel narcissism is undoubtedly a hazard inherent in becoming a Nobel laureate. It must be resisted like a plague. And most of the time, these brilliant men successfully overcome it.

This illustrious group of 19 Nobel laureates had every right to expect *instant obedience* to their will. Though the National Academy of Sciences was earlier called "the most important and prestigious scientific body in the world," Kornberg's collection of celebrities was entitled to even greater honor — man for man — than the National Academy.

After all, only a tiny fraction of the Academy ever receive a Nobel Prize whereas all 19 of Kornberg's prestigious prodigies (with the exception of Richard P. Feynman) were members of the National Academy of Sciences at the time Kornberg convened his cast of colleagues to castigate the crazy. (Feynman had been elected to the Academy in 1954 but was not listed in the 1972 roster of members). And, of course, the Nobel Prize is an *international* award whereas the Academy is technically a *national* body.

Not only did Kornberg conceive a masterful stroke of strategy. His *timing* couldn't have been better! Capitalizing on the massive mobilization of denunciatory resolutions from every quarter of scientific officialdom, he timed the release of his "bomb" perfectly.

It was like a red maraschino cherry carefully placed at the peak of a delicious ice cream sundae. Nothing could top that!

Since the State Board was scheduled to make their decision on Thursday, 14 December 1972, Kornberg dated his missile on the previous *Sunday*, 10 December. In that way, it would be the last and crowning blow to bowl the Board over.

Evidently not trusting the United States Post Office to deliver this unprecedented achievement the 83 miles from Stanford University to Sacramento in the 4 days between Sunday and Thursday, Kornberg was forced to use the news media instead of a postman to convey it to the Board.

Of course, it would just so happen that the entire world would get hit by the weapon at the same time. And what *leverage* world opinion thereby would bring on the State Board! Not *all* correspondence addressed to the State Board comes via the Associated Press and United Press International. But after all, Kornberg was entitled to a little extra privilege since he was a notable Nobel nobleman. Further, where time is critical, the wire services are generally faster than the Post Office.

What was this stupendous weapon formed by 19 Nobel laureates? A 3-sentence resolution (see Exhibit 53).

The key word in these 3 sentences is "appalled." This ad hoc coalition of world-renowned barons of science was *horrified*, *shocked* and *dismayed* that the State Board of Education was "considering" a requirement that the Board had unanimously adopted more than 3 years earlier.

Where had these famed fellows been all this time? Did they just wake up? Had they not read newspapers and scientific periodicals for 3 years? Surely, they would carefully examine the statement that Kornberg was urging them to sign — rather than sign it blind.

Or would they?

The State Advisory Committee on Science Education, that the Nobel resolution urged the State Board to support, had *ceased to exist* several years earlier. Only 3 of the 10 members of the State Board of Education to whom the Nobel resolution was addressed were on the Board when the Committee died.

So regardless of the repute of the signatories, the resolution was *irrelevant*!

Kornberg wisely kept the resolution brief. The longer a statement, the more likely the signers will read before signing. Because of its inaccuracies, it would have been prudent to have shortened the Nobel writ even further. But Kornberg had some additional "deploring" to dump on the Board.

So he wrote a separate letter to release concurrently with his brainchild (see Exhibit 52). The thrust of his letter was directed at a phrase I had written in the unanimous Curriculum Development Commission resolution calling for editing of science textbooks (Exhibit 45).

Similar to Jukes' paranoia "that *removal* of Genesis is the first step to *introducing* Genesis into science textbooks," Kornberg foresaw sly trickery in the Commission's recommendation to the State Board. The editing urged by the Commission to "change dogmatism to conditional statements where speculation is offered as explanation of origins" might *appear* to be innocuous, Kornberg said (in Exhibit 52).

However, apparently considering himself astute at discerning the Commission's intent, he declared the editing requirement to be *clearly tailored* to make room for the "creation theory" as an alternative to evolution. This keen perception set the stage for the remainder of Kornberg's letter that established why the dogmatism of *evolution* should not be changed to a conditional statement.

There is no doubt that Kornberg perceived what he honestly felt was a mortal danger to the future direction of science. Acting according to his conscience, he attacked that danger with what he conceived to be his most devastating resource — his membership in the most exclusive fraternity in his world. No one should fault him for that. I certainly do not.

However, there are two aspects of Kornberg's efforts — both the *resolution* (Exhibit 53) that he wrote and persuaded 18 other Nobel laureates to co-sign and his personal *cover letter* (Exhibit 52) — that should disturb all persons concerned about achieving objectivity about origins:

1. Kornberg allowed *belief* to thwart *truth.*

2. Kornberg confused *prestige* with *expertise.*

These are serious charges — especially against a man of Kornberg's scientific eminence. The reader is invited to weigh the evidence and reach a verdict.

First, regarding Kornberg's permitting *belief* to defeat *truth*, note the first sentence of his cover letter (Exhibit 52). What did Kornberg and his elitist colleagues *believe?* They felt obliged to join "in deploring the attack on evolution and beclouding its significance in the science textbooks." They *believed* that there was an *attack* and a *beclouding.*

Was there?

Only if distinguishing between Evolution I and Evolution II attacks and beclouds. Only if calling for alternatives for primordial events that are non-repeatable extrapolations beyond physical evidence is attacking and beclouding. Only if unmasking the fact that Evolution II and Creation II stand equally and side-by-side *outside* the realm of science constitutes an attack and a beclouding.

The average age of the 19 Nobel signers was 61 years. Seven of the 19 were over 65 years of age — Harold Urey being the eldest at 4 months less than 80. Darwinism has repeatedly been dubbed an "old man's belief." Basic beliefs are seldom discarded or revised past middle age — especially cosmological ones like Evolution II or Creation II. With little imagination, the signing of the Nobel resolution could be viewed as a pact until death —19 old men huddled together, defending a faith they had held too long to change.

If their faith was in vain, it was already too late to recant. The only course left was to declare that "there are *no alternatives* to our belief and any statement to the contrary is an *attack* on our belief."

There is no evidence whatever that Kornberg or his illustrious co-signers seriously examined or studied the issue of origins before lashing out at the State Board. They provided no indication that they understood the difference between Evolution I and Evolution II.

Likewise, they showed no differentiation between creation as a *scientific theory* and creation as a *religious belief*. Their own faith in Evolution II blinded their minds to the truth that origins are *scientific unknowlables.*

What about Kornberg's confusion of *prestige* with *expertise*? Ponder the supposed significance of the Nobel resolution. Did the headlines scream, "19 of World's Top Experts on Origins Attack State Board?" No, they noted that 19 winners of the top scientific *honor* in the world were attacking. What's the difference? The former speaks of scientific *expertise* while the latter denotes *prestige* of the attackers.

An Olympic Gold Medal for athletes is analogous to a Nobel Prize in science. What if 19 Olympic Gold Medalists in track, field, gymnastics and wrestling all signed a joint resolution deploring the chemical purity of the swimming pool at the Olympic games? What significance should be given their testimony? None. Why? Because they are not experts on swimming pool chemistry. Holding a Gold Medal does not automatically give them license to speak on *everything* at the Olympic Games.

Evolution II (the belief that origins occurred without intervention of a Creator) is a non-verifiable extrapolation of primordial history. It is postulated most frequently by scientists in geology or its subordinate branch of paleontology. Nobel Prizes in science have not been awarded in the field of geology or paleontology. Physics, chemistry, biology and medicine are generally the sciences for which Nobel Prizes are bestowed.

Of the 19 Nobel signers of Exhibit 53, 8 won their Nobel Prize in physics, 7 were awarded theirs in chemistry, and the remaining 4 came in the field of medicine.

Significantly, all 3 of these fields are devoted to an already-existing *universe*, an already-existing phenomenon of *life*, and the already-existing creature of *man*. Therefore, not one of the signers of the Nobel resolution had a Nobel Prize conferred on them for work they had done in *origins!*

Why then did anyone pay attention to what they said about origins? Because their *prestige* was substituted for *expertise.* What proof is there for this claim? Read the lead editorial in the *San Francisco Chronicle* on the day before the State Board's "day of decision":[447]

> NINETEEN NOBEL PRIZEWINNERS have pooled their *scientific prestige* in support of Darwinism's claim to tell the true story of the origin of all earthly species, including man, and the State Board of Education is going to have to respond to their challenge at tomorrow's meeting.
>
> Evolution, the Nobelists submit, is the sole explanation of man's origins that belongs in science textbooks...
>
> We have no intellectual respect for those who would ask science textbook publishers to bowdlerize their account of evolution's role in the story of all life in order to make way for balancing it off against the Bible's account...(emphasis added)

The unfortunate but unmitigated hypocrisy of these 19 Nobel laureates is best illustrated by one of the 19 himself — William Shockley. His affliction with Type B scientific heresy was discussed extensively in Chapter 2. Shockley's Nobel Prize in physics came in 1956 in recognition of his development of the transistor. The universal persecution that he has undergone at the hands of his fellow scientists — even within the National Academy of Sciences — is legendary.

How do Shockley's scientific persecutors justify their illiberal intolerance? *On Shockley's speaking outside his area of expertise.* As Stebbins says in Exhibit 26, "Scientific opinion can be regarded as informed only when the scientists in question have acquired first-hand information about the subject being discussed."

Kornberg's solicitation of Shockley's signature on the famous Nobel resolution raises some very interesting questions:

1. Did Kornberg want Shockley to sign because of his *expertise* on the origin of the universe, life, and man, or because of his *prestige* of being a Nobel laureate?

2. Since Kornberg and 17 other signers were fellow members of the National Academy of Sciences that had repeatedly denied Shockley's appeal for 6 previous years to resolve the role that heredity might play in determination of intelligence, did Kornberg offer Shockley any solace or compromise in order to secure his signature?

3. Why would Kornberg and the remaining signatories *deplore* Shockley's transgressions into genetics from transistors and yet *join* with him on a statement in a field equally distant from transistors?

There is equal intrigue when looking at Shockley. Being an outcast as long as he had, he might have acted expediently. In other words, he might have jumped at the chance to rejoin as an equal the community that had rejected him. By such prostitution (if he realized his total incompetence on origins), he might have hoped to gain support for his "dysgenics" theories at a later date.

In April 1974, I was teaching at NASA's Johnson Spacecraft Center in Houston. One evening in my motel, I tuned in a TV interview show, *Black Journal*, on which Shockley was pitted against a black woman sociologist from Howard University, Frances Welsing. The topic of this show, moderated by Tony Brown, also a black, was: "Black or White Superiority."

It was the first time I had seen Shockley, although I was aware of his aberrant views. Shockley was treated with undisguised contempt from the moment the show started until its completion. He was not allowed to answer questions that were asked of him or to even finish many of his sentences. Regardless of how you viewed his theories, you would have needed a heart of stone to avoid empathizing with Shockley in his inhumane humiliation.

As I watched this unbelievable scene, I soon perceived a similarity between Shockley's treatment and that which I had received during the Science Textbook Struggle. The unwillingness to listen, the prior prejudice, the arrogance perpetrated by a righteous cause, and the lack of respect for honest dialogue were identical to what I had experienced.

I wrote to Shockley a short time later, describing the book I was writing about scientific prejudice. I asked him if he would be willing to spend approximately 30 minutes in private discussion with me concerning this subject.

Offering to fly up to Stanford and meet at any time of his convenience, I even proposed several dates to him. He never even acknowledged receipt of my letter.

I know that Shockley received my letter, however, because nearly 4 months later, the postman brought a mimeographed postcard. It read:

> Dear Writer (of a letter in my office files):
>
> Your correspondence suggests that your face in the studio audience might add a much needed expression of sincere interest — perhaps even approval — rather than dogmatic hostility, while David Sachs M. D. interviews me for taped TV broadcast in L.A. on Saturday 14 Sep 74 from 11 PM to midnight. Probably thirty stations at the same hour may present Sachs, the unconventional heart surgeon who now seeks a fuller life on TV.
>
> To make your appearance support my campaign for objective research on dysgenics, come to *KTTV, 5746 Sunset, at 6:20 PM Friday, 23 Aug 74*, to be seated at 6:30 for taping at 6:55. Use the Van Ness entrance, 100 ft.

from corner at Sunset, saying if asked, that you are a guest of Hal Parets, Producer of the Sachs Show.

With appreciation for your interest,

Professor W. Shockley

Although I was not able to attend, I would have loved to — for two reasons. First, I would have liked to support Shockley in obtaining objective research for his project. Secondly, I would have liked to ask him why he signed a Nobel resolution of 10 December 1972 to defeat objectivity on the question of origins.

Do scientists — even Nobel Prize-winning scientists — want objectivity only for their own pet projects?

Without belaboring the now-expected inaccuracies of the news media, press reports of the Nobel resolution fell into the same old rut.

An Associated Press report in the *Los Angeles Times* erroneously claimed that the Nobel laureates had called for a rejection of a requirement for "religious doctrine on man's origins to be given equal prominence alongside the theory of evolution in science textbooks."[448] The *San Francisco Chronicle* — equally inaccurate — twisted the resolution's wording to say that the Nobel laureates were appalled because "the board is considering including religions as well as scientific explanations in science books."[449]

The religious news services got the message even more garbled. One article, "Nobel Laureates Oppose 'Equal Time' for Biblical Creation in Education," stated:[450]

> The Nobel laureates (commented) on a "compromise" measure suggested by the state Curriculum Commission a week earlier, which would have changed the wording of the textbooks to suggest that Darwin's theory is still speculative, and omit all mention of God and the Book of Genesis. The measure was defeated.

The dropping of the atomic bomb on 6 August 1945 proved to be the clincher in getting the Japanese to surrender unconditionally. The dropping of the Nobel resolution on 10 December 1972 proved...

The sentence would not be finished until 14 December in Sacramento. Would Kornberg's "trump card" sweep the game?

REBUFF OF NOBILITY

On 6 August 1945, the B-29 *Enola Gay* headed for Hiroshima with the fateful atomic bomb nicknamed "Little Boy" tucked in its belly. At 9:15 AM, "Little Boy" dropped through open bomb bay doors at 30,000 feet over the city. Instead of falling free, "Little Boy" descended by parachute. This allowed a deliberate delay of 43 seconds between the drop and detonation some 5 miles below so that the *Enola Gay* could escape the fearful blast.

There was also a deliberate delay between drop and detonation of the Nobel resolution in 1972. But it was longer than 43 seconds. Arthur Kornberg dropped his bomb on Sunday, 10 December 1972, so that 4 days of news coverage could build up massive public pressure on the State Board of Education to capitulate on 14 December.

Undoubtedly, those 4 intervening days were suspenseful for Kornberg as he measured his impact by reading news reports of his resolution. He must have repeatedly wondered whether he had sufficiently embarrassed the Board that they would crawl snivelingly under a rock to escape his attack.

Frankly, I was apprehensive myself. Kornberg had clearly played the trump card of science. All press coverage favored the Nobel laureates. The die seemed to be cast.

A United Press International release in *The Seattle Times* the day before the Board's decision reported:[451]

> Scientists seemed today to have won their struggle to keep the "creationist theory" of man out of the books.
>
> But they were battling on the eve of a decisive State Board of Education meeting to prevent the evolutionary theories of Charles Darwin from being presented in the book as mere speculation.

Meanwhile, my work on the Governor's Select Committee on Law Enforcement Problems was becoming increasingly demanding.

Although the Curriculum Development Commission was also scheduled to meet in Sacramento on Wednesday, 13 December, the Select Committee had to meet with the Governor's cabinet from 7-9 that morning. Then, we had a separate meeting alone with Governor Reagan at 3:30 PM. My hopes for getting the pulse of battle were dashed.

Leaving our meeting with Governor Reagan in the Capitol, I returned to the Cosmopolitan Hotel just across the street. In the lobby, I ran into State Board member Clay Mitchell. He proposed that we have dinner together.

During the meal, we agreed that it might be well to meet with Board members John Ford and Eugene N. Ragle later that evening. Mitchell called them on the house telephone, and they agreed.

From 9 PM until midnight, Mitchell, Ragle, Ford, Curriculum Commissioner Bud Creighton and I met in Ford's hotel room. Some of us were quite concerned about the effect the news blitz of the past month might have had on Board members. However, Ford was confident that the science textbooks would be edited.

As we discussed strategy for the next morning's meeting, 3 key points seemed to emerge.

First, Ford would have to get the floor just as soon as President Steward called for the science adoption on the agenda. That would preclude any other Board member from introducing a motion to kill editing of the books. All three Board members in the hotel room felt that Ford's position as Vice President and as chief spokesman on the origins issue would favor his getting first recognition — especially if he spoke with Steward about it before the meeting.

Second, it was critical to utilize the unanimous resolution of the Curriculum Development Commission (Exhibit 45) as a foundation argument to demand editing of the books. The State Department of Education, that reports to Wilson Riles, had prepared a standard adoption resolution that could be — with minor revisions — the vehicle to accomplish editing.

The *third* point was that the *Board* — not Riles' Department of Education — should control the editing. This perhaps would call for the appointment of an editing committee of five — three of whom would have to be people who understood the difference between Evolution I and Evolution II.

Awaking on the big day — 14 December 1972 — from a good but short sleep, I could feel the tenseness that precedes battle beginning to mount within me. I rushed downstairs to the coffee shop for breakfast. As I passed a stack of morning newspapers, I bought the *Sacramento Union* and took it to my table. Opening it, my eyes fell on headlines — *Adam, Eve and Darwin*!

What I read began to amaze me:[452]

> Evolutionists — including 19 Nobel prizewinners — were losing their battle to prevent the downgrading of Charles Darwin's theories in California public school texts.
>
> On the eve of a decisive State Board of Education Meeting, a UPI poll showed that a majority of the board members were leaning toward a compromise move to leave Adam and Eve out of the texts and to reduce evolutionary theory to simple speculation.
>
> But a couple of swing votes on the seven-man, two-woman panel said Wednesday that they might change their mind in favor of creationists before today's vote to adopt textbook standards.

Maybe Kornberg's crowning blow wasn't as effective as I had feared! Maybe the State Board members had recalled a saying by Thomas H.

Huxley (grandfather of Sir Julian and Aldous Huxley): "If a little knowledge is dangerous, where is the man who has so much as to be out of danger?"

Speedy Showdown

It was like an instant replay. Same day of the week, same time of day, same jammed auditorium of the State Resources Building in Sacramento, same bank of TV cameras all lined up...and same tension in the air.

The 34 days that had elapsed between 9 November and 14 December 1972 hadn't seemed to change anything.

As I walked in the door of the auditorium, there was a stack of news releases — *News in Education*, Number 126 — published by Wilson Riles' State Department of Education. Under a caption "For Immediate Release," the announcement described in cryptic terms — similar to those used to relate the execution of a criminal — what was now only minutes from happening:

> The State Board of Education will decide whether or not to include the creation theory in school science texts...At approximately 10:30 a.m., on December 14 the board will receive a routine resolution proposing the adoption of the kindergarten through eighth grade series as suggested by the board's curriculum commission. None of the recommended books contains the creation theory.
>
> The board can either adopt or reject the resolution, or adopt it with modifications. The resolution comes at the end of a year-long process of advisory comments, public bids, public displays and a public hearing.

I took a seat on the opposite side of the auditorium from where I had sat on 9 November. My role, like all other Curriculum Commissioners, was passive this time — simply available to answer questions of the Board.

The meeting started on time. There were actually 3 textbook adoptions scheduled for the day — music, health and science. The time allotted on the agenda for each adoption gave some indication of its controversiality. Music got 15 minutes, health 30 minutes, and science 90 minutes.

The Board's routine business items dragged a bit at the outset of the meeting. The music adoption, however, took less than the allotted 15 minutes.

So the health adoption was called on time at 10 AM. But there had been no public hearing on health books as there had been on science. Sex education, another controversial subject, is part of the health adoption. Because of the sex controversy, the Board allowed public testimony prior to entertaining an adoption resolution on health. Instead of taking the scheduled 30 minutes, the health adoption ran about an hour.

The TV cameramen and crews had come to this meeting for only one purpose — to witness the end of the 3-year-old Science Textbook Struggle. The outcome of this war — even with the super-weapon of Nobel nobility thrown in — was not yet as predetermined as General MacArthur's signature ceremony on the battleship Missouri had been for World War II.

In addition to the earlier-mentioned UPI Poll of Board members reported in the *Sacramento Union* as withstanding the Nobelist attack, reports in Sacramento's other newspaper, *The Sacramento Bee*, added to the uncertainty:[453]

> Some veteran board watchers say that if you are inclined toward betting, place your money on the side of creation.

As the scheduled 10:30 AM hour for science came and went, the TV personnel became increasingly restless. They had been lollygagging around since 9 AM waiting for the excitement to break open. The boring health adoption dragged on…

Finally, just a minute or two after 11 AM, President Steward intoned dryly: "Now we come to the science adoption."

My heart leapt within me as I anxiously scanned the Board members. This was the most critical moment of time. If anyone but Ford got the floor, the outcome of the Science Textbook Struggle could obliterate everything I had fought so hard to accomplish for over 3 years.

Although it seemed like ages, a tape recording reveals that only about a second elapsed between Steward's announcement and his next sentence — "The chair recognizes Dr. Ford."

Steward had not even glanced up to recognize anyone. Evidently, Ford had previously gotten Steward's agreement to allow him to speak first.

Immediately, there was a flurry of activity among the TV crews. They ran, jumped, and pushed furniture aside to get to their cameras. TV lights came on. Throughout the auditorium, everyone strained eyes and necks to get focused on the scene. The final showdown was now going to begin…

"Although there has been some controversy in relation to the proposed science adoption," Ford began with unusually subdued tone of voice, "I feel that the resolution that was prepared for us — as well as the unanimous decision of the Curriculum Commission that certain editorial changes must be made in many areas of concern — would, with a few minor changes, satisfy the Board's *Framework* and criteria. I would like to move that the proposed resolution, as prepared for us, be adopted with 2 minor changes."

Ford had total control at this point as he listed his revisions to the routine resolution of adoption that Riles' State Department of Education had prepared for the Board.

His first change was to delete the adjective "minor" from the phrase, "...subject to any minor editorial changes that may be needed prior to contracting for these instructional materials..."

Ford's second modification to the resolution was far more significant. The original wording of the resolution read: "The revision of the books will be at the request of the Department of Education..." To this sentence, Ford tacked on an extension, "...and the State Board of Education, and such revisions will be under the direction of the State Board of Education."

That revision was a power play, to be sure. However, Wilson Riles had methodically undercut the State Board's position as the primary policy source for California education ever since he took office. Further, he and his entire State Department of Education had been unsupportive of — if not hostile to — the origins issue. By this revision, Ford grabbed the ball away from Riles. Had he not done so, Riles would have "pocket-vetoed" the Board's desires on editing the book for origins.

As soon as Ford had listed his two changes, Steward asked if there was a second to Ford's motion. Virla R. Krotz quickly seconded it.

Steward, after repeating Ford's modifications, called for discussion. Unbelievably, *there was no discussion!* It was *unreal!*

Once more, Steward asked if there was any discussion. Again, *no response.* Mrs. Krotz immediately called for a roll-call vote.

A hush fell over the jammed auditorium. The secretary began to call the roll...Mr. Mitchell? *Aye.* Mrs. Krotz? *Aye.* Mrs. Drinker? *Aye.* Mr. Ragle? *Aye.* Like the last few grains of sand pouring through an hourglass, the roll call seemed to speed up...Mr. Steward? *Aye.* Dr. Ford? *Aye.* Dr. Hubbard? *Aye.* Mr. Gates? *Aye.*

That completed the roll call of the 8 members present. (Member Tony N. Sierra was absent, and Governor Reagan had not yet filled one vacancy.)

Almost in unbroken rhythm with the roll-call cadence, Steward quietly announced, "The motion is unanimously carried." He slammed his gavel down.

Pandemonium broke loose. No one could believe it! Many of the TV cameramen had not yet started to film the scene by the time it was all over. From the time Steward had called for discussion of Ford's motion until he rapped the gavel ending the war, only *25 seconds* had elapsed!

People began to stand throughout the auditorium. Newsmen ran out to get to telephones. Steward really had no choice but to adjourn. What had been scheduled as a 90-minute deliberation on science textbooks had lasted a total of 3 minutes and 20 seconds!

Wilson Riles left his seat on the dais and walked down the aisle on which I was sitting. Like the Pied Piper, he had an entourage of news reporters trailing him, all clamoring for his reaction to what had just happened. As he approached me, I shook hands with him.

He had a dazed look and turned around to look behind him at the crowd of reporters. "Does anyone here understand what just took place?" he asked rhetorically. It was a cinch that *he* didn't.

Not getting any satisfaction from Riles, many of the reporters left him and converged on me. My excitement at this victory was not disguised, I'm sure. Trying to remain responsibly calm, I explained that this motion had accomplished what I had been fighting to achieve for 3 years. *Science textbooks were going to be edited concerning origins!*

"But what about *Genesis*?" one reporter yelled above the roar around us.

"I never wanted Genesis at *any* time!" I shouted back. "You people are the ones that invented that straw man."

As I walked out of the auditorium for lunch, it was difficult to rationally recap what had just transpired. Two major points, however, seemed to be indelibly impressed on my mind:

1. The State Board of Education had heard the Darwinian mobilized message — in mail, meetings and media — and *denied* its validity.

2. The State Board of Education had heard my appeal on 9 November 1972 for *removal of dogmatism* — even if the other speakers had ignored that message — and *affirmed* its validity.

The analogy that I had foreseen between the Hiroshima atomic bomb and the noxious Nobel notification had ultimately broken down. Why?

It was not because I or anyone opposed to the Nobel laureates' imperative caveat were more *authoritative, qualified, or prestigious*. Nor was it because the general public or the State Board members did not hear the Nobel decree.

On the contrary, Kornberg's "ultimate weapon" turned out to be a farce because the public and the Board simply did not agree that *dogmatism about unknowables* — however prestigious its advocates — has any place in science.

The war was over. There are special headlines that newspapers reserve for cataclysmic events like the end of a war. On this eventful day, the *Los Angeles Times* was to break out such headlines to announce the end of the Science Textbook Struggle (see Figure 17).

Truman Condition Very Serious

THURSDAY

LATE FINAL

COMPLETE N.Y. STOCKS

Los Angeles Times

VOL. XCI ELEVEN PARTS—PART ONE ★★★ THURSDAY, DECEMBER 14, 1972 198 PAGES DAILY 10c

VICTORY FOR ADAM AND EVE

Market Off

Astronauts Leaving Moon, Will Rejoin Command Vessel

Public Schools to Downgrade Evolution

Figure 17. Victory for Adam and Eve

VICTORY FOR ADAM AND EVE, they proclaimed. Of course, the headlines are absolutely untrue. But who would buy a newspaper with headlines — however large — that read, "VICTORY FOR OBJECTIVITY?"

The Board's decision had come like a stroke of lightning — no warning, no preliminary buildup. Just a simple motion by Ford — without so much as one word of discussion — had wiped out the massive onslaught of the barons of science.

Dazed Darwinians didn't disguise despair at their disastrous defeat. The early adjournment for lunch allowed them some extra time to plot how they might minimize the magnitude of the editing changes. They had a strong and able advocate on the State Board — attorney Mark T. Gates.

Darwinians began their effort to scuttle the editing immediately — during lunch in the eighth-floor cafeteria of the State Resources Building. A small knot of worried people gathered around Mark Gates as he ate his lunch.

Meanwhile, the next step facing the State Board after lunch was deciding on *who* would edit the books. During the lunch adjournment, I suggested a possible candidate to Dr. Ford: Bob Fischer, Dean of the School of Natural Sciences and Mathematics at California State College in Dominguez Hills. I felt Fischer had both the proper credentials and the understanding between philosophy and science that was crucial in recognizing dogmatism in the textbooks.

Just before the Board reconvened, I also proposed Fischer's name to Board member Dave Hubbard. While we were discussing this matter at Hubbard's desk on the dais, he asked if I also knew Richard Bube, Chairman of the Department of Materials Science and Engineering at Stanford. Hubbard had been impressed by some things Dick had written.

Of course, I explained that Dick had been one of the first people I had contacted before getting involved in 1969.

Ford, Hubbard and I supposed that either the State Department of Education or other Board members would also have candidates for an editing committee. As it turned out, the results were astounding.

In naming a committee, Ford's name was proposed first. Then someone on the Board nominated Hubbard. They had been spokesmen from different orientations on the origins question. In turn, Ford asked that Fischer be named, and Hubbard requested Bube. Those 4 were considered an adequate committee, and all nominations were closed!

The night before in Ford's hotel room, we had hoped to get 3 out of a possible 5 who would be able to differentiate between Evolution I and II. Now all 4 editors would meet that criterion. So far, our victory had exceeded our greatest expectation.

I was obliged to return to the Governor's office in the Capitol, a few blocks away, shortly after lunch. Therefore, I missed the maneuvering by Mark Gates to destroy the editing process. As recounted to me later, it was a masterful attempt — but met with little success.

First, the Darwinians must have been so overwhelmed by Ford's incisive stroke that they failed to foresee the next necessary step — *appointment of an editing committee.* It was amazing that they didn't at least nominate Richard M. Lemmon or Thomas H. Jukes — both of whom had offered their eminently-qualified opinions on who should edit science books (see Exhibits 50 and 51). Instead, the editing committee ended up devoid of a dedicated Darwinian dogmatist.

Apparently, the Darwinists decided to attack first the *criteria* of editing rather than the editing *personnel.* Tactically, that was a poor choice, as the latter are far more consequential. Gates' first proposal was to limit editing to only what could be considered "dogmatic" about evolution. Obviously, both Ford and Hubbard objected.

It was Hubbard who showed the fallacy of Gates' proposal. Restating what I had called "the rudimentary issue at stake" in my address to the Board the previous month (Exhibit 47), Hubbard said that the line between science and philosophy needed to be clarified. The problem was much larger than just softening dogmatic language about one philosophic belief (Evolution II).

"That boundary has been transgressed and a great deal of what is said in science is also in the area of philosophy," Hubbard pointed out. He went on to say that if evolution is stated as a *fact*, we "may put students in conflict between what they think they are learning in science and what they learn spiritually, morally and even politically."

Striking at the heart of the controversy, Hubbard drove home two points. Evolution, as currently presented in science textbooks, appears to rule out any divine intervention in the universe. Secondly, it views man as a biological accident. These two points, if accepted by the schoolchild, Hubbard feared would make the pursuit of human rights and dignity moot.

"The really harmful thing," Hubbard concluded, "is the *factual tone* given to evolution taught as a world view and the *arrogance* scientists have adopted in their writing on evolution."

Having lost that attempt, Gates next tried to force a showdown on whether creation would be allowed as an alternative to evolution. He offered a motion to remove dogmatism on evolution but also to exclude creation. It also failed.

The only ground Gates gained was an agreement by the Board to forego editing on the third point of the unanimous recommendation of the Curriculum Development Commission (Exhibit 45). Thus, the children of California became the losers. Evidently, Darwinians feared that if "questions yet unresolved in science" were "presented to the science student to stimulate interest in inquiry processes," Darwinism might be doomed.So, the last day that man may ever walk on the Moon in our lifetime (and also my father's 76th birthday) — 14 December 1972 — signaled the end of an era of unchecked Darwinian dogmatism.

Keeping its readers completely off-balance, the *Los Angeles Times* published a large front-page story, INCLUSION OF RELIGIOUS CREATION THEORY IN TEXTBOOKS REJECTED,[454] the day after it carried the headlines, VICTORY FOR ADAM AND EVE (Figure 17).

Which way was it? Actually it was neither — as the facts recounted here should prove. But the only way a straw man can be kept alive is by error and confusion.

Figure 18.

An Associated Press release, datelined Sacramento, with the more accurate story that the pleas of 19 Nobel laureates and members of the National Academy of Sciences had been rejected by the California State Board of Education was carried nationwide.[455] It even appeared in the newspaper that I once delivered door-to-door in Spokane, my hometown.[456]

The cartoonist Renault chose to revive one of his earlier caricatures in *The Sacramento Bee* and heap even greater scorn on the State Board for its repudiation of the Nobelists. Figure 18 is his update of Figure 14.

National news coverage, in addition to Associated Press releases, included another article in *Time* magazine.[457] This one quoted my suggested revisions (from Exhibit 49) as a capitulation to "conservative objections." A lengthy article in *The National Observer*, on the other hand, presented arguments on both sides of the issue. It was in this report that I first learned what Stebbins thought of me:[458]

> Responses from some scientists opposing the guidelines have been terse, direct, and not always polite. Geneticist Ledyard Stebbins of the University of California at Davis refers to Grose as a "trickster."

But it didn't really matter what the news media had to say from then on. The war over dogmatism had been won. My first confirmation of victory came the next day — 15 December.

As part of my assignment on the Governor's Select Committee on Law Enforcement Problems, I was at the Los Angeles Police Department's Parker Center developing a systematic model for criminal apprehension.

"There's a long-distance call for Vern Grose," someone shouted across the room. Picking up the receiver, I recognized Clarence Hall's voice in Sacramento. Hall was Wilson Riles' Assistant Superintendent of Public Instruction.

"Vern, the 4-man *ad hoc* committee appointed yesterday by the State Board to edit the science books has asked me to contact you. I have been assigned from the State Department of Education to work with them, and we need all the specific suggestions for editing that you may have compiled. No one has examined the books as thoroughly as you have, so your help will be invaluable to us."

I appreciated the spirit of his request. But here was another demand for a gift of my free time. No compensation — just a lot of hard work. Yet, if I shirked the request, my 3-year investment in objectivity for origins might evaporate.

So, my family lost me for another weekend, as I prepared Exhibit 54 for the State Board.

A total of 136 specific changes were attached to Exhibit 54 — 19 for American Book Company, 32 for Harcourt Brace Javonovich, 33 for Harper & Row, 10 for Holt, Rinehart and Winston, 15 for Laidlaw Brothers, 10 for Charles E. Merrill, and 17 for Silver Burdett.

The number of suggested changes was not a true measure of dogmatism, however. The *magnitude* of each change varied greatly. By far, the books by Harcourt Brace Javonovich contained the most flagrant examples of converting speculation into "scientific facts" and of unqualified statements favoring the "amoeba-to-man" belief.

Meanwhile, as the *ad hoc* Editing Committee started their project, the *Los Angeles Times* article by McCurdy,[459] that twisted the victory of 14 December 1972 into a "defeat of religious creationism," provoked widespread reaction in Los Angeles. The straw man simply would not die.

William C. O'Donnell, Vice President and General Manager of KNXT (Channel 2) — the CBS-owned television station in Los Angeles, wrote an editorial on the subject, "Teaching Creation in the Schools" (see Exhibit 55). O'Donnell read this editorial personally on television, and it was repeated several times over a 2-day period.

When I later requested a copy of this editorial, the Editorial Director at KNXT said, "Although it was brief and did not attempt to analyze the many ramifications of the argument but merely expressed a general point of view, we've been surprised at the amount of response. You have a topic for discussion on the Commission which creates more stir than we guessed."

The response in favor of creation being taught alongside evolution that surprised KNXT is a *real force* in our society. The general public has shown greater perception of the critical *philosophic* issue at stake than has the scientific community.

Certainly there is confusion of terminology for both evolution and creation as shown in Figure 5 in Chapter 4. However, the confusion is as pervasive in the scientific community as it is among the general public.

The same concern that provoked the massive public response favoring KNXT's position caused State Board member Eugene N. Ragle to react at the Board's next meeting.

On 12 January 1973, Ragle introduced the following motion:[460]

> All science books (should) contain the following statement as a part of any treatment of the subject of the origin of life: "There are two basic conflicting theories of the origin of life. Neither is fully supported by scientific fact. The older of the two is the theory of divine or special creation. Of more recent date is the theory of evolution, to which many scientists subscribe."

That motion failed to pass. It should be obvious by now that I also would have opposed Ragle's motion on several points. First, it was too restrictive in scope — discussing only the origin of *life*. Second, it perpetrated the *religion-versus-science* straw man. Third, it lent credence to the idea that creation is an antiquated belief that has recently been superseded by evolution. Finally, it fell far short of *removing from science books all beliefs* about how origins may have occurred.

Ragle's action, of course, gave the news media an opportunity to confirm the straw man they had built over the years. In addition to another lengthy front-page story, MOVE TO PUT RELIGIOUS CREATION THEORY IN SCIENCE BOOKS KILLED,[461] the *Los Angeles Times* also published an editorial, KEEP PRESSURE OFF TEXTBOOKS.[462]

"THOU SHALT HAVE NO OTHER THEORIES BEFORE US!"

Figure 19.

Two days later, Figure 19 appeared in another Los Angeles newspaper, *The Valley News*, apparently reacting to the arrogance of the *Times* editorial position.

Ron Moskowitz, Education Correspondent for the *San Francisco Chronicle*, continually vied with Jack McCurdy of the *Los Angeles Times* for honors in sustaining the straw man.

In *Saturday Review of the Sciences*, Moskowitz published an account filled with the same errors he had been publishing for years. Only *this* time, he called the State Board "all-Christian!" Some of the Board members I knew would resent that label for sure. Lamenting the emphatic editing that was taking place, Moskowitz described the situation:[463]

> The State Board of Education agreed to alter the wording of new science textbooks so that evolution will be taught as speculative theory, not "scientific dogma." From now on children in the fourth and sixth grades will learn that Darwin's theory of evolution is simply science's way of guessing how man was created. Modified textbooks will expose what the board has decided are holes in the theory — areas in which scientists have no hard proof for evolution.

Wouldn't that be a shame to tell children that, where there was no proof for their beliefs, scientists were only *speculating?* It might even ruin the child's image of scientists as those who know *everything!*

Remarkable Revisions

The news media didn't give up easily. After all, it was probably disheartening for them to see their child, the straw man "God-versus-Darwin" die at the age of 3. So they did their best to pump life back into the phantom.

After the historic decision of 14 December 1972 to edit the science textbooks but before the *ad hoc* Editing Committee could meet for the first time, the *Los Angeles Times* devoted a major portion of their Sunday "Opinion" section to the subject, "Creation vs. Evolution — Battle Over What to Teach."[464] This section contained 8 separate articles. The collective bias against editing was obvious. Two of the 8 articles were later reprinted throughout the nation.[465]

In Washington, D.C., the National Education Association's daily bulletin, *What's Happening Today*, carried an article on 2 January 1973 with the phony headline, "Genesis Given Equal Time in California Texts." Naturally, it pitted religionists against scientists. But the most curious thing about the article was that it contained no mention of Genesis being given equal time!

Instead, it talked about the State Board of Education's elimination of scientific dogmatism in textbooks.

The first meeting of the Editing Committee was on 3 January 1973. They accomplished two things. First, they prepared 39 specific changes in the Harcourt Brace Jovanovich series of science books. In addition, they gave Professor Bube the assignment of drafting an introductory statement on origins that could be placed in *all* science textbooks. The purpose of such a statement was to avoid numerous specific editorial changes.

The Harcourt Brace Jovanovich revisions, together with the introductory draft, were presented to the State Board the next week in Sacramento as a progress report.

Bube's draft included the declaration, "Where the first matter and energy came from, for example, is not a question that can be answered scientifically. At this ultimate question of origin, science can only be silent." This true but direct remark provoked the Board's prime Darwinian defender, attorney Mark Gates.[466]

"This statement seems to treat science and scientists in a rather derogatory manner," Gates remonstrated.

The Editing Committee pointed out to the Board that the study of origins also occurs in other disciplines including philosophy, religion and history. This observation influenced the Board to unanimously adopt a resolution to discuss the philosophy of origins in an upcoming adoption of new social science textbooks. That decision was to come back and haunt the Board later.

Education-Training Market Report, published bi-weekly by New York Institute of Technology, was quick to note the State Board's decision to include a discussion of both creation and evolution in social science textbooks as a *compromise.*

On one hand, this report showed mild alarm that creation might be discussed in history, geography, economics, ecology, philosophy, psychology, and other social problem's texts. On the other hand, the compromise, as they perceived it, would keep science books limited to discussing "how man evolved, not ultimate causes of man's origins."[467]

All four members of the Editing Committee — Ford, Hubbard, Bube and Fischer — were very busy people. And I knew it. I also knew only too well how much work was involved in reviewing and editing textbooks.

The news reports of the Committee's reception at the January Board meeting weren't encouraging. I feared that the editors might yield to pressure from the news media, the State Board, or their full-time professions to do only a superficial editing job.

On 18 January, I wrote to all 4 of them. With delayed hindsight, my letter has an admittedly passionate tone. Some of the key points included:

> There is a "blurring" of that area (in the textbooks) where facts are overtaken by beliefs regarding origins. This blurring occurs in the minds of scientists as well as in written materials on science. Since the facts or data of science are nearly universally endorsed, the blurring between facts and beliefs has permitted beliefs to enjoy the same credibility with the public and scientists that the facts enjoy...
>
> The State Board in December 1972 properly called for the downgrading of dogmatism to speculation. However, this is *not* the total editing that must be performed in the textbooks.
>
> Even though the evolutionary concept may be proposed as speculation rather than fact, there is no reason why it should be afforded *exclusive* status in interrelating the facts of science. It simply is *not* the only conclusion that could be drawn from the facts...
>
> I am adamant in stressing that "watering down" or correcting the error of dogmatism in macroevolution, while necessary and proper, does not solve the issue for the child who believes in God. If evolution is retained as the *exclusive* explanation for origins, science cannot be neutral...I feel most strongly that there must be an explanation that differentiates between facts and beliefs. It must point out that, while scientific facts are universally endorsed, so few facts concerning origins exist that no certain conclusion can be drawn. Therefore, there are a number of belief systems that will fit those few facts...
>
> Regarding the action of the State Board to discuss origins in social science texts, I have checked this possibility out with other Commissioners. There are no "social science" books in K-8 where origins are even alluded to today or where origins might be appropriately inserted. Certainly geography, history and social studies textbooks seem inappropriate for such discussion. Therefore, the motion to avoid discussion of philosophy of origins in science and to do so in social science texts could appear as a capitulation to pressures ...

Included with this unsolicited letter was a proposed statement regarding origins for insertion in science textbooks (see Exhibit 56). Considerably shorter than the preliminary draft Bube had prepared for the January Board meeting, my proposal incorporated language from both the *Science Framework* and curriculum criteria that had been unanimously adopted by the State Board.

The second Editing Committee meeting took place on 24 January.

Changes in books published by Harper & Row, Holt Rinehart & Winston, Laidlaw Brothers, and Harcourt Brace Jovanovich were prepared. These changes were submitted in a written progress report to the Board during their February meeting and drew no comment. However, the Board made an important and critical decision affecting the editing process in their February meeting.

Vice President Ford missed the January Board meeting when Eugene Ragle's motion to include creation in the discussion of the origin of life failed. In February, Ford offered a motion that science textbooks be edited "to provide examples of unresolved questions in science as a means of stimulating the inquiry process, including a theory of creation."[468]

Only 2 Board members — President Steward and theologian Hubbard — voted against Ford's proposal. Ironically, since only 7 of the 10 Board members were present, 2 negative votes killed the motion because it required 6 affirmative votes to pass.

Since the news media had persistently said that *religionists* were the driving force behind the inclusion of creation, Hubbard's consistent voting *against* creation was a mystery. He was the only *professional religionist* on the Board! Whereas Ragle's motion in January had been both limited in scope and sectarian in tone, Ford's February motion overcame both objections.

How then did Hubbard justify his vote to defeat a motion that would have accomplished the "neutralizing" of Evolution II (that was absolutely necessary if Evolution II were allowed to remain)? The *Los Angeles Times* reported Hubbard's position this way:[469]

> The reason was, (Hubbard) said, that time is now too short to make the necessary revisions to include the creation theory in the textbooks in an adequate manner...
>
> Because time is lacking, Mr. Hubbard said, he prefers to see the books edited to make certain that evolution is stated as theory rather than fact.

Hubbard's decision to forego principle under a time constraint evidently encouraged the scientific officialdom to exert additional pressure on the State Board.

Apparently under the illusion that the State Board was still wavering over the validity of my two paragraphs they had unanimously adopted into the *Science Framework* over 3 years earlier, the President of the Society of Vertebrate Paleontology wrote an open letter to the world on 23 February 1973 (Exhibit 57). Its message? Simpson's "singularly sensitive six" words that I had plagiarized — "the regular absence of transitional forms" — are *not true!* The AAAS even resurrected the resolution its Commission on Science Education had passed 4 months earlier (Exhibit 40) and published it in the February 1973 issue of *AAAS Bulletin*. Exhibits 60 and 61 are two irate reactions that AAAS received to that reprinted resolution.

The State Board had given the Editing Committee a deadline for its work — the Board's March meeting. So the third and final time that the Committee met was on 21 February. Ready or not, the 4 editors compiled their efforts and braced themselves for the upcoming presentation to the Board.

The Day of Reckoning was Thursday, 8 March 1973. The place? The very same auditorium in Los Angeles where my two paragraphs had been unanimously incorporated into the *Science Framework*.

Though it seemed like an eternity, only 3 years, 3 months and 23 days had passed.

The Editing Committee made its report to the Board. First, there was the standard introductory statement to be inserted in all science textbooks concerning origins (see Exhibit 58). This statement was significant in several respects. Most importantly, *it limited science.*

"Considerations extending beyond a natural description of the physical universe" it said do not belong in science. It also declared that science has no way to determine whether or not a supernatural reality exists. Further, it differentiated, albeit in rather obtuse language, between Evolution I (science) and Evolution II (belief).

The primary weakness of the introductory statement was that it failed to address the *origin* of the universe, life, and man — the primary points of contention. The "theory of organic evolution" was allowed to remain in the textbooks on the excuse that it accounted for "the complex forms of life in the past and present." But obviously, it did *not* account for the *origin* of life or the universe.

Even worse, it tacitly proposed that the evolutionary explanation for the origin of man was *scientific* rather than *an unprovable belief.*

In addition to the introductory statement, the Editing Committee offered a list of definitions for various forms of the word "adapt." Whether the decision to clarify this elusive term resulted from my warning that 24 unique definitions were used in the science textbooks (see Exhibit 54), I do not know.

However, there is no question that this term, so critical to the evolutionary belief, needs all the definition it can get. Even the definitions submitted by the Editing Committee and that were adopted by the State Board (Exhibit 59) are nothing more than *broad, unfocused statements* about "structure, physiology or behavior."

The most remarkable revisions proposed by the Editing Committee were some 219 specific changes in the science textbooks. Each of these changes varied as to magnitude, of course. Some required major rewriting of whole pages while others needed only the addition or subtraction of a sentence. Nonetheless, the number of changes were distributed among publishers as follows:

Publisher	No. of Changes
Harcourt Brace Jovanovich, Inc	149
Laidlaw Brothers	29
Charles E. Merrill Publishing Co.	16
Silver Burdett Company	7
Harper & Row, Publishers	3
Rand McNally & Company	3
Holt, Rinehart & Winston, Inc.	1
	219

Evidently, the Editing Committee agreed with my earlier-stated conclusion that Harcourt Brace Jovanovich books required the greatest revision for dogmatism. Over two-thirds of all the changes were made in those books.

News media reaction to this large number of changes was nearly mute. Large headlines were missing this time. What little coverage they gave the ignominious disappearance of their straw man — God-versus-Darwin — was devoted to trivia. Perhaps it was too much to expect them to shake hands and openly acknowledge their defeat.

The implementation of 219 specific changes in science textbooks on the subject of origins must certainly rank among the most significant events in science education — perhaps surpassing the Scopes Trial of 1925.

Any measurement of its significance must consider the unprecedented and massive resistance that was mounted by the scientific elite to *one single change!* Likewise, the news media bombarded the general public *for over three years*, trying to whip up reaction against any change at all. Therefore it is astounding that revisions were put into effect.

In view of this near-miraculous accomplishment by the State Board of Education in the face of unsurpassed odds, it is amazing that *the changes were opposed only by Board members who favored creation over evolution!*

The vote was 7 to 3. The 3 members who voted against the 219 changes were Clay Mitchell, Gene Ragle, and Tony Sierra — all of whom had consistently fought for the inclusion of creation.

How can this unexpected twist of irony be explained? Easier than one might think. While all 3 of these opponents of the editing motion undoubtedly favored every one of the 219 changes, they also saw that the editing had fallen short of the *objective* — to either cleanse science textbooks of *all beliefs* or else provide an *alternative belief* to Evolution II.

Board member Ragle observed that this significant milestone of editing, noteworthy as it was, still "confirmed and continued the dismal tradition of mandating scientific atheism as a required portion of the curriculum."[470] And he was absolutely right.

However, the key to correcting that critical error had already passed. Dave Hubbard's vote during the February 1973 Board meeting, in favor of *meeting a publication schedule* instead of *holding to the principle of balancing* anti-theistic belief with a theistic position destroyed all hope of accomplishing objectivity about origins in science textbooks.

Hubbard, of course, found himself in a most difficult position at the conclusion of the editing. He was a prominent theologian — President of Fuller Theological Seminary. The primary purpose in the Governor's appointment of a theologian to the State Board of Education is to assure moral, ethical and religious sensitivity in education. Of all Board members, a theologian should be most concerned about *avowed anti-theism* — especially in the name of objective science. And Hubbard was.

He was the most articulate spokesman on the Board for exposing the *factual tone* of Evolution II and the *arrogance* it exhibited in science textbooks. He served as one of the 4 editors on the Editing Committee, so he knew as much as anyone about the dogmatism and indoctrination concerning origins that ruled out the possibility of a Creator. Therefore, he should have voted in February with the majority of the Board to carry out the very theme he had so eloquently espoused for months. But he didn't.

A lengthy account of Hubbard's reasoning was published a week after the historic editing was mandated.[471] Missing from this report, however, is any explanation of why he either changed his mind or failed to perceive the crucial nature of his February vote.

In fairness to Hubbard, he may have been influenced by several other factors that arose at the time of editing. One was a revision in the law under which textbooks were selected. This law (AB 531), passed halfway through the science textbook adoption process, required that as many as 15 (instead of 4) textbooks in a given subject be approved by the State Board.

The *intent* of this law was to give local school districts much more flexibility in selection of materials. Its *practical effect* to publishers, however, was that it reduced the number of books they were likely to sell, even if their books were approved by the State Board. Harcourt Brace Jovanovich, probably because they were so severely edited, tried to argue that they would lose money if their books were adopted. A legal opinion obtained by the Department of Education held Harcourt Brace Jovanovich to their contractual obligation, however.

A second factor that arose was that rescheduling of the science adoption would have been required if either excising Evolution II or balancing it with Creation II were to be done in a professional manner. Hubbard was correct in making this observation. Such an effort could not have been accomplished by a part-time committee of otherwise busy people.

The constantly changing membership of the State Board of Education was a third factor that influenced the editing of science textbooks. Contrary to news media reports, the State Board of Education was *less* inclined to editing for origins in 1973 when *all* the members had been appointed by Governor Reagan than they were in 1969 when 30% of the Board were yet appointees of Governor Edmund G. (Pat) Brown! This is but another example of fact failing to validate the contention of the press that editing of textbooks was a *conservative conspiracy*.

The massive editing of 219 specific changes in science textbooks is now history. The fact that hundreds of additional changes could and perhaps should have been made cannot detract from the importance of the editing.

Beyond the direct rewriting, there is no doubt that textbook publishers have quietly noted the impact and are instituting private corrective measures within their firms to assure that dogmatism is checked in the future. In addition, the general public has been openly exposed to the bigotry and hypocrisy of the scientific aristocracy that would have never occurred if the editing had not been demanded.

The State Board of Education, as a result of their March 1973 decision to have origins discussed in *social science* rather than in *science*, ran into subsequent headwinds. As could have been expected, there was no appropriate niche in social science for such discussion. Yet the State Board had shut the door (so they thought) on origins being henceforth discussed in science.

My last unsolicited advice to the Board on the subject of origins was dated 24 February 1974 (see Exhibit 64). In this statement, I clearly showed the Board, by means of an excerpt from *Science* (Exhibit 63), that origins are discussed by scientists in other than religious or metaphysical terms.

This position paper prompted Board member Marion W. Drinker to raise an objection to my observation that "the State Board of Education chose (not from scientific accuracy but from expedience of publishing science textbooks on an arbitrary schedule) in 1973 to avoid addressing origins in science textbooks and to defer discussion of origins to social science books." She and other Board members apparently resented the truth, but Board member Hubbard's admission to this fact has already been discussed.

Mrs. Drinker (who was ultimately to become President of the State Board) and I later exchanged cordial letters of understanding on this matter.

However, the State Board's efforts to discuss origins intelligently in social science textbooks has never materialized.

As a postlude, an observation concerning Darwinians during the Science Textbook Struggle seems appropriate. Perhaps due to overconfidence or inability to understand the real issue at stake, Darwinians always seemed to lag a step behind what was happening.

Therefore, they were perpetually in a *reactive* mode. One can seldom win anything by reacting, of course. Yet Darwinians were forever claiming great *victories* by their reactions — that were actually *losses*!

One illustration of this pattern serves to close the recounting of this lengthy but successful odyssey on origins. My two paragraphs in the *Science Framework* were a perpetual irritant to Darwinians. Their reaction against those paragraphs always followed the same sequence: *Attack — be repulsed — declare victory — realize defeat — attack again.*

At least 3 distinct iterations of this series occurred:

1. They demanded total withdrawal of the paragraphs from the *Science Framework* in November 1969 (failed).

2. They then demanded publisher noncompliance to them by a footnote in February 1970 that disavowed scientific authenticity (failed).

3. They then demanded disregard of them by the State Board of Education via massive public pressure from news media and scientific elite in the fall of 1972 (failed).

Ironically, Darwinians were so certain of triumph at every step that they would declare a victory before checking. Their final desperate attack consisted of demanding replacement of my offensive paragraphs as a part of the normal 5-year revision cycle for the *Science Framework*.

But I had been publicly declaring my own dissatisfaction with the paragraphs from the day they were adopted! So I led the move to replace them. Exhibit 65 was substituted for my 2 paragraphs in the *Science*

Framework in March 1974.

The Darwinians once more declared a great victory — *the two paragraphs had finally been obliterated!* But, just as in the 3 previous attacks, they had failed. They didn't realize that the war was over — and they were *defeated.* The purpose for which the two paragraphs had been inserted had already been accomplished!

By the time Exhibit 65 was adopted, *science books had been edited...*

Chapter 10

ADDRESS THE ISSUE?

"Tarde quae credita laedunt credimus
(We are slow to believe what hurts when believed)." — *Ovid*

It's a bit uncomfortable...For a war that *officially* lasted only about 3 years, it just doesn't seem to have quite ended yet.

Both parties in the Science Textbook Struggle — the Godly and the Darwinians — still appear to be uneasy...like they're in an undeclared truce.

In the Prologue, the Science Textbook Struggle was described as "the current *form* of an unresolved conflict among scientists which has alternately smoldered and blazed ever since Charles Darwin published his *The Origin of Species* in 1859."

"Unresolved conflict" it *is*...and will apparently *remain*.

Yet this "unresolved conflict" was never *defined* in the Prologue. What *is* it? Why didn't the Huxley-Wilberforce debates *resolve* it? Over 50 years later, it erupted in the famous Scopes Trial. Seemingly settled in that 1925 courtroom, it burst forth in about 50 more years in still another forum — the science textbook.

So there is *real cause* for the uneasiness between the Godly and Darwinians. The track record for resolving their conflict is miserable. The long trail of failures doesn't deter *everyone* from attempting resolution, however. Some academicians are still trying...

The lead article in *Scientific American*, 3 years after the tacit truce in the Science Textbook Struggle, was entitled, "The Science-Textbook Controversies."[472] Despite many serious errors, a Cornell University professor produced an interesting synthesis — even a *bizarre tapestry*.

Eastern mysticism, intuitive expectation, anti-evolutionism, obscenity concern, creationism, cults, fundamentalism, patriotism, sects, absolute ethical values, theism, nonprofessional participation, and egalitarianism were some of the threads this professor wove together into an imaginary textile that supposedly threatens to smother science.

There were very few interests or activities *outside* of science, in fact, that the author *excluded* from being poised and ready to lunge at science's jugular vein. Grasping wildly to detect a connecting link between all the disparate *provocateurs* she had assembled, for example, she even twisted an observation I had made in a Curriculum Development Commission meeting regarding egalitarian social activism in public schools into a statement about *evolution!*[473] Yet, for all her effort, she completely missed the *cause celebre* that is yet unresolved.

As apparently impossible as it has been in the past, let's try once again to define the issue that must ultimately be identified if science is not to be torn asunder by frustrating and bedeviling conflict.

ONE MORE TIME!

Sir Karl R. Popper, Professor Emeritus at the University of London as well as a leading philosopher of our day, recently endorsed the book, *Darwin Retried.*[474] "I regard the book as most meritorious and as a really important contribution to the debate," Popper's review says on the colorful jacket of the book.

What debate? Surely Darwin's niche in science is secure after 115 years! Searching for an answer to that question led Tom Bethell, Editor of *The Washington Monthly*, to discover that there actually *is* a debate — still alive, still boiling, still unresolved — over Darwin's renowned theory. As Bethell dug into scientific literature, he uncovered a startling statement by Nobel laureate Thomas H. Morgan.

> "It may appear little more than a truism to state that the individuals that are the best adapted to survive have a better chance of surviving than those not so well adapted to survive," Morgan said.[475] No *theory* — just a *tautology*. "Selection, then," Morgan went on to say, "has not produced anything new, but only more of certain kinds of individuals. Evolution, however, means producing new things, not more of what already exists."

Another geneticist, C. H. Waddington, confirmed Morgan's conclusion when he addressed the Darwin Centennial at the University of Chicago in 1959. "Natural selection," he said, "which was at first considered as though it were a hypothesis that was in need of experimental or observational confirmation turns out on closer inspection to be a tautology, a statement of an inevitable although previously unrecognized relation. It states that the fittest individuals in a population (defined as those that leave most offspring) will leave most offspring."

Bethell discovered that the debate revolved about the very *core* of Darwin's supposed contribution to science — *natural selection* or "survival of the fittest." And, Bethell concluded in a *Harper's* article, the debate is fast going against Darwin! Here is his summation:[476]

> The machinery of evolution that he (Darwin) supposedly discovered has been challenged, and it is beginning to look as though what he really

> discovered was nothing more than the Victorian propensity to believe in progress...
>
> I think it should now be abundantly clear that Darwin made a mistake in proposing his natural selection theory, and it is fairly easy to detect the mistake. We have seen that what the theory so grievously lacks is a criterion of fitness that is independent of survival. If only there were some way of identifying the fittest beforehand, without always having to wait and see which ones survive, Darwin's theory would be testable rather than tautological.

But is this revelation of a chink in Darwin's armor by such notables as Popper, Morgan and Waddington essential to uncovering that "unresolved conflict" that goads scientists into combat? Not really.

In fact, Darwin may have been subject to an even greater oversight:[477]

> A scientist is a person seeking insight into the harmony of things. The harmony and the human spirit seeking to comprehend it are there first. They are prescientific. Darwin seems never to have grasped the implications of this fact. He had a profound intuition of the harmony of nature, of her "endless forms most beautiful and most wonderful," but he distrusted his intuitions. He distrusted them, his autobiography shows, because he feared that they could be explained scientifically as holdovers from man's animal past. Having doubted the reality of spirit, he suffered the spiritual consequences of his doubt. There is no escape from reality, least of all from spiritual reality.

Yet there is no way that Darwin's shortcomings could be more than *symptomatic* of the "unresolved conflict." That conflict is far more than a single person could invent. The Science Textbook Struggle recounted thus far, while exposing all kind of Darwinian aberrations, is not basically a struggle over mistakes made by Darwin or his disciples.

Then maybe the "unresolved conflict" is due to *irrationalism* — that popular but undefined label hung on anything that appears threatening to the scientific *status quo*. Irrationalism was a topic at the annual meeting of the American Association for the Advancement of Science in December 1972. Edward Shils of the University of Chicago said at that meeting that some roots of irrationalism come from science itself:[478]

> From an original position of exploring God's will, science became secularized first to a largely esthetic pursuit of cognitive understanding of the universe and then to a pragmatic development of knowledge useful for man's comfort or defense. The latter effort has brought vast financial support but has given science an association with political power leading to distrust of a scientist's motivation.

At the same AAAS meeting, Charles Frankel of Columbia University proposed 2 fundamental principles of irrationalism:[479]

1. The universe is divided into 2 realms — *appearance* and *reality*.
2. People mistake appearance for reality because of *presuppositions* imposed by culture, class or practical concerns.

Frankel claimed that science "draws a sharper distinction between appearance and reality than does anything in the scheme of philosophical irrationalism." On the other hand, Frankel freely admitted that no inquiry of any kind is possible without making *assumptions*.

Therefore, Frankel's line between irrationalism and science's rationalism is not all that distinct — certainly not as obvious as the popular use of the 2 terms might suggest. Irrationalism seems to connote *stupidity*, *absurdity*, *foolishness* and *nonsense* while rationalism implies *logic*, *reason*, *wisdom* and *intelligence*. The two terms, when placed in juxtaposition to defend science, are unjustifiably pejorative.

Even so, is the root of the "unresolved conflict" to be found in *irrationalism versus rationalism*? Hardly.

Where else can we seek? Possibly the root lies in whatever prompted Lord Alfred Tennyson to pen the mystical words now inscribed in the magnificent rotunda of the Library of Congress in Washington — "One God, One Law, One Element, and One Faroff Divine Event to Which the Whole Creation Moves."

The *unity* — the *singleness* — the *oneness* — the *simplicity* of Tennyson's lyric verse is something that scientists and laymen alike inherently seek and desire.

In a complex world growing with the exponential complexity Alvin Toffler documents in his *Future Shock*, we are all caught up at a dizzying pace in a direction we try to resist. Saul Bellow, 1976 Nobel laureate for literature, confirmed this yearning to return to less complexity in his Nobel lecture at the Swedish Academy. He said that the world is sick of intellectual bosses and has an immense desire for a return to "what is simple and true."

Bellow's observations — especially his attack on organized intellectualism — startled the world:[480]

> A group of mummies, the most respectable leaders of the intellectual community, has laid down the law...We must not make bosses of our intellectuals...
>
> At the moment, neither art nor science but mankind is determining, in confusion and obscurity, whether it will endure or not. The whole species — everybody — has gotten into the act. At such a time it is essential to lighten ourselves, to dump encumbrances, including the encumbrances of education and all organized platitudes, to make judgments of our own, to perform acts of our own . . .

> The struggle that convulses us makes us want to simplify, to reconsider, to eliminate the tragic weakness which prevented writers — and readers — from being at once simple and true.

This deep yearning that Bellow describes is for what is more *basic* or, as he defines it, "an immense, painful longing for a broader, more flexible, fuller, more coherent, more comprehensive account of what we human beings are, who we are, and what this life is for." Could this profound yet simple cry be the "unresolved conflict" we must identify? It may be closer than the other alternatives we have examined, but it's not quite on target...

There is a rudimentary common denominator of each of the three candidates — Darwin's shortcomings, public irrationalism, and personal yearning for simplistic truth — just proposed for the "unresolved conflict." It is a persistent, unrelenting factor that, while common to all three, is more basic than any of them.

It is the bi-polar alternate for *cosmology* — **NATURALISM versus THEISM**. *This is the ultimate source of conflict.* Reducing the conflict to simpler terms is impossible.

Attempts to understand the Science Textbook Struggle (and generic exhibitions such as the Huxley-Wilberforce debates and the Scopes Trial) are doomed to despair and confusion unless there is recognition of the basic issue — NATURALISM versus THEISM.

Robert B. Fischer, who played such a significant role in my own thinking, as well as in the Science Textbook Struggle itself, placed science textbook editing in perspective with this basic issue:[481]

> In retrospect, it appears that the issue which really underlies the public conflict over these books is the conflict between two basic ways of looking at all of life and reality: (1) a purely naturalistic world-view and (2) a theistic world-view. This basic issue is surely not a problem for science education as such, even though it is one of considerable concern to human beings, including those who are scientists and students of science.
>
> Over the years, various specific conflicts have arisen by superimposing this basic issue upon other issues which do not coincide. The present situation appears to be such a case.
>
> One result of this superimposing of a basic issue (purely naturalistic vs. theistic world-views) upon a specific issue (theories of origins in science books) when the two issues do not fully or clearly coincide, has been a polarization of people that has stood in the way of full, valid understandings, both of the nature and content of science, and of the overall world-views and commitments of persons.

Note that Fischer is blaming neither the Godly nor the Darwinians by his observations. It is a *mutual* misunderstanding. *Both* parties have failed to perceive the radical root of rancor.

If this premise — that the "unresolved conflict" is due to NATURALISM versus THEISM — can be *granted*, there is hope for (1) mutual understanding between the Godly and Darwinians, (2) peaceful resolution of antagonistic alternatives, and (3) restoration of objectivity among scientists regarding origins.

If the premise is *denied*, then future eruptions of the basic issue can be expected — though their *form* may continue to surprise, frustrate, confuse and bewilder whoever holds the majority power position (e.g., the Godly in the Scopes Trial and the Darwinians in the Science Textbook Struggle).

Obviously, everyone should favor *acknowledgement* over *denial* of the premise — just for their respective results. Consider then 2 points of evidence that validate the premise.

First, all scientists *believe* — in either NATURALISM or THEISM. Second, current science writing, whether in textbooks or professional literature, exclusively and dogmatically supports only one of these beliefs – NATURALISM.

Scientists Are Believers!

"Let me give *you* an example!" Porter M. Kier, Director of the National Museum of Natural History of the Smithsonian Institution in Washington, D. C. was about to describe how scientists react when a pet belief is challenged.

Kier and I had just been introduced to each other by a mutual friend, Glenn S. Pfau. Pfau, it will be recalled, arranged for me to meet in his National Education Association office in June 1972 with nationally prominent science educators. However, the setting for Kier and me to meet one another was far different.

It was a cold December night in 1973 in Washington. Pfau had made dinner reservations at the Luau Hut 3 blocks north of the Capitol. As soft Hawaiian music wafted about our table set in South Sea decor, Pfau pulled off the "introduction of the century." To have a world-renowned paleontologist agree to meet the non-scientist who had raised all the ruckus in California about evolution was an accomplishment a la Henry Kissinger.

Pfau's associate, David A. Spidal, rounded out a foursome as we dined on Polynesian food. Kier and I entered into dialogue from the moment we sat down.

Graciously, he listened as I described how the issue in California had differed from press reports about it. I had just mentioned the emotional tirade unleashed by the scientific elite over the State Board's decision when Kier offered a similar example.

"Arthur Meyerhoff, a petroleum geologist, wrote an article in a geological journal about 6 months ago which attacked continental drift. It got a *terrible* reaction! Just like your attack on evolution! Then there was a *counter-reaction*. A lot of thoughtful people wrote in and said, 'What is going on here? Continental drift isn't a *proven fact!* Have we closed our minds to the possibility that it may *not* have occurred?' It rather startled everybody. But we've done the same thing with evolution. As far as most scientists are concerned, evolution is an accepted *fact* — things have been changing over time. But you wouldn't argue with that either..."

"No, I don't disagree with that point at all!" I responded.

Kier continued. "As far as the *origin* of life goes, I know *some* paleontologists — when pressed — would say that they are unsure about it. They don't want to take a Supreme Being out of their ideas. Probably they would say, 'Well, evolution occurred, but I think that there was a divine spirit guiding things along.' I've no argument with that kind of thinking as long as it doesn't make them do unscientific work — as long as they don't refuse evidence that's available."

I sensed Kier's fairness as he systematically laid out his arguments. His next point was a masterpiece.

"I've got a friend who does exactly the *opposite* — sees evidence that isn't even there! He can find an evolutionary climb — where you get one fossil going to another to another — every time he turns around! I've been with him in the field — in the Middle East, in Arabia where we have been going up a section. He'll say, 'Look at these clams! See them changing? This one is a little bit longer than that one.' Then he'll write a paper with all his 'data' laid right out. I know that it's just a joke. I know he didn't see that evidence, but he has his brain turned in that direction...and he's not a good scientist!"

This "ability to see evidence that isn't there" that Kier noticed in his colleague can be defined as *belief*. That scientist "believed" (in error) that he had seen something he had not seen. There are 4 logical possibilities for coupling evidence with a scientist's observation:

Case 1. Evidence exists and is observed by the scientist.

Case 2. Evidence exists and is overlooked by the scientist.

Case 3. Evidence does not exist but is "seen" by the scientist.

Case 4. Evidence does not exist and is not observed by the scientist.

Two of the four possibilities (Cases 1 and 4) are *desirable* and the other two are undesirable. However, Case 3 is far more *undesirable* than Case 2. Why? Because Case 3 involves *belief* whereas Case 2 is probably due only to *inadequate observation.*

While belief is not *forbidden* in science, it must be *validated* by repeated experiments if it is ultimately to be incorporated into the body of science. Obviously, since evidence in Case 3 is only imaginary and has no real existence, belief in that case must be ruled out of science. That's why Kier said his friend was not being *scientific.*

Note that the belief held by Kier's co-worker is not a *religious* belief. Beliefs are simply convictions that something is true based on faith. And, of course, we can have faith in *anything* — not just in a *deity.*

It can and will be propounded that beliefs in science are most *useful* as starting points for the interrelationship of observations by scientists. However, as the 19th century Swiss philosopher Henri-Frederic Amiel once said: "*A belief is not true because it is useful.*" The great temptation to confuse a belief's *truth* with its *usefulness* is one that all scientists must resist.

Porter Kier, who had just completed a term as President of the Paleontological Society, earned his doctorate in paleontology as a Fulbright Scholar at Cambridge. He is probably the top authority in the world on the fossils of echinoderms — starfish, crinoids, and sea urchins. Kier held me spellbound as he described his frequent deep-sea expeditions around the world to collect echinoderm fossils. As he related his tedious method of laying out collected echinoid specimens in a laboratory, I was intrigued by one of his comments.

"It's very hard for us to find in echinoids species-to-species evolution — change from one to another. We don't very often find, when going through a rock section, species A evolving slowly into species B.

"It would be a lot easier to believe that every single species was independently created, as far as the record goes. This is something that has puzzled paleontologists for a long time. It has been a source of a lot of concern to paleontologists. We *do* have gaps between species."

Kier pointed out, on the other hand, that in some coral there is a continuum of fossils. By laying these fossils side-by-side, a scientist can detect a "climb" or a continuous change over a period of millions of years without any gap.

"Alright," I said, "when you look at these specimens or fossils which you have arranged in a time-ordered sequence that indicates continuous change, you also see a construct, skeleton or framework that ties them interdependently all together. You call that framework 'evolution.' Is that correct?"

"That's right."

"That's obviously a *preconditioned* set of eyes through which you are viewing the specimens or fossils, though, isn't it?"

"Certainly. I have heard previously about evolution, so I look for that possibility."

"Okay. But would you also grant the possibility that there could be other scientists looking — as you look for evolution — for the handiwork or contribution of a Designer? Do you think that they could see Him if they chose to have that predisposition?"

"Of course. They can say that He made life. He made selection occur. He had a certain approach in mind, and it went that way because He set the machine up. He could also individually create every one of the fossils along the line. In fact, anything is possible, if you believe in the supernatural."

"Would you personally *deny* scientists the right to look at things that way?"

"No, I would not do so — not at all."

"Should they be allowed to *say* that they see it that way?"

"Of course they should! *Absolutely*. And I think it's perfectly all right to say it in science textbooks. I think that textbooks should describe interrelated observations as a *theory* — not as a fact. Science textbooks should *never* say that life began 3 billion years ago in the sea. That's just bad writing. It isn't good science. It isn't *science*, I can assure you of that!"

Though Glenn Pfau may have expected a clash of ideologies by introducing Porter Kier and me, our 3-hour dinner conversation was amazingly congenial. We parted real friends and have remained so. My personal friendship with Kier has grown through subsequent visits with him at the Smithsonian.

On several occasions, he has interrupted his busy day to take me, Phyllis, or some of our children on private tours of the Museum of Natural History.

Our acquaintance also led to an extensive exchange of correspondence with Norman D. Newell, Chairman and Curator, Department of Invertebrate Paleontology at the American Museum of Natural History in New York. This exchange started when Kier sent me a copy of Newell's "Special Creation and Organic Evolution"[482] — another attempt to resolve the "unresolved conflict" between evolution and creation.

My initial letter to Newell was brief. Enclosing Exhibits 47 and 54, I proposed that they might provide middle ground to bridge the 2 poles of thought Newell had described in his treatise. "Any *error* in my position which you could point out would be particularly appreciated," was how I concluded my letter.

Newell called Kier after receiving my letter. He was curious how Kier and I could possibly be friends. However, Kier not only assured him that we were friends, but that my viewpoint was worthy of his attention and response. Thereupon, Newell wrote to me.

Newell replied with 6 single-spaced pages of *criticism of the Creation Research Society* — a group with whom I had never been affiliated! Not one comment addressed *my* position! Much of what Newell wrote I had also publicly said in agreement with him.

Therefore, I was disappointed.

I waited a month before answering Newell. One reason for delaying was that I thought he'd spent so much time criticizing the Creation Research Society (while believing he had addressed my position) that he would never take any more time with me. Yet, I really wanted to know any *specific errors* he could find in *my* writing.

So, as delicately as possible, I wrote that my views and those of Creation Research Society were not synonymous. Then I repeated my request for Newell's criticism:

> My intent, from the outset of my involvement in the issue in California, has been to provide calm, rational, studied, and objective review of what science knows and does not know about *origins* — not mechanisms of *change* subsequent to such origins; e.g., evolution. I have additionally attempted to elicit (and without much success, I must confess) acknowledgement of various "belief systems" that exist in the minds of *all* scientists, regardless of their persuasion regarding *causes* whose effects we observe and correlate in science.
>
> Although re-examination of our basic presuppositions, beliefs, and constructs of thought is painful for every person — scientist or anyone else, I do believe that there is great advantage to such re-examination on rare occasions. I trust that you will still find it profitable to examine my thesis once more from your position of insight, experience and wisdom. Again, I am not desirous of accolades but of constructive criticism leading toward mutual understanding instead of polarization.

Newell got my point — that I was disassociated from the Creation Research Society. His next letter gave me credit for excluding reference to religion or to the Holy Scriptures. But he still claimed that many of my arguments were identical to those of the Creation Research Society. Nonetheless, I was heartened by his promise to comment more specifically on my writing at a later date.

When Newell wrote 2 weeks later, I was excited as I opened his 4-page letter that had an additional 4 pages of specific comments on Exhibits 47 and 54. Granting that my reports were "interesting and persuasive," Newell did not accept my conclusions nor the rationale on which they were based. Several of what Newell perceived as basic differences between us really were only semantic in character. However, I did disagree with his thinking on a few points, two of which are significant:

1. Newell evidently had difficulty differentiating between Evolution I and Evolution II. On one hand, he clearly acknowledged that even atheistic scientists cannot really eliminate "first causes" and since such causes cannot be studied by science, they are not relevant to science curricula. Yet, he then stated that creationism (1) has been rejected by the scientific community over the world, (2) has outlived its intellectual value, and (3) "has absolutely no meaning for 20th century morals and ethics." Since creationism deals exclusively with "first cause," how could *science* have rejected it?

2. While granting the need for making clear distinction between what is *known* and what is only *suggested*, Newell was more concerned about the need "to avoid eroding the layman's confidence in the scientific method." He felt that since the general public intuitively expects definitive answers to complex problems, they would become disillusioned if scientists acknowledged uncertainty. Obviously, my position is quite the opposite. To me, it is both arrogant and dishonest to appear certain when one is uncertain just to protect one's method from eroded confidence.

The biggest disappointment for me in the protracted exchange of letters with Newell was my inability to persuade him to look beyond what he frighteningly saw as an attack on science. He seemed to be so overwhelmed at a *religious* threat that he could not see any *scientific* issue. His central argument illustrates his fixation of belief:[483]

> We are in the midst of an atavistic revival of the primitive idea of the supernatural creation of individual species, *especially* man. This revival is riding a more general wave of antirationalism characterized by emotionalism, mysticism, astrology, and other occult arts, and is

> accompanied by wide-spread mistrust of the objectivity which is the foundation of science and reason...
>
> It is a massive attack on the methods, assumptions and limitations of scientific knowledge in biology and geology. Such an attack could be highly beneficial if it were an objective critique of scientific errors accompanied by constructive suggestions for improving scientific knowledge. Unhappily, science is presented to the unwary reader as an entrenched and ultraconservative establishment.

But how are "constructive suggestions for improving scientific knowledge" developed? Perhaps *attack* is too strong, but certainly error (if it exists) should be *identified.* Likewise, if science is "an entrenched and ultraconservative establishment," that is undesirable for science and should be *remedied.*

On the other side of the coin, there *is* an anti-science mood abroad in the land. Much of the thrust behind the "creationist cause" definitely *favors* the destruction of science — at least a decimation of its power. But Newell "throws the baby out with the bathwater." Why? Because he *believes* that scientific naturalism wholly answers all questions about the physical universe. Thereby, he equates *science* with only one pole of the "unresolved conflict" — NATURALISM.

Newell is entitled to his belief. But it must be recognized as *belief* — not as a *scientific* statement or position. As uncomfortable as it might be for him to admit, the current effort (to secure recognition that origins and associated "first causes" must remain outside scientific determination) is "an objective critique of scientific errors accompanied by constructive suggestions for improving scientific knowledge." This effort does have spurious offshoots, but the gravamen is still valid.

I totally agree with Newell when he says, "The essence of the scientific method is to assess probability from empirical evidence." When assessing probability from empirical evidence — particularly evidence for origins and primordial events, *the role of assumptions is overpowering.*

Harris says in *The Foundations of Metaphysics in Science*:[484]

> When we talk of assumptions the epistemological and interpretative character of the process is inescapable. Strictly an assumption is a proposition, but of course it may be operative in behavior without being formulated...This activity involves the assessments and weighting of probabilities, which inevitably involves past experience, and involves it not just as having occurred but as having significant bearing on the interrelation of sensory and motor events. The past experience here useful must be organized experience of objects in spatio-temporal and causal relations, which can provide evidential foundations for the estimate of probabilities. This is a highly complex body of already

acquired knowledge, and it is the bearing of this upon the immediately presented sensory clues that constitutes the percept. Such a process of reference of clues to a system of interrelated objects is precisely what is meant by interpretation...

Thomas Kuhn in *The Structure of Scientific Revolutions* claims that scientists are limited in their observations to viewing phenomena from a pre-theoretical framework, belief or paradigm.[485] Likewise, Herbert Butterfield in *The Origins of Modern Science* points out that choice of scientific experiment is based on *presupposition* or *belief*:[486]

> Butterfield argues that the observations or evidence do not themselves thrust upon the scientists conceptual patterns of interpretation that are univocal and necessary. Instead the scientist needs to choose deliberately which alternative conceptual framework to use for his interpretation...This is a choice the scientist makes that is not dictated by the observations and experiments. We can't even say that after we have the data at hand we can choose our conceptual framework.

Harvard's Gerald Holton has proposed 8 distinct facets of scientific work that function as themata. The first of these is "understanding of the scientific event." To obtain this understanding, Holton says:[487]

> For this we need to establish the awareness, within the area of public scientific knowledge at the time of the event, of the so-called scientific facts, data, laws, theories, techniques, lore. I would include under this heading the larger part of historical research on what are called *scientific world views*, paradigms, and research programs (emphasis added).

Holton uses Einstein's example of *believing* in an indefinable "reality" that serves as a useful concept in scientific work:[488]

> In the end, Einstein came to embrace the view which many, and perhaps he himself, thought earlier he had eliminated from physics in his basic 1905 paper on relativity theory: that there exists an external, objective, physical reality which we may hope to grasp — not directly, empirically, or logically, or with fullest certainty, but at least by an intuitive leap, one that is only guided by experience of the totality of sensible "facts."

Robert K. Merton of Columbia University is among those who have evaluated Holton's thematic analysis. He summarized the contribution of Holton's thematic approach as "a distinctive effort to deal with tacit knowledge:"[489]

> The themata of scientific knowledge are tacit cognitive imagineries and preferences for or commitments to certain kinds of concepts, certain kinds of methods, certain kinds of evidence, and certain forms of solutions to deep questions and engaging puzzles.

What does this mean in layman's language? Scientists *believe* silently and subconsciously. Belief influences their analysis, conclusions — even their "facts." And they are seldom aware of the influence of belief on their work.

When A. I. Oparin said, "Matter perpetuates its own organization,"[490] was he stating a *fact* or a *belief*? "The first form of life, or self-duplicating particle, did arise spontaneously from chemical inanimate substances."[491] Is that a *fact* or a *belief*?

Many scientists are fearful of admitting their presuppositions, bias, prejudice, or "intuitive leaps" because such an admission would supposedly make them *sub-scientific*. They have been brainwashed, in a sense, by their educational upbringing (and later by the sterile system of academic respectability) into thinking that *beliefs* belong only to others. Holton describes this frame of mind:[492]

> The very institutions of science — the methods of publication, the meetings, the selection and training of young scientists — are designed to minimize attention to this element. The success of science itself as a shareable activity seems to be connected with this systematic neglect of what Einstein called the "private struggle." Moreover, the apparent contradiction between the often "illogical" nature of actual discovery and the logical nature of well-developed physical concepts is perceived by some as a threat to the very foundations of science and rationality itself.

Science education stresses *inductive* reasoning — as contrasted with the *deductivism* in mathematical literature. "At the very heart of induction," Sir Peter Brian Medawar writes, "lies this innocent-sounding belief: that the thought which leads to scientific discovery or to the propounding of a new scientific theory is logically accountable and can be logically spelled out."[493] He goes on to show that, *supposedly* in the inductive scheme, *discovery* and *justification* form an integral act of thought. *But it's a false belief or assumption.*

While inductive theory demands primacy of *facts* — of propositions that put on record the simple and uncomplicated evidence of the senses, it simply doesn't happen that way.

As Medawar says:[494]

> Deductivism in mathematical literature and inductivism in scientific papers are simply the postures we choose to be seen in when the curtain goes up and the public sees us. The theatrical illusion is shattered if we ask what goes on behind the scenes. In real life discovery and justification are almost always different processes.

Charles Sanders Peirce, who has been widely acclaimed as the most original thinker in American philosophy,[495] has been credited with a third type of inference — beyond induction and deduction. He, like Medawar, did not agree that hypothesis formation was inductive. Instead, he called it *abduction*. All three — abduction, deduction and induction — "must be taken as an integrated whole in order to complete any scientific enterprise."[496]

T.C. Schneirla points out that abduction (the circumstances, processes or *beliefs* which generate an explanatory hypothesis — that very beginning of scientific endeavor) is most neglected in scientific consciousness. Schneirla views *the preconceptions and predilections an investigator brings to his work as being equal in importance to any other factor.* To merely take hypotheses for granted and to ignore the conditions, intentions, beliefs, and processes of their formation is a serious error.[497]

For those who may be unfamiliar with Peirce's abduction, there is apparently a rediscovery of his thought now underway. One account of his development of abduction describes how he had formerly confused "hypothesis" and "induction" until he realized that "probability had nothing to do with abduction..."[498]

Peirce's perception of the role of *belief* in science is germane to the point that *scientists are believers*. Peirce shows that belief virtually *predetermines* how a scientist will react to observations or data. Here is W. B. Gallie's summary of Peirce's position:[499]

> "While belief lasts," Peirce tells us, "it is a strong habit, and as such, forces a man to believe until surprise breaks up the habit. The breaking of a belief can only be done to some novel experience, whether external or internal..." What is the precise force of the word "experience" in this connection? Peirce does well to remind as, in one passage, that "experience which could be summoned up at pleasure would not be experience": on the contrary, experience is the main thing that constrains our thinking, and, at any rate on some occasions, compels us to a "forcible modification of our ways of thinking." In a word, belief, like action, must accommodate itself to that which it finds thrust upon it: that is, to the broad course of experience.
>
> Once we are constrained to abandon a given belief, however, a new condition of mind, Doubt, ensues, which provides in almost every aspect the sharpest possible contrast to belief. Doubt is an "uneasy and dissatisfied state" from which we "at once struggle to free ourselves and pass into the state of belief." Peirce compares it to the irritation of a nerve and the reflex action produced thereby, whereas the physiological analogue of belief would be provided by nervous associations. The struggle to *re*-fix belief, by removing the irritation of doubt, Peirce names Inquiry: and the sole purpose of inquiry, he tells us in his 1878 paper, is the settlement of belief. "We may fancy that this is not enough for us, and that we seek, not merely an opinion, but a true opinion. But put this

> fancy to the test, and it proves groundless; for as soon as a firm belief is reached we are entirely satisfied, whether the belief be (*sc.* in fact) true or false...The most that can be maintained is, that we seek for a belief that we shall *think* to be true...But we think each one of our beliefs to be true, and, indeed, it is a mere tautology to say so."

Of all beliefs that scientists hold, the most critical is a *Weltanschauung* or an outline system of metaphysics that Peirce says those sciences that deal with matter *take for granted.*[500] It is this very point that has caused the unresolved conflict since Darwin.

One *Weltanschauung* — NATURALISM — says that the universe and all it contains originated and continues to exist entirely without any external (supernatural) agency. The other *Weltanschauung* — THEISM — says that the universe and all it contains came into being and is continuously sustained by a supernatural agency. These two *Weltanschauungs* are mutual exclusives. Most importantly, *science has absolutely nothing to say concerning the validity of either one!*

But what was the prime Darwinian argument during the Science Textbook Struggle? "The vast majority of scientists have agreed to believe in NATURALISM." *Ipso facto*, no other belief shall be allowed expression in science. (Note that the *inverse* was true in the Scopes Trial.) Of course, the Darwinian spokesman never took an actual vote to see if NATURALISM was the majority belief. They simply *believed* it was true.

If scientists were polled for their personal *Weltanschauung*, the results might surprise Darwinians. A 1976 Gallup Poll conducted for the Charles F. Kettering Foundation showed that 94% of all Americans believe in the existence of God.[501] And the percentages elsewhere in the world were just as startling — 89% in Canada, 98% in India, sub-Sahara Africa 96%, Latin America 95%, and even 78% in Western Europe!

Now, unless it were assumed that scientists are most anomalous, wouldn't it be a bit absurd to say that "the vast majority of all scientists" believe in the NATURALISM *Weltanschauung*? While those who set themselves up as spokesmen for "the vast majority of all scientists" may feel qualified to make that presumptuous claim, it would be more scientific to actually poll scientists for their *Weltanschauung*.

One reason the Science Textbook Struggle generated worldwide interest may be that the scientific elite expressed a position so contrary to the THEISM *Weltanschauung* held by the general public. Opinion surveys across the nation overwhelmingly indicate that taxpayers and parents want schoolchildren exposed to a creationist viewpoint whenever evolution is taught in the science classroom. Acknowledging that the lay public fails to understand that neither Creation II nor Evolution II belong in the science classroom, Darwinians might reconsider their familiar complaint that the public view is a "reversion to medieval superstition." *No reversion has occurred.* Adherence to a THEISM *Weltanschauung* by most human beings *has been there all along*. Darwinians have simply ignored it.

The NATURALISM *Weltanschauung* has many articulate advocates within the scientific hierarchy. Several have been quoted earlier. To aid the reader who might otherwise have difficulty in identifying this philosophic predisposition, a few examples are offered. The important point to remember about the following statements is that they are *extra-scientific*. Even though distinguished scientists are saying them, the assertions have *no scientific validity* or *significance*.

Remember these two declarations by George Gaylord Simpson in Chapter 2?

> Adaptation (by natural selection) is real, and it is achieved by a progressive and directed process. This process is natural, and it is wholly mechanistic in its operation. This natural process achieves the aspect of purpose, *without the intervention of a purposer*, and it has produced a vast plan, *without the concurrent action of a planner*.[502] (emphasis added)

> Man is the result of a *purposeless* and *materialistic* process that did not have him in mind. He was *not planned*. He is a state of matter, a form of life, a sort of animal.[503] (emphasis added)

Harold C. Urey, one of the 19 Nobel laureates who signed Exhibit 53, is more candid about his belief in the NATURALISM *Weltanschauung*:[504]

> All of us who study the origin of life find that the more we look into it, the more we feel that it is too complex to have evolved anywhere. *We all believe as an article of faith* that life evolved from dead matter on this planet. It is just that its complexity is so great it is hard for us to imagine that it did. (emphasis added)

The late Sir Julian Sorell Huxley probably espoused the NATURALISM *Weltanschauung* more clearly and prolifically than any other Darwinian. When delivering the keynote address at the Darwin Centennial Convocation held at the University of Chicago in 1959, Huxley said:

> Evolutionary man can no longer take refuge from his loneliness by creeping for shelter into the arms of a *divinized father figure whom he himself has created* (emphasis added).

But Huxley not only decreed God out of existence. He promulgated a new vision for the universe, with man as the supreme being:

> A new vision has been revealed by post-Darwinian science and learning. It gives us a new and an assured view of ourselves. Man is a highly peculiar organism. He is a single joint body-mind, not a body plus a separate mind or soul, but with mind on top, no longer subordinate to body, as in animals. By virtue of this, he has become the latest dominant type in the solar system, with three billion years of evolution behind him...His role, whether he wants it or not, is to be the leader of the evolutionary process on earth, and his job is to guide and direct it in the general direction of improvement.[505]
>
> In the light of our new and comprehensive vision, we must redefine religion itself. Religions are not necessarily concerned with the worship of a supernatural God or gods, or even with the supernatural at all; they are not mere superstition nor just self-seeking organizations exploiting the public's superstitions and its belief in the magical powers of priests and witch doctors.
>
> The ultimate task will be to melt down the gods, and magic, and all supernatural entities, into their elements of transcendence and sacred power; and then, with the aid of our new knowledge, build up these raw materials into a new religious system that will help man to achieve the destiny that our new evolutionary vision has revealed. Meanwhile, we must encourage all constructive attempts at reformulating and rebuilding religion. My personal favorite is Evolutionary Humanism, but there are many others tending in the same general direction, like Yoga and Zen, ethical and meditative systems, and the cults of release through psychedelic drugs or bodily rituals.[506]

Just as Simpson, Urey and Huxley have believed in the NATURALISM *Weltschauung*, there are many other scientists who believe in THEISM. Certainly Kepler, Newton, Pascal, Faraday, Maxwell and other giants of science declared their predisposition to THEISM as openly as Simpson, Urey and Huxley have their beliefs in NATURALISM.

The important factor is neither how many scientists are lined up on one side or the other, which viewpoint has the loudest and most articulate disciples, the prestige or accomplishment of individual believers on either side, nor which belief seems most attractive or persuasive.

The noteworthy point is that *scientists do believe!* Further, scientists, as a group, are divided in their beliefs between THEISM and NATURALISM — they have not lined up on only one side.

Lest one draw the conclusion that either *Weltanschauung* can be represented by a simple creed, it must be explained that both worldviews contain a wide spectrum of belief. NATURALISM embraces a broad range of beliefs from *agnosticism* ("I think it's impossible to know whether there is a God") through *atheism* ("I am convinced there is no

God") to *anti-theism* ("I am adamantly opposed to the idea of a God").

Likewise, THEISM would incorporate a variety of beliefs including *pantheism* ("God is everything and everything is God"), *polytheism* ("There are many individual gods"), *deism* ("God exists and created the universe but assumed no control over the creation"), and *theism* ("God not only created the universe but also is actively governing it"). Obviously, there are subsets of belief within these major positions, just as there are within *agnosticism*, *atheism*, and *anti-theism*.

One of the prime objectives of this book is to have the general public understand the role that *belief* plays in scientists' work. The first step toward that objective obviously would be to get *all* scientists — not just Simpson, Urey, Huxley and Einstein — to openly declare their predisposition or cosmological belief. This openness would accomplish at least four things:

1. Demonstrate to laymen an *honesty* and *maturity* among scientists.
2. Help scientists to become more *objective* (by self-recognition of specific bias).
3. Eliminate the "science-*versus*-religion" syndrome (for scientists and laymen alike).
4. Prove the inappropriateness of including Evolution II and Creation II in science textbooks.

As a gesture toward open acknowledgement of private belief by all scientists, I will declare my own position. Thus far, I have attempted (probably not successfully) to conceal my personal beliefs because revelation of them during the Science Textbook Struggle seemed to be counterproductive to objectivity, as discussed in Chapter 3.

I have since changed my mind. I now see that exposure — at least of my *conscious* predispositions and beliefs — is required if I expect others to do the same.

On 17 August 1945, I encountered Jesus Christ in terms of a living person rather than a historical figure. As a result, I surrendered my whole being — life, values, property, reputation — to Him. It was a personal, conscious act of my will.

To clarify the nature of this commitment, it may be helpful to restate it in negative terms. It was *not* the agreement or subscription to a creed. *Nor* was it the act of joining an organization — like a church. *Neither* was it an experience inherited from parents, defined by an organization, or resulting from living in a certain cultural setting.

My personal relationship with Christ produced a life-changing *reorientation* of my values, my outlook on life, and also my understanding of ultimate causes. In other words, I adopted a new *Weltanschauung* — THEISM.

For those anxious to classify or label me, I will continue to resist categorization. It is far more important that my *beliefs*, per se, be identified than for me to be properly slotted into a rigid, preconceived niche. The most likely tag others will wish to hang on me is *fundamentalist.* My response to that stigma is that a fundamentalist is committed to a *book* — the Bible — while my commitment is to a *Person*, Jesus Christ.

While I could properly be called a *Christian* because of my relationship to Christ, I tend to shun that title as well since there is great confusion in its public usage and interpretation. For example, I feel absolutely no spiritual kinship to "Christian" armed forces battling Moslems in Lebanon, those engaged on either side of the "Christian" civil strife in Ireland, or those Nazis (considered "Christians" by many of my Jewish friends) who slaughtered 6,000,000 Jews during World War II. Once again, labels are counterproductive if my beliefs are to be understood.

How do my beliefs influence my interpretation of evidence for origins of the universe, life, and man? Quite drastically. First of all, I believe there is a Creator. Why? Because I have met Him personally and enjoy a daily, vital relationship with Him.

For anyone unfamiliar with who Jesus Christ is, these statements explain His role in creation:

> Christ is the visible expression of the invisible God. He existed before creation began, for it was through Him that everything was made, whether spiritual or material, seen or unseen. Through Him, and for Him, also, were created power and dominion, ownership and authority. In fact, every single thing was created through, and for, Him. He is both the first principle and the upholding principle of the whole scheme of creation...Life from nothing began through Him, and life from the dead began through Him, and He is, therefore, justly called the Lord of all.[507]

> When in former times God spoke to our forefathers, He spoke in fragmentary and varied fashion through the prophets. But in this the final age He has spoken to us in the Son (Jesus Christ) whom He has made heir to the whole universe, and through whom He created all orders of existence: the Son who is the effulgence of God's splendour and the stamp of God's very being, and sustains the universe by His word of power.[508]

It follows then that I view God's creative work to encompass far greater scope of activity than simply *initiating* the material universe. I concur in Denis Alexander's perspective in *Beyond Science*:[509]

> (God) has not only created everything in the past, but is actively creating everything now and will continue to do so in the future. Everything is held together and consists by his power. God is not viewed as an explanation for anything — rather there would be nothing for the scientist to explain if God had not first willed it to exist. He is not like the gardener who occasionally potters around in his garden. He is like the playwright who is actively creating all the structures and situations necessary for a continuing drama...
>
> So, in this view, even the atoms in the hands of the biochemists synthesizing life artificially in the laboratory would be held together by the continuing creative work of God. To suggest that God did not exist because man had created life would be as stupid as saying that Picasso did not exist because someone had made a copy of one of his paintings. To suggest that scientific descriptions of creation made God unnecessary would be rather like one of the actors in a play maintaining that the author of the play was not really necessary.

Because of this intimate involvement of the Creator, the following argument establishes the significance of the physical evidence used by scientists to postulate their laws:[510]

> The holy anger of God is disclosed from Heaven against the godlessness and evil of those men who render truth dumb and inoperative by their wickedness. It is not that they do not know the truth about God: indeed He has made it quite plain to them. For since the beginning of the world the invisible attributes of God, for example, His eternal power and divinity, have been plainly discernible through things which He has made and which are commonly seen and known, thus leaving these men without a rag of excuse. They knew all the time that there is a God, yet they refused to acknowledge Him as such, or to thank Him for what He is or does...
>
> These men deliberately forfeited the truth of God and accepted a lie, paying homage and giving service to the creature instead of to the Creator.

The material universe thus becomes the proof in the eyes of God for His *existence* — to the point of being overwhelming! Yet, despite the logic of this argument that proposes that failure to acknowledge the Creator is inexcusable, I would not expect anyone who has not personally met Jesus Christ to be committed to the concept of a Creator. The Genesis account, by itself, would be insufficient argument for me.

(Should a reader view a personal encounter with Jesus Christ as a rare, unusual, mystical event, another recent Gallup Poll showed that about one-third of all American adults — about 47,000,000 — have experienced "a moment of sudden religious insight or awakening" similar to mine in 1945.[511])

It is now fashionable to ridicule the mythological language of Genesis as an unworthy and outmoded explanation for origins in light of the vast understanding and precise terminology provided by modern science. Critics of Genesis mistakenly assume that it is the focal point, if not the essence, of all argument favoring a Creator.

However, it should be obvious that the Genesis account is *not* the primary basis of my personal conviction that there is a Creator. Genesis only *corroborates* the personal relationship I have with the Creator.

Lest someone misinterpret my commitment to the authority of the Genesis account, I must clarify its *scientific* significance to me. Its primary message involves *who* and *why* rather than *how* or *when* of origins. The writings are devoid of any description of the mechanisms that science seeks to understand; e.g., chemical constituents or energy sources. Yet, it clearly stipulates that origins were planned with a purpose by an intelligent Designer.

So there is my prejudice — briefly but openly laid out for all to see. Though it is a bit unnerving to allow everyone to know my intimate and private thoughts, the recounting is viewed as an investment in the future of science. It is a small price to pay — if it will crack open and expose the beliefs that feed scientific *abduction* — that fountainhead of science.

Remember Ralph Gerard's protest that *scientists* do not believe in the same sense that *religionists* do? I'm still not convinced…

Dogmatic Demand to Deny

Darwinians claim to be mystified by the Science Textbook Struggle. They cannot figure out why it occurred. It makes no sense to them. Because they believe themselves to be most rational, they attribute the struggle to *irrationalism* — to the *occult* — to the *religious* — to anything and anyone but themselves. The whole world is apparently out-of-step with true science.

But is this the case? Are Darwinians the purists they imagine themselves to be? If not — if they contributed even one *smidgen* of impurity to the conflagration, they could help themselves immensely by ferreting it out — looking at it — studying it. Since predictability is a prized objective in science, Darwinians — if they could construct a model of society — might correlate this "smidgen of impurity" in such a way as to predict the next public eruption — instead of being caught off-guard as they were in the Science Textbook Struggle.

Of course, there has been a little breast-beating by scientific elitists, to wit, that they themselves might have failed to adequately communicate "the true nature of science" to the ignorant masses. This condescending admission, however, is clearly protective of the scientific ego and fails to mask the arrogance and paternalism that prompted it. Such pseudo-humility will never allow Darwinians to predict the next challenge to their authority.

On the other hand, there is at least one clue that Darwinians could use to determine the next "flash point" of confrontation between NATURALISM and THEISM.

Consider once more the morphology of conflict between the Godly and Darwinians. Have not the 3 major collisions between these forces — the Huxley-Wilberforce debates, the Scopes Trial, and the Science Textbook Struggle — all involved *extremism*? It wasn't the lukewarm believers — *agnostics* in the NATURALISM camp and *pantheists* in the THEISM crowd — who tangled, was it? No, it was the hotheads — the *hyper*-believers — who squared off against each other. There never would have been war if it had been up to agnostics and pantheists.

But let the extremists — the *theists* and *anti-theists* — catch sight of each other, and there is an automatic explosion. Both Huxley and Wilberforce were extremists. The same can be said for Clarence Darrow and William Jennings Bryan in the Scopes Trial. What about the antagonists in the Science Textbook Struggle ? Same song, third verse.

In addition to extremism, however, there is another factor to consider if brawls are going to be *predictable*. There must be a *power imbalance*. One of the extremists must hold a decided edge over the other. In the Huxley-Wilberforce debates and the Scopes Trial, *theists* held high ground. In the Science Textbook Struggle, the power positions were reversed. *Anti-theists* had the upper hand.

Finally, the *magnitude* of the detonation will vary directly with the power differential existing between the two forces. The more extreme either or both become, the bigger the bang.

One reason that the Science Textbook Struggle lasted over 3 years, reached the highest rung on the scientific hierarchy, and became an international issue was that the dominant opponent — the *anti-theists* — had become *excessively* extreme. They — indifferently or deliberately — removed themselves so far from the theist position that a major outburst was inevitable.

(Incidentally, this focus on the NATURALISM — THEISM bipolar tension should expose the foolish conclusion Darwinians reached when they perceived their opposition coming, as Newell described it, from "emotionalism, mysticism, astrology, and other occult arts accompanied by widespread mistrust of objectivity."[512] Their opponents were *theists.* There is no evidence that Newell's aberrant elements played any role in the Science Textbook Struggle.)

It can be concluded, then, that Darwinians themselves must bear the brunt — if not the *totality* — of culpability in the Science Textbook Struggle. To blame others, they would only compound the insensitivity that provoked the struggle in the first place. With the blame placed squarely on the Darwinians, what did they *specifically* do to warrant this indictment? *They imposed their anti-theistic belief exclusively and dogmatically on science teaching.*

Think of it this way. In the Scopes Trial, *theists* held a dominant and dogmatic position and fought to retain supremacy. In the Science Textbook Struggle, *anti-theists* held a dominant and dogmatic position and fought to retain supremacy.

Can this analysis be justified? Did dominant Darwinians dogmatically demand the denial of all views but their own anti-theistic dogma? If so, how did they gain the upper hand?

Historians someday may recognize the 100th anniversary of Darwin's *The Origin of Species* as a key juncture in the power shift from theism to anti-theism. It may have been the crossover point. At the Darwinian Centennial Celebration at the University of Chicago in 1959, Sir Julian Huxley, as the keynote speaker, set the stage for revolution by his clarion call:[513]

> Darwinism removed the whole idea of God as the creator of organisms from the sphere of rational discussion. Darwin pointed out that no supernatural designer was needed; since natural selection could account for any known form of life, there was no room for a supernatural agency in its evolution.

By Darwinian fiat, God no longer existed. No classical liberalism in that decree! No toleration — or even *recognition* — of a contrasting view. In the name of *science* yet, theism was annihilated — not compromised — not reduced to insignificance — not even jeopardized, but *totally obliterated.*

Obviously, if the sole opposition no longer exists, caution may be thrown to the wind — and *was.* As in *The Wizard of Oz,* Darwinians danced with delight as they sang, "Ding dong, the Witch is dead — the wicked old Witch — the Witch is dead..."

The next year, 1960, the Biological Science Curriculum Study (BSCS) was established at the University of Colorado to produce 3 versions of a high school biology textbook that would stress Darwinism as the only acceptable dogma. That there remained a twinge of conscience about being so blatant in espousing their *belief* is clear from this account:[514]

> In the mid-1960's the censoring power of the school boards proved to be less strong than had previously been assumed. Word was spreading that BSCS biology was the "new thing," and there were community pressures on school boards to be up to date, even if a little wicked, rather than behind the tinges and fully virtuous. Once this situation was understood, nearly every newly published biology book included an explicit discussion of evolution.
>
> Despite the fact that many teachers continued to ignore what was clearly a contentious subject in their communities, many professional biologists interpreted these brave doings as "victory." In 1959, the well-known geneticist H. J. Muller had written a forceful article, "A Hundred Years without Darwinism Are Enough." But now, apparently, biology teaching was ready to emerge from its century in darkness.

Before long, Darwinian pronouncements became so bold that *Darwinism* and *science* became interchangeable terms in their minds. The anti-theistic tone of their dictums was obvious, like in Ernst Mayr's pompous but paralogistic pontification:[515]

> For the devout of past centuries, such perfection of adaptation seemed to provide irrefutable proof of the wisdom of the Creator. For the modern biologist it is evidence for the remarkable effectiveness of natural selection.

Meanwhile, nothing had changed the minds of those who believe there is a God — 94% of all Americans. While the Darwinians increasingly indulged in intellectual incest, a storm was brewing. But they remained oblivious of anyone but themselves — as though they were entranced with their own navel — as they continued to issue alienating statements in the name of science. Only a short time later, Darwinians established a new "anti-theology:"[516]

> Human "goodness" and behavior considered ethical by many societies probably are evolutionary acquisitions of man and require fostering...An ethical system that bases its premises on absolute pronouncements will not usually be acceptable to those who view human nature by evolutionary criteria.

This was the apex of ascendance. After all, one cannot go higher than to become God. History is replete, of course, with those who have attained this lofty summit in their own mind — Caesar, Napoleon, Hitler. But it is a precarious perch . . .

Some may consider this description of development of Darwinian dominion a bit myopic. If so, the London University biochemist, Denis Alexander, provides a broader historical perspective from which to view the long-standing tension between NATURALISM and THEISM:[517]

> Science and rationality have always gone hand in hand. Yet as science became more and more successful, so its roots in the Christian world-view began to be forgotten. First, in the deistic view, as we have seen, God was relegated to the position of a First Cause, the one who started the whole thing off, but who could now be comfortably relegated to the side-lines because science was quite adequate to give a fully satisfactory mechanistic explanation, thank-you-very-much. God was occasionally brought in to explain otherwise inexplicable phenomena...But for all practical purposes God was unnecessary.
>
> In one sense, the early deists were right. If a satisfactory scientific explanation could be found for something, then it was absurd to try to make God into a rival explanation, since after all he was the one who was responsible for the whole process. The eventual mistake of the deists was to make their scientific explanation autonomous and to suggest that the whole cosmic system could be explained by science without reference to God at all.
>
> The next stage was inevitable. If science could explain everything and God was the distant creator who set the whole system going, then he could be quite readily dispensed with altogether. The rationality derived from a world-view where belief in God was almost universal was kept. But the God himself who had made this rationality possible was discarded. Rationality became rationalism, the theory that man's unaided reason could by itself deal with all man's problems without any recourse to outside help.

"Motives by excess reverse their very nature," English critic Samuel Taylor Coleridge once said. This was what happened to Darwinians in the 1960's. Without doubt, they were entitled to a voice in science textbooks in 1925 — for which they fought in the Scopes Trial. Their motive to attain freedom of expression for their belief — if the Genesis belief was allowed — is one that I would endorse and defend.

But by 1969 they had replaced one tyrannical monopoly (theism) with another (anti-theism). They shut the door against *theism* as tightly as it had been shut against *anti-theism* in the Scopes Trial.

Being blind to what they had done, it was natural for Darwinians to be confused when theists played Darwinian music from the Scopes Trial back to them. "What is the reason for the revival today of such fierce fundamentalism?" *Time* magazine asked regarding the Science Textbook Struggle.[518] Answering its own question, *Time* said:

> Perhaps the cause is an increased need for spiritual security in a troubled world. It may also derive from the current distrust of science and disillusionment with rationalism.

It was obviously the wrong answer to an improper question. Americans, in a most traditional sense, were simply demanding a basic freedom that had been dogmatically denied them.

An interesting and perceptive article appeared in the *Washington Star and News* which, in contrast to *Time* magazine and most other news reports, zeroed in on the role of *dogma* in the Science Textbook Struggle. Entitled, "Science Vs. Religion: Man's Best Friend Is His Dogma," it pointed out that both scientists and religionists are afflicted by their own set of dogmas.[519]

The spark that ignited the Science Textbook Struggle in 1969 then was the realization by theists that Darwinians, in a monopolistic position of power, had dogmatically demanded denial of the Deity. It was blackmail! The insulation between the anti-theistic and theistic high-voltage poles broke down, and there was a massive flashover.

Hopefully, it will never happen again. It *can* be avoided. Both poles, however, must remain aware of and sensitive to each other. Acting as though the other has no right to exist is almost certain to provoke another altercation.

Those who most passionately call for reason over irrationality have not always practiced what they preach. *Physics Today* recently published a letter that commented on the reaction of physicists to the unorthodox ideas of Immanuel Velikovsky:[520]

> Some of Velikovsky's claims are hard to accept. Nevertheless, many of his predictions have come true, although they originally contradicted scientific belief. Eventually, some of his major predictions. may be found to be untrue, and then, and not until then, Velikovsky's theories should be rejected or amended.
>
> But the real point of this conflict stretches far beyond any argument about the validity of Velikovsky's theories. There is no question that his theories are unorthodox, that his use of historical references as a basis was unusual, and that he presented his theories in an unscientific manner. Nevertheless, why should so much anger have been vented on Velikovsky? Why are we so eager to put down divergent and unorthodox theories? Why do we wish to punish transgressors?
>
> We can calmly and rationally reject a theory after it has been proven to be incorrect. The outburst of passion from the scientific community should serve as a lasting example of how physicists should *not* behave.

Scientists, if science is to be protected from the dogmatism that would destroy it, must not only *avoid* such passion, they must also take positive steps to *correct error* in public opinion that fosters dogmatism — even when the dogmatism favors their private beliefs.

Figure 20. 14 December 1972:
The Day Darwinian Dogmatism was Defeated

The three-year Science Textbook Struggle ended decisively on 14 December 1972 when massed forces of scientific illuminati were ignored by the *unanimous* decision of the California State Board of Education to defeat Darwinian dogmatism (see Figure 20).

As earlier noted, the alarming headline in the *Los Angeles Times* on 14 December 1972 was far from accurate – and obviously intended to create public outcry. On the other hand, its accompanying story correctly reported what had happened:

> Turning aside pleas from scientists across the nation, the state Board of Education voted today to downgrade Charles Darwin's theory of evolution in new science texts for millions of California schoolchildren. The board voted 8 to 0 to adopt the books with editorial changes stressing that Darwin's theory of the origin of man is *speculative.* (emphasis added)

Had the Darwinian dogmatic demand to deny a Deity not been challenged by the Science Textbook Struggle, science in America could well have reached the anti-theistic state demanded of it in the Soviet Union.

Did that challenge *help* or *hinder* science? Is science *threatened* by exposing theistic as well as anti-theistic predispositions among scientists? Is there a higher probability of scientific breakthrough and advance in the Soviet Union due to its official dogma of anti-theism?

Historically (and ironically for anti-theism), modern science has flourished where theism is prevalent.

REVOKE, REPAIR OR REVOLT?

The *problem* is clear, but the *solution* is not.

All scientists are afflicted with beliefs that influence their work. None of their beliefs are *scientific.* Although there is a wide dispersion of such beliefs among scientists, only one extreme belief (anti-theism) is being allowed expression. All other beliefs are denied recognition. Worst of all, suppression of rival beliefs is justified as being "in defense of science."

What solutions to this problem are available? One cure might be to *revoke* all scientific work that is supported exclusively by belief — hypotheses and theories that are neither provable nor disprovable by testing and observation. Another answer could be to *repair* all scientific literature — particularly textbooks — by inserting and discussing other cosmological beliefs where anti-theism alone prevails today. A third possibility — certainly less desirable — is to *revolt* — intellectually, physically or politically — violently overthrowing the power structure ala 1776 or the French Revolution.

These three alternatives — *revocation*, *repair* or *revolution* — can be implemented several ways. The most desirable method, obviously, would be for the scientific community to *solve the problem themselves*. The record of self-policing is dismal, however. The American Medical Association doesn't do well for physicians. The American Bar Association fails miserably with attorneys. American industry seems to deliberately invite government interference in corporate life by their insensitivity. Labor unions are tyrannical. Even the Congress is spineless when it comes to disciplining its own members. So, the prospect of scientists correcting an entrenched perversion in science — regardless of its severity — is not encouraging.

Another approach would be to use *legal action* to force objectivity among scientists regarding their private beliefs. And it has been and is being done with increasing frequency. Probably the most vocal advocate for settling the issue in the courtroom is Mrs. Nell Segraves, the California housewife mentioned earlier. Her first success came in 1963 when she secured an opinion from the California Attorney General that set the stage for challenging anti-theism in science textbooks. Since that time, she has traveled extensively throughout the nation urging lawsuits and providing guidelines and information required for legal action.

Even *The Humanist*, whose anti-theism is noteworthy, admits that Mrs. Segraves has constitutional grounds for advocating legal confrontation concerning anti-theism. Her claim that science textbooks are not neutral and that neutrality is essential under the First Amendment to the Constitution of the United States has merit in the eyes of two prominent humanists, Paul Blanshard and Edd Doerr:[521]

> This struggle (is not) a constitutional struggle in the literal sense. Both sides are operating within the limits of fairly good California statutes on religious freedom and they are acting within the boundaries of the First Amendment as interpreted by the Supreme Court. Mrs. Segraves has a constitutional right to ask for an objective description of Adam and Eve in public textbooks if she can find anybody in California who can win acceptance as an objective authority in this field. The great Supreme Court decisions outlawing religious instruction and promotion in public schools — *McCollum* v. *Board of Education*, 33 U.S. 203 (1948); *Engel* v. *Vitale*, 370 U.S. 421 (1962); and the combined *Abington* v. *Schempp* and *Murray* v. *Curlett*, 374 U.S. 203 (1963) — confirm the constitutionality of objective summaries *about* religion in public classrooms and guarantee religion protection against hostile classroom attacks. As Justice Tom Clark said in the *Schempp-Murray* decision, "the State may not establish a religion of secularism in the sense of affirmatively opposing or showing hostility to religion." And he added that the public schools may treat

> religion objectively"... as part of a secular program of education ... (our society) is firmly committed to a position of neutrality"...

There is no doubt that Mrs. Segraves has come to the right forum for her attack since it is generally admitted that school boards have the legal right to determine the content of textbooks so long as they do not defy criminal laws or curtail any citizen's rights under the Constitution.

Mrs. Segraves' motives in securing constitutional rights for schoolchildren whose religious beliefs are impugned by anti-theism in science textbooks would be less suspect if she were not also in the business of marketing "creationist" materials for the science classroom. Nonetheless, lawsuits to obtain equal protection under the First Amendment for *all* religious beliefs (since anti-theism has been recognized by earlier court decisions as a religion) can be expected in the future.

A third avenue of implementing revocation or repair of private belief in science textbooks has been by *new legislation*. Among the states that have introduced new laws aimed at balancing anti-theism with theism have been Colorado, Michigan and Tennessee. In April 1973, for example, the Tennessee Senate (28 to 1) and House (54 to 15) passed a new statute that read:[522]

> Any biology textbook used for teaching in the public schools which expresses an opinion of, or relates to a theory about origins or creation of man and his world shall be prohibited from being used as a textbook in such system unless it specifically states that it is a theory as to the origin and creation of man and his world and is not represented to be scientific fact. Any textbook so used in the public education system which expresses an opinion or relates to a theory or theories shall give in the same text book and under the same subject commensurate attention to, and an equal amount of emphasis on, the origins and creation of man and his world as the same is recorded in other theories including, but not limited to, the Genesis account in the Bible...

The bill, as originally proposed, would have required equal *space* to non-evolutionary accounts of origins, but was modified to read "equal *emphasis*."[523] It should be obvious that I would have opposed this law because it called for the inclusion of the Bible's Genesis account. That account is clearly inappropriate in science textbooks.

I urged the removal of all references to Genesis in proposed science textbooks in November 1972 (see Exhibit 49). *However*, had the wording of the second sentence of the Tennessee statute stipulated that *anti-theism* (Evolution II) must be balanced by the general concept of *theism* (Creation II as shown in Figure 5), I would have supported the law.

Tennessee's new law was held to be unconstitutional by the Nashville Chancellery Court a year after its passage on grounds of placing the Biblical account above other theories, thereby "establishing a religion" and violating the Constitutional doctrine of separation of church and state.[524] This ruling was appealed in a variety of ways that resulted in rulings by the Sixth District U.S. Court of Appeals on 10 April 1975 and the Tennessee Supreme Court on 20 August 1975. Both courts concurred with the earlier ruling by the Nashville Chancellery Court.

It has been contended in this book that Darwinians, consistently and in a manner totally unscientific, demonstrate the following behavior:

1. They fail to recognize Evolution II as a *belief* (blind faith).
2. They refuse to consider any *alternative* cosmological perspective to Evolution II (illiberality).
3. They demand *exclusive adherence* to Evolution II (dogmatism).
4. They view any alternative cosmological belief as a *threat* against Evolution II (paranoia).

The latter point could not be better confirmed than in this account from *Science*, the journal of the American Association for the Advancement of Science:[525]

> A *serious threat* to the teaching of evolution in schools has been dissipated, or at least blunted, by a ruling of the U.S. Court of Appeals for the Sixth Circuit. The ruling, issued on 10 April, strikes down as unconstitutional a law passed by the state of Tennessee which requires text-books to give "equal time" to the Darwinian and biblical explanations of man's origins.
>
> The importance of the ruling transcends the boundaries of Tennessee. It possibly marks the end to a nationwide campaign by fundamentalists *to adulterate the teaching of evolution* (emphasis added).

This systematic delusion of persecution by Darwinians will probably do more to stimulate public reaction against Evolution II than any other behavior they might exhibit.

Prior to his recent death in the Soviet Union, Trofim D. Lysenko — a plant biologist who is denounced in American scientific circles for his dictatorial enforcement of a politically favorable scientific theory — acted much like Darwinians have in the Science Textbook Struggle. We can be thankful that a democratic form of government prevails here, or Darwinians would undoubtedly adopt Lysenko's tactics and thereby remove all future threats to Evolution II by declaring it official state policy. Announcements of Lysenko's death described this charlatan's behavior:[526]

> At the peak of his power, Lysenko accused his opponents of being reactionary, bourgeois scientists and "class enemies" and he sent dozens to Siberian labor camps, where many died.
>
> For 30 years, from 1934 to 1964, only one article seriously critical of Lysenko's works was allowed to appear in the Soviet Press. Now his schemes are universally recognized as wild quackery and there is suspicion that his data were falsified as well.

Though the general public may not fully understand science, they do know that scientific theories are never "threatened" by other theories. *Theories* may be *disproven* — but only *beliefs* are *threatened.* Lysenko, "a fraud, a nincompoop and a vicious careerist" as the *Los Angeles Times* described him, dominated Russian biology most of his professional life.[527] He succeeded in getting his aberrant views established as Holy Writ. His approach was alarmingly similar to the 4 points of Darwinian behavior just listed.

Lysenko resorted to faking laboratory results to bolster his edicts. A recent report in the British journal *New Scientist* says that, in a questionnaire sent to 200 scientists, 90% reported knowledge of faked scientific reports.[528] When Darwinians insist that any belief other than theirs is a *threat*, how can the public be assured that faking will not be employed? Remember Porter Kier's colleague who "saw evidence that wasn't there?"

A fourth alternative for achieving scientific objectivity about origins — if *self-policing*, *legal action*, and *legislation* all fail — is *public uprising.* In a law-abiding society, revolution is never advocated. History constantly reminds us, however, that revolt is the ultimate resort for justice.

There is nationwide discontent over school textbooks that is spearheading citizen involvement. Public debate is springing up in most unlikely forums. Scientists would be wise to recognize the root causes of such discontent — among them being the arrogant reordering of values contrary to those of taxpayers and parents.

U.S. Commissioner of Education T. H. Hell recently said:[529]

> Parents have a right to expect that the schools, in their teaching approaches and selection of instructional materials, will support the values and standards that their children are taught at home. And if the schools cannot support those values, they must at least avoid deliberate destruction of them.

> Parents have the ultimate responsibility for the upbringing of their children. The school's authority ends where it infringes on this right. *We must pay more attention to parents' values and seek their advice more frequently* (emphasis added).

The scientific barons immediately respond by saying that science is *elitist* rather than *egalitarian*. Therefore, science should not be subject to parental (layman) pressure. And they are correct — as long as they don't allow *beliefs* like Evolution II to be confused as being a part of science. Evolution II is legitimately subject to parental pressure because it can neither be proven nor disproved by the scientific method.

Without question, the preferable means of assuring that science textbooks remain objective and free from personal bias is through self-policing by the scientific community. The haunting question is whether or not scientists *will* exercise control over beliefs.

Remember, first of all, the onslaught mounted by the National Academy of Sciences, the American Association for the Advancement of Science, and nearly all other professional societies of science together with 19 Nobel laureates to defeat *any editing whatsoever* of science textbooks. They not only opposed every single one of the 219 changes made in California science textbooks, but also saw any *reduction of dogmatism* as a *threat* to science. Obviously, the 219 changes were symbolic rather than complete — both in number and in scope. For example, none of the changes address the *dominance of anti-theism*. That task still lies ahead.

Will scientific authorities wake up and cleanse textbooks on their own? The *need* for modification is clear. What will the scientists' *response* be?

If scientists fail to respond, they can expect every other alternative to be exercised. It has been alleged that the late Sir Cyril Burt — whose theory that intelligence is largely inherited had a huge impact on British education — used fakery in his results. Ian St. James-Roberts, who conducted the survey on faked laboratory results mentioned in the *New Scientist* report, said:[530]

> "My own feeling is that the data here and the implications of the Burt case together form a sufficient argument for developing more stringent controls," he said.
>
> "Perhaps the most important consideration of all is whether, *if science doesn't develop its own controls, some new revelation of the Burt sort will cause them to be imposed from outside*" (emphasis added).

What must scientists revoke or repair if they expect to avoid revolution? Three factors – *anti-theism*, the *arrogance of elitism*, and the *use of ultimatums* to express cosmological beliefs — are good starters.

Anti-theism — Ruthless Sovereign

"In science, one should never accept a metaphysical explanation if a physical explanation is possible or, indeed, *conceivable*," says George Gaylord Simpson.[531] Is that a *scientific* statement? Not really. Since science is concerned only with the material universe, it is entirely appropriate to limit science to statements about physical phenomena. However, Simpson betrays a bias that is not scientific when he suggests that the wildest unverifiable pipe dream (just so it's expressed in physical terms) is superior to any other non-demonstrable phenomenon like the intervention of a Purposer or Creator in the universe.

Simpson's "quantum evolution" (see Exhibit 48) is clearly one such wild unverifiable pipe dream that he postulates to explain what might otherwise be explained as the unrepeatable activity of a Creator. This fantasy Simpson calls a "most controversial and hypothetical" attempt to establish the existence of some process (that he never has described) that yields a "relatively rapid shift of a biotic population in disequilibrium to an equilibrium distinctly unlike an ancestral condition."

In very metaphysical language that carefully skirts mention of a Creator, Simpson says that his imagery "is believed to include circumstances that explain the mystery that hovers over the origins" of taxonomic units of relatively high rank, such as families, orders, and classes.

Simpson's right to invent such an unfounded fiction to jump over the obvious gaps in the fossil record is unquestioned. However, until he can demonstrate how it could have occurred physically, it has no more scientific validity than saying, "God visited the earth and created some new living beings." Using physical instead of metaphysical *terminology* does not entitle it to scientific status.

Why does Simpson feel so strongly biased against the *metaphysical?* Not because he is a *scientist*, you may rest assured. It is entirely a *personal* choice. As one Nobel biologist was quoted about the origin of life: "The reasonable view was to believe in spontaneous generation; the only alternative, to believe in a single primary act of supernatural creation. There is no third position..." The choice then is between a miracle by NATURALISM or one by THEISM — and it's *personal*, not *scientific*.

In the Old Testament, God issued this edict: "You may worship no other god than me...I, the Lord your God, am very possessive. I will not share your affection with any other god!"[532] Anti-theism is even more possessive. It declares that there is no *alternative* — no other *choice*. God at least admits to *other* gods!

When a person becomes anti-theistic, he must close his mind tightly against all evidence of a Creator. Since scientists pride themselves on being objective and open-minded, taking an anti-theistic stance must be painful for them.

G. Ledyard Stebbins, who played a significant role in the Science Textbook Struggle, is committed to an anti-theistic position on origins. He requested that the following statement be included in this book as an example of his thinking. Pleased to honor his request, I suggest that special attention be paid to Stebbins' logic on the role of chance:

> A teacher of evolution must often respond to two questions. The first one is "How can you imagine that the elegant design of a butterfly's wing, a spider's web, or the intricacy of the human brain could have arisen by chance?"
>
> This question is spurious and irrelevant. No evolutionist who really knows the facts believes that any elaborate design has suddenly emerged out of a state of disorder, by purely chance events. The oft-repeated statement that because mutations are random relative to the adaptive demands of the environment, therefore evolution depends upon chance, is a dangerous bit of false logic. It ignores many of the most important facts about the evolutionary process. Although mutations are the ultimate sources of variability, most of them individually make only a tiny contribution to visible evolutionary change. Does anyone believe that because the molecules of the air move at random when the air is still, the direction and force of the wind is a random affair? This analogy can be extended even further. The wind blows in a particular direction and at a certain velocity because of differences in the environment between one part of the earth and another. Similarly, organisms have evolved in certain directions and at various rates not because of the directions and rates of mutations but because of differences in the environment between one locality and another, as well as changes in the environment of the same region from one geologic epoch to another. These differences have brought about differential natural selection of genes present in their populations. The tendency of chance mutations to produce disorder is counterbalanced by the antichance factor of natural selection...
>
> The second question is "Can you imagine the existence of a design without an intelligent designer?" The evolutionist's answer to this question is twofold: first, some designs exist solely because of the symmetry of the molecules that compose them; second, the functional designs of modern animals and plants have evolved from preexisting designs. Furthermore, circumstantial evidence, which is being obtained in ever-increasing volume, strongly indicates that this succession of evolving functional designs will eventually be traced back to primordial designs that are based upon the properties of inorganic molecules.[533]

Stebbins, like most other anti-theistic scientists, also admits to holding a cosmological belief outside of science — Evolution II — that he described in a statement prepared for a panel discussion on 24 May 1973:

> If we believe that our kind and every other kind of living being got here by a miracle of special creation, then we would logically conclude that the best way out of our present difficulties would be to pray for another miracle and wait for it to happen...
>
> If, on the other hand, we recognize that our present situation has come about through the biological and cultural evolution of our own kind, and that one result of our evolution is our ability to modify the environment to suit our needs, then we have a different outlook. The evolutionist says: we can't solve our problems by doing nothing except praying to God in the same spirit as children writing letters to Santa Claus. We must analyze and understand the processes of evolution that have taken place and have brought the world to its present state. We must work together to modify these processes in such a way that we can right past wrongs, and make the world a better place for our children, grandchildren, and all humanity in the future. For myself, I can think of no more meaningful purpose in life.

Anti-theism is a pervasive belief that influences far more than the subject of origins or ultimate causation. The sociological application of anti-theism is generally described as *scientific humanism.* It requires an intelligent being to replace God in cosmology. So, as Stebbins has shown, man becomes the ultimate entity in the universe. As such, much is expected of him — maybe more than he can deliver. T. M. Kitwood puts it in this context:[534]

> What government by the proletariat is to the communist, and the coming of the kingdom of God is to the Christian, the forward surge of evolution to a better future is to many a humanist. It is the ultimate hope behind his creed.

NATURALISM, anti-theism, Evolution II, and scientific humanism are parts of each other — if not interchangeable in many contexts. Humanism is sometimes confused with *humanitarianism.* As is true with many words today, the popular usage of the two terms is exactly the opposite to their definitions. In the dictionary, *humanism* is "any system or way of thought or action concerned with the interests and ideals of people" while *humanitarianism* is "the doctrine that man's obligations are limited to the welfare of mankind and that man may perfect his own nature without the aid of divine grace."[535]

Most theists are involved in what the general public would call "humanitarian" activities — concern for the poor, nurture of the underprivileged, healing of the sick. On the other hand, no theist could subscribe to this statement by Paul Kurtz, Editor of *The Humanist*:[536]

> Although humanists share many principles, there are two basic and minimal principles which especially seem to characterize humanism. First, there is a rejection of any supernatural conception of the universe and a denial that man has a privileged place within nature. Second, there is an affirmation that ethical values are human and have no meaning independent of human experience; thus humanism is an ethical philosophy in which man is central.

Likewise, the anti-theistic nature of humanism is clearly stipulated in "A Humanist Manifesto" published in 1933 and subsequently elaborated in a second manifesto in 1973. The original manifesto had 15 points, most of which interrelated anti-theism and Evolution II with the term *humanism*. Some of these points (in order of appearance in the Manifesto are relevant to this point:[537]

1. Religious humanists regard the universe as self-existing and not created.

2. Humanism believes that man is a part of nature and that he has emerged as the result of a continuous process.

3. Holding an organic view of life, humanists find that the traditional dualism of mind and body must be rejected.

5. Humanism asserts that nature of the universe depicted by modern science makes unacceptable any supernatural or cosmic guarantees of human values...

6. We are convinced that the time has passed for theism...

8. Religious humanism considers the complete realization of human personality to be the end of man's life and seeks its development and fulfillment in the here and now. . .

9. In place of the old attitudes involved in worship and prayer the humanist finds his religious emotions expressed in a heightened sense of personal life and in a cooperative effort to promote social well-being.

10. It follows that there will be no uniquely religious emotions and attitudes of the kind hitherto associated with belief in the supernatural.

Thus, when prominent scientists like Hudson Hoagland, former President of the American Academy of Arts and Sciences, urge support of humanist manifestos, one must not take their statements to be *scientific*. They are statements of personal persuasion or *belief*:[538]

> The supernatural world created by primitive man (was) extended by medieval theologians into religious dogmas, resulting in views of the nature of man quite at variance with what we know from science. . . The biological sciences today furnish a magnificent perspective of life over the past 300 million years, beginning with the spontaneous emergence of self-producing molecules in the pre-Cambrian slime, and evolution thence by natural selection to man, with his imagination and achievements.

Even classical *socialism*, that today enjoys a rather pervasive influence among Western intellectuals, is founded on the premise of scientific humanism. As Peter L. Berger, Rutgers sociology professor, says, "The socialist program is based on all the standard assumptions of modernity — history as progress, the perfectibility of man, scientific reason is the great liberator from illusion, and man's ability to overcome all or nearly all of his afflictions by taking rational control of his destiny."[539]

The teasing possibility that man can cure all ills through rational thought is what fuels scientific humanism. Stebbins believes desperately in this thesis as does Kurtz, Hoagland and all other humanists. Yet the following statement expresses the prospects of success succinctly:[540]

> The gospel of Humanism invites men to make the best of a bad job. The world is ultimately pointless. All we can do is to try to give it a few temporary points before we pass into the abyss of nothingness. The ancients said: Eat, drink, and be merry, for tomorrow we die. The Humanism of the sixties has become more sophisticated. It allows man to be serious in his merry-making. The image is new, but the basic idea is the same. Life has no meaning than that which we give it ourselves, and no other end by death. This in itself is not an objection to Humanism. For this view of life is perhaps the best we can do, if we assume that God does not exist and that he does not care for the world. It is better to face the facts and recognize that this is the only valid alternative to Christianity than to pretend that there are various intermediate options open to as.

Thus ageless tension between NATURALISM and THEISM underlies the sporadic outbursts — such as the Science Textbook Struggle — that bedevil science. Failure to not only *acknowledge* these two diametric poles of belief but also to *admit that they influence science* (through scientists) will continue to produce irrational conflict. The anti-theism of NATURALISM is a ruthless sovereign — unwilling to give up the throne upon which it temporarily sits. It must either be *revoked* or *repaired* in science or *revolution* is inevitable.

Elixir of Elitism

"Science is hierarchial and elitist. For the same reason that automobile mechanics are not permitted to carry out abdominal surgery; we must not permit individuals who are illiterate in the biological sciences to insert their opinions and prejudices into science textbooks," Thomas H. Jukes wrote during the Science Textbook Struggle (see Exhibit 51). And he is correct. In a specialized world, specialists should surely specialize in their specialties. No argument.

On the other hand, there is also timeless wisdom in Hans Christian Andersen's allegory, "The Emperor's New Clothes," in which the least-qualified — a small child — alone spoke the truth. The "hierarchial elitists" in *that* story somehow failed. Of course, it was only a fairy tale...

Jukes, and all others who think as he does, allude to a rather insidious, unspoken but fearsome premise — that they have vast superiority over all other people. Why was it "an *intellectual* affront" to Jukes (in Exhibit 51) that I was serving on the Curriculum Development Commission?

On what *basis* did he "object to the presence of Mr. Grose" on the Commission? Maybe the State Board of Education slipped up and violated the law when they appointed me. Is that what upset Jukes? Not at all.

Well, maybe he considered me rude, a poor dresser, or dull of mind. Even if true, that is not *really* what irritated Jukes. He was incensed because *he found me insufferably inferior to himself.*

To understand Jukes' vexation, we must look closely at how he described science. The two adjectives he used are crucial to comprehending his frustration with my presence on the Curriculum Development Commission.

HIERARCHIAL	"Having an organization of persons arranged one above the other according to grade, order or class."
ELITIST	"Advocating or favoring rule by an elite — the choice or distinguished ones, the best people."

Just like people, science has both a *public* and a *private* image. *Publicly*, science appears open-minded, objective, receptive to all propositions (that are hopefully verifiable but not necessarily so; e.g., natural selection), and tolerant of all types of proposers. The *private* side of science is almost a mirror-image of the public one. Jukes dropped the public mask enough to let us glimpse the private side.

The international journal of science *Nature* published an editorial on the Science Textbook Struggle.[541] A Scottish scientist, responding to that editorial, described the *private* world of science thus:[542]

> There are more anti-Darwinists in British universities than you seem to realize. Among them is a friend of mine who holds a chair in a department of pure science "in a field bearing on the evolutionary question," to use your phrase. If his friends ask why he keeps silent about his unorthodox views, he replies in words very like those used recently in another connexion by Professor Ian Roxburgh:
>
> "Science is only concerned with the truth, and as the way to truth is by conjecture and refutation, why don't they (orthodox scientists) listen? Science is not like that at all. There is a powerful establishment and a belief system. There are power seekers and career men, and if someone challenges the establishment he should not expect a sympathetic hearing."
>
> The majority of biologists accept the prevailing views uncritically — just as a great many competent Russian biologists were once brainwashed into accepting Lysenko's quackery. Others have thought for themselves and have come to realize the flaws in contemporary Darwinism. But for them to speak out would be to invite ridicule, and probably ruin their careers. Can you blame them for keeping silent?

Scientists are human — and subject to all the frailties of humanity. Just because they work in science, they are not freed from bigotry and inordinate pride. In no way can Jukes' outburst be justified as being in the best interest of science. The normal methods of science have ample provision for eliminating the spurious. There is no need or excuse for using suppression, threat and personal attack as Jukes did.

Whenever scientists revert to imperialism — stressing *hierarchial elitism*, here are some interesting observations that could be pointed out to them:

1. They obviously consider *themselves* to be the elite — otherwise they wouldn't be raising elitism as a topic.
2. While they expect the general public to *support* their scientific work, they obviously want no *interference* from their supporters. (Fortunately, we have maintained civilian control over military elitists and averted monarchy — inherited political elitists — even though scientific elitists think themselves above answerability.)
3. Elitism, by definition, is "minority rule" — only the *best* should rule. Therefore, scientific elitists detest and denounce the application of egalitarianism to science. Yet, they hypocritically

use *majority* rule when they feel threatened — speaking as they did during the Science Textbook Struggle of the "vast majority of scientists" in terms of a bludgeon to subdue anyone disagreeing with them. It is an *elite* "majority rule" of course — and the rest of us are *unelites*.

4. What is truly "elitist" about science? Is it not the *scientific laws* — those postulates that have withstood all types of challenge without yielding an exception — that rule science? Or is it, as Jukes seems to believe, *scientists* themselves who form the elite — looking down their noses at everyone they believe to hold lesser rank or station in the hierarchy?

The fourth observation above, if answered in favor of *scientists* instead of *scientific laws*, immediately raises the specter of tyranny. Elitism of the past, whether religious, political, social, or economic — has consistently led to tyranny. One example of scientific tyranny illustrates this danger.

I have a long-standing interest and involvement in aviation and space technology. My fondness for studying aircraft, missiles and spacecraft often draws me to the Smithsonian Institution during the many weeks I spend in Washington D.C. each year. In numerous visits, I have spent literally hours studying my favorite exhibit there — the Wright brothers "Kitty Hawk" aeroplane. It has hung suspended inside the main entrance to the Arts and Industries Building until very recently. Since our Nation's bicentennial birthday on 4 July 1976, the "Kitty Hawk" has occupied a permanent place in the beautiful new National Air and Space Museum of the Smithsonian.

There is something about the "Kitty Hawk" that you will never learn by visiting the Smithsonian, however. It involves the tyranny of scientific elitism.

Very few people know that the "Kitty Hawk" was denied a place at the Smithsonian Institution until after Orville Wright, one of the co-inventors and the first man in history to fly, died on 30 January 1948. From 1928 to 1948, the "Kitty Hawk" was displayed in the Science Museum in South Kensington, London. Why? Because of a dispute over the scientific aspect of the first flight by man.

Samuel P. Langley, while Director and Secretary of the Smithsonian, had unsuccessfully attempted to build a man-carrying flying machine. After several failures, the Wright brothers successfully flew at Kitty Hawk, North Carolina in 1903. Shortly thereafter, Langley died and was succeeded at the Smithsonian by Charles B. Wolcott.

Under Wolcott's supervision, the Smithsonian went so far as to issue false and misleading statements about who had discovered or produced the inventions necessary for flight. From 1910 until Wolcott's death in 1927, the Smithsonian did everything it could to discredit the Wright brothers and favor Langley as the inventor of flying. The story of political intrigue that involved Chief Justice William Howard Taft, Senators, Congressmen and other directors of the Smithsonian Institution makes very interesting reading.[543]

The Smithsonian tampered with data, modified Langley's 1914 model of the aeroplane, and issued annual reports from 1915 through 1918 containing untruths that would make it look as though the Wright brothers had not been the originators of a successful flying machine:[544]

> Altogether here had been something probably unique in scientific procedure. A test was made purporting to determine if the original Langley plane was capable of flight; but the test was not made with the machine as designed and built by Langley, nor with an exact copy of it. No disinterested official observer was present. Misstatements were published about the results, and no information was furnished, regarding the changes made, to enable anyone to learn the truth. To have made one more honest test of the Langley plane that had immediately crashed each time it was launched over the Potomac would have been permissible. But for a scientific institution officially to distort scientific facts, and in collaboration with a man who stood to gain financially by what he was doing, has been called worse than scandalous.
>
> After the Langley machine had been restored as nearly as possible to its original state, it was placed on exhibition by the Smithsonian. Soon afterward it bore a label that falsely proclaimed it to be "the first man-carrying aeroplane in the history of the world capable of sustained free flight"...
>
> By omitting from its published reports at the time and for many years afterward, the facts about the changes in the Langley machine, the Smithsonian Institution succeeded in deluding the public.

The Science Textbook Struggle witnessed many exhibits of scientific elitism beside Jukes' outburst. The resolution signed by 19 Nobel laureates is another one — since it was clearly based on *prestige* (the mark of elitism) rather than on *expertise*.

Of course, the pompous interrogation of me by Richard M. Lemmon, following the 9 November 1972 public hearing, is yet another classic manifestation of elitism. Lemmon must also be given credit for his *persistence* in trying to enforce elitism. He not only took it upon himself to apprise the State Department of Education of detailed reasons why I was unqualified to serve on the Curriculum Development Commission (Exhibit 50). He even organized a letter-writing campaign that triggered Jukes, among others, to vent his spleen (Exhibit 51).

But Lemmon, in his overwrought passion to destroy all noncompliance to elitist demands, continued to carry his grudge long after the Science Textbook Struggle had ended. And he got carried away — to a point that may be his undoing.

Publishers require authors to secure permission to publish private correspondence in a book — even for letters addressed to the author. Because I wished to reproduce, verbatim in Exhibit 50, two of Lemmon's letters to me, I wrote to him requesting his permission. I mentioned that I had a contract with Macmillan Publishing Company for the book. That presented Lemmon with a temptation he could not withstand — a chance to use his elitist status to kill my book.

Using precisely the same tactics that astronomers Harlow Shapeley and Fred Whipple used against Immanuel Velikovsky in 1950 — and against the *same publisher*, Lemmon wrote to Macmillan in an obvious attempt to break my contractual relationship. Here's what he said:

> My reason for writing to you is to express my disappointment that you have chosen Mr. Grose to author a contemplated book on the recent California textbook controversy. Mr. Grose was a prime mover in the effort to introduce religious natters into the State's science textbooks, and I can hardly expect anything but a biased account from him.
>
> The essence of what Mr. Grose achieved, temporarily, in California was, in my opinion, the debasement of some of the language of the State's "Science Framework." I enclose a copy of two versions of page 106 of that "Framework." On the left are delineated three very bad paragraphs, all written by Mr. Grose. Posing as some kind of a scientist, and manipulating the religious-political pressures then on the State Board of Education, he succeeded in getting the Board, in 1972, to adopt paragraphs 2, 3, and 4 of that page into the "Framework". It was a long step toward getting California the title of "Tennessee-West." You note that the first two paragraphs were adopted over the objections of the State Advisory Committee on Science Education. Mr. Grose's achievement was a political victory over the Board of Education's own advisors on what should be, and what should not be, science education.
>
> Fortunately, it was only a temporary victory. Faced with the volume of complaints from America's scholarly scientific societies, the State Board of Education threw out Mr. Grose's language in March of 1974. The new version is shown on the right side of the enclosure.
>
> I am disappointed in the Macmillan Publishing Company's plans for a book on the California textbook controversy. I feel that, like Mr. Grose's earlier efforts, it will be another burden on California teachers who wish to teach science objectively.

Academicians like Lemmon can be ignorant of subjects outside their narrow specialty. Perhaps he was unaware that it is against the law to commit "tortious interference with a contractual relationship." On the other hand, he may have deliberately violated the law. Either way, Lemmon transgressed the law.

First, Lemmon wrote his letter to Macmillan on "University of California, Berkeley" letterhead, implying that it represented the policy of that organization. Second, he had an employee of the University type his letter. As a taxpayer, I pay both Lemmon's salary and that of his secretary. Taxpayers normally do not intend that their taxes be spent on illegal attempts to break contractual relationships, especially ones in which they are involved.

Lemmon's third offense was the most serious. His letter contained 2 direct errors of record (I was the author of only 2, not 3 paragraphs, and those paragraphs were adopted in 1970, not 1972) as well as 4 allegations of intent that are defamatory. Building on these false statements, he used his title and affiliation to attempt to force Macmillan to break our contract and not publish this book.

When I received a copy of Lemmon's letter, I called the California Attorney General's Office in Sacramento, and it was confirmed that Lemmon's letter, indeed, constituted a violation of the law. Legal action against Lemmon is pending.

The humanist's vision and faith in the perfectibility of man through mankind's own control of evolution as the only hope for the universe was described earlier by Stebbins. H. J. Muller concurs with Stebbins:[545]

> I believe, (as) an outgrowth of the thesis of modern humanism, as well as of the study of evolution, that the primary job for man is to promote his own welfare and advancement, both that of his members considered individually and that of the all-inclusive group in due awareness of the world as it is, and on the basis of a naturalistic, scientific ethic ...Through the unprecedented faculty of long-range foresight, jointly serviced and exercised by us, we can, in securing and advancing our position, increasingly avoid the missteps of blind nature, circumvent its cruelties, reform our own natures, and enhance our own values.

Who is "we" in Muller's grand scheme? Knowing what has occurred through scientific elitism in the past, I fear that Muller, along with all other scientific humanists, does not intend to allow mankind's future control of evolution to be determined by a *democratic* process. Rather, the *scientific elite* will decide for us where evolution should take us all — since only they possess the necessary knowledge. How do you get to be an elite?

The elixir of elitism is a heady brew indeed. But it has no place in science — of all places, in the preparation of science textbooks for young children. The California State Board of Education successfully withstood, during the Science Textbook Struggle, those who had imbibed of this intoxicant. However, as scientific elitism continues to flourish, "eternal vigilance" must be our watchword...

Ultimates by Ultimatum

The American Chemical Society (ACS), one of the oldest and largest professional societies in science, has about 120,000 members. Patrick P. McCurdy, Editor of the ACS journal — *Chemical & Engineering News*, published a startling editorial to coincide with the 165th National ACS meeting. After 164 previous national meetings, the need for McCurdy's message seemed *incredible*! He called on scientists to have an *open mind*:[546]

> Some nonscientists seem to feel that scientists, almost by definition, are endowed with an unusual amount of this quality. After all, isn't the "job" of a scientist to seek truth objectively? The truth is, though, that scientists can be just as cerebrally opaque as anyone else.
>
> Scientific theories can be useful devices. As a frame of reference to help order our thinking and a model against which we can measure and evaluate apparent reality, they are fine. But let's not forget that they are indeed just theories. Completely captive to your theories, you cannot be open-minded.

McCurdy then discussed two major issues in the scientific world that were apparently disturbing members of the ACS — the Science Textbook Struggle in California and the renewed interest in the theories of Immanuel Velikovsky. Regarding the California issue, McCurdy provocatively asked why scientists become so defensive about evolution, and in particular, why evolutionists refuse to acknowledge the contradictions and disagreements both in and outside their own ranks. Calling once more for an open mind, he reasoned:[547]

> Do we *really* know? We sometimes like to pretend that we've just about figured it all out. Yet we still cannot fully and adequately explain such fundamental phenomena as gravity, magnetism, light, even electricity. All we have really done is work out certain rules that work in a perhaps limited and local way. But they don't explain ultimate cause...

I found his viewpoint refreshingly candid — and quite atypical of those I had confronted in the Science Textbook Struggle. But McCurdy had apparently not reckoned with one ACS member in California — Richard S. Lemmon. Infuriated by the editorial, Lemmon railed that it had effectively belittled “the efforts that have been made to keep superstition, masquerading as science, out of science classes.”[548]

Recounting with obvious pride his personal crusade to keep “religious dogmas” out of public school science, Lemmon chided *Chemical & Engineering News* for failing to publish the rash of resolutions by professional science societies that had been mobilized in the final days of the Science Textbook Struggle. “Mr. McCurdy seems to think that evolution is 'just a theory.' I find this appalling,” he cried.

Editors always have the last word. McCurdy did, too. And it was a *classic*:[549]

> Reader Lemmon apparently feels that our scientific framework is completely in place and beyond any question. The editor seriously questions his proposition. But in any event, an open mind is essential to scientific thought. As for teaching religious theory in science classes, *which the editorial did not deal with*, the editor's personal feeling is that it should not (emphasis added).

McCurdy's analysis struck at the heart of Lemmon's rationale. Lemmon not only believes that *ultimates* have already been established by science but that those ultimates must be enforced by *ultimatum* from the scientific elite. Precisely how does his position differ from the medieval theological position that Galileo faced?

A social science text recently adopted in California for eighth grade students describes "the scientific revolution." Contrary to Lemmon's rigidity of thought, this text discusses *pros* and *cons* of the effect of Darwinism on man's view of the universe.[550] The teacher's manual for this text rightly points out that Darwinism fails to answer many basic questions that all people ask.

Among the yet-unanswered enigmas of Darwinian evolution listed and discussed in the book are these:[551]

1. Biological evolution does not account for a first cause.
2. Evolutionists have not established a link between the higher apes and *Homo sapiens*.
3. Evolution appears to be haphazard, denying the existence of divine order in the universe.
4. Survival of species can be explained through the *cooperation* within and between species rather than through the *struggle* that Darwin envisioned.
5. There is no proof that different major species are related — rather they appear to be the result of parallel development in the fossil record.

6. Biological evolution does not specifically account for a moral sense.

Most of these points involve *ultimates* — fundamental causative premises that are unknowable and unprovable by scientific methodology. They are not only unresolved *now*. They will *remain* unanswered — at least as far as science is concerned.

Shakespeare once said, "Will your answer serve to fit all questions?" Obviously, Darwinism is not and never will be an answer that fits all questions. In fact, it raises many unique questions of itself — not the least of which is "social" Darwinism. This same social science text discusses Darwin's "survival of the fittest" in the social context:[552]

> It seems to describe a brutal, extravagant process of development of living matter. Man himself seems to be a "chance" result of millions of coincidences. All life is a "struggle for survival," not only between species, but even more between varieties of species and between individuals of a species. In this endless struggle, the "fittest" will survive. Thus it could be argued that Darwin's theory would tend to make men ruthless and unethical in their dealings with one another. For example, cut-throat tactics would be acceptable in business or political affairs if they ensured survival or success — in fact, any means, so long as they secured the desired end, would be approved. The doctrine of "survival of the fittest" seemed to make everyone view his fellow man as a rival in the struggle to survive — a rival who must be subdued, if not destroyed...
>
> Many sociologists, political scientists, politicians, economists, and businessmen used Darwin's arguments to defend ruthless economic competition, the elimination of the unfit by allowing "nature to take its course," the use of war and conquest to prove the superiority of this or that race or nation, and similar brutal acts. This school of thought came to be known as "social Darwinism."

It cannot be stated too often that science is *probabilistic* — and it is *never* certain — that *all* its laws are constantly subject to challenge and revision. Thus, it would be improper to expect Darwinism to answer all questions. The fact that it doesn't provide explanations at critical junctures is not an indictment.

On the other hand, the *certainty* — the *closing* of all debate — the adamant *insistence* that no other idea or hypothesis can be supported in science, that elitists like Lemmon exhibit is totally unjustifiable. Once a hypothesis, theory or law is postulated in science, the true scientist is the one seeking to find an *exception* — not to shut the book and demand universal belief by ultimatum.

Therefore, the scientific elite — if they are the ruling class — should lead a crusade to *elicit exceptions* to all theories — including Darwinism.

Instead, these elites have done just the *opposite.* They are a "protectionist platoon," viewing all challenges to Darwinism as *threats!*

One of the most interesting phenomena in the Science Textbook Struggle was the frequent reference, always parenthetical, to "the many unsolved details of Darwinism" by the scientific elite. It was the universal caveat. By sprinkling it throughout their pontifications on the invincibility of Darwinism, they could retain a degree of apparent humility, avoid appearance of intolerable certainty, and in general be gentlemanly about their dogmatic dictums.

"In spite of nearly a century of work and discussion there is still no unanimity in regard to the details of the means of evolution," the noted geneticist Richard B. Goldschmidt said. But it's only a parenthetical statement because, in the same breath, he continues: "This statement should not be misunderstood; all biologists who have mastered the available facts are agreed upon the main points, the big outlines of the explanation of evolution...The differences in opinion mentioned...refer only to the technical details within this framework."[553]

"If some details (of Darwin's thesis) are still in dispute, the main thesis is not," humanists Paul Blanshard and Edd Doerr write.[554] The editor of *Nature* joins them:[555]

> Although there are many important questions about the details of the evolutionary process to be understood, especially at the molecular level, Darwinism occupies a place in science at least as strong as that of Newton's laws — no doubt there are many reinterpretations and refinements to come, but nobody in his senses can deny that the doctrine of evolution is an exceeding powerful means of relating such a variety of phenomena that it deserves to be called the truth.

Even the resolution by the Academic Senate of the University of California (Exhibit 43) that was solicited to pressure the State Board of Education into bowing to scientific elitism says: "There are many facets of the evolution picture that are not yet thoroughly understood . . . But virtually all biological scientists are agreed on the broad features of the theory of evolution of life forms, the evidence for which is completely overwhelming."

Is it insolent or irrelevant to ask all these illustrious elitists to *identify* the "many details of evolution that are not understood?" Is it bad for science if the *public* is allowed to know these trade secrets? Is the "right to know" limited to the scientific elite? If not, maybe the National Academy of Sciences could commission a study to identify and discuss the unknowns of Darwinism.

Since these critical "building blocks" from which the wall of Darwinism is constructed are acknowledged but mysteriously kept secret, any science textbook that proposes Darwinism (even Evolution I) as a scientific theory cannot be objective. Objectivity awaits full disclosure of uncertainties, unknowns, unsustainables, and undeterminables.

William Robin Thompson was ranked, at his death in 1972, as one of the world's greatest authorities on entomology — the branch of zoology that deals with insects. He was one of the pioneers of the scientific study of entomology and also studied the mathematical theory of population growth in connection with parasite-host relationships and its bearing on population control. His extensive research on the taxonomy and systematics of tachnids continued until his death.[556]

For many years, Thompson was Director of the Commonwealth Institute of Biological Control in Ottawa. Many scientific honors were bestowed on Thompson during his working life. He was a Fellow of both the Royal Society and the Canadian Royal Society. An author of note, he was also responsible for the *Catalogue of the Parasites and Predators of Insect Pests*, an invaluable reference work to the literature of biological control.

However, Thompson may be best remembered for an honor he accepted with reservation. When Everyman's Library decided to produce their latest edition of Darwin's *The Origin of Species*, they invited Thompson to write an introduction replacing the one prepared a quarter of a century earlier by Sir Arthur Keith. Thompson's response was guarded:[557]

> I felt extremely hesitant to accept the invitation. I admire, as all biologists must, the immense scientific labours of Charles Darwin and his lifelong, single-hearted devotion to his theory of evolution . . . But I am not satisfied that Darwin proved his point or that his influence is scientific and public thinking has been beneficial.
>
> I therefore felt obliged to explain to the editors of the Everyman's Library, that my introduction would be very different from that of Sir Arthur Keith . . . I am of course well aware that my views will be regarded by many biologists as heretical and reactionary. However, I happen to believe that in science heresy: is a virtue and reaction often a necessity, and that in no field of science are heresy and reaction more desirable than in evolutionary theory.

Thompson then proceeded to write one of the finest analyses of Darwin's theory that I have ever read. If I had to recommend one single, brief critique of Darwin's work, it would be Thompson's Introduction. This 18-page examination of those "many details of evolution that are not yet understood" that everyone admits but won't identify should be required reading for all biology students, if not every science scholar.

It is time to question the previously unquestioned "details." If the *building blocks* don't exist, maybe the *wall* doesn't either!

The scientific hierarchy, then, insists that they alone have the *ultimate* truth about origins. They enforce their "truth" by *ultimatum*. Will this dysfunction of science be *revoked*, *repaired* or *overthrown by revolution*? Is there any alternative to the unfortunate specter of another uprising by noble Nobel notables to "defend the faith?" Can science tolerate a free marketplace of ideas? I hope so.

Let all — theists and anti-theists alike — pursue objectivity together.

ONWARD TO OBJECTIVITY

Long before the dawn of modern science — in the 5th century B.C., the Greek tragic dramatist Euripedes said, "In a case of dissension, never dare to judge till you've heard the other side." Things haven't changed in 2500 years. It's still good advice.

Since both NATURALISM and THEISM have in the past occupied the dominant or majority position of power in science, it should be obvious that the occupation cannot be permanent. As George Bernard Shaw opined, "Man can climb to the highest summits, but he cannot dwell there long." Therefore, it is in the interest of both viewpoints to respect the right of the other to be heard.

In particular, the *ultimate* right of anti-theists (Darwinians) — who today hold the superior status — to express their belief about origins in a scientific context is directly linked to their recognition of the right of their opposition — theists — to also be heard. Why? Because *neither viewpoint can be proven or disproven by scientific methodology.*

Open, reasoned dialogue — rather than rhetoric — between anti-theists and theists is essential if both are to respect each other's right to be heard. Both must seek dialogue.

The first step for *theists* toward dialogue appears to be a recognition that their theistic predisposition can be an *additional* rather than an *alternative* perspective to anti-theism. John A. McIntyre, Professor of Physics at Texas A&M University, has pointed out that evidence for evolution — the fossil record — should be regarded as pages of history. They are, in one sense, no different than the pages of history written by ancient historians. Neither give a complete enough picture that, by setting up the original conditions described, we could guarantee the events that followed would reproduce the events of history. Just as there is both secular and Biblical history, the fossil record is but another view of what has happened in the past. It *supplements* rather than *replaces* the Biblical account. Here's how McIntyre develops the parallelism:[558]

> We find in the Bible a history in which God plays the leading role. Yet, the Biblical history in no way disagrees with secular history. The facts of history are correct; however, a new dimension is added showing God at work as history unfolds. This feature of the Bible appears particularly clearly in one of the earliest historical accounts, the Book of Judges (4:1-3). The secular historian would say that the people of Israel were defeated because their enemies had mastered the art of forging iron. The Biblical writer acknowledges this too but also includes God's activity (6:1, 5) . . .
>
> Returning, then, to the history of life described by the evolutionary account, the Christian can also incorporate the Biblical account of God's activity in the development of life as described in the early chapters of Genesis. The evolutionist may strongly oppose the Biblical account since his materialist philosophy will not tolerate God; however, the Christian should have no problem with the evolutionist's account (in so far as it restricts itself to the facts), any more than the Christian is troubled by secular history.

The first stop for *anti-theists* toward open, reasoned dialogue seems to be twofold: (1) to recognize and declare their own cosmological presuppositions and (2) to gain an up-to-date understanding of theistic viewpoints. Both parts of this step are in the best tradition of liberalism that most anti-theists prize highly.

Admittedly, an up-to-date understanding of theistic viewpoints is not readily obtained. Anti-theists can legitimately claim that the loose term "creationism" has been poorly defined and articulated in recent years. In a sense, the only thinking available to anti-theists has been the outmoded arguments of Bishop Samuel Wilberforce and William Jennings Bryan. By that set of standards, it is easy to understand how Darwinists underestimated any opposition during the Science Textbook Struggle.

Anti-theists can also point to the nearly hopeless task of refuting every pseudo-scientific whim that masquerades as science.[559] But seriously studying theism's influence on science is not to be confused with debunking parascience cultism. If either theists or anti-theists consider each other less than equals, the type of dialogue required to attain objectivity regarding origins will not be obtained.

J. T. Fraser of the Department of Physics and Astronomy at Michigan State University has pinpointed "the almost complete absence of unhurried, unfrivolous, formal dialogue between the sciences and humanities about man's many ways of perceiving reality." Fraser's solution to this problem is to propose a non-trivial common theme about which men of different professional backgrounds may speak with confidence without transgressing the limits of their fields of specialization.[560] The subject of origins could be such a non-trivial common theme *provided* both parties respect each other.

Since the anti-theist's position on origins is well documented and more available for theist review than is the converse, desired dialogue could be accelerated if some references on current theistic thought about origins become known. To that end, some suggested writings are offered.

A background study on scientific presuppositions held by theists might be a good departure point. W. Stanford Reid of the University of Guelph (Ontario) has prepared such a paper.[561] Reid points out that Christians (a specific subset of theists) possess two levels of presuppositions:

1. *Religious* presuppositions that are constant, based on revelation rather than observation, and free from historical development.
2. *Phenomenal* presuppositions that are in a *perpetual state of flux or change* due to the growth of knowledge of the universe and increased understanding of man, whether as scientist or as object of investigation.

It may well be that Darwinians (a specific subset of anti-theists) also possess two similar levels of presuppositions. Their belief in the "amoeba-to-man continuum" is undoubtedly analogous to a *religious* belief that is invariant. On the other hand, they may also possess presuppositions about evolution that are *phenomenal* — frequently revised clues to phenomena uncovered in scientific research.

Part of the anti-theistic misunderstanding of the theistic position that emerged in the Science Textbook Struggle was centered about the role that the Bible — specifically the Book of Genesis — plays in theistic presupposition. Surprising as it may be to many anti-theists, theistic scientists hold a wide variety of theories based on Genesis. Some more prevalent ones (with overly simplified descriptions) include:[562]

1. *Progressive Creative Catastrophism Theory* — Presumes a gap of an indefinite but vast expanse of time between Genesis 1:1 and Genesis 1:2.
2. *Day-Age Catastrophism Theory* — The "days" of Genesis literally mean geologic "ages."
3. *Progressive Creationism Theory* — Proposes that God in a continual process, created suddenly at different times throughout the geological ages.

4. *Alternate Day-Age Theory* — Each of the Genesis "days" were actually 24-hour periods but were separated by vast geological ages.
5. *Eden-Only Theory* — The Genesis account has nothing to do with the rest of the earth except for the Garden of Eden.
6. *Concurrent (Overlapping) Ages Theory* — A total independence of time so that the Genesis "days" could refer to creative acts overlapping in occurrence without reference to time.
7. *Split Week Theory* — Each of the Genesis "days" may have varied in length — from a moment to millions of years.
8. *Revelation Day Theory* — The Genesis "days" refer to days in Moses' life in which he received God's revelation of the creation.

As can be seen by the diversity of these, interpretations, the important point to theists is not the mechanistic details that can be either literally or figuratively deduced from Genesis. For one thing, Genesis is significant only to those of Judeo-Christian persuasion, and they are only one type of theist. The crux for *all* theists, however, is centered in the *existence* and *involvement* of a Creator as an initiating and sustaining force in the universe.

On the other hand, since the formation of hypotheses in the mind of the scientist *precedes* his observation of data (as Peirce has shown in his development of *abduction*), Genesis provides, for Judeo-Christian theists, a predictive model of what might be expected as the scientist observes the physical universe. In the same sense that Darwin offered a possible unifying theme for living things, Genesis does also.

Thompson summarized the essence of Darwin's model this way:[563]

> The view that natural selection, leading to the survival of the fittest, in populations of individuals of varying characteristics and competing among themselves, has produced in the course of geological time gradual transformations leading from a simple primitive organism to the highest forms of life, without the intervention of any directive agency or force.

A similar but contrasting sketch or outline is provided in Genesis. Specific deductions can be made that should not only correlate scientific observations but also allow a scientist to predict what he should find in the future. Among the premises or presuppositions that a theistic scientist might deduce from Genesis could be the following (for which I acknowledge the insight of James D. Sales):

1. It would be predicted that the ultimate reality in nature is *singular* rather than diverse — we should observe a *uni*verse not a *multi*verse.

2. It would be predicted that the universe is *not self-ordering* — we should observe the universe proceeding from a state of order into disorder rather than the inverse.
3. It would be predicted that the universe has not always existed but that it had a *beginning.*
4. It would be predicted that at one point in time, the earth was *uninhabitable.*
5. It would be predicted that light preceded life.
6. It would be predicted that plants came before mankind.
7. It would be predicted that permanent clear *distinctions* exist between living beings — we should not observe an unbroken continuum of living things.
8. It would be predicted that living things would reproduce *similar* creatures.
9. It would be predicted that a principle of "*life producing life*" would be observed.
10. It would be predicted that man's physical body would be composed of materials found in the *earth's crust.*
11. It would be predicted that man shares with animals the same principle of *life* in a physical body.
12. It would be predicted that man would be not only the climax but the *completion* of created life.
13. It would be predicted that mankind would be *unified*, possessing a common origin.
14. It would be predicted that mankind would be *distinctly different* from other animal life — we should observe man to be rational, moral, and a being who has certain freedoms as well as a spiritual character greater than matter.
15. It would be predicted that mankind would be *religious*, with the desire to worship something outside of itself.
16. It would be predicted that mankind would have *dominion* over all other living things on the earth.

Anti-theists could object to the *source* of these predictions just as some theists have irrationally objected to Darwin as the author of evolution. However, the important element in a hypothesis used for predictive purposes in science is not its source but how well it correlates what we observe. As Weisberg says:[564]

> There are many cases in physics where correct laws were formulated on an ad hoc basis, or even on incorrect grounds. Examples of the former are the Lorentz equations, Planck's radiation theory, and perhaps Schrodinger's equation. A causative basis is certainly an advantage for a theory, but its absence does not necessarily refute a theory. *The validity of a theory is based on how well it correlates events, and how well it predicts future findings* (emphasis added).

The 16 candidates for predictions from Genesis have continued to be useful — from the rise of modern science until the present. They not only were the presuppositions of the founding fathers of science but serve that purpose for many scientists today.

The fact that these Genesis predictions have never been disproven, however, does not make them *scientific*. The anti-theistic view postulated by Darwin — that predicts antitheses for almost half of the 16 Genesis predictions — has never been disproven either. The purpose of briefly comparing both presuppositions is to prove several points:

1. Both presuppositions are used by scientists today and have been for over 100 years.

2. Both presuppositions can be used to correlate the findings of science today.

3. Both presuppositions are subject to further testing to determine the validity of the predictions they yield.

4. Neither presupposition can be ultimately proven or disproven by scientific methods.

5. If objectivity is to be retained in science textbooks, *both* beliefs must be presented when either one is discussed.

There *are* unknowables in science — subjects about which scientists, as scientists, have no way of determining truth. Objectivity about these unknowables is a worthy goal. To obtain and then retain objectivity, anti-theists and theists alike must understand and respect each other's presuppositions. Further, there must be mutual agreement that such cosmological predispositions have no place in science — unless openly acknowledged as such and presented in a balanced manner.

Although we are far from the goal of objectivity in science today, the path to that goal is fairly obvious. First, science must be cleansed from all religious belief – anti-theistic as well as theistic. Secondly, knowns must increasingly be balanced by unknowns and unknowables when science is presented to either student or layman. Finally, a method of rational resolution of cosmological issues that are certain to arise in science must be established in order to avoid another outburst between NATURALISM and THEISM.

Science Sans Religion

There is an anti-science mood that ebbs and flows in society — but today it seems to be flowing instead of ebbing.

"Just as I thought when I first heard about it," one scientist says to another.

"What do you mean?"

"All this ballyhoo about *origins* is just another devious means for jumping on the anti-science bandwagon."

It could certainly appear that way. If you propose taking something (e.g., religion) away from something else (e.g., science), there is certainly bound to be less of the something else. Assuming that *less* means *anti*, then maybe "cleansing science of all religion" is an anti-science maneuver.

Before jumping to that conclusion, however, listen carefully to the argument. What appears as a *weakening* of science by removing an element could actually be *strengthening* — especially if the removed element had *deluded* and *diffused* science. So, for the moment, set aside fears of anti-science and consider what is being proposed.

The first step toward purifying science by elimination of religious ingredients would be to differentiate between *science* and *scientists*.

Fischer says it this way:[565]

> There are long-standing inadequacies in the teaching of the nature of science...
>
> First, very little attention is paid to the limits and presuppositions of science, either in general discussions or in dealing with specific topics. This neglect results, for example, in a confusion of scientific *evidence* with scientific *interpretation*. It results in a lack of recognition of the proper role of speculation in scientific work. Science is permitted to come through to students (and at times to scientists, as well) with a much greater degree of certainty and finality than is warranted.
>
> Second, there is a frequent blurring of the distinctions between science and scientists and of the inevitable relationships between the two. Science involves knowledge and methods for the acquisition of knowledge. Science deals with the realm of matter and energy, for this is the arena in which the hypotheses and theories of science must be tested.
>
> But scientists are not necessarily limited thus. Scientists are people, not unlike other human beings. One consequence of the blurring of this distinction is a confusing of "science" with "what scientists think" or even with "what scientists believe."

So we propose to separate *science* from scientists. Then we would make another separation — science from *scientism*. Fischer, noting that there is a strong tendency to extrapolate from science to scientism, defines the latter as "the worldview that all of life and all of reality are totally naturalistic and non-theistic." As he further points out:[566]

> Yet science is not scientism, and scientism is not science and the scientific integrity of science books and of science education is weakened by permitting the two to be confused with each other.

Having separated science then from both scientists and scientism, we would be ready next to cull religion out of science. Before doing so, however, it may be helpful to recognize that the origins issue isn't the only forcing factor for "science sans religion." What is done to purify or (as antagonists might say) limit science could yield additional benefits.

The United States Congress helped the National Science Foundation (NSF) celebrate its 25th birthday in 1975 by severely criticizing it for an elementary school behavioral science course developed with NSF funds.

Why was this course — "Man: A Course Of Study (MACOS)" — so harshly censured by Congress? Because it had established — in the name of science — a religion of secularism, particularly as it related to radically reorienting values and beliefs about moral law. The *Congressional Record* for 8 April 1975 described MACOS as:[567]

> A course for 10-year-olds mainly about the Netsilik Eskimo subculture of Canada's Pelly Bay Region. Student materials have repeated references in stories about Netsilik cannibalism, adultery, bestiality, female infanticide, incest, wife-swapping, killing old people, and other shocking condoned practices.

So the subject of origins is not the only place where science has violated its purity by meddling in religion. The NSF criticism points up the guilt of *science educators* along with *scientists* in this matter.

Culling out religion will make some scientists and science educators edgy because it seems to resurrect another worn-out cliché — the *limitations of science*. Certainly science is and should be limited as discussed in Chapter 2. However, purging religion from science would not be nearly as restrictive as the limitations championed by Holton's "New Apollonians."[568]

One avenue of achieving religion-free science education might be to pattern after one of the latest methods for teaching science. Believing that experience with science must begin with the learning of language and continue throughout the educational process, the Commission on Science Education of the American Association for the Advancement of Science, with financial support from the National Science Foundation, embarked upon the preparation of a science program for kindergarten through grade six in 1963. The program, known as *Science — A Process Approach*, was adopted for California public schools while I was a Curriculum Development Commissioner.

Traditional science education programs assume that science is mostly a collection of facts and concepts about the natural world. In contrast, the AAAS approach assumes that science is *what scientists do*. Therefore, there is no apparent way that the AAAS approach could ever introduce belief systems — theistic or anti-theistic — and thereby befoul science.

The biggest danger of ejecting religion from science would be to interpret "science *sans* religion" as "science *versus* religion." To avoid that danger, some guidelines for the extrication effort are offered. Two are negative (actions to shun) and three are positive (measures to incorporate):

1. Do *not* polarize science and religion as typified by the *Los Angeles Times* editorial position[569] or the National Academy of Sciences resolution in Exhibit 41. While their approaches to truth differ, science and religion are not mutual exclusives.

2. Do *not* limit the definition of "religion" to theism when revoking or revising science materials. *All* belief systems must be eradicated, whether based on NATURALISM or THEISM.

3. *Recognize* that what Edwin G. Boring calls "cognitive dissonance" — the existence of incompatible beliefs or attitudes held simultaneously by a human being is a problem for scientists.[570] Two of such dissonances that bear on beliefs in science are *ambivalence* and *personal bias*.

4. *Force* constant exposure of previously-hidden assumptions, problems, unresolved questions and imponderables in science. This honesty by scientists should enhance credibility with the public and attenuate anti-science thinking.

5. *Resist* consciously and continuously the temptation to discuss origins in science textbooks. Direct the inquisitive student to the appropriate discipline for this subject — philosophy, metaphysics and religion.

Should these guidelines be followed, science should be *strengthened* —not weakened. Most importantly, the first step toward objectivity will have been taken.

Knowns and Unknowns

It was one of those rare occasions when I was alone with only one of our six children in the car. We were driving on the Ventura Freeway near our home in the San Fernando Valley. Suddenly, completely out of context with our previous conversation, she reflected aloud, "Dad, I wonder if there will be anything left to discover when I grow up."

"Oh *sure.* There'll be *lots* of things," I responded almost by reflex.

"How *can* there be?" Her voice had a sincere urgency. "Everything's already been *discovered!*"

Our ensuing discussion covered lots of subjects — the possibility of her visiting Mars and other planets, of new cures in medicine, of new breakthroughs in materials . . . But somehow I was *sick.*

It really hurt me to think that her bright young mind had been stifled by half-truth. Scientific discovery, to her, was a virtually buttoned-up frontier — like applying for a homestead would be for me.

Scientific closemindedness need not be vocalized in order to be real. There is an *implied assurance* — discernible to the youngest of children — that all the facts are in and that the case is closed. Scientific notables in the Science Textbook Struggle, both in writing and in oral statements, did their dead level best to propagate this message.

Evolution was declared by Nobel laureates to be just as certain as electricity or the law of gravity. Yet it is not at all scientifically demonstrable. It is a *belief system.*

Any time that the phrase "the vast majority of scientists believe" is used, it is a veiled threat that anyone who disagrees with this statement is disagreeing with something that is already resolved by science. It also turns out to be the death knell to stimulating the inquiry processes in young children's minds.

What is the answer to this bad situation? Simply to carry out the curriculum criterion that I proposed in 1970 and that was unanimously adopted by the Curriculum Commission, its Science Committee, and State Board of Education — "Learner materials shall provide examples of unresolved questions in science as a means of stimulating the inquiry process."

You may recall that I also inserted a similar statement in the resolution to the State Board of Education passed unanimously by the Curriculum Development Commission on 8 November 1972 (see Exhibit 45) concerning editing of science textbooks. It read "that questions yet unresolved in science be presented to the science student to stimulate interest and inquiry processes."

State Board member Mark Gates, apparently trying to protect the sanctity of Darwinism, successfully persuaded the Board to kill that possibility for schoolchildren on 14 December 1972. Yet the question continues to boil, "How else do science educators hope to interest the young student in studying science?"

There is real irony in the suppression of unresolved scientific problems. On one hand, science educators are giving lip service to stimulating young minds in science while refusing to identify and discuss problems that they fully know are unsolved. On the other hand, even the public press regularly prints information that reveals the true situation — that in science all the answers aren't in yet. The exploration of Mars by Viking spacecraft produced headlines like SCIENTISTS SEEK CHEMICAL CLUE TO MAPS PUZZLE[571] and SCIENTISTS WARY OF "LIFE" ON MARS.[572]

Even the subject of controversy in science is treated in a manner foreign to true science. Why do scientific journals print articles entitled EVOLUTION CONTROVERSY FLARES ANEW[573] *as though controversy is unnatural or undesirable in science*? Pliny the Elder (Caius Plinius Secundus) said 2,000 years ago that "The only certainty is that nothing is certain." Of all fields of study that should sham the rigidity of certainty, science would be expected to be foremost.

The obscurant tendency of the scientific elite to patronizingly claim that the masses cannot understand the true nature of science is nothing more than a subtle means of maintaining elitism. It is not unlike the snobbery of the law — lawyers writing laws that can only be interpreted by lawyers.

The only way to overcome this malady is to teach children, even in the earliest years of education, that (1) science is only one method of looking at the universe, and (2) science is a self-limited field of study (it can never know all the truth). Thus, controversy will be readily shown as *normal*, *expected*, and even *desired.*

Controversy can only occur when there is a difference of opinion. Difference of opinion can only occur when something is unresolved or unknown. If open acknowledgement of controversy in science is desirable, then *open acknowledgement of unknowns is mandatory.*

Some scientists believe that *admitting* unknowns can be dangerous. Elitists like Norman D. Newell fear that openness about unknowns will erode "the layman's confidence in the scientific method." But that's only because he's never been a layman. If he had been, he would realize that credibility among laymen is built on the honesty of "telling it like it *is* — unknowns right along with knowns."

But *denying* unknowns is also hazardous. It is a sure-fire way of destroying curiosity — what von Braun calls "the mainspring of science." Mystery must be kept alive if science is to progress. *Ask...probe...criticize...seek...inquire...search.* Aren't these the watchwords of science? As Francis Bacon said, "He that questioneth much shall learn much."

Probably the most disturbing aspect of the Science Textbook Struggle to me personally — aside from news media refusal to print truth — was the concerted denial by the scientific hierarchy that Darwinism raised any questions about the fundamental laws of science. The standard attack by the barons of science consisted of either discrediting the person who raised the questions or ridiculing the questions. Seldom was there any indication that the questions had been seriously examined.

Of course, foolish questions can be raised by anyone on any subject. Ignoring those questions may be entirely appropriate. However, the type of questions to which I refer are *persistent* ones — they've been asked by many knowledgeable scientists over a long period of time. That still does not make the questions necessarily "proper," but a studied and serious answer seems warranted — if only to pinpoint the impropriety of the question.

Specific questions, per se, were less important than the *topic* — the supposed conflict between Darwinism and a fundamental law of science — that provoked the questions. A few of these classic *topics* include:

1. Abiogenesis (Life arising *spontaneously* — or *inevitably* — from inanimate matter).
2. The mathematical concept of randomness (and its role in *favorable* genetic mutations throughout geologic history).
3. Uniformitarianism (the belief that all physical laws have remained constant from the origin of the universe).
4. The Second Law of Thermodynamics (with its concept of increasing entropy or disorder in the universe).

Most of these topics were brought to the attention of the science textbook publishers in October 1972 (see Exhibit 38). I do not wish to defend the sophistication — or lack thereof — with which I and others formulated questions about the topics. The critical issue is that the topics have either *imagined* or *actual* significance for Darwinism.

Yet, Darwinians continue to ignore the topics. Why can they not be openly debated? Why not either *eliminate* all argument by irrefutable proofs, or *acknowledge irresolution* and/or *violation* of scientific laws?

I was consistently castigated for linking Pasteur's spontaneous generation experiment with the Darwinian requirement for spontaneous generation in the origin of life. Yet, science textbooks adopted by the State Board of Education (and for which all the barons of science demanded no revision — even though they hadn't even seen them) made the same comparison! Here is one example:[574]

> The concept of biogenesis is built on the assumption that only living organisms can produce more living organisms. Although we do accept Pasteur's results and reject spontaneous generation, the problem of the origin of life remains. Even if we were absolutely certain about biogenesis, we would have to explain how the first living things came into being. Thus, it appears possible that sometime during the early history of the earth spontaneous development of life took place.

Or compare this college-level textbook explanation with what I wrote in Exhibit 38:[575]

> We have been unceasingly taught not to believe in spontaneous generation, the view that living things can originate from nonliving objects. The classical experiments of Francesco Redi, Lazarro Spallanzani, and Louis Pasteur provided proof that life can only come from preexisting life. However, these experiments revealed only that life cannot arise spontaneously under conditions that exist on earth today. Conditions on the primeval earth billions of years ago were assuredly different from those of the present, and the first form of life, or self-duplicating particle, did arise spontaneously from chemical inanimate substances.

Is that statement *scientific*? Why are the scientific hierarchy silent on *that* violation of science? Because it favors their belief. Is the obvious collision between Darwinism and Pasteur's work *illusionary* or *real*?

The same castigation by scientific notables greeted any mention of the inadequacy of uniformitarianism to account for discontinuities in historical records, inverted geological formations, or abruptness versus gradualism in the fossil record. Why can't uniformitarianism be a widely-debated premise in science textbooks, if not scientific symposia? Why must it always be "ridiculed out of sight?"

Does the recent observation that our sun, long regarded as an exceptionally stable star, actually is a more variable, more dynamic object than ever imagined — that its rotational velocity varies greatly over short time spans — have any effect on the doctrine of uniformitarianism?[576] What about the recent Caltech observations that the earth's temperature has oscillated or fluctuated throughout geological history rather than cooling uniformly?[577] These observations, along with many others, seem to suggest an impact on uniformitarianism, yet questions about it are greeted with arrogant sarcasm. Why?

Alfred Adler, the Austrian psychiatrist, observed that "exaggerated sensitiveness is the expression of the feeling of inferiority." By far, the most sensitive topic — the one that draws the most vociferous venting of vitriolic venom — is the Second Law of Thermodynamics. This book is not an appropriate place to wrestle with highly technical definitions, interpretations and proofs of the celebrated Second Law. Further, it is immaterial to the point I am making — to wit, that Darwinism is far more often thought by qualified scientists to violate the Second Law of Thermodynamics than it does the Law of Gravity. My proof?

Show me one paper in a scientific journal that attempts to explain why Darwinism and the Law of Gravity are compatible, and I'll match it 10:1 with papers that are desperately dedicated to defending Darwinism's compatibility with the Second Law of Thermodynamics.

I completed some graduate work in statistical mechanics and information theory at Dartmouth under Myron Tribus, author of *Thermostatics and Thermodynamics*.[578] Though not claiming expertise in thermodynamics, I understand the major theme of the Second Law — that the universe is in a running-down (disordering) state when considered in its entirety. I am also aware that exceptions to disorder do occur. In fact, order can and does increase in certain rare and restrictive situations (called "open" systems). However, there is no known proof that order increases in *every* "open" system. As Tribus points out:[579]

> Since the second law is a statistical law, we must interpret (it) in the following way: It is true that occasionally there may occur a net decrease in the entropy of a system — such a decrease must be viewed as an improbable event. It is impossible to construct a machine that will *consistently* violate the second law. The odds against the success of such a machine are so great that, in the long run, any backer must go bankrupt. As a matter of fact, his chases of success on any one trial are so small (say of order 10^{-10} to 10^{-100}) that, even if he is of an adventurous spirit and inclined to gamble, he would be better off betting that tomorrow all men will become virtuous.

But even this persuasive argument of improbability, together with the appeal by Darwinians "to view the earth as an open system" (which is meaningless because *everything* is an open system) are irrelevant to my central point. If Darwinism did not strongly imply a violation of the Second Law, why would a scientific journal like *Physics Today* waste its readers' time by publishing, as it recently did, an extensive 2-issue series,

"Thermodynamics of Evolution," to try one more time to explain away the obvious?[580]

Not all the unknowns of science are equally consequential if ignored. The unknowns discussed thus far will only kill *science* if they continue to be neglected. There are some other unknowns in science that threaten the *entire human race* — possibly all of *life* itself.

Recent advances in biology have made it possible for scientists to manipulate the genetic material that governs the physical destiny of every living cell. This process is called "recombinant DNA." While it may offer great benefits, this process also poses great peril. There are many unknowns associated with it.

Recombinant DNA technology has popularly been called "genetic engineering." Clifford Grobstein, Professor of Biology at the University of California in San Diego, believes the process involves hazards of uncertain magnitude. "A new genie has emerged from the bottle of scientific research," Grobstein writes.[581] The unknowns of genetic engineering, however, undoubtedly will be more openly acknowledged than have the earlier-discussed unknowns of Darwinism because of their high leverage on humanity's survival.

The head of the Biology Division at Caltech, Robert L. Sinscheimer, paints a sober picture of scientific unknowns immediately ahead:[582]

> Mankind is about to extend its dominion by redirecting the course of biological evolution...great advances in molecular and cellular biology have provided the basis of a technology capable of reshaping the living world — of restructuring man's fellow life forms into projections of the human will. Naturally, many are profoundly troubled by this prospect.
>
> As science pursues this synthetic biology, we will be leaving the secure web of evolutionary nature that, blindly and strangely, has supported all living creatures for some 3 billion years. With each step we will be increasingly on our own. How do we prevent grievous, inherently irreversible, errors? Can we in truth foresee the ultimate consequences of intervention into this ancient process?

Although the world's leading molecular biologists agreed to impose a voluntary moratorium on their research into recombinant DNA long enough for the National Institutes of Health (NIH) to issue a set of guidelines intended to regulate their research, they have already resumed their research. The NIH guidelines do not have the force of law and are obviously unenforceable throughout the world.

The role of the scientific elite in determining their own guidelines has been severely criticized because those who had the greatest voice in formulating the guidelines were scientists *committed to continuing* recombinant DNA research. Urgent calls for more stringent guidelines to control the unknowns of genetic engineering could have the effect of driving research underground.[583] If the history of science concerning open admission of unknowns is applicable to this process, all life on our planet may be endangered.

This fearsome possibility obviously raises the question about what can be done if scientists *refuse* to expose unknowns. It is clear, by the massive Darwinian reaction during the Science Textbook Struggle, that scientists do not wish anything other than self-policing. One account was most explicit on this point:[584]

> The suggestion that questions of scientific fact and scientific education should be settled by public debate has left most scientists amazed...Concepts of pluralism, of equity and of participatory democracy, as they are defined in the political context, are incongruous in science. The internal standards of science may run counter to egalitarian principles. Hence scientists are particularly distressed by the proposal that laymen participate in defining the nature of an appropriate education in science.

But even a spoiled brat feels exactly the same way. Something clearly must be done to yield greater objectivity in science. What should it be? Should the public *revolt* and cut off all financial support to its incorrigible dependent?

Hopefully, there is a *rational* path to resolving the impasse...

Rational Resolution

I admired Winston Churchill from my childhood. As a young teenage boy, I remember — as though it were yesterday — his rasping, gravelly voice alternately fading and surging across the Atlantic via the BBC as I sat glued to the radio.

It was 10 November 1942 — 2 days after Allied forces had landed rather easily on the beaches of North Africa. Millions throughout the world thrilled at the amazing success. To many, the smell of victory wafted past their nostrils . . . I too caught the aroma . . .

Then the voice of the ages spoke. "This is not the *end*. It is not even the *beginning* of the end. But it is, perhaps, the *end of the beginning*."

Winston Churchill, at Mansion House had delivered another telling phrase that would ring on — not just through a few more years of war, but throughout millennia. Over three decades later, Churchill's maxim applies again. For both the Godly and the Darwinians, the struggle for objectivity in science is, perhaps, at "the end of the beginning."

Four significant milestones — the persecution of Galileo, the publication by Charles Darwin, the trial of John Scopes, and the editing of the Science Textbook Struggle — mark overt collisions between the forces of NATURALISM and THEISM.

None of these battles produced more than a temporary truce. The time between each affray is shortening exponentially — 227, then 66, and finally only 44 years have separated them. The time of the next outbreak can be predicted — unless there is a rational resolution to the underlying issue.

Scientists must learn to live with the eternal tension between NATURALISM and THEISM. They cannot wish it away — abolish it by edict — shame it into oblivion. It is here to stay. The balance of power between the two poles will continue to shift back and forth. Science, as a discipline, must not be thrown off balance by such shifts. More importantly, *science must not take sides*, even though individual scientists obviously do.

But science has been thrown off balance by power shifts between NATURALISM and THEISM. In fact, it is presently in such a state. Who should *referee* science? How can it *regain* its balance? Does science have a built-in *gyrocompass* to restore balance? Apparently not.

Throughout the Science Textbook Struggle, there were statements by scientific barons that "if religious concepts about origins were to be discussed in science textbooks, science would be set back into the Middle Ages." What is *actually* meant by that rhetoric? Without reading too much into it, one could conclude that science is *extremely vulnerable.* If it could be tipped so far so easily, then it must be highly unstable. Why else the irrational fear?

If science is truly built upon consistently repeatable observations of nature, how can it be so *fragile* — so subject to 300-year reversals? It would certainly appear probable that science is in dire need of *protection* — of independent *adjudication* — of a *court of justice.*

That's exactly what *Physics Today*, journal of the American Institute of Physics proposed recently — *a scientific court.*[585] To overcome the inability of the general public to distinguish *facts* from *opinions* in statements by scientific experts, a panel of *scientific judges* would be established. These judges would replace the traditional *panel of experts.* The experts, in turn, would change functions and be given the role of *advocates* arguing the issues before the judges, much as in a court of law.

At least four new and important dimensions to achieving objectivity would occur if such a scientific court were instituted. First, it would separate the *judging* from the *advocacy* of an issue. This separation would heighten the distinction between *fact* and *opinion*, while also providing a means to reduce bias.

The second advantage would come from the selection process for judges. While the judges would be scientists, they would *not* be allowed to have expertise in fields relevant to the issue under review.

Third, the expert advocates, of course, would present the strongest case possible for their own position. However, they could go further. They could *cross-examine* and *criticize* their opponents' position. Only after this open and interactive discussion would the judges publish what they decided were the *scientific facts* in the issue.

Here's how the fourth advantage over the familiar "board of reputable scientists" is described:[586]

> Perhaps an equally important advantage of this approach is that bringing together the opposing views in critical debate offers a means of further refining the scientific understanding of the issue at hand. This kind of public confrontation and cross examination of rival positions is not encouraged in the panel-of-experts approach where the emphasis is more on negotiating a consensus.

Nearly four years previous to the *Physics Today* editorial, Allen D. Allen, a theoretical physicist, published a comparison between scientific and judicial fact-finding in a technical journal.[587] It was later reviewed at length in *Psychology Today*.[588] Allen's paper showed that, while in many respects both scientific and judicial methodologies for ascertaining truth (or "facts") were analogous, there were two points at which they differed significantly.

First, the traditional scientific method for judging a case (e.g., a publication) is based on the adequacy of its listed references. In a court of law, this type of testimony is called "hearsay" because the researcher fails to have *first-hand* evidence.

The second difference is that, in a court of law, the "finders of fact" must be shown to be free from bias. Jurors and judges alike can be dismissed for bias and the location of the trial changed to guarantee this principle. Not so in science! Much of the theme of this book — that scientists are both consciously and unconsciously biased in favor of what they "observe" — demonstrates this striking difference.

If a scientific court — patterned after the points discussed herein —were to be formed to judge Darwinism, for example, the advocates would be few indeed. By limiting testimony exclusively to those who (1) had *observed first-hand* the mechanisms of evolution at work and (2) had *absolutely no bias* as to whether or not the mechanisms they observed had occurred with or without the involvement of an outside agency, there would be no case for Darwinism. Certainly, the appeal used in the Science Textbook Struggle — to wit, that "the vast majority of the scientific elite *believe*" — would be thrown out.

However, my purpose in suggesting a radical reorientation for the determination of scientific truth lies not in how Darwinism — or any other belief system in science — would fare. Rather, it is based on the dismal historical record in science of being philosophically unstable — so open to every wind of doctrine — so vulnerable to the persuasion of elitists rather than fact. Ask Velikovsky, Jensen, or Shockley which system of fact-finding they would prefer. Their answer might be enlightening.

A final summary is in order at this point. If we are to address the issue of how science should deal with ingrained, unconscious, and irreconcilable presuppositions in the minds of all those scientists on whom science is dependent to produce the objective, verifiable, value-free, incontrovertible "facts" shown in Figure 2, what should be done? Can there be *rational resolution* of the inherent dissonance between *subjective scientists* and *objective science*? After all, if science does not produce a pure distillate, has it not, in essence, lost its reason for being?

Here then are some conclusions bearing on a rational solution:

1 Origins are and will remain important, intriguing enigmas for humanity. Even the origin of *death* — a phenomenon that is available for first-hand observation — is every bit as puzzling as the origin of *life*. How did the universe, life, man — yes, and death — come into existence? Don't ask scientists for a *scientific* answer — because there is no way for an answer to be found in science.

2. Scientists, like all other human beings, are desirous of answers for origins. They also suffer the temptation of passing their private beliefs on to the public under the renowned cloak of scientific certainty. *Forgive* them first — and then almost simultaneously *ignore* them. Unfortunately, their knowledge of ultimate causation is no more reliable than that of a cab driver or a chimney sweep.

3. Scientific laws must be better understood — by scientists as well as by laymen. Scientists are inclined to overestimate the sanctity of scientific laws to the same degree that laymen may underestimate it. Donald M. MacKay has succinctly pointed out that scientific laws are founded on the realization that if one observes nature carefully and intelligently enough, regular and causal precursors for many physical events can be found. These

precursors occur in a precedence that can be codified and called a *law*. So long as everyone understands that things do not happen because a scientific law *says* they will — or worse yet, because a scientific law *determines* that they will, we are relatively safe from scientism — that spurious religion of naturalistic determinism.

4. The conclusion that *only one Weltanschauung* is acceptable in science, as manifested by Ernst Mayr's criteria for the "Darwinian revolution,"[589] must be rejected if objectivity is to be obtained. Such a conclusion is the essential antithesis of objectivity because it establishes and enforces a creed — a belief system — a non-rational allegiance at the very moment that it is advocating a value-free mind for the scientist.

5. At the conclusion of Chapter 1, questions that might be asked of scientists were grouped into 4 classifications. The humility that Wernher von Braun in the Foreword called "a seemingly natural product of studying nature" needs to be restored and nurtured so that scientists decline answers to 2 of the 4 classes of questions — the *unanswerable* and those lying *outside* the limits of science.

6. Whereas science has had to struggle and fight for a place in the sun for a few centuries among so many other older disciplines, the time has arrived for scientists to realize that science is now accepted as a full-fledged, co-equal member in the family of disciplines. This realization will then permit scientists to seek out and explore — in far less defensive fashion than has been shown thus far in history — the boundaries or interfaces of science with other disciplines in the family. In particular, there appears to be a critical need for understanding the interplay between *philosophy* and *science*. The sterility of studying science in a philosophic vacuum has produced most of the ammunition for the sporadic open battles between NATURALISM and THEISM.

Recounting and documenting the struggle in which so much of one's physical, mental and spiritual resources were expended is bound to produce an unscientific, subjective bias that cannot be hidden. As sincerely as I tried to be objective so as to lead the way to the goal of scientific objectivity, I undoubtedly failed more often than I succeeded. The wonder of it all is that my account of this confrontation between NATURALISM and THEISM has been granted a forum.

I can personally answer Walt Whitman's profound question — "Have you not learned great lessons from those who reject you?" — in the affirmative. The debt I owe those who sincerely opposed me is inestimable. I would honestly hope that my esteemed opponents — all of them far more qualified than I — will someday join me in my answer to Whitman.

The most fitting finale to this lengthy chronicle that I could imagine is a resolution prepared in April 1976 by the National Academy of Sciences — that very organization which I harshly criticized for an earlier resolution they passed attempting to suppress a viewpoint in the Science Textbook Struggle that they had not objectively analyzed! However, by urging that every reader endorse the "Affirmation of Freedom of Inquiry and Expression" (Figure 21), I trust that it will be taken as a gesture of conciliation between people of good will — regardless of cosmological predisposition.

Even more importantly, the freedoms that this resolution advocates — freedom from religious, political or ideological restriction — freedom to search where inquiry leads — freedom to publish without censorship — freedom from retribution for unpopularity of conclusions, should preclude another cataclysm in science over the beliefs of scientists.

Throughout the Science Textbook Struggle, John Scopes and I were pictured as antagonists. Just the opposite was true. We were fighting for exactly the same principle.

I regret that he and I will never be able to meet because, if we could have, John Scopes and I would have linked arms and repeated in unison what he *alone* had earlier said: "The basic freedoms defended at Dayton are not so distantly removed; each generation, each person must defend these freedoms or risk losing them forever..."

'An Affirmation of Freedom of Inquiry and Expression'

This resolution was approved on April 27 by National Academy of Sciences members at their annual meeting in Washington, D.C. Signed statements should be directed to the Commission on International Relations, National Research Council, National Academy of Sciences, 2101 Constitution Avenue NW, Washington, D.C. 20418, U.S.A.

WE, THE MEMBERS PRESENT at the 113th Annual Meeting of the National Academy of Sciences of the United States of America, hereby support the following affirmation. We invite others at home and abroad to join with us in this declaration, and we offer the offices of the Academy as a repository for such individual statements of affirmation.

AN AFFIRMATION OF FREEDOM OF INQUIRY AND EXPRESSION

I hereby affirm my dedication to the following principles:

- That the search for knowledge and understanding of the physical universe and of the living things that inhabit it should be conducted under conditions of intellectual freedom, without religious, political or ideological restriction.
- That all discoveries and ideas should be disseminated and may be challenged without such restriction.
- That freedom of inquiry and dissemination of ideas require that those so engaged be free to search where their inquiry leads, free to travel and free to publish their findings without political censorship and without fear of retribution in consequence of unpopularity of their conclusions. Those who challenge existing theory must be protected from retaliatory reactions.
- That freedom of inquiry and expression is fostered by personal freedom of those who inquire and challenge, seek and discover.
- That the preservation and extension of personal freedom are dependent on all of us, individually and collectively, supporting and working for application of the principles enunciated in the United Nations Universal Declaration of Human Rights and upholding a universal belief in the worth and dignity of each human being.

Signed

Date

Figure 21. Affirmation of Freedom of Inquiry and Expression signed by nearly 3,000 scientists worldwide

EPILOGUE

By now, you likely recognize that this book was *written* in 1976. And it was neither updated nor edited before being *published* over 30 years later.

Why? For at least three reasons.

First, *scientific knowledge* about origins has not *progressed* in 30 years — since such knowledge has *never existed.* Second, the *struggle for objectivity* concerning origins has not changed in 30 years because the combatants — naturalism and theism — will *always* be at odds regarding origins. Third, this first-hand account is *unique* — occurring only once in history — because the conflict about origins moved radically from its last eruption in a small, remote Tennessee courtroom in 1925 to an educational forum in the largest state in the Union in 1969.

Much changed in the world after Chapter 10 was finished in 1976. The Cold War ended, radical Islamic terrorism exploded across the globe, and technology continued to surprise us with its marvels.

After living a quarter century in southern California, I ceased my transcontinental commuting and moved with Phyllis to the Nation's capital for nearly another quarter century.

When this book was written, there were no cell phones, FAX machines or home computers. In fact, *Science But Not Scientists* would have remained un-published had technology not have created electronic scanners.

Consider this unlikely story ...

You will recall that this book was written under contract to The Free Press, a division of The Macmillan Company. My gratitude to Macmillan for their courage to publish *Science But Not Scientists* was expressed in Chapter 2, since Macmillan had capitulated in 1950 to blackmail by prominent scientists concerning Velikovsky's *Worlds in Collision.*

But my praise for Macmillan was premature.

The day after attending Jimmy Carter's inauguration in Washington as 39th President of the United States, I carried the manuscript for *Science But Not Scientists* to New York City and personally presented it to Edward W. Perry, President of The Free Press. Admittedly, the number of pages did exceed the stipulated length in the contract, but I had met the submission deadline.

Mr. Perry cordially accepted the manuscript, giving me no indication that the book would not be published. However, I was to learn a short time later that G. Ledyard Stebbins and Richard M. Lemmon — my two chief antagonists in the Science Textbook Struggle — had threatened

Macmillan just like Harlow Shapley and Fred Whipple did in 1950 against Velikovsky as recounted in Chapter 2. Recall Lemmon's intimidating letter to Macmillan in Chapter 10?

So Macmillan refused to publish *Science But Not Scientists* after all!

The California Attorney General confirmed that Stebbins and Lemmon had committed "tortuous interference with a contractual relationship" — thus giving me legal basis to take action against this obvious illegal maneuver to prevent publication of a book that had taken me about one year to write.

But ultimately I did not pursue retribution — for two reasons.

First, the book was neither germane nor contributory to my profession – only documenting an avocational diversion. Second, I lacked both time and finances to take such a major digression from my career.

My secretary Shirley Markus had dedicated hundreds of hours in typing the manuscript on an IBM Selectric typewriter to meet Macmillan specifications. The book understandably represented a major personal achievement for her. So the rejection by Macmillan was very painful for both of us.

Nonetheless, the manuscript sat untouched in 3-ring binders for 30 years ...

My professional life continued successfully in spite of Macmillan's capitulation – hardly impacted by the three-year battle over science textbooks and an additional year to write *Science But Not Scientists*.

When President Ronald Reagan appointed me as a Member of the National Transportation Safety Board in 1983, Phyllis and I moved to Washington. There I wrote another book – a best-seller published by Prentice Hall, *MANAGING RISK: Systematic Loss Prevention for Executives*, that is currently in its third printing.

Following my time in public office, I founded Omega Systems Group Incorporated in 1986 to provide systems management methodology for diverse business and government applications. A technique known as *SMART™ (Systems Methodology Applied to Risk Termination)* that I created has become widely implemented in the fields of telecommunications, health care, criminal justice, offshore oil drilling, coal mining, commercial space, petrochemicals, environmental protection, mass transit, food processing, nuclear power, aviation, and terrorism control.

For well over a decade, my expertise has been solicited in over 400 interviews on CNN, NBC, ABC, CBS, FOX, MSNBC, CNBC, PBS and international networks such as BBC, ITN, Phoenix TV Hong Kong, CTV, CBC, and Nine Network Australia.

On 9/11, Jon Scott and I were discussing on FOX News Channel the odd crash of American Airlines Flight 11 into the World Trade Center north tower at the moment that United Airlines Flight 175 smashed into the south tower. I spent the next 15 hours awaiting more interviews in the FOX Green Room in Washington — along with Reagan's Secretary of State Alexander Haig, former Speaker of the House Newt Gingrich, and Senators John Breaux, Chuck Hegel, and James Inhofe among others — as the city went into virtual siege.

Despite my very busy professional life, however, there remained one residual *aftereffect* from the Science Textbook Struggle. It persisted like a silent shadow in my subconscious ...

The news media — from the outset — referred constantly to that conflict and my role in it as a reincarnation of the Scopes Trial. Such perpetual reference to that historic event continued to pique my interest to know more about it — to confirm whether there was any truth to the correlation.

In 1969, I had little knowledge about that 1925 trial — beyond knowing that Clarence Darrow defended John Scopes using the same argument for academic freedom that I used before the State Board of Education (Exhibit 4). I knew that it had been called "The Monkey Trial" because the legal issue centered on teaching whether humans had descended from apes via evolution.

Also I became aware of the 1955 Broadway play, *Inherit the Wind*, that later became a theatre movie in 1960 and an NBC TV movie in 1988 — all of which had parodied the theme of the Scopes Trial.

But that submerged curiosity persisted. I spoke to no one about it. However it led me to read John Scopes' 1967 book *Center of the Storm*, meet and interview John's son, and even take two trips to Dayton, Tennessee to see and photograph the setting for the trial. These personal encounters involving the Scopes Trial were to influence the publication of *Science But Not Scientists* after a 30-year delay — but in a stranger-than-fiction fashion.

Definite *parallels* between John Scopes and me emerged as I studied his life:

- Neither of us allowed the issue of evolution to interrupt our professional lives. John, according to his son Bill, was the most proud of his scientific accomplishments as a reserve geologist in the petroleum industry and never even spoke about the trial until after he retired.[590]

- We viewed our roles *identically* – and we both supposedly *lost* the case for academic freedom we were defending. We were on the same side of an issue where, in a truly objective climate, we should have *won*. John expressed it just as I would have:[591]

> I believe that the files (of the trial) would reveal that the defense had on its side all of the facts, logic, and justice, but these weapons are often ineffective in a battle against bigotry and prejudice.

- We were both committed in our involvement to the idea of academic freedom and creating a climate for understanding divergent points of view.[592]

- We both joined in the hope "that religion and science may now address one another in an atmosphere of mutual respect and of a common quest for truth."[593]

- Both of us made two trips to Dayton, Tennessee after our involvement. Twice in early 1982 I drove there while spending about a month in eastern Kentucky addressing coal mining risk with a client. John's two visits to Dayton he described this way in 1965:[594]

> I fully intended to be a regular visitor to Dayton since (1925) had been one of the most pleasant years of my life. I had met many people I liked and wanted to continue to call friends. But the realities of a career alter the course for many of us. I have been able to visit Dayton only twice in forty years.

- We shared interests in the *issues* raised by the trial but did not continue any role in the *outcome* of our respective involvement. John's reflection on this point — after 40 years — summarizes it well:[595]

> I have had a continuing interest in the issue of the trial but never as a participant. Many times I have been asked why I have had no further role to play relative to the issues – even why I did not at least capitalize on my publicity and reap the monetary harvest that was close at hand. Perhaps my best answer is to paraphrase Calvin Coolidge's "I do not choose to run;" for me it would be, "I did not choose to do so."

On the other hand, there were *contrasts* between John Scopes and me. We differed in at least three ways:

- On the topic of faith in God, I have already declared my strong commitment to Jesus Christ and His resurrection. John, according to his son Bill, was an agnostic who was not sure about life after death.

- John was a very heavy smoker all his life, quitting a few months before he died of esophagus cancer at age 70 in October 1970. I have never smoked.
- His son Bill said that John drank alcohol "probably like a lot of people. We'd like to every now and then go out and have a little spree, getting drunk as everyone does to a certain extent."[596] I have been a teetotaler all my life.

I cannot recall how I learned that John Scopes' son, William C. (Bill) Scopes, was in the insurance business with an office only about a mile from my California home. However on 11 October 1979, I met him there and had a wonderful time interviewing him on tape about his father. (See Figure 22)

We discussed many topics like his Dad's physical characteristics, education, professional scientific work – especially during World War II, family life, boredom following retirement, fondness for children, rare recollections of the trial, political views, religious beliefs, and his intent in writing his book *Center of the Storm*.

Some surprising facts for me included John's almost belligerent refusal to be involved in anything to do with the trial. For instance, when the playwrights for the Broadway show, *Inherit the Wind*, mailed him a copy for his comment, "He just bundled it back up and sent it right back to them, signed a release, and said, 'I don't want anything to do with it.'"[597]

John Scopes was tight-lipped about the trial throughout Bill's life. As John grew older and particularly after he retired from the petroleum business, Bill was able to get him to relate some details:[598]

> As a conversation, we really never did sit down and have, say, a long drawn-out conversation concerning the trial itself. Dad, on a few different occasions, related certain things to me. One in particular, he told me that if he were ever'd been called (to testify) – the reason he never testified . . . because he never taught evolution. That it was just something that came up. He was asked by the school board president to help test this law (the Butler Act), and he volunteered, not realizing exactly how far it was going to get out of hand. And at the same time, he was a football and basketball coach. And he substituted every now and then when a teacher was not in, and he had no recollection of ever teaching evolution.

This first-hand and intimate insight continued to stoke my interest in knowing more about the correlation that the news media had made between the Scopes Trial and the Science Textbook Struggle. Discovering this uncanny relationship with John Scopes caused me to want to visit (and photograph) the *scene* where the famous trial had occurred.

Dayton, Tennessee isn't even easy to *locate*. It isn't on an Interstate highway. A tiny town of less than 5,000 between Knoxville and Chattanooga, one must deliberately desire to go there.

Figure 22.
Vernon L. Grose interviews William (Bill) Scopes, 11 October 1979

On two weekends in January and February 1982 during the implementation of our SMART™ methodology at Sierra Coal Company in eastern Kentucky, I drove many miles on secondary roads to finally reach Dayton. (See Figure 23)

Once there, of course, I wanted to see the Rhea County Courthouse where the trial was held, the house where William Jennings Bryan had died only five days after the Scopes Trial ended, and Frank Robinson's drugstore where John Scopes had been recruited to stand trial for teaching evolution in violation of the newly-passed Butler Act that forbade such teaching.

Figure 23
Vernon L. Grose in Dayton, Tennessee, 9 January 1982

Bryan College, a private co-educational school founded in 1930 to memorialize Bryan, is nearby. Even on a Saturday afternoon, I was able to meet there some people who helped me to locate, interview, and photograph Harry Sheldon, Daisy Morgan, Giles Ryan, Earl Morgan, and Virgil Wilkey – all of whom had attended the famous 11-day trial.

Robinson's drugstore had moved from the location it occupied in 1925, but I found the original building. Its significance for *Science But Not Scientists* will be discussed later.

Obviously, the Rhea County courthouse stimulated lots of wonderment as I wandered about the grounds and explored the courtroom itself. (See Figure 24) It was so much smaller than I had imagined. I wished that the walls could have spoken. *If only ... What if ...* ran through my mind.

There was another element of the Scopes Trial that also caught my interest. Though I favored Clarence Darrow's position in the trial because he argued for openness and freedom regarding science, I was intrigued by William Jennings Bryan as a person.

Figure 24
Rhea County Courthouse in Dayton, Tennessee, 9 January 1982

John Scopes had met Bryan in 1919, six years before the trial. Bryan delivered the commencement address when John graduated from Salem, Illinois high school. It was also the same high school from which Bryan had graduated. So the two had a common root.

When Bryan arrived in Dayton for the trial, a banquet was held for him. John attended it. They exchanged pleasantries that evening – before the trial started. Even though Scopes never did testify in the Scopes Trial, he was impressed by Bryan's reputation as an orator as well as the fact that he had been a Congressman, the Democratic candidate for President three times, and Secretary of State under Woodrow Wilson. My own father, a life-long Democrat, often spoke of Bryan's stirring and famous "Cross of Gold" address.

On the 5th day after the trial ended, Bryan — at age 65 years – attended church in Dayton with his wife, offering the morning prayer. He laid down for a Sunday afternoon nap and never awoke.

His death triggered an outpouring of grief from the "common" Americans who felt they had lost their greatest champion. A special train carried him to his burial place in Arlington National Cemetery. Thousands of people lined the tracks.

Historian Paul Boyer says, "Bryan's death represented the end of an era. This man who had loomed so large in the American political and cultural landscape for thirty years had now passed from the scene."

Given my interest described in Chapter 2 in visiting cemeteries to photograph graves, it should not surprise you that Bryan's grave attracted me. When I learned that he was buried in Arlington National Cemetery – a favorite site that I had visited often for years, I decided to locate his grave one evening in 1979 after a class I was teaching at The George Washington University. See Figure 25.

Ironically, only four years later we moved from California to accept the White House appointment in Washington. We purchased a condominium penthouse that overlooks Bryan's grave only about 800 yards away – and have lived there for 23 years!

So how did all these details about the Scopes Trial influence the publication of *Science But Not Scientists* – after a 30-year delay? Believe it or not, there is an almost unbelievable progression that runs between that Dayton event and my decision to publish.

I teach an adult class at The Falls Church, a historic Anglican house of worship established the year that George Washington was born – 1732. George even attended and served the church as a vestryman. The city of Falls Church, Virginia is named after the church.

Figure 25
Vernon L. Grose at grave of William Jennings Bryan in Arlington National Cemetery, 19 September 1979

Early in 2006, a Royal Australian Navy officer, Captain Michael Anthony Houghton, began attending the adult class. He, with his family, had moved to Washington as Deputy Project Manager on a joint heavyweight torpedo program with the US Navy. Offhand one day in class, I evidently mentioned that I had written a book concerning the origin of the universe, life and man. I have no recollection of doing so, however.

Captain Houghton approached me later and asked about the book, since the topic had been one of high interest with him for several years. Learning that I had never published my book, though having written it 30 years earlier, he asked if I would mind if he found a publisher for it. He even declared that he was going to *pray* for the book to be published!

Another attendee in the adult class, John B. Mumford whom I'd known for many years, became aware of Captain Houghton's interest in the book. I explained to John that all I had to offer a publisher was a massive *typewritten manuscript* of the book in four huge 3-ring binders. And I was certain that no publisher would be interested in working with such a clumsy document. Publishers today no longer work with paper manuscripts – everything is processed *electronically*.

John founded and chairs *The Washington Group*, a business management firm. A former White House Fellow (along with Colin Powell) holding a Harvard MBA, John graduated from the US Military Academy at West Point in 1962. He offered the use of a high-speed scanner in his office to convert the paper version into an electronic one. So he and I took a full day to scan the book. What a wonderful – almost miraculous — transformation!

However, the IBM Selectric font and the 8.5"x11" format of the original manuscript required that it be transformed and re-formatted into a structured configuration that involved new font, reading style, and printed adaptation to the size of a 6"x9" book before it could be published. I was unaware of this need. The skill for accomplishing this major step is rare indeed, I was soon to learn.

A third person in the adult class, Keith A. Godwin, was visiting in our home one evening shortly after John and I had scanned the manuscript. He had never heard of the book's existence. Keith, also a West Point graduate – Class of 1979, is an information systems engineer with MITRE Corporation working on missile defense.

Almost as an afterthought, I expressed my excitement of having just scanned a useless paper manuscript of a book I had written 30 years ago. Frankly, I had no understanding that the scanned document was not yet useful for a publisher. I was simply thrilled to have it on my computer harddrive instead of sitting on a bookshelf.

"What's the book *about*?" Keith asked.

"Oh, it describes a battle I had in California years ago regarding how the origins of the universe, life and man are described in science textbooks. But the struggle was generally viewed as the resurrection of the Scopes Trial – pitting evolution against creation," I replied.

"That's a subject that has interested me for *years*," he replied.

Keith has an extensive library, and he had been sorting and assembling all his books related to the history of evolutionary thought – even prior to Darwin. Immediately, he began to ask me questions about my book.

That evening took an unplanned turn, as he began explaining computer work he had done 15 years earlier on document conversion. That was the missing piece in the puzzle of publishing *Science But Not Scientists*!

Would it surprise you to know that Keith and his wife once lived in Maryville, Tennessee – only about 60 miles from Dayton – when he worked as an independent computer consultant on bank marketing?

But the tapestry of threads that interlace the Scopes Trial, evolution, and Dayton, Tennessee — ultimately leading to publication of *Science But Not Scientists* continues on ...

Frank Robinson's drugstore in Dayton was not only the setting for recruiting John Scopes to stand trial as earlier recounted. William Jennings Bryan, on the day he arrived in Dayton to prepare for the trial, "downed an ice cream sundae at Robinson's drugstore."[10] More than likely, it was a popular location on Main Street for homefolks to meet and chat.

"Did I ever tell you that I was born in Dayton?" John Mumford asked me while we were scanning the manuscript for *Science But Not Scientists* in his office. Somehow I had mentioned the Scopes Trial connection to what I had written.

"No, I never knew *that*!" I exclaimed.

"Yes, my Dad was the trainmaster for the Southern Railway in Dayton."

"Do you know *where* you were born in Dayton?"

"Yes, the hospital was on the second floor above Robinson's drugstore."

However, Robinson's original drugstore – where John Scopes had agreed to stand trial and that I had photographed in 1982 — had moved two blocks in the 1930's to the site above which John was born. Yet it was an uncanny coincidence.

There is an amazing *interconnectedness* to life — the viability that emerges when vision, people, activity, events, and insight all work together. And the delayed publication of *Science But Not Scientists* seems to illustrate that phenomenon.

Thirty years have passed since Macmillan refused to publish this book – because a couple scientists feared itss content. The intellectual freedom to share my ideas was denied. Yet at the time I finished the book – in Chapter 10 – I was unaware of this outcome. In fact, it appeared that the future for controversial thinking about science had *brightened* due to the Science Textbook Struggle.

The greatest evidence for my enthusiasm – touted as the fitting finale for the book — was an *"Affirmation of Freedom of Inquiry and Expression"* approved by the National Academy of Sciences in April 1976, soliciting signed endorsement from individual scientists (see Figure 21).

How thrilled I was to learn recently from the Archivist at the Academy that within six months after issuance, 2,622 scientists from around the world had signed it! The *Affirmation* had been distributed by such professional societies as the American Anthropological Association, American Association of Immunologists, American Chemical Society, American Society of Biological Chemists, American Society of Plant Physiologists, Association of Professional Geological Scientists, Operations Research Society of America, and the Society for Industrial and Applied Mathematics.

What a victory for objectivity! It validated and confirmed my conviction that science must always remain open to searching for "knowledge and understanding of the physical universe and of the living things that inhabit it ... without religious, political or ideological restriction."

Finally, this epilogue would be incomplete if I failed to acknowledge my two chief protagonists who – as far as they knew – successfully blocked publication of this book. My greatest wish is that they might have had an opportunity to read it. Why? Because I sincerely believe that they *might* have changed their minds about its primary thesis: ***that science will NEVER have a role in explaining the origin of the universe, life, and man.***

G. Ledyard Stebbins died in his home in Davis, California on 19 January 2000 at 94 years of age. His obituary stated that he was one of the leading evolutionary biologists and foremost botanists of the 20th century.

Richard Millington Lemmon, an award-winning Lab scientist and an authority on the chemical origins of life, died on 22 February 2001 in an Oakland hospital from complications following surgery for a broken hip. He was 81 years old.

They were both very bright men and great contributors to their respective fields of science. Though I earnestly and fiercely opposed them during the Science Textbook Struggle, they were worthy opponents indeed.

Could it be that this saga was *intended* to be revealed after all — as one far grander in significance than any of the individuals who played roles in it?

Vernon L. Grose
September 2006

Endnotes

Prologue

[1] Scopes, John T., and James Presley, *Center of the Storm* (New York: Holt, Rinehart and Winston, 1967), p. VI.
[2] Standen, Anthony, *Science is a Sacred Cow*, (New York: E. P. Dutton & Co., Inc., 1950), p. 13.
[3] Grose, Vernon L., "Deleterious Effect on Astronaut Capability of Vestibulo-Ocular Disturbance During Spacecraft Roll Acceleration," *Aerospace Medicine*, Volume 38, Number 11 (November, 1967).

Chapter 1

[4] Bremner, Donald, "How Did Man Get Here? Replies Vary Around the World," *Los Angeles Times*, (31 December 1972) Part C, p, 4.
[5] Anonymous, "Will the Universe Expand Forever?" *Science News*, Washington, D.C. Science Service Incorporated, (8 November 1975) Volume 108, No. 19, p. 293.
[6] *Ibid.*, p. 294.
[7] Alexander, George, "Our Universe: Is It Expanding or Is It Closed?" *Los Angeles Times*, (2 January 1974) Part I, p. 1.
[8] Anonymous, loc. cit.
[9] *Ibid.*
[10] Anonymous, "'Supernova Explosion: A Computer Model," *Science News*, Washington, D.C.: Science Service Incorporated, (8 November 1975) Volume 108, No. 19, pp. 295-6.
[11] Allen, Allen D., "The Big Bang Is Not Needed, " *Foundations of Physics*, New York: Plenum Publishing Corporation, Volume 6, No. 1, pp 59-63.
[12] Getze, George, "'Big Bang' Hit as 'Act of Faith'," *Los Angeles Times*, (30 January 1975) Part II, p. 1.
[13] Alexander, loc. cit.
[14] Getze, George, "Earth Born In 10,000-Year 'Instant', Two Scientists Say," *Los Angeles Times*, (30 June 1972) Part I, p. 1.
[15] Getze, George, "Moon Too Big, New Planet Theory Claims," *Los Angeles Times*, (1 February 1972) Part I, p. 3.
[16] Miles, Marvin, "New Evidence of Moon Split From Earth Told," *Los Angeles Times*, (12 June 1975) Part II, p. 1.
[17] Anonymous, "Evolution Controversy Flairs Anew," *Chemical and Engineering News*, American Chemical Society, (11 December 1972) Volume 50, No. 50, p. 13.
[18] Cowen, Robert C., "Biological Origins: Theories Evolve," *The Christian Science Monitor*, Boston, Mass.: Christian Science Monitor, (4 January 1962) p. 4.
[19] Rosenfeld, Albert, "Did 'Someone Out There' Put Us Here?", *Los Angeles Times*, (29 November 1973) Part H, p. 7.
[20] Toth, Robert C., "Hunt For Other Life Held Top Space Goal," *Los Angeles Times*, (27 March 1965) Part 1, p. 1.
[21] Cowen, loc. cit.

[22] Cowen, loc. cit.
[23] Cowen, loc. cit.
[24] Getze, George, "Doubt Raised On Theory of Life Origins," *Los Angeles Times,* (1 April 1971) Part 2, p. 4.
[25] Anonymous, "Violent Prelude to Life Theorized," *Los Angeles Times*, (22 April, 1974) Part 1, p. 5.
[26] Bengelsdorf, I. S., "Idea Sheds New Light On Origin of Life," *Los Angeles Times,* (5 July 1970) Section 2, p. 1.
[27] Anonymous, "Discovery Redates Life A Billion Years, Russ Says," *Los Angeles Times*, (5 July 1975) Part 1, p. 4.
[28] Wilder Smith, A. E., *The Creation of Life - A Cybernetic Approach to Evolution,* Wheaton, Illinois: Harold Shaw Publishers, (1970) Chapters 2 through 6.
[29] Washburn, Sherwood L., "Man: On the Origin of the Species," *California Monthly*, Berkeley, California: California Alumni Association, (June-July 1972) Volume 82, No. 8, pp. 12-3.
[30] *Ibid.*, p. 12.
[31] *Ibid*, pp. 12-3
[32] *Ibid.*, p. 13.
[33] *Ibid.*
[34] *Ibid.*
[35] *Ibid.*
[36] *Ibid.*
[37] Howell, F. Clark, *Early Man*, Sacramento, California, California State Department of Education, (1965).
[38] *Ibid.*
[39] *Ibid*, pp. 98-99.
[40] *Ibid.*
[41] *Ibid.*
[42] *Ibid*, p. 7.
[43] Alexander, Denis, *Beyond Science*, Philadelphia and New York: A. J. Holman Company - Division of J. B. Lippencott Company, (1972) pp. 41-72.
[44] *Science Framework for California Public Schools*, Sacramento, California, California State Department of Education, (1970), p. 8.

Chapter 2

[45] Grose, Vernon L., "Constraints on Application of Systems Methodology to Socio-Economic Needs," *Proceedings of First Western Space Congress*, Santa Maria, California (October 1970).
[46] Anonymous, "It Would Seem Incredible," *Los Angeles Times*, (29 December 1975) Part 11 p. 6.
[47] Tournier, Paul, *The Healing of Persons*, New York: Harper and Row (1965) pp. 225, 226.
[48] Nielsen, H.A., *Methods of Natural Science (An Introduction)*, Englewood Cliffs, N.J.: Prentice-Hall (1967) Chapter V, pp. 50-53
[49] *Ibid.*, p, 53.

[50] Standen, Anthony, *Science Is A Sacred Cow*, New York: E. P. Dutton & Company, Incorporated (1950).
[51] Kerkut, G. A., *Implications of Evolution*, Oxford, England: Pergamon Press Ltd. (1960).
[52] Macbeth, Norman, *Darwin Retried -- An Appeal To Reason*, Boston, Mass.: Gambit, Incorporated (1971) p. 5.
[53] *Ibid.*, pp. 6, 7.
[54] Holton, Gerald, "On Being Caught Between Dionysians and Apollonians," *Daedalus*, Cambridge, Mass.: American Academy of Arts and Sciences (Summer 1974) Volume 103, No. 3, p. 66.
[55] *Ibid.*, p. 67.
[56] *Ibid.*
[57] *Ibid.*, p. 81.
[58] Allen, Allen D., "Does Matter Exist?" *Intellectual Digest*, New York: Ziff-Davis Publishing Co. (June 1974) p. 60.
[59] Ibid.
[60] Luyten, Willem J., "Astrofantasies and Contracts," *Science*, Washington, D.C.: American Association for the Advancement of Science (17 July 1964) Vol. 145, No. 3629, p. 231.
[61] Yerges, Lyle F., "Science – Friend or Foe?" *Sound and Vibration*, Bay Village, Ohio: Acoustical Publications, Inc. (April 1972) Volume 6, No. 4, p. 9.
[62] Tournier, Paul, *The Whole Person In A Broken World*, New York: Harper and Row (1964) p, 2.
[63] *Ibid.*, pp. 2-12.
[64] *Ibid.*, pp. 2-3.
[65] *Ibid.*, pp. 3-4.
[66] *Ibid.*, pp. 8.
[67] *Ibid.*, pp. 10.
[68] Long, T. Dixon, "Communication with the Humanities," *Science*, Washington, D. C.: American Association for the Advancement of Science (7 July 1964) Volume 145, Number 3629, pp. 231, 232.
[69] Kornberg, Arthur, "Cut in Basic Research Funds Called 'Tragic'," *Los Angeles Times,* (17 June 1974) Part 2, p. 5.
[70] Anonymous, "Inadequate or Biased Research," *The Montgomery Advertiser*, Montgomery, Alabama: Harold E. Martin (9 December 1974) p. 4.
[71] Hebrews 11:1 (J. B. Phillips translation).
[72] Alexander, Denis, *Beyond Science*, Philadelphia and New York: A. J. Holman Company, Division of J. P. Lippincott Company (1972).
[73] Neyman, Jerzy, "Copernicus," *News Report -- National Academy of Sciences*, Washington, D. C.: National Academy of Sciences (December 1974) Volume XXIV, No. 10, p. 6.
[74] Anderson, Alan, Jr., "Gods and Devils in Science," *Saturday Review of the Sciences*, San Francisco, California: Saturday Review Company (27 January 1973) Volume 1, No. 1, p. 24.

[75] Cousteau, Jacques-Yves, "The Next Billion Years," *PSA Magazine*, Los Angeles, California: East/West Network, Incorporated (February 1974) Volume 9, No. 2, pp. 57-59.

[76] Asimov, Isaac, "Your Spare Parts," *American Way*, New York, N.Y., American Way (November 1975) Volume 8, No. 11, pp. 8, 9.

[77] Warshofsky, Fred, "When The Sky Rained Fire: The Velikovsky Phenomenon," *Readers Digest*, Pleasantville, New York: Reader's Digest Press (December 1975) Volume 107, Number 644, pp. 220-240.

[78] *Pensee*, Portland, Oregon: Student Academic Freedom Forum, (May 1972) Volume 2, Number 2.

[79] Kallen, Horace, "Shapley, Velikovsky And the Scientific Spirit," *Pensee*, Portland, Oregon. Student Academic Freedom Forum, (May 1972) Volume 2, Number 2, p. 36.

[80] *Ibid.*, p. 37.

[81] *Ibid.*

[82] *Ibid.*, p. 38.

[83] *Ibid.*

[84] *Ibid.*, p. 40.

[85] Hibbs, Albert R., "Inquisition, Repression and Ridicule," *Engineering and Science*, Pasadena, California: California Institute of Technology and the Alumni Association (December 1974 - January 1975) Volume XXXVIII, No. 2, p. 28.

[86] *Ibid.*, p. 29.

[87] *Ibid.*, p. 6.

[88] *Ibid.*, p. 29.

[89] Buckley, William F., Jr., *Execution Eve – And Other Contemporary Ballads*, New York: G. P. Putnam's Sons (1972) pp. 471, 472.

[90] Hibbs, loc. cit.

[47] Anonymous, "When Not to Pull Punches," *Nature*, London: Macmillan Journals Limited (1 September 1972) Volume 239, Number 5366, p. 1.

[48] *Ibid.*

[93] Koestler, Arthur, *The Case Of The Midwife Toad*, New York: Random House, Incorporated (1971).

[94] Alexander, George, "Experts Detect Monopole--- the 'Unicorn' of Science," *Los Angeles Times* (15 August 1975) Part 1, p. 1.

[95] Gaylin, Willard and Gorovitz, Samuel, "Academy Forum: Science And Its Critics," *Science* , Washington, D. C.: American Association for the Advancement of Science (25 April 1975) Volume 188, No. 4186, p. 315.

[96] Darwin, Charles, *The Origin of Species*, New York: Everyman's Library, Dutton: New York (1967) Sixth Edition, p. xxii.

[97] Abelson, Phillip H., "Bigotry in Science," *Science*, Washington, D. C.: American Association for the Advancement of Science (24 April 1964) Volume 144, Number 3617, p. 371.

[98] Jesseph, John E., "Bigotry in Scientists," *Science*, Washington, D. C.: American Association for the Advancement of Science (26 June 1964) Volume 144, No. 3626, pp. 1529, 1531.

[99] Mayr, Ernst, "The Nature of the Darwinian Revolution," *Science*, Washington, D.C., American Association for the Advancement of Science (2 June 1972) Volume 176, p. 982.

[100] Anonymous, "Second Thoughts About Man," *Time*, New York: Time, Inc. (23 April 1973) p. 83.

[101] Anonymous, "Evolution Not a Fact, Says Professor," *Los Angeles Times* (24 October 1973) Part IA, p. 5.

[102] Sieghart, Paul, "A Corporate Conscience for the Scientific Community?", *Nature*, London: Macmillan Journals Limited (1 September 1972) Volume 239, No. 5366, pp. 15-18.

[103] *Ibid.*

[104] Anonymous, "Second Thoughts About Man," loc. cit.

[105] Anonymous, "Strong X-ray Source Found in Deep Space," *Los Angeles Times* (20 March 1971) Part 1, p. 16.

[106] Bengelsdorf, Irving S., "Sixties Added Amazing Scientific Advances," *Los Angeles Times* (6 January 1970) Part 2, p. 12.

[107] Anonymous, "Scientists Unexpectedly Find Unknown Elementary Particle," *Los Angeles Times* (17 November 1974) Part 1, p. 3.

[108] Anonymous, "Second New Atom Particle Found By Stanford Lab," *Los Angeles Times* (23 November 1974) Part 3, p. 16.

[109] Meagher, Ed, "Galaxy Reported Traveling Faster Than Speed of Light," *Los Angeles Times* (24 October 1972) Part 1, p. 1.

[110] Harris, Errol E., *The Foundations of Metaphysics in Science*, New York: Humanities Press (1965) p. 23.

[111] *Ibid.*, p. 31.

[112] Getze, George, "Lunar Rocks Pinpoint Moon's Creative Stage, " *Los Angeles Times* (20 November 1969) Part 1, p. 28.

[113] Anonymous, "Studies Indicate Moon May Be More Complex Than Believed," *Los Angeles Times* (1 December 1969) Part 1, p. 1.

[114] Getze, George, "Scientists Doubt Real Age of Moon is Known," *Los Angeles Times* (9 January 1970) Part 1, p. 16.

[115] Anonymous, "Rocks Upset Theories on Moon's Age," *Los Angeles Times* (27 June 1971) Section A, p. 19.

[116] Anonymous, "'Genesis Rock' From Moon a Misnomer," *Los Angeles Times* (18 September 1971).

[117] *Ibid.*

[118] Campbell, Sam, "Hoaxes Among The Bone Finds," *The Register Focus*, Santa Ana, California: The Register (26 November 1972) P. 2.

[119] Anonymous, "Establishment of a Science of Genetics," *Los Angeles Times* (21 December 1972) Section C, p. 5.

[120] Miles, Marvin, "Amino Acids in Meteorite Hint Non-Earth Life," *Los Angeles Times* (2 December 1970) Part 1, p. 3.

[121] Getze, George, "Mythology of Aging Thwarts Longer Lives," *Los Angeles Times* (21 April 1970) Part 1, p. 1.

[122] Anonymous, "Triumph For a Heretic," *Newsweek*, Dayton, Ohio; Newsweek (20 July 1970) Volume LXXVI, No. 3, pp. 56, 57.

[123] Anonymous, "Morals Make A Comeback," *Time*, New York: Time, Incorporated (15 September 1975) p. 94.
[124] Burton, Alan C., "The Human Side of the Physiologist, Prejudice and Poetry," *The Physiologist*, American Physiological Society (1957) Volume 1, No. 1, pp. 1, 2.
[125] *Ibid.*, p. 2.
[126] Chein, Isidor, "Some Sources of Divisiveness Among Psychologists," *American Psychologist*, Washington, D. C.: The American Psychological Association, Incorporated (April 1966) Volume 21, No. 4, pp. 337-338.
[127] *Ibid.*, p. 337.
[128] Simpson, George Gaylord, "The Problem of Plan and Purpose in Nature," *Human Evolution*, New York: Henry Holt and Company (1959) p. 104.
[129] Simpson, George Gaylord, *The Meaning of Evolution*, New Haven, Conn.: Yale University Press (1952) p. 344.
[130] Nelson, Harry, "New Biology: Has It Become Too Confident?" *Los Angeles Times* (22 November 1970) Part 1, p. 1.
[131] *Ibid.*
[132] Kirsch, Robert, "Scientific Papers on Psychic Study," *Los Angeles Times* (17 January 1975) Part IV, p. 6.
[133] Porter, Sir George, "The Relevance of Science," *Engineering and Science*, Pasadena, California: California Institute of Technology and the Alumni Association (December 1974 - January 1975) Volume XXXVIIi, No. 2, pp. 22, 23.
[134] Porter, George, "Science: What Is Its Function?" *Los Angeles Times* (6 February 1975) Part II, p. 5.
[135] Ford, James H.,"Science: What Is Its True Purpose?" *Los Angeles Times* (15 February 1975) Part II, p. 4.
[136] Hoagland, Hudson, "Science and the New Humanism,"' *Science*, Washington, D. C.: American Association for the Advancement of Science (10 January 1964) Volume 143, No. 3602, p. 113.
[137] Ibid, p. 114.
[138] Schmitt, Roman A., "Birth Control: Science and Values," *Science*, Washington, D. C.: The American Association for the Advancement of Science (12 June 1964) Volume 144, No. 3624, pp. 1293, 1294.
[139] Heller, Walter W., "Why Can't Those (#$"%*!) Economists Ever Agree?" *TWA Ambassador*, St. Paul, Minnesota, Trans World Airlines (May 1975) Volume 8, No. 5, pp. 14, 15 and 40.
[140] Weinberg, Steven, "Reflections of a Working Scientist," *Daedalus -- Journal of the American Academy of Arts and Sciences*, Harvard University, Cambridge, Mass.: American Academy of Arts and Sciences (Summer, 1974) Volume 103, No. 3, pp. 42, 43.
[141] Roszak, Theodore, "The Monster and the Titan: Science, Knowledge and Gnosis," *Daedalus -- Journal of the American Academy of Arts and Sciences*, Harvard University, Cambridge, Mass.: American Academy of Arts & Sciences (Summer, 1974) Volume 103, No. 3, p. 18.
[142] Oliver, Bernard M., "The Search for Extraterristrial Intelligence," *Engineering and Science*, Pasadena, California: California Institute of Technology and the Alumni Association (December 1974 - January 1975) Volume XXXVIII, No. 2, pp. 7-11.

Chapter 3

[143] Hayakawa, S.I., (Foreward), *The Use and Misuse of Language*, Greenwich, Connecticut: Fawcett Publications, Inc. (1943) p. viii.
[144] Volpe, E. Peter, *Understanding Evolution*, Dubuque, Iowa: Wm. C. Brown Company Publishers (1967).
[145] *Ibid.,* p. 15.
[146] Macbeth, Norman, *Darwin Retried: An Appeal to Reason*, Boston Massachusetts; Gambit Incorporated (1971) p. 3.
[147] Moore, John A., "Creationism in California," *Daedalus – Journal of the American Academy of Arts and Sciences,* Cambridge, Massachusetts: American Academy of Arts and Sciences (Summer 1974) Volume 103, Number 3, pp. 177, 178.
[148] Holton, Gerald, "On the Duality and Growth of Physical Science," *American Scientist,* New Haven, Connecticut: The Society of the Sigma Xi and The Scientific Research Society of America (January 1953) Volume 41, Number 1, pp. 89-99.
[149] *Ibid.*, p. 96.
[150] *Science Framework for California Public Schools*, Sacramento, California; California State Department of Education (1970) p. 106.
[151] Sears, Francis Weston, "The Nature and Propagation of Light," *Principles of Physics III - Optics*, Cambridge, Massachusetts: Addison-Wesley Press, Inc. (148) p. 3.
[152] *Science Framework for California Public Schools*, loc. cit.
[153] Simpson, George Gaylord, *Tempo and Mode in Evolution*, New York, New York: Hafner Publishing Company, Inc, (1944) p. 107.
[154] Buswell, James O., III, "A Creationist Interpretation of Prehistoric Man," *Evolution and Christian Thought Today*, Grand Rapids, Michigan: Wm. B. Eerdmans Publishing Company (1959) p. 180.
[155] Miles, Marvin, "Comets Displaying Process of Creation, New Theory Claims," *Los Angeles Times* (9 January 1973) Part 2, p. 1.
[156] Alexander, George, "The Universe. Is the Edge of it Now in Sight?" *Los Angeles Times* (29 January 1973) Part 1, p. 1.
[157] Anonymous, "First Artificial Gene Created by Scientists," *Los Angeles Times* (3 June 1970) Part A, p. 1.
[158] Mayer, William V., "The Nineteenth Century Revisited," *BSCS Newsletter*, Boulder, Colorado: Biological Science Curriculum Study (November 1972) Number 49, pp. 7-13.
[159] Davis, Percy Roland, "State Publication of Textbooks in California," Thesis Submitted for Doctorate of Education, University of California, Berkeley, California: California Society of Secondary Education (1930) p. 10.
[160] *Ibid.*, p. 11.
[161] *Ibid.*, p. 12.
[162] Ryan, Dent and Stull, *Assembly Bill #531* (as amended in Assembly May 9, 1972) Chapter 2, Article 1, 9400 (a), p. 10.
[163] McCurdy, Jack, "Last of Liberals Leave State School Board," *Los Angeles Times* (10 January 1970) Part 2, p. 1.
[164] McCurdy, Jack, "State Education Board Backs Teaching of Evolution as Theory," *Los Angeles Times* (14 November 1969) Part 1, p. 1.
[165] Mayer, op. cit., p. 10.

[166] Simpson, George Gaylord, *The Meaning of Evolution*, New Haven, Connecticut: Yale University Press (1952) p. 344.
[167] Moore, op.cit., pp. 173, 174.
[168] *Ibid.*, p. 176.
[169] *Ibid.*, p. 174.
[170] Anonymous, "Survey Finds Physicists On The Left," *Physics Today*, Easton, Pennsylvania: American Institute of Physics, Inc. (October 1972) Page 61.
[171] Shils, Edward, "Faith, Utility, and Legitimacy of Science, " *Daedalus - Journal of the American Academy of Arts and Sciences*, Cambridge, Massachusetts: American Academy of Arts and Sciences (Summer 1974) Volume 103, Number 3, p. 1.
[172] Seidenbaum, Art, "Lock-Ear Liberals," *Los Angeles Times* (11 January 1973) Part 2, p. 1.
[173] *Ibid.*
[174] Ladd, Everett Carl1, Jr. and Lipset, Seymour Martin, "Politics of Academic Natural Scientists & Engineers," *Science*, Washington, D. C.: American Association for the Advancement of Science (9 June 1972) p. 1097.
[175] Hardy, Kenneth R., "Social Origin of American Scientists and Scholars," *Science*, Washington, D.C.: American Association for the Advancement of Science (9 August 1974) pp. 497-505.
[176] Ladd and Lipset, op. cit. , p. 1094.
[177] Clark, James V., "Motivation in Work Groups: A Tentative View," *Psychology in Administration* (edited by Timothy W. Costello and Sheldon S. Zalkind): Englewood Cliffs, N. J. (1963) pp. 61-63.
[178] Kent, Francis B., "Criminologist Moonlights as Policeman," *Los Angeles Times* (14 August 1975) Part 1, p. 24.
[179] Mathias, Charles McC., Jr., "Can the Political Parties Survive?" *The Washington Post* (8 November 1975) p. A-27.
[180] Hallett, Douglas L., "'New Progressivism' Links Brown, Reagan," *Los Angeles Times* (19 October 1975) Part 4, p. 1.
[181] Schrag, Peter, "Brown Skimps on Specifics," *Los Angeles Times* (11 January 1976) Part V, p. 1.
[182] Efron, Edith, "There *Is* A Network News Bias," *TV Guide*, Radnor, Pennsylvania: Triangle Publications, Inc. (28 February 1970) p. 8.
[183] Kilpatrick, James J., "Pot, Conservatively Speaking," *Los Angeles Times* (5 December 1974) Part II, p. 7.
[184] *Webster's New World Dictionary of the American Language*, College Edition, Cleveland & New York: The World Publishing Company (1959) p. 843.
[185] Gerard, Ralph, Form letter distributed to the public at large, dated 2 December 1969, pp. 1, 2.
[186] Larsen, Rebecca, "California's Evolution War: Should Genesis Get Equal Time?" *The Christian Century*, Chicago, Illinois: Christian Century Foundation (25 February 1970) Volume LXXXVII, Number 8.
[187] *Ibid.*, p. 252.

Chapter 4

[188] Anonymous, "Agnew vs. Press: Round Two," *Los Angeles Times* (23 November 1969) Section G, p. 6.
[189] McCurdy, Jack, "State Education Board Backs Teaching of Evolution as Theory," *Los Angeles Times* (14 November 1969) Part I, p. 1.
[190] Anonymous, "Palo Alto Trustees Challenge State Bible Ruling," *Palo Alto Times* (2 December 1969).
[191] *Ibid.*
[192] *Ibid.*
[193] McCurdy, Jack, "School Boards Oppose Shift on Evolution Guide," *Los Angeles Times* (10 December 1969) Part II, p. 6.
[194] Anonymous, "California's 'Evolution' Controversy," *Los Angeles Times* (9 December 1969) Part II, p. 6.
[195] Anonymous, "Genesis — Teaching Decision Stirs Arguments," *Santa Barbara News Press* (21 December 1969) Section A, p. 18.
[196] *Ibid.*
[197] Hooykaas, R., *Religion and the Rise of Modern Science*, Grand Rapids, Michigan: William B. Eerdmans Publishing Company (1972).
[198] Darwin, Charles, *The Origin of Species*, New York, New York: Everyman's Library (Sixth Edition: latest reprint 1967) Introduction by W. R. Thompson.
[199] Mathews, Linda, "Justices Narrow Definition of Public Figures in Libel Cases," *Los Angeles Times* (3 March 1976) Part I, p. 16.
[200] Ibid.
[201] Anonymous, "Biblical Creation Account to be Included in Texts," *Pentecostal Evangel*, Springfield, Missouri: Assemblies of God (11 January 1970) p. 26.
[202] Larsen, Rebecca, "California's Evolution War: Should Genesis Get Equal Time?" *The Christian Century*, Chicago, Illinois: Christian Century Foundation (25 February 1970) pp. 251-253.
[203] *Ibid.*
[204] *Ibid.*
[205] Plowman, Edward, "California's Fourth 'R': Board of Education Flap," *Christianity Today*, Washington, D. C.: Christianity Today (30 January 1970) p. 37.
[206] Gerard, R. W., "Some Things Should Not Be Mixed," *Los Angeles Times* (6 December 1969) Part II, p. 4.
[207] Brown, Nicholas E., "More on Science Teaching," *Los Angeles Times* (7 December 1969) Section G, p. 6.
[208] Grispino, Joseph A., "The Ghost of the Scopes Trial!" *Los Angeles Times* (6 December 1969) Part II, p. 4.
[209] Gerard, R. W., loc. cit.
[210] Anonymous, "Creation vs. Evolution; Battle Over What to Teach," *Los Angeles Times* (31 December 1972) Section C, p. 4.
[211] McCrea, W. H., "Origin of the Solar System, Review of Concepts and Theories," *On The Origin of the Solar System*, Paris, France: Centre National de la Recherche Scientifique (1972) p. 2.
[212] *Ibid.*

[213] McCurdy, Jack, "State Evolution Guide Stirs Education Dispute," *Los Angeles Times* (23 November 1969), Section B, p. 1.

[214] Gerard, R. W., "Some Things Should Not Be Mixed," *Los Angeles Times* (6 December 1969) Part II, p. 4.

[215] Anonymous, "Reshaping County Government," *Los Angeles Times* (15 February 1976) Part VI, p. 1.

[216] Gerard, R. W., loc.cit.

[217] Anonymous, "Equal Time for Eden," *Time*, Chicago, Illinois: Time-Life Publishers (12 December 1969) Volume 94, Number 24, p. 14.

[218] Personal correspondence of the Author, letter from Ralph D. Winter, 18 November 1969.

[219] Moore, John A., "Creationism in California," *Daedalus-- Journal of the American Academy of Arts and Sciences,* Cambridge, Massachusetts: American Academy of Arts and Sciences (Summer 1974) Volume 103, Number 3, pp. 179, 180.

[220] McCurdy, Jack, "Misconceptions Blamed in Furore in Evolution," *Los Angeles Times* (12 December 1969) Part I, p. 3.

[221] *Ibid.*

[222] Personal correspondence of the Author, letter from Ralph W. Gerard, 8 December 1969.

[223] Anonymous, "Ministers Meet," *The Valley News*, Van Nuys, California: The Valley News and Greensheet (22 January 1970).

[224] Binder, Otto O, and Flindt, Max H., "Is Man a Hybrid 'Developed' by a Super Space Civilization?" *Saga -- The Magazine for Men,* Brooklyn, New York: Gambi Publications, Inc. (June, 1970) Volume 40, Number 3, pp. 22-25; 62-64.

[225] Binder, Otto O. and Flindt, Max H., "Is Man a Hybrid 'Developed' by a Super Space Civilization?" *Saga -- The Magazine for Men*, Brooklyn, New York: Gambi Publications, Inc. (1972) Volume 1, Number 4, pp. 12-15, 100-103.

[226] Anonymous, "Downgrading Darwin in Class," *Medical World News*, New York, New York: McGraw-Hill, Inc. (3 April 1970) Volume 11, Number 14, pp. 20-21.

[227] Holles, Everett R., "Science Teaching at Issue on Coast," *The New York Times* (22 October 1972) p. 76.

[228] Anonymous, "Darwin Gains in Bible Fight," *Pentecostal Evangel*, Springfield, Missouri: Assemblies of God (25 January 1970) p. 26.

[229] Irish, Peter, "Teach Creation in the Schools?" *Discovery*, Palo Alto, California. Peninsula Bible Church (15 February 1970) Volume 1, Number 1, p. 1.

[230] News report, *KNX News Radio*, Los Angeles, California, 13 November 1969.

[231] Scopes, John T., and Presley, James, *Center of the Storm -- Memoirs of John T. Scopes*, New York, New York: Holt, Rinehart & Winston (1967) p. 152.

[232] *Ibid.*, p. 153.

[233] Dart, John, "Textbook Fight Partly Laid to Bible Scholars," *Los Angeles Times* (21 January 1973) Part IV, p. 5.

[234] Anonymous, "New Voices Heard Against Evolution," *Los Angeles Times* (21 January 1973) Part IV, p. 5.

[235] Scopes, John T., and Presley, James, loc. cit., Preface p. VI.

Chapter 5

[236] Gurko, Miriam, *Clarence Darrow*, New York, New York; Thomas Y. Crowell Company (1965) p. 220.
[237] Anonymous, "Evolution and the State Board," *Los Angeles Times* (14 October 1969) Editorial Page.
[238] Alexander, George, "Celestial X Rays a Fresh Mystery," *Los Angeles Times* (19 March 1976) Part II, p. 5.
[239] Proverbs 29:11.
[240] Gerard, Ralph, Form letter distributed to the public at large, dated 2 December 1969, pp. 1, 2.
[241] Romans 1:20, 21, 25 (Phillips Translation).
[242] Anonymous, "Evolution Just Theory, Three Experts Agree," *Los Angeles Times* (9 December 1972) Part II, p. 6.
[243] Moore, John A., "Creationism in California," *Daedalus -- Journal of the American Academy of Arts and Sciences,* Cambridge, Massachusetts American Academy of Arts and Sciences (Summer 1974) Volume 103, Number 3, pp. 179, 180.
[244] Mayer, William V., "The Nineteenth Century Revisited," *BSCS Newsletter*, Boulder, Colorado: The University of Colorado, The Biological Sciences Curriculum Study (November 1972) Volume No. 49, p. 10.
[245] *Ibid.*, pp. 10, 11.
[246] McCurdy, Jack, "Guidelines for Science Texts Okayed by Board," *Los Angeles Times*.(14 March 1970) Part I, p. 3.
[247] Mayer, William V., op, cit., pp, 7-13.

Chapter 6

[248] Holton, Gerald, "On the Duality and Growth of Physical Science," *American Scientist*, New Haven Connecticut: The Society of the Sigma Xi and the Scientific Research Society of America (January 1953) Volume 41, Number 1, p. 96.
[249] *Ibid.*
[250] Thompson, W. R., *Science and Common Sense*, Albany, New York: Magi Books, Incorporated (1965) p. 229.
[251] Burnam, Tom, "Facts We Know That Are Not So", *Readers Digest*, Pleasantville, New York: The Reader's Digest Association, Inc. (April 1976), Volume 108, Number 648, pp. 169-174.
[252] Roszak, Theodore, "The Monster and the Titan: Science, Knowledge, and Gnosis," *Daedalus -- Journal of the American Academy of Arts and Sciences*, Cambridge, Massachusetts: American Academy of Arts and Sciences, Harvard University (Summer 1974) Volume 103, Number 3, p. 20.
[253] Harris, Errol E., *The Foundations of Metaphysics In Science*, New York, New York: Humanities Press (1965) p. 260.
[254] *Ibid.*, p. 261.
[255] Brush, Stephen G., "Can Science Come Out of the Laboratory Now?" *Bulletin of the Atomic Scientists*, Chicago, Illinois: The Educational Foundation for Nuclear Science (April 1976) p. 41.
[256] *Ibid.*, p. 43.

[257] Hoagland, Hudson, "Science and the New Humanism," *Science*, Washington, D.C. : American Association for the Advancement of Science (10 January 1964) Volume 143, Number 3602, p. 113.
[258] Edwards, Paul, "Why I am Not a Christian," *Why I am Not a Christian: and Other Essays on Religion and Related Subjects*, New York, New York: Simon & Schuster (1957) p. 22.
[259] Kerkut, G. A., *Implications of Evolution*, New York, New York: Pergamon Press, Incorporated (1960) pp. 4, 5.
[260] Hoagland, Hudson, loc. cit.
[261] MacKay, Donald M., *The Clockwork Image*, Downers Grove, Illinois: InterVarsity Press (October 1974) p. 43.
[262] Anonymous, "Has America Had It?" *Firing Line*, Columbia, South Carolina: Southern Educational Communications Association (20 August 1973) pp. 6, 7.
[263] Dobzhansky, Theodosius, "Evolution at Work," *Science*, Washington, D. C.: American Association for the Advancement of Science (9 May 1958) p. 1091.
[264] Vanderkooi, Garret, "Evolution as a Scientific Theory," *Christianity Today*, Washington, D. C. ; David R. Rehmeyer (7 May 1971) Volume XV, Number 16, p. 13.
[265] Wilder Smith, A. E., *Man's Origin, Man's Destiny*, Wheaton, Illinois: Harold Shaw Publishers (1968) pp. 152, 153.
[266] Vanderkooi, Garret, loc. cit.
[267] Ayala, Francisco J., "Genotype Environment and Population Numbers," *Science*, Washington, D.C.: American Association for the Advancement of Science (27 December 1968) Volume 162, p. 1456.
[268] Waddington, C. H., *The Nature of Life*, New York, New York: Anthenium (1962) p. 98.
[269] Harris, Errol E., op.cit. , p. 259.
[270] *Ibid.*
[271] *Ibid.*, p. 261.
[272] *Ibid.*, pp. 268-273.
[273] Lodge, George Cabot, "Business and the Changing Society," *Harvard Business Review*, Boston, Massachusetts: Harvard.University (April 1974) Volume 52, Number 2, p. 63.
[274] Harris, Errol E. op. cit., p. 280.
[275] *Ibid.*
[276] Dirksen, Everett McKinley and Prochnow, Herbert, *Quotation Finder*, New York, New York: Harper and Row (1971).
[277] Dart, John, "Origin of Man: Creation Theory Far From Dead," *Los Angeles Times* (25 December 1969), Part 1, p. 1.
[278] Dart, John, "Textbook Fight Partly Laid to Bible Scholars," *Los Angeles Times* (21 January 1973) Part IV, p. 5.
[279] Viner, Jacob, *The Role of Providence in the Social Order*, Philadelphia, Pennsylvania, American Philosophical Society (1972) Volume 90, p. 18.
[280] Roszak, Theodore, op. cit., p. 18.
[281] Dubos, Rene, *A God Within*, New York, New York: Charles Scribner's Sons (1972) pp. 42, 43.

[282] Mayr, Ernst, "The Nature of the Darwinian Revolution," *Science*, Washington, D.C. : American Association for the Advancement of Science (2 June 1972) Volume 176, pp. 981-989.
[283] Kriehn, David A., "Evolution or Creation?" *The Black and Red*, Watertown, Wisconsin: Northwestern College (May 1973) Volume 77, Number 1, pp. 7-9.
[284] Darwin, Charles, *The Origin of Species*, New York, New York: Everyman's Library No. 811; E. P. Dutton & Co., Inc. (Latest Reprint 1967) Sixth Edition, p. 455.
[285] Schmidt, Oscar, *The Doctrine of Descent and Darwinism*, New York, New York: D. Appleton & Company (1896) Introduction v-vi.
[286] *Ibid.*, p. 195.
[287] *Ibid.*, p. 221.
[288] *Ibid.*, p. 310.
[289] Anonymous, "Minority Still Hoping to Discredit Darwin," *Los Angeles Times* (17 December 1971) Part 1 D, p. 8.
[290] Smith, Ned, "No Easy Answers," *American Way*, New York, New York: American Way (November 1975) Volume 8, Number 11, p. 12.

Chapter 7

[291] Monod, Jacques, *Chance & Necessity*, New York: Vintage Books, A Division of Random House (October 1972) pp. 146, 147.
[292] *Ibid*, p. 118.
[293] Simpson, George Gaylord, "The Problem of Plan and Purpose in Nature," *Human Evolution*, New York: Henry Holt and Company (1959) pp. 118, 125.
[294] Yourno, Joseph, Kohno, Tadahiko, and Roth, John, "Enzyme Evolution: Generation of a Bifunctional Enzyme by Fusion of Adjacent Genes," *Nature*, London. Macmillan Journals, Ltd. (28 November 1970) pp. 820-823.
[295] Getze, George, "Laboratory Feat with Genes May Help Explain Evolution," *Los Angeles Times* (29 November 1970) Section C, p 1.
[296] Toth, Robert C. , "Hunt For Other Life Held Top Space Goal," *Los Angeles Times* (27 March 1965) Part 1, p. 1.
[297] Anonymous, "Universe No Accident," *Pentecostal Evangel*, Springfield, Missouri: Assemblies of God, (11 February 1973).
[298] Anonymous, "Space-Life Evidence on Upswing," *Los Angeles Times* (16 April 1971) Part 1 B, p. 19.
[299] Barnhart, Clarence L. (Ed.), *The World Book Dictionary*, Chicago, Illinois: Field Enterprises Educational Corporation (1967) p. 324.
[300] Monod, Jacques, op. cit., p. 118.
[301] Getze, George, "Chemist Says Other Planets' Life Would Be Like Earth's;" *Los Angeles Times* (9 November 1971).
[302] Anonymous, "Carrots Don't Help Vision, Expert Says," *Los Angeles Times* (22 February 1971) Part 1, p. 19.
[303] Miles, Marvin, "New Evidence of Moon Split From Earth Told," *Los Angeles Times* (12 June 1975) Part 11, p. 1.
[304] Getze, George, "Moon Too Big, New Planet Theory Claims," *Los Angeles Times*, (1 February 1972) Part 1, p. 3.

[305] Hibbs, Albert R., "Inquisition, Repression and Ridicule," *Engineering and Science*, Pasadena, California: California Institute of Technology and the Alumni Association (December 1974-January 1975) Volume XXXVIII, No. 2, p. 6.
[306] Medawar, Sir Peter Brian, *Induction and Intuition in Scientific Thought*, Philadelphia, Pennsylvania: American Philosophical Society (1969) Volume 75, pp. 10, 11.
[307] Crosby, Kent M., "A Futile Argument," Sacramento, California: *The Sacramento Bee* (13 December 1972).
[308] Birx, H. James, "Teaching Evolution: The Relevance of Philosophical Literature," *Bioscience*, Washington, D.C.: The American Institute of Biological Sciences (15 June 1971) Volume 21, No. 12, p. 573.
[309] Huxley, Julien, "Evolution and Genetics," *What Is Man?* (Edited by J. R. Newman), New York: Simon and Schuster (1955) p. 278.
[310] Dobzhansky, Theodosius, "Changing Man," *Science*, Washington, D. C.: American Association for the Advancement of Science (27 January 1967) Volume 155, p. 409.
[311] Dubos, Rene, "Humanistic Biology," *American Scientist*, New Haven, Connecticut: The Society of the Sigma Xi and The Scientific Research Society of America (March 1965) Volume 53, p. 6.
[312] Vaeck, S. V., "Scientific Responsibility," *Nature*, London: Macmillan Journals, Ltd. (28 November 1970) Volume 228, Number 5274, p. 888.
[313] Birx, H. James, loc. cit.
[314] Goldschmidt, Richard B., "Evolution, as Viewed by One Geneticist," *American Scientist*, New Haven, Connecticut: The Society of the Sigma Xi and the Scientific Research Society of America (January 1952) Volume 40, Number 1, p. 84.
[315] *Ibid.*
[316] *Ibid.*, p. 93.
[317] *Ibid.*
[318] *Ibid.* , pp. 97, 98.
[319] Gore, Rick, "The Awesome Worlds Within a Cell," *National Geographic*, Washington, D.C.: National Geographic Society (September 1976) Volume 150, Number 3, p. 390.
[320] Moorhead, P. S. and Kaplan, M. M., *Mathematical Challenges to the Neo-Darwinian Interpretation of Evolution*, Philadelphia, Pennsylvania: Wistar Press (1967) p. 109.
[321] Wilder Smith, A. E., *Man's Origin, Man's Destiny*, Wheaton, Illinois: Harold Shaw Publishers (1968) p. 117.
[322] Alexander, George, "Man In New World 50,000 Years Pair Say," *Los Angeles Times* (14 May 1974) Part II, p. 1.
[323] Bada, Jeffrey L., Schroeder, Roy A. and Carter, George F., "New Evidence for the Antiquity of Man in North America Deduced from Aspartic Acid Racemization," *Science*, Washington, D. C.: American Association for the Advancement of Science (17 May 1974) pp. 391-383.
[324] Cromie, William J., The Moon's Unfolding History," *Science Year*, Chicago, Illinois: Field Enterprises Educational Corporation (1972) p. 41.
[325] Anonymous, "Science File - Geoscience," *Science Year*, Chicago, Illinois: Field Enterprises Educational Corporation (1975) p. 295.
[326] 36 Anonymous, "'Genesis Rock' from Moon a Misnomer," *Los Angeles Times* (18 September 1971).

[327] Howell , F. Clark, "Our Earliest Ancestors," *Science Year*, Chicago, Illinois: Field Enterprises Educational Corporation (1973) p. 229.
[328] Ibid.
[329] Anonymous, "Science File - Geoscience," *Science Year*, Chicago, Illinois. Field Enterprises Educational Corporation (1974) pp. 299, 300.
[330] Wilder Smith, A. E. , op. cit. , p. 127.
[331] Kerkut, G. A., *Implications of Evolution*, New York; Pergamon Press, Incorporated (1960) p. 153.
[332] Anonymous, "Science File - Geology," *Science Year*, Chicago, Illinois: Field Enterprises Educational Corporation (1971) p. 313.
[333] Getze, George, "Earth Born in 10,000-Year 'Instant', Two Scientists Say," *Los Angeles Times* (20 June 1972) Part 1, p. 1.
[334] Hanks, Thomas C. and Anderson, Don L., "Formation of the Earth's Core," *Nature*, London, Macmillan Journals, Ltd. (16 June 1972) Volume 237, Number 5355, pp. 383-388.
[335] Wilder Smitb, A. E., op. cit., p. 146.
[336] Sagan, Carl, "'Dec. 31 – The First Cosmic Year Is Ending About Now," *Los Angeles Times* (31 December 1975) Part II, p. 3.
[337] Ibid.
[338] Dalrymple, Richard, "This Existence: Creation or Evolution?" *Los Angeles Herald-Examiner* (14 November 1970) Part A, p. 7.
[339] Ibid.
[340] Anonymous, "Science File – Astronomy," *Science Year*, Chicago, Illinois: Field Enterprises Educational Corporation (1973) p. 273.
[341] Sagan, Carl, loc. cit.
[342] Howell, F. Clark, *Early Man*, Sacramento, California: California State Department of Education (1965) p. 176.
[343] Anonymous, *Science Framework for California Public Schools* (Preliminary Draft) Sacramento, California: California State Department of Education (1969) p. 125.
[344] Anonymous, "Discovery Redates Life a Billion Years, Russ Says," *Los Angeles Times* (5 July 1975) Part I, p. 4.
[345] Alexander, George, "Scientists Dig in Africa for Primitive Man," *Los Angeles Times* (12 November 1972) Section A, p. 14.
[346] Anonymous, "Scientist Calls Pyrenean Skull a Historic Find," *Los Angeles Times* (5 November 1971).
[347] Alexander, George, "Fossils Reveal 'Man' That Didn't Survive," *Los Angeles Times* (25 February 1974) Part 11, p. 2.
[348] *Ibid.*
[349] *Ibid.*
[350] Ottaway, David B., "Human Fossil Find Believed Oldest Yet,"
Los Angeles Times (26 October 1974) Part I, p. 1.
[351] *Ibid.*
[352] *Ibid.*
[353] Torgerson, Dial, "Team Seeks Positive Proof of Human Fossil," *Los Angeles Times* (24 November 1974) Part 7, p. 4.

[354] Anonymous, "Hominid Bones: Old and Firm at 3.75 Million," *Science News*, Washington, D. C.: Science Service Incorporated (8 November 1975) Volume 108, Number 19, p. 292.

[355] Alexander, George, "Fossils 600 Million Years Old Found," *Los Angeles Times* (29 November 1974) Part 11, p. 1.

[356] Anonymous, *Science Framework for California Public Schools*, Sacramento, California: California State Department of Education (1970) p. 8.

[357] Darwin, Charles, *The Origin of Species*, New York: Everyman's Library, New York: Dutton (1967) Sixth Edition, pp. 315, 316.

[358] Newell, Norman D., "The Nature of the Fossil Record," *Proceedings of the American Philosophical Society*, Philadelphia, Pennsylvania: The American Philosophical Society (1959) Volume 103 (2) pp. 264-285.

[359] *Ibid.*

[360] Oleyar, Rita D., "'In The Beginning…': A Case for Myths," *Los Angeles Times* (19 November 1972) Section F, p. 1.

[361] *Webster's New World Dictionary of the American Language*, College Edition, Cleveland & New York: The World Publishing Company (1959) p. 843.

[362] Leakey, Richard E., "Skull 1470," *National Geographic*, Washington, D. C.: National Geographic Society (June 1973) Volume 143, Number 6, p. 819.

[363] Ibid.

[364] Mullin, Robert, "The Making of a Textbook: McGuffey's Days Are Gone," *Los Angeles Times* (31 December 1972) Section C, p. 5.

[365] Bruckner, D. J, R., , "Research Into Origins of Life Is Too Vital to Remain Private Study," *Los Angeles Times* (28 December 1970) Part II.

Chapter 8

[366] Anonymous, *Science Framework for California Public Schools,* Sacramento, California: California State Department of Education (1970) p. 106.

[367] Anonymous, "Alfred Romer, Evolutionary Authority Dies," *The New York Times,* (7 November 1973).

[368] Anonymous, "Awards and Prizes," *Science Year,* Chicago, Illinois: Field Enterprises Educational Corporation (1966) p. 425.

[369] Simpson, George Gaylord, *Tempo and Mode in Evolution,* New York: Hafner Publishing Company, Inc. (1944, reprint 1965) p. 107.

[370] Newell, Norman D., "The Nature of the Fossil Record," *Proceedings of the American Philosophical Society*, Philadelphia, Pennsylvania: The American Philosophical Society (1959) Volume 103 (2) pp. 264-285.

[371] Cowen, Robert C., "Biological Origins: Theories Evolve," *The Christian Science Monitor*, Boston Mass.: Christian Science Monitor, (4 January 1962) p. 4.

[372] Cowen, Robert C., "Tracking Down the Dawn of Life," *Science Year,* Chicago, Illinois: Field Enterprises Educational Corporation (1973) p. 137.

[373] Jukes, Thomas H., Open Letter to Dr. Junji Kumamoto distributed to the public at large, dated 29 August 1972.

[374] Calvin, Melvin, Open Letter to Dr. Junji Kumamoto distributed to the public at large, dated 6 September 1972.

[375] Kumamoto, Junji, Letter to Professor J. Ledyard Stebbins, Herman T. Spieth and 14 others, dated 16 September 1972, p. 3.
[376] Carpenter, Don, "Loving God Will be Understanding," *The Ledger,* Montrose, California: Donald T. Carpenter (31 January 1973) Section 2, p. 2.
[377] McCurdy, Jack, "Science Academy Hits Texts' Creation Theory," *Los Angeles Times* (18 October 1972) Part 11, p. 1.
[378] *Ibid.*
[379] *Annual Report of the National Academy of Sciences, Fiscal Years 1973 and 1974,* 94th Congress, 1st Session: Senate Doctunent No. 94-41, Washington, D.C., pp, 146, 147.
[380] *Ibid.*, p. 147.
[381] *Ibid.*, p. 170.
[382] Dubos, Rene, *A God Within*, New York: Charles Scribners Sons (1972) p. 255.
[383] Barnett, Lincoln, *The Universe and Dr. Einstein*, New York: Signet Science Library Books (1964) p. 108.
[384] Anonymous, "Alternative Hypothesis," *Scientific American,* New York: Scientific American, Incorporated (August 1972) Volume 227, No. 2, pp. 43, 44.
[385] McCurdy, Jack, "Bible Creation Theory in School Texts OKd," *Los Angeles Times* (13 October 1972) Part II, p. 1.
[386] *Ibid.*
[387] Anonymous, "Science Textbook Ruling," *Sacramento Bee*, Sacramento, California: Sacramento Bee (13 October 1972).
[388] Anonymous, "Evolution Issue Stirs Furore on State Textbooks," *Los Angeles Times* (15 September 1972) Part 1II, p. 1.
[389] Fischer, Robert B., *Science, Man and Society,* Philadelphia, Pennsylvania: W. B. Saunders Company (1971).
[390] Anonymous, "Alternative Hypothesis," *Scientific American*, New York: Scientific American, Incorporated (August 1972) Volume 227, No. 2, pp. 43, 44.
[391] Forbes, Cheryl, "God and Darwin in California," *Christianity Today,* Washington, D.C., Wilbur D. Benedict (22 December 1972) p. 35.
[392] Anonymous, "Evolution Issue Stirs Furore on State Textbooks," *Los Angeles Times* (15 September 1972), Part II, p. 1.
[393] McCurdy, Jack, "Bible Creation Theory in School Texts OKd " *Los Angeles Times* (13 October 1972), Part 11, p. 1.
[394] McCurdy, Jack, "Science Academy Hits Texts' Creation Theory," *Los Angeles Times* (18 October 1972) Part II, p. 1.
[395] Anonymous, "NAS Eyes California Controversy, Urges That Science Texts Deal Only With Science,"*News Report - National Academy of Sciences*, Washington, D.C.; National Academy of Sciences (November 1972) Volume XXII, Number 9, p. 1.
[396] McCurdy, Jack, "Special Creation Theory in School Books Opposed," *Los Angeles Times* (2 November 1972) Part II, p. 1.
[397] Anonymous, "Creation in California," *Nature*, London, England: Macmillan Journals Limited (20 October 1972) Volume 239, Number 5373, p. 420.
[398] Lucas, E. C., "Creationism and Evolutionism," *Science*, Washington, D.C.: American Association for the Advancement of Science (9 March 1973) Volume: 179, Number 4076, p. 953.

[399] *Ibid.*
[400] *Ibid.*
[401] Martz, D.C., "Creationism and Evolutionism," *Science*, Washington, D. C.: American Association for the Advancement of Science (9 March 1973) Volume 179, Number 4076, p. 953.
[402] Tinkle, William J., "Creationism and Evolutionism," *Science,* Washington, D.C.: American Association for the Advancement of Science (9 March 1973) Volume 179, Number 4076, p. 954.
[403] Anonymous, "Neoclassicism Challenges QED," *Physics Today*, New York: American Institute of Physics (October 1972) Volume 25, No. 10.
[404] Payne, Robert, *The Life and Death of Adolf Hitler*, New York: Praeger Publishers, inc. (1973) p. 589.
[405] Mayer, William V., "The Nineteenth Century Revisited," *BSCS Newsletter*, Boulder, Colorado: The University of Colorado, The Biological Sciences Curriculum Study (November 1972) Volume 49, pp. 7-13.
[406] *Ibid*, p. 9.
[407] Payne, Robert, loc. cit.
[408] Mayer, William V., "The Nineteenth Century Revisited," *BSCS Newsletter*, Boulder, Colorado: The University of Colorado, The Biological Sciences Curriculum Study (November 1972) Volume 49, p. 13.
[409] Lightner, Jerry P., "The National Association of Biology Teachers ...where it stands," *BSCS Newsletter*, Boulder, Colorado: The University of Colorado, The Biological Sciences Curriculum Study (November 1972) Volume 49, p. 14.
[410] *Ibid.* p. 15.
[411] *Ibid.*
[412] Anonymous, "Evolution Debate Resumes in California," NSTA *News-Bulletin*, Washington, D.G.: National Science Teachers Association (September 1972).

Chapter 9

[413] Revelation 16:16.
[414] Hansen, Earl, "Nov. 9 Decisive Day for State's School Textbooks," *Seattle Post-Intelligencer,* Seattle, Washington: Seattle Post-Intelligencer, (28 October 1972).
[415] Ruppert, Ray, "Ex-Seattleite Wages War Against 'Scientific Religion'," *The Seattle Times,* Seattle, Washington: The Seattle Times (2 November 1972) Section B, p. 10.
[416] Anonymous, *Call For Bids For Textbooks and Reusable Educational Materials -- Science Health Music,* Sacramento, California: California State Board of Education (1 March 1971) p. 27.
[417] Anonymous, "Scopes Revisited," *California Journal,* Sacramento, California: California Center for Research & Education in Government (December 1972) p. 350.
[418] Plowman, Edward, "California's Fourth 'R': Board of Education Flap," *Christianity Today,* Washington, D. C.: Wilbur D. Benedict (30 January 1970) p. 37.
[419] Remmel, Ronald S. , "Randomness in Quantum Mechanics and Its Implications For Evolutionary Theory," *Journal of the American Scientific Affiliation,* Elgin, Illinois: American Scientific Affiliation (September 1974) p. 96-98.

[420] McCurdy, Jack, "Evolution or Creation? The Fight's Revived," *Los Angeles Times* (10 November 1972) Part I, p. 3.

[421] *Ibid.*

[422] *Ibid.*

[423] *Ibid.*

[424] Moskowitz, Ron, "Move Over, Darwin," *San Francisco Chronicle,* San Francisco, California: San Francisco Chronicle (10 November 1972) P. 1.

[425] *Ibid.*

[426] Speich, Don, "Board Will Decide Text Content," *The Sacramento Bee,* Sacramento, California: The Sacramento Bee (10 November 1972) Section A, p. 1.

[427] *Ibid.*

[428] Whiteside, Beverlei, "Creation Vs. Darwin Dispute In Schools," *Los Angeles Herald-Examiner* (19 November 1972) Section A, p. 18.

[429] Anonymous, "What Is Man's Origin?" *The Register,* Santa Ana, California: The Register (26 November 1972) Focus, p. 1.

[430] Hansen, Earl, "Equal Time for God in Science Classes?" *Seattle Post-Intelligencer,* Seattle, Washington: Seattle Post-Intelligencer (25 November 1972) Section A, p. 8.

[431] Anonymous, "Scopes Revisited," *Newsweek,* Dayton, Ohio: Newsweek, Inc. (20 November 1972) p. 86.

[432] Wade, Nicholas, "Creationists and Evolutionists: Confrontation in California," *Science,* Washington, D. C.: The American Association for the Advancement of Science (17 November 1972) pp. 724-729.

[433] *Ibid.*

[434] Hoover, Arlie J., "Science Joins Religion in Ranks of Prejudice," *Los Angeles Times* (10 November 1972) Part II, p. 11.

[435] Hoover, Arlie J., "Science Joins Religion in Ranks of Prejudice," *Christianity Today,* Washington, D. C.: Wilbur D. Benedict (22 December 1972) pp.12, 13.

[436] Spears, Larry, "State Board to Tackle Evolution Problem," *Oakland Tribune,* Oakland, California: Oakland Tribune (10 December 1972) p. 14.

[437] Anonymous, "Evolution Controversy Flairs Anew," *Chemical & Engineering News, American* Chemical Society (11 December 1972) Volume 50, Number 50, p. 13.

[438] Anonymous, "Letters On Science and Religion," *Chemical & Engineering News,* American Chemical Society (12 February 1973) Volume 51, Number 7, p. 32.

[439] Hoppe, Arthur, "A Simply Divine Theory of Evolution," *San Francisco Chronicle,* San Francisco, California: San Francisco Chronicle (14 November 1972) p. 41.

[440] Dart, John, "Biblical Creation In Textbooks Urged," *Los Angeles Times* (17 November 1972) Part II, p. 2.

[441] Anonymous, "Creation Theory and Empirical Science," *Los Angeles Times* (18 November 1972).

[442] Moskowitz, Ron, "The Textbook Evolution Fight," *San Francisco Chronicle,* San Francisco, California: San Francisco Chronicle (27 November 1972) p.15.

[443] Anonymous, "How A Publisher Handled The Issue," *San Francisco Chronicle,* San Francisco, California: San Francisco Chronicle (27 November 1972) p. 15.

[444] Moskowitz, Ron, "Evolution Debate Evolves An Answer," *San Francisco Chronicle,* San Francisco, California: San Francisco Chronicle (8 December 1972) p. 4.

[445] Spears, Larry, loc. cit.

[446] Spears, Larry, "State Text Ruling Near," *Oakland Tribune,* Oakland, California: Oakland Tribune (11 December 1972) p. 1.
[447] Anonymous, "The Schoolbook Controversy," *San Francisco Chronicle,* San Francisco, California. San Francisco Chronicle (13 December 1972) p. 62.
[448] Anonymous, "19 Nobel Laureates Join Creation Fight," *Los Angeles Times* (12 December 1972) Part II, p. 11.
[449] Anonymous, "Scientists' Plea In Evolution Debate," *San Francisco Chronicle,* San Francisco, California: San Francisco Chronicle (12 December 1972).
[450] Anonymous, "Nobel Laureates Oppose 'Equal Time' for Biblical Creation In Education," *Pentecostal Evangel,* Springfield, Missouri. Assemblies of God (18 February 1973) p. 25.
[451] Anonymous, "Adam and Eve Evicted from Calif. School Books," *The Seattle Times,* Seattle, Washington: The Seattle Times (13 December 1972).
[452] Anonymous, "Adam, Eve and Darwin," *The Sacramento Union,* Sacramento, California: The Sacramento Union (14 December 1972) p. A4.
[453] Speich, Don, "State Decision Pends On Creation, Evolution (Or Both) In School Texts," *The Sacramento Bee,* Sacramento, California: The Sacramento Bee (10 December 1972) Section A, p. 22.
[454] McGurdy, Jack, "Inclusion of Religious Creation Theory in Textbooks Rejected," *Los Angeles Times* (15 December 1972) Part I, p. 1.
[455] Anonymous, "Calif. Downgrades Darwin's Theory," *The Seattle Times,* Seattle, Washington; The Seattle Times (15 December 1972).
[456] Anonymous, "Darwin Demoted," *Spokane Daily Chronicle,* Spokane, Washington: Cowles Publishing Company (15 December 1972) p. 24.
[457] Anonymous, "Darwin Who?" *Time,* New York, New York: Time Magazine (25 December 1972) pp. 34-5.
[458] Spears, Lawrence M., "'Creationists' Win Right to Revise Science Texts," *The National Observer,* Washington, D. C.: The National Observer (30 December 1972) p. 6.
[459] McCurdy, Jack, loc. cit.
[460] McCurdy, Jack, "Move to Put Religious Creation Theory in Science Books Killed," *Los Angeles Times* (13 January 1973) Part I, p. 1.
[461] *Ibid.*
[462] Anonymous, "Keep Pressure Off Textbooks," *Los Angeles Times* (17 January 1973) Part II, p. 8.
[463] Moskowitz, Ron, "Eden and Evolution," *Saturday Review of The Sciences,* San Francisco, California: Saturday Review Company (27 January 1973) Volume I, Number 1, pp. 58, 59.
[464] Anonymous, "Creation vs. Evolution -- Battle Over What To Teach," *Los Angeles Times* (31 December 1972) Section C, p. 5.
[465] McCurdy, Jack, "Controversy Swirls Again Over Evolution Teaching: Other Clashes Loom," *The Spokesman-Review,* Spokane, Washington: The Spokesman-Review (7 January 1973) p. 2.
[466] McCurdy, Jack, "Silent Stand Urged On Creation Issue in Science Textbooks," *Los Angeles Times* (12 January 1973) Part II, p. 3.

[467] Anonymous, "California Text Compromise," *Education-Training Market Report, Old* Westbury, New York: New York Institute of Technology (5 February 1973) Volume IX, Number 3, p. 8.

[468] McCurdy, Jack, "God-as-Creator Theory Nearly OKd for Texts," *Los Angeles Times* (9 February 1973) Part 1, p. 26.

[469] *Ibid.*

[470] *Ibid.*

[471] Hubbard, Harold N., "Fuller Chief Tells Position On Evolution," *Pasadena Star-News*, Pasadena, California (14 March 1973) p. 4.

[472] Nelkin, Dorothy, "The Science-Textbook Controversies," *Scientific American,* New York, New York: Scientific American, Inc. (April 1976) Volume 234, No. 4, pp. 33-39.

[473] *Ibid.*, p. 35.

[474] Macbeth, Norman, *Darwin Retried -- An Appeal To Reason,* Boston, Mass.: Gambit, Incorporated (1971).

[475] Bethell, Tom, "Darwin's Mistake," *Harpers,* New York, New York: Harper's Magazine Company (February 1976) Volume 252, No. 1509, p. 72.

[476] *Ibid.*, pg. 72, 75.

[477] Greene, John G., *Darwin And The Modern World View,* Mentor Books (1963) p. 116.

[478] Anonymous, "Science Tilts With Irrationalism," *Chemical and Engineering News,* Washington, D. C.: American Chemical Society (15 January 1973) p, 18.

[479] *Ibid.*

[480] Anonymous, "Bellow Assails the Intellectual Bosses," *Los Angeles Times* (13 December 1976) Part 1, p. 1.

[481] Fischer, Robert B., "The Evolution of a Policy on Textbooks," *Los Angeles Times* (15 April 1973) Part VII, p. 7.

[482] Newell, Norman D., "Special Creation and Organic Evolution," *Proceedings of the American Philosophical Society,* Philadelphia, Pa.: The American Philosophic Society (August 1973) Volume 117, No. 4, pp. 323-331.

[483] *Ibid.*, p. 323.

[484] Harris, Errol E., *The Foundations of Metaphysics in Science,* New York: Humanities Press (1965) p. 408.

[485] Vander Vennen, Robert E., "Is Scientific Research Value-Free," *Journal of the American Scientific Affiliation,* Elgin, Illinois: The American Scientific Affiliation (September 1975) Volume 27, No. 3, pp 108.

[486] *Ibid.*

[487] Holton, Gerald, *Thematic Origins Of Scientific Thought: Kepler to Einstein,* Cambridge, Massachusetts: Harvard University Press (1973) p. 245.

[488] *Ibid.*

[489] Merton, Robert K., "Thematic Analysis in Science: Notes on Holton's Concept," *Science,* Washington, D.C.: American Association for the Advancement of Science (25 April 1975) Volume 188, No. 4186, p, 335.

[490] Oparin, A.I., "Introduction by S. Morgulis, Translator," *The Origin Of Life,* New York: Dover Publications, Incorporated (1953) P. XXII.

[491] Volpe, E. Peter, *Understanding Evolution,* Dubuque, Iowa: Wm, C. Brown Company Publishers (1967) p. 129.
[492] Holton, Gerald, "On The Role of Themata in Scientific Thought," *Science,* Washington, D. G.: American Association for the Advancement of Science (25 April 1975) Volume 188, No. 4186, p. 329.
[493] Medawar, Sir Peter Brian, *Induction and Intuition in Scientific Thought,* Philadelphia, Pennsylvania: American Philosophical Society (1969) Volume 75, p. 24.
[494] *Ibid.*, p. 26.
[495] Herbenick, Raymond M., "Peirce On Systems Theory," *Transactions Of The Charles S. Peirce Society,* Binghamton, New York: Department of Philosophy of the State University of New York (Spring 1970) Volume VI, No. 2, p. 84.
[496] Harris, F. Gentry, "The Comparative Psychology of T. C. Schneirla," *Psychotherapy and Social Science Review,* New York, New York: Dr. Jason Aronson, Editor (19 March 1973) p. 19.
[497] *Ibid.*, p. 20.
[498] Haas, William Paul, *The Conception Of Law And The Unity Of Peirce's Philosophy,* Friborg, Switzerland: The University Press (1964) P. 73.
[499] Gallie, W. B., *Peirce And Pragmatism,* New York, New York: Dover Publications, Inc. (1966) p. 86.
[500] Herbenick, Raymond M., op. cit., p. 87.
[501] Dart, John, "Americans 'Extraordinarily Religious,' Poll Finds," *Los Angeles Times* (9 August 1976) Part 11, p. 1.
[502] Simpson, George Gaylord, "The Problem of Plan and Purpose in Nature," *Human Evolution,* New York: Henry Holt and Company (1959) p. 104.
[503] Simpson, George Gaylord, *The Meaning of Evolution,* New Haven Connecticut: Yale University Press (1952) p. 344.
[504] Cowen, Robert C., "Biological Origins: Theories Evolve," *The Christian Science Monitor,* Boston, Massachusetts: Christian Science Monitor, (4 January 1962) p. 4.
[505] Huxley, Sir Julian, "The Crisis in Man's Destiny," *Playboy,* Chicago, Illinois: Playboy Magazine (January 1967) p. 2.
[506] *Ibid.*, p. 217.
[507] Colossians 1:15-20 *(J. B. Phillips version)* New York: The Macmillan Company (1963).
[508] Hebrews 1:1-3 *(The New English Bible)* Oxford University Press (1961).
[509] Alexander, Denis, *Beyond Science,* Philadelphia and New York: A. J. Holman Company - Division of J. B. Lippencott Company (1972) p. 64.
[510] Romans 1:18-25 *(J. B. Phillips version)* New York: The Macmillan Company (1963).
[511] Anonymous, "3 in 10 Adults Report Having Had Religious Experience At Some Point, Survey Shows," *Los Angeles Times* (18 December 1976) Part I, p. 30.
[512] Newell, Norman D, loc. cit.
[513] Tax, Sol -- Editor, "At Random: A Television Preview," *Issues in Evolution (Evolution After Darwin),* Chicago, Illinois: University of Chicago Press (1960) Volume III, p. 45.
[514] Moore, John. A., "Creationism in California," *Daedalus – Journal of the American Academy of Arts and Sciences,* Cambridge, Massachusetts: American Academy of Arts and Sciences (Summer 1974) Volume 103, No. 3, p. 175.

Chapter 10

[515] Mayr, Ernst, "Behavior Programs and Evolutionary Strategies," *American Scientist,* New Haven, Connecticut: The Society of the Sigma Xi And The Scientific Research Society of America (November-December 1974) Volume 62, p. 650.
[516] Motulsky, Arno G., "Brave New World?" *Science,* Washington, D.C.: American Association for the Advancement of Science (23 August 1974) Volume 185, p. 654.
[517] Alexander, Denis, op. cit., pp. 102, 103.
[518] Anonymous, "The Bible: The Believers' Gain," *Time,* Chicago, Illinois: Time, Incorporated (30 December 1974) Volume 104, No. 27, p. 37.
[519] Willoughby, William, "Science vs. Religion: Man's Best Friend," *The Star and News,* Washington, D.C. (17 March 1973) Section A, p. 8.
[520] Weisberg, Leonard R., "Velikovsky Again," *Physics Today,* Easton, Pennsylvania: American Institute of Physics, Inc. (September 1972) Volume 25, No. 9, p. 13.
[521] Blanshard, Paul and Doerr, Edd, "Darwin vs. Adam and Eve In California," *The Humanist,* Hoffman Printing Co. (March/April 1973) Volume XXXIII, No. 2, pp. 11, 12.
[522] Wade, Nicholas, "Evolution: Tennessee Picks A New Fight with Darwin," *Science,* Washington, D.C.: American Association for the Advancement of Science (16 November 1973) Volume 182, No. 4113, p. 696.
[523] Anonymous, "Teaching of Darwin's Theory as Fact Barred," *Los Angeles Times* (19 April 1973) Part I, p. 19.
[524] Anonymous, "Tennessee's New 'Genesis' Law Ousted By Court," *Palo Alto Times,* Palo Alto, California: Palo Alto Times (10 September 1974) p. 4.
[525] Wade, Nicholas, "Fundamental Setback for Fundamentalists," *Science,* Washington, D.C.: American Association for the Advancement of Science, (2 May 1975) Volume 188, No. 4187, p. 428.
[526] Toth, Robert C., "Lysenko, Soviet Charlatan, Dies," *Los Angeles Times* (24 November 1976) Part I, p. 1.
[527] Anonymous, "Lysenko: He Played Stalin's Tune," *Los Angeles Times* (25 November 1970) Part II, p. 4.
[528] Anonymous, "Many Fake Findings by Scientists Seen," *Los Angeles Times,* (26 November 1976) Part I, p. 4.
[529] Busch, Noel F., "The Furor Over School Textbooks," *Reader's Digest,* Pleasantville, New York. The Readers Digest Association (January 1976) Volume 108, No. 645, p. 128.
[530] Anonymous, "Many Fake Findings by Scientists Seen," *Los Angeles Times,* (26 November 1976) Part I, p. 4.
[531] Simpson, George Gaylord, "The Problem of Plan and Purpose in Nature," *Human Evolution,* New York: Henry Holt and Company (1959) p. 94.
[532] Exodus 20: 3, 5 (*The Living Bible*) Wheaton, Illinois: Tyndale House Publishers (1971).
[533] Stebbins, G. Ledyard, "The Evolution of Design," *The American Biology Teacher,* (February 1973) Volume 35, No. 2, p. 57.
[534] Kitwood, T. M., *What Is Human?,* London, England: InterVarsity (1970), p. 46.
[535] *Webster's New World Dictionary of the American Language,* College Edition, Cleveland & New York: The World Publishing Company (1959) p. 707.

[536] Kurtz, Paul, "What is Humanism?" *Moral Problems In Contemporary Society,* Englewood Cliffs, New Jersey: Prentice-Hall, Incorporated (1969) pp. 1-7.
[537] Anonymous, "A Humanist Manifesto (1933)," *The Humanist,* Hoffman Printing Co. (January/February 1973) Volume XXXIII, No. 1, p. 13.
[538] Hoagland, Hudson, "Toward A New Humanist Manifesto," *The Humanist,* Buffalo, New York: Hoffman Printing Company (January/February 1973) Volume XXXIII, No. 1, p. 16.
[539] Berger, Peter L., "Oh So Seductive Socialism," *Los Angeles Times* (22 August 1976) Part IV, p. 5.
[540] Brown, Colin, *Philosophy and The Christian Faith*, London, England: Tyndale Press (1969) p. 230.
[541] Anonymous, "Creation in California," *Nature,* London, England: Macmillan Journals Limited (20 October 1972) Volume 239, No. 5373, p. 420.
[542] Hayward, A.T.J., "Creation in California," *Nature,* London England: Macmillan Journals Limited (29 December 1972) Volume 240, No. 5383, p. 577.
[543] Kelly, Fred C., *The Wright Brothers,* New York, New York: Ballantine Books, A Division of Random House, Inc., (1943) p, 191, 193.
[544] *Ibid.*
[545] Muller, H. J., "Human Values in Relation to Evolution," *Science,* Washington, D.C.: American Association for the Advancement of Science (21 March 1958) Volume 127, p. 629.
[546] McCurdy, Patrick P., "The Open Mind, ACS, and Velikovsky," *Chemical and Engineering News,* Washington, D.C.: American Chemical Society (9 April 1973) Volume 51, No. 15, p. 1.
[547] *Ibid.*
[548] Lemmon, Richard M., "Religious and Scientific Dogmas," *Chemical and Engineering News,* Washington, D.C.: American Chemical Society (14 May 1973) Volume 51, No. 20, p. 50.
[549] *Ibid.*, p. 39.
[550] *Technology: Promises and Problems,* Belmont, California: Allyn and Bacon, Inc. (1972) pp. 48-53.
[551] *Technology: Promises and Problems (Teachers' Guide),* Belmont, California: Allyn and Bacon, Inc. (1972) pp. 82, 83.
[552] *Ibid.*
[553] Goldschmidt, Richard B., "Evolution, as Viewed by One Geneticist," *American Scientist,* New Haven, Connecticut: The Society of the Sigma Xi and the Scientific Research Society of America (January 1952) Volume 40, No. 1, p. 84.
[554] Blanshard, Paul and Doerr, Edd, "Darwin vs. Adam and Eve in California," *The Humanist,* Hoffman Printing Go. (March/April 1973) Volume XXXIII, No. 2, p. 12.
[555] Anonymous, "Creation in California," *Nature*, London, England: Macmillan Journals Limited (20 October 1972) Volume 239, No. 5373, p. 420.
[556] Anonymous, "Obituary -- Dr. W. R. Thompson, " *Nature,* London, England: Macmillan Journals Limited (14 July 1972) Volume 238, No. 5359, p. 116.
[557] Darwin, Charles, *The Origin of Species,* New York: Everyman's Library, Dutton: New York (1967) Sixth Edition, p. vii.

[558] McIntyre, John A., "Creation, Evolution and the ASA," *Journal of American Scientific Affiliation,* Elgin, Illinois: The American Scientific Affiliation (December 1973) Volume 25, No. 4, p. 172.

[559] Frazier, Kendrick, "Science and the Parascience Cults," *Science News,* Washington, D.C.: Science Service Incorporated (29 May 1976) Volume 109, pp. 346-350.

[560] Fraser, J. T., "Communication with the Humanities," *Science,* Washington, D.C.: American Association for the Advancement of Science (17 July 1964) Volume 145, No. 3629, p. 231.

[561] Reid, W. Stanford, "The Historical Development of Christian Scientific Presuppositions," *Journal of the American Scientific Affiliation,* Elkin, Illinois: American Scientific Affiliation (June 1975) Volume 27, No. 2, pp. 69-75.

[562] Key, Thomas D. S., "The Influence of Darwin on Biology," *Evolution and Christian Thought Today,* Grand Rapids, Michigan: William B. Eerdmans Publishing Company (1959) pp. 30, 31.

[563] Darwin, Charles, *The Origin Of Species,* New York: Everyman's Library, Dutton: New York (1967) Sixth Edition, p. ix.

[564] Weisburg, Leonard R., "Velikovsky Again," *Physics Today*, Easton, Pennsylvania: American Institute of Physics, Inc. (September 1972) Volume 25, No. 9, p. 13.

[565] Fischer, Robert B., "The Evolution of a Policy on Textbooks," *Los Angeles Times* (15 April 1973) Part VII, p. 7.

[566] *Ibid.*

[567] Walsh, John, "NSF: Congress Takes Hard Look At Behavioral Science Course," *Science*, Washington, D.C.: American Association for the Advancement of Science (2 May 1975) Volume 188, No. 4187, P. 426.

[568] Holton, Gerald, "On Being Caught Between Dionysians and Apollonians," *Daedalus*, Cambridge, Mass.: American Academy of Arts and Sciences (Summer 1974) Volume 103, No.3.

[569] Anonymous, "Science and the Creation Theory," *Los Angeles Times* (16 November 1972) Part II, p. 6.

[570] Boring, Edwin G., "Cognitive Dissonance: Its Use in Science," *Science*, Washington, D.C.: American Association for the Advancement of Science (14 August 1964) Volume 145, No. 3633, pp. 680-686.

[571] Alexander, George, "Scientists Seek Chemical Clue to Mars Puzzle," *Los Angeles Times* (9 August 1976) Part I, p. 3.

[572] Alexander, George, "Scientists Wary of 'Life' on Mars," *Los Angeles Times* (11 August 1976) Part II, p. 1.

[573] Anonymous, "Evolution Controversy Flairs Anew," *Chemical and Engineering News*, American Chemical Society (11 December 1972) Volume 50, No. 50, p. 13.

[574] Schwartz, George I. and Troost, Cornelius J., *Patterns of Life*, New York, New York: American Book Company (1972) p.598.

[575] Volpe, E. Peter, *Understanding Evolution*, Dubuque, Iowa: Wm. C. Brown Company Publishers (1967) p. 129.

[576] Alexander, George, "Scientist Detects Puzzling Boost in Sun's Rate of Spin," *Los Angeles Times* (3 January 1977) Part II, p. 1.

[577] Alexander, George, "Primitive Earth Seen as Caldron," *Los Angeles Times* (11 October 1976) Part II, p. 2.

[578] Tribus, Myron, *Thermostatics and Thermodynamics*, Princeton, New Jersey: D. Van Nostrand Company, Inc. (1961).
[579] *Ibid.*, pp.148-9.
[580] Prigogine, Ilya; Nicolis, Gregoire, and Babloyantz, Agnes, "Thermodynamics of Evolution," *Physics Today*, Easton, Pennsylvania: American Institute of Physics (November 1972) Volume 25, No. 11, pp. 23-28; (December 1972) Volume 25, No 12, pp 38-44.
[581] Grobstein, Clifford, "A New Genie Is Out of the Bottle," *Los Angeles Times* (29 August 1976) Part IV, p. 5.
[582] Sinsheimer, Robert L., "Caution May Be an Essential Scientific Virtue," *Los Angeles Times* (29 August 1976) Part IV, p. 5.
[583] Denike, L. Douglas, "More Stringent Guidelines Are Vitally Needed," *Los Angeles Times* (29 August 1976) Part IV, p. 5.
[584] Nelkin, Dorothy, "The Science-Textbook Controversies," *Scientific American*, New York, New York: Scientific American, Inc. (April 1976) Volume 234, No. 4, p. 39.
[585] Davis, Harold L., "A Scientific Court?" *Physics Today*, Easton, Pennsylvania: American Institute of Physics (January 1976) p. 120.
[586] *Ibid.*
[587] Allen, Allen D., "Scientific vs. Judicial Fact-finding in the United States," *IEEE Transactions on Systems, Man, and Cybernetics*, New York, New York: Institute of Electrical and Electronics Engineers, Inc. (September 1972) Volume SMC-2, No. 4, pp. 548-550.
[588] Allen, Allen D., "Fact-Finding in the Lab And the Courtroom," *Psychology Today*, (April 1973) p. 85.
[589] Mayr, Ernst, "The Nature of the Darwinian Revolution," *Science*, Washington, D.C.: American Association for the Advancement of Science (2 June 1972) Volume 176, pp 981-989.

Epilogue

[590] Tape-recorded interview with William C. (Bill) Scopes by Vernon L. Grose in Woodland Hills, California on 11 October 1979
[591] Scopes, John Thomas, "Reflections – Forty Years After," unpublished notes, *Famous Trials in American History: Tennessee vs. John Scopes – The "Monkey Trial"* available at www.law.umkc.edu
[592] *Ibid.*
[593] *Ibid.*
[594] *Ibid.*
[595] *Ibid.*
[596] Tape-recorded interview, op.cit.
[597] *Ibid.*
[598] *Ibid.*

Exhibits

EXHIBIT 1: *LOS ANGELES TIMES* EDITORIAL - I

This editorial was published on 14 October 1969 and evoked a letter to the Editor by Vernon L. Grose on that date

EVOLUTION AND THE STATE BOARD

ISSUE: Some State Board of Education members doubt the accepted theory of evolution. Will they impose their views on schools?

Educators could hardly believe their ears at the recent meeting of the California Board of Education.

Some of the board members actually complained that "evolution appears to be stated as fact" in new guidelines for science instruction in state elementary schools.

There were unpleasant echoes of the 1925 Scopes "monkey trial" as members questioned a theory long accepted as fact by virtually the entire scientific community.

"I believe in the creation theory, not evolution," declared board member Dr. Thomas Harward, a Needles physician. "You people should try to find out more of a scientific background of creation ..." he told authors of a 205-page framework for elementary science textbooks.

Educators at the meeting "were almost speechless," reported *Times* writer Jack McCurdy.

They could hardly be blamed.

We apparently have not advanced as far as we thought from that day 44 years ago when John Scopes, a biology teacher, was fined $100 for violating a Tennessee law prohibiting the teaching of evolution.

A similar Arkansas law was thrown out last year by the U. S. Supreme Court, which ruled that it breached the constitutional separation of church and state. And the California Attorney General's office declared in 1963 that the teaching of evolution in the schools was constitutional if it did not involve indoctrination of the idea.

The creation concept, board members were told, would be presented to students in other study matter. But in the science guidelines the authors stuck to supportable scientific conclusions.

That seems to us to be the only proper approach to the teaching of science — as opposed to religious instruction. We hope that the State Board of Education will concur when they reconsider the science guidelines next month.

After all, one need not be an atheist to accept the theory of evolution and the mass of scientific evidence that supports it.

EXHIBIT 2: LETTER TO EDITOR, *LOS ANGELES TIMES*

14 October 1969
The Editor
Los Angeles Times
Los Angeles, California

Dear Sir:

Your editorial of 14 October 1969, "Evolution and the State Board," appears to fall below the standard of excellence I have come to expect from the *Los Angeles Times.*

Both as an educator and a member of the scientific community, I was shocked that you could so completely twist an issue out of context. Dr. Harward was being most scientific in pointing out that evolution is a *theory* — not a fact. Further, it is but *one* theory. Another one which he happens to consider more valid, is the creation theory. He was asking that ample consideration be given to all theories (certainly a prime thesis of the scientific method).

To compound your narrow view, you allude to the Scopes trial as though the theory of evolution was again on trial when, in actuality, just the opposite is true. Creation appears to be ruled out as a scientific theory and limited to "religious instruction." Therefore, creation (or any other theory) must not be taught — only evolution! You are correct in saying that we have not advanced in 44 years because we are prohibiting the teaching of all theories but one.

Ironically, the evolution theory does leave the student without any supportable rationale with regard to "first causes," i.e., under what stimulus and by what means did the very first bit of matter originate? After all, Louis Pasteur conclusively disproved abiogenesis (spontaneous generation) over 100 years ago, thereby leaving the theory of evolution without an explanation for "first causes." Yet this very question is the most likely question for a student to raise!

Creation, on the other hand, does provide rationale for "first causes." In the objectivity that the scientific method demands, how can we afford not to examine all theories, particularly when some have explanations for major premises that a currently popular theory does not?

Please understand that this is not a vote for creation over evolution as a theory. Rather, I am disturbed if educators are so insecure regarding the validity of the evolution theory that they are afraid to allow any other theory to be taught or compared with what they may have already decided is a "fact."

I agree with you that "one need not be an atheist to accept the theory of evolution." I would however also favor the converse: "One need not be a religious person to accept the theory of creation." As a father of six children, I would demand that the schools permit students to examine *all* the theories about the origin of life — not just a pet one.

Very truly yours,
TUSTIN INSTITUTE OF TECHNOLOGY

Vernon L. Grose Vice President

Copies: California Board of Education members

EXHIBIT 3: LETTER TO CALIFORNIA STATE BOARD OF EDUCATION BY DIVISION OF BIOLOGY. CALIFORNIA INSTITUTE OF TECHNOLOGY

CALIFORNIA INSTITUTE OF TECHNOLOGY
Pasadena, California 91109
Division of Biology
October 15, 1969

Mr. Howard Day, President
State Board of Education
State Education Building
721 Capitol Mall
Sacramento, California 95814

Dear Mr. Day:

Biologists consider evolution to be a fact, not a theory. Statements to the contrary attributed to members of the State Board of Education (*L.A. Times*, October 10, 1969) are not correct. When scientists speak of the "theory of evolution" they are not questioning the reality of evolution, but are saying that we do not yet understand in detail the mechanisms by which evolution comes about. There are theories about these mechanisms, and these constitute the "theory of evolution." But there is no doubt that evolution as such has occurred and is occurring. This is a fact that can be demonstrated in nature and in the laboratory.

Yours truly,

NHH: pe
cc: Dr. Max Rafferty

(The above letter was signed by the following persons:)

N. H. Horowitz Professor	Max Delbruck Professor	F. Strumwasser Professor	R. D. Owen Professor
S. Benzer Professor	James Olds Professor	R. L. Sinsheimer Professor	Jerome Vinograd Professor
R. S. Edgar Professor	R. W. Sperry Professor	W. B. Wood Assoc. Professor	D. McMahon Asst. Professor
	C. J. Brokaw Professor	G. Attardi Professor	James Bonner Professor

---

EXHIBIT 4: EVOLUTION AND CREATION

Presentation to the California State Board of Education on 13 November 1969 by Vernon L. Grose

For my appearance today before the State Board of Education, I am deeply indebted to the *Los Angeles Times.* It was first their biased reporting of the Board's October meeting regarding evolution and then their distorted and illogical thinking in an editorial entitled, "Evolution and the State Board," which appeared on 14 October 1969 that piqued my personal concern in this issue. Had they been either scientific, logical or fair, I certainly would not have become involved. On the other hand, even though they undoubtedly had not intended to have someone defend theories other than evolution, I believe that they accomplished one of the primary aims of a great newspaper — to cause people to think and to act as a result of their thought.

At the outset, I would like to make my position clear. I do favor the teaching of evolution in the public schools. I myself believe in the various theories of evolution where scientific data have been shown to verify them. Further, since the *Times* raised the issue of the Scopes trial of 1925, I would have opposed William Jennings Bryan and upheld Clarence Darrow's point of view at that trial that evolution should be taught. Finally, I am entirely favorable to the separation of church and state and thereby the separation of scientific and religious instruction.

The 205-page Science Framework for California Public Schools under consideration here today I have read. I find no particular statement in that Framework to be offensive. Rather, I consider it overall to be an excellent piece of work. My only comments concerning it would be in the sense of additions by way of improvement rather than deletions or criticism of what is said.

The Framework presents an excellent rationale for the philosophy of teaching science. For example, the Prologue states:[1]

---

[1] *Science Framework for California Public Schools (preliminary), California State Department of Education, Sacramento (1969), ix.*

The development of inquiry processes was identified as a major purpose of science instruction.

This idea of developing inquiry processes continues to be acknowledged throughout the Framework until one reaches Appendix A.

However, Appendix A appears to become more dogmatic than I believe it should be if we are to maintain the flexibility necessary for development of inquiry processes of children. In other words, most, if not all, the fascinating mystery, conflict, and disagreement of various scientific theories is eliminated in Appendix A. Instead, many if not most of the statements are in the form of unqualified facts. Let me illustrate. On page 125, we read,[1]

> A soup of amino acid-like molecules, formed in pools some 3 billion years ago, interacted with oxygen and other elemental constituents of the earth, probably giving rise to the first organization of matter which possessed the properties of life.

As a father of six children of school age, I believe that a child would have no reason to question that statement although there is a wide diversity of opinion in scientific circles regarding its validity. A more humble and realistic statement might have been, "One hypothesis of evolution suggests that a soup of amino acid-like molecules, formed in pools some 3 billion years ago, interacted with oxygen and other elemental constituents of the earth, possibly giving rise to the first organization of matter which possessed the properties of life."

OBJECTIVES

There are four reasons why I accepted the gracious invitation extended to me by the Board today. I will refer to them as objectives, not only for consideration by the Board and the State Advisory Committee on Science Education but also by all teachers of science in California.

Perspective of Reservation — The first objective I would propose involves the attitude we hold regarding the so-called accomplishments of science. I believe that many of us in the scientific and technological community are responsible for the widespread false image we have created of being cold, calculating, precise, always right, never subjective, and the sole possessors of true facts. This almost chauvinistic attitude that many of as have shown toward the non-scientific world certainly tempts educators, for example, to put undue faith in science or to take too seriously the statements of scientists. After all, scientists put on their trousers one leg at a time every morning just like the rest of the world. They even still have erasers on the end of their pencils, too!

The *Framework*, on page 6, points out that science is an organization of knowledge. Our knowledge can be categorized several ways, based on how much faith we can put in it. Perhaps the loosest category is the *hypothesis*, defined as something assumed because it seems likely to be a true explanation. Something a bit more certain could be called a *theory* which is explanation based on observation and reasoning. Only when we have removed all doubt do we dare call something a "*fact*," because a fact is something known to be true or known to have really happened. A collection of facts, then, permits us to postulate laws. A scientific *law* is a statement of a relation or sequence of phenomena invariable under the same conditions.

I apologize for this tutorial review of definitions, but it is germane to my first objective — having a perspective of reservation about science. The almost hysterical response a few weeks to the fallacious issue raised by the *Los Angeles Times* that some Board members had "questioned a theory long accepted as fact by virtually the entire scientific community" illustrated this misunderstanding.

As the *Framework* points out on page 6, even facts do not constitute science. Science exists only when relationships are discovered. Try as we will in science, we cannot always be objective when we are establishing these relationships. Human bias, prejudice, and even bullheadedness creeps into these relationships or theories. Over a period of time, these

---

[1] *Ibid.* 125.

undesirable traits tend to be exposed and rejected. However, some error persists for years.

To protect, then, our young people both from error and dogmatists (who would make science a religion if you will), it is imperative that we show them the unknowns right along with the knowns. As I read Appendix A, I sensed a "ring of certainty" that I feel must be balanced with a corresponding acknowledgement of gaps in our knowledge. There could be a popular misconception pervasive in our whole society that it is only a matter of time, effort, and possibly dollars until science will have all the answers. Unfortunately, just the opposite appears to be more likely. As we uncover more and more physical evidence, we are amazed by the complexity in back of these discoveries. Instead of the search being a *converging* one where we are finding all the answers, it is a *diverging* one where we are raising more questions! Instead of *simplicity* emerging from investigation, it is increasing *complexity* which emerges.

Agreeing with the *Framework* when it says, "The progress of science requires the human imagination as an active agent," I suggest that less dogmatic language than appears in Appendix A would be a step in the right direction. Perhaps this quotation from the latest textbook used at Massachusetts Institute of Technology and written by Dr. Edwin C. Kemble, Chairman Emeritus of the Physics Department at Harvard University sheds light on what I would consider in better taste:[1]

> The scientist is confronted by an aggregation of baffling interconnected problems as he seeks to understand ... as he struggles to fit the stories ... Here we are at the frontier of knowledge, where partial information must be supplemented by conjecture, where rival theories are battling for supremacy, and where ideas are in continual flux. No one knows the answers yet.

Need for Plurality - My second objective, toleration of plurality, is in tune with much that is going on in our country at this time. Rather than emphasizing uniformity as we have in the past, we have greater recognition of the need to maintain various points of view. This is borne out by a recent decision in Laredo, Texas to treat both English and Spanish as permanently American languages in the classroom. Likewise, the courts have recently ruled that reading, writing or understanding English cannot be a prerequisite to voting. A further recognition of plurality is the establishment of black studies in universities.

Contrary to popular thought, science too has operated in a pluralistic environment for years. If I may be permitted a personal illustration, I would point out that when I studied optics in my undergraduate physics curriculum using a textbook written by that eminent physicist, Dr. Frances Weston Sears from Massachusetts Institute of Technology, we were confronted with this statement:[2]

> The present standpoint of physicists, in the face of apparently contradictory experiments, is to accept the fact that light appears to be dualistic in nature. The phenomena of light propagation may best be explained by the electromagnetic wave theory, while the interaction of light with matter, in the processes of emission and absorption, is a corpuscular phenomenon.

Frankly, I was not particularly disturbed by this statement. Rather, I was challenged that there was yet a frontier for me and that I might even contribute to a reconciliation of these two apparently contradictory theories.

This idea of pluralism in science is not foreign to the *Framework* either, since the pluralistic value system of our culture is acknowledged in Chapter 1. Furthermore, the need to analyze society prior to deciding on the nature of science curriculum reform is correctly stressed on Page 2. The first of five questions in the *Framework* regarding analysis of society is: "What are

---

[1] *Kemble, Edwin C., Physical Science, Its Structure and Development, Cambridge, Massachusetts and London, England: The MIT. Press (1966), 148.*

[2] *Sears, Francis Weston, Principles of Physics III - Optics, Cambridge, Massachusetts: Addison-Wesley Press, Incorporated (1948), 3.*

the fundamental beliefs, values and moral principles basic to the American tradition?"

Certainly the Judeo-Christian heritage of our country is a fundamental set of principles. Implicit in this heritage is the concept of creation, and it is far more than a religious concept. For example, it is basic to our political and judicial philosophy. Thomas Jefferson, in writing the Declaration of Independence, referred to "the laws of Nature and of Nature's God" as well as stating, "that all men are *created* equal and that they are endowed by their *Creator.* "

Can you imagine the impact on the logic required for justice in our courts if we were forced to amend the Declaration of Independence to read:

> We hold these truths to be self-evident, that all men arose as equals from a soup of amino acid-like molecules and that they, by virtue of this common molecular ancestry, are endowed with certain unalienable Rights such as Life, Liberty, and the pursuit of Happiness?

Just as dramatically as the concept of creation is ingrained in the warp and woof of American philosophy, so the idea of evolution is inexorably a part of other political philosophies such as those of Karl Marx and Adolf Hitler. Quite apart from their scientific validity, these two groups of theories should be exposed and discussed in the classroom so that children may see the effect that scientific theories have on other areas of life.

Before leaving this objective of pluralism, permit me to refer to my favorite *Times* editorial once more. It mentioned that the California Attorney General had declared in 1963 that the teaching of evolution in the schools was constitutional providing it did not involve indoctrination of the idea. When one and only one concept is presented to a child, how can we avoid indoctrination?

Elimination of Evolution-Versus-Creation Syndrome — This discussion of the political impact of the theories of evolution and creation leads me to my third objective — the elimination of the long-standing antagonism by non-scientists concerning these two groups of theories on grounds other than science. I have been careful thus far to indicate no preference between evolution and creation because I firmly believe that both have a rightful place in explaining scientific observations to date. Neither explain, in entirety, what we have found thus far by experimentation.

Unfortunately in some religious circles, evolution is viewed as an atheistic plot, and conversely, some who hold no religious convictions view creation with equal suspicion as a persistent but unwelcome myth. I would hope that the time is near when these two schools of thought can be removed from the realm of religion and be examined, in the cold light of scientific objectivity, for their validity in explaining the relationship of observations we have made.

I do agree with the *Times* editorial in its statement that "one need not be an atheist to accept the theory of evolution." However, they stopped short by omitting the corollary, "One need not be a religious person to accept the theory of creation."

To lend weight to this corollary, I quote the *New York Times* of February 8, 1962:

> Within the most advanced echelon of Soviet science there is emerging a tendency to seek a non-materialist, spiritual concept of the universe ... there is reason to believe that some of the most eminent figures in the galaxy of Soviet physicists, astronomers and mathematicians are involved ... these men have not become believers in a formal religion or dogma. Their faith is more akin to that appearing among many of their western scientific colleagues. But they are no longer atheists.

Once these two schools of thought are examined in the cold light of objectivity and thereby removed from irrational and biased bigots from both sides, it becomes evident that neither is an alternative for the other. Certainly there are some points where they may reach conflicting conclusions from the same observation, but in the main, they compliment one another far more than they conflict. Since they are not primarily competitors, I become disturbed when it appears, as it did in the recent *Los Angeles Times* articles, that proponents of one theory (evolution in this case) are afraid to allow any other theory to be taught or compared with what they may have already decided is a "fact."

If we limit ourselves to the theories of evolution, the student in the classroom is left without any supportable rationale with regard to "first causes," e.g., under what stimulus and by what means did the very first bit of matter originate? After all, Louis Pasteur is considered to have conclusively disproved abiogenesis (spontaneous generation) over 100 years ago, thereby leaving the theories of evolution without an explanation for "first causes." Yet this very question is the most likely question for a student to raise!

Creation, on the other hand, does provide rationale for "first causes," and here the complimentary feature of the two concepts is illustrated.

Our most recent and spectacular product of science and technology, Project Apollo, provides us with another illustration for including creation in the curriculum. Having met a number of the astronauts personally in connection with my work in all three manned space programs, I have a deep respect not only for their courage but also their no-nonsense, technical understanding. When Frank Borman returned from orbiting the Moon, I met with him in Madrid, Spain at length, and I am convinced of his sincerity in selecting the Genesis account of creation as the most appropriate commentary for what he, Jim Lovell and Bill Anders were viewing. I doubt whether you could get them to agree that they were dipping back into mythology or indulging in a mystic reverie. If creation should be omitted from the science curriculum, however, what choice would a schoolchild have concerning the astronauts' judgment during this momentous event?

There are literally thousands of men, holding some of the most responsible teaching positions in leading universities, who not only endorse the principle of creation where it is applicable but who also teach this concept. A few of these men whom I happen to know are:

Dr. Irving W. Knobloch, Professor of Botany, Michigan State University

Dr. Walter R. Hearn, Professor of Biochemistry, Iowa State University

Dr. V. E. Anderson, Assistant Director, Dight Institute for Human Genetics, University of Minnesota

Dr. Wilbur L. Bullock, Associate Professor of Zoology, University of Tennessee

Dr. Mark Biedebach, Assistant Professor of Biology, California State College at Long Beach

Understanding of the Scientific Method — My fourth and last objective takes the form of an appeal — that the Framework's noble purpose of developing inquiry processes in each student be carried beyond the philosophic stage right into the classroom. This implies a wider understanding of the scientific method than is exhibited in Appendix A.

Implicit in the development of the inquiry process must be the presentation of alternatives — the continuing emphasis on asking "why?" and "how do we know?" — the clear identification of gaps in our knowledge.

We fall short of the goal if we simply elaborate what science has decided to be factual. It is even a tragedy if we shut the door to all alternatives for an idea like evolution whose prime constituent at certain crucial points is conjecture. In defense of this latter statement, I quote an evolutionist, Dr. W. R. Thompson:[1]

> As we know, there is a great divergence of opinion among biologists, not only about the causes of evolution but even about the actual process. This divergence exists because the evidence is unsatisfactory and does not permit any certain conclusion. It is therefore right and proper to draw the attention of the non-scientific public to disagreements about evolution. But some recent remarks of evolutionists show that they think this unreasonable. This situation, where men rally to the defense of a doctrine they are unable to defend scientifically, much less demonstrate with scientific

---

[1] *Darwin, Charles, Origin of the Species, Darwin Centennial edition, J. M. Dent & Sons LTD, No. 811 Everyman's Library (1959), xxii.*

rigor, attempting to maintain its credit with the public by suppression of criticism and the elimination of difficulties is abnormal and undesirable in science.

Dr. Thompson's warning should be heeded. If we choose to ignore him and insist on teaching only one pet concept thereby denying our youngsters the thrill of the mysterious, the perplexing, and the conflict of scientific viewpoints, we may well preclude them from ever really understanding science. As Albert Einstein said:[1]

> The most beautiful and profound emotion we can experience is the sensation of the mystical. It is the power of true science. He to whom this emotion is a stranger, who can no longer wonder and stand rapt in awe, is as good as dead.

PROPOSALS

What I have said thus far is a preface for two specific proposals I would like to have the Board consider.

Revision for Insertion in Framework — I have incorporated the material in the last three paragraphs of section *G-2: Interpendence and interaction with the environment are universal relationships*, on page 124-5 into a rewrite which includes a description of the principle of creation. I propose that it be considered for incorporation into the *Framework*. (See Exhibit 5 for this revision)

Offer of Focal Point — Inasmuch as the present version of the *Framework*, in limiting itself to the single idea of evolution, is only perpetuating the current method of teaching in California public schools, there may be a shortage of research texts and materials which address creation from the purely scientific viewpoint. Inserting a few words about creation in the *Framework* without providing adequate study materials in the classroom would be a waste of time. Therefore, I hereby offer my services, not as an expert in creation, but as a focal point for securing bibliographies as well as personal contacts for the various scientific disciplines involved in creation curriculum formulation.

CONCLUSION

To conclude my thoughts, I have selected a quotation by Charles A. Siepmann which seems to summarize what I had to say to the Board today:[2]

> For those who go far with it, education is, on one side, a lonely business. For to be eminent in mind or in imagination is, as with all eminence, to stand alone and apart.
>
> Education is also awful, for it leads us inexorably beyond the known to the unknowable — so that, like Socrates, we end with the recognition that we know nothing. Consequently, the chief attributes of a truly educated man are humility (defined as courageous insight that refuses to seek simple answers to complicated questions) and a certain suspension of judgment (not to be confused with skepticism) in recognition, always, of 'that reserve of truth beyond what the mind reaches but still knows to be behind.' This quality of mind is common both to great scientists and great men of faith.

Thank you so much for the invitation to express my views.

---

[1] *Barnett, Lincoln, The Universe and Dr. Einstein, New York; Signet Science Library Books (1964), 108 .*

[2] *Drake, William E., The American School in Transition, Prentice-Hall, Incorporated (1955), frontispiece.*

EXHIBIT 5: SUGGESTED REVISION OF SCIENCE FRAMEWORK

A revision of pp. 124-5 in Science Framework for California Public Schools, *California State Department of Education, Sacramento (Preliminary version) 1969. Prepared by Vernon L. Grose for the California State Board of Education — 13 November 1969*

Original Version

Another order of interactions are evolutionary events which produce predictable changes in certain kinds of objects over long periods of time. One theory claims that even atoms, interacting with one another, evolved over eons of time, give rise to the present assemblage of different kinds of elements. Another evolutionary thesis describes the progress of stars all the way from young gaseous nebulae to pulsating dying stars. Still another interacting series of events has produced the evolution of rocks from igneous to sedimentary and metamorphic.

Perhaps the best known evolutionary product of interactions are changes which occur in living organisms over long periods of time. From the origin of the first living particle, the evolution of living organisms was probably directed by environmental conditions and the changes occurring in them. A soup of amino acid-like molecules, formed in pools some 3 billion years ago, interacted with oxygen and other elemental constituents of the earth, probably giving rise to the first organization of matter which possessed the properties of life.

Evidence indicates that nearly 2 million living species, and millions of extinct species, are descendents of this early form of life. This diversity among living organisms is the result of natural selection, preserving characteristics which have allowed adaptation to the many kinds of environments on this planet. Long-term adaptation is evolution. Evolution results from mutations and genetic recombinations in the organism which, through natural selection, have produced a more efficient relation with the changing environment than less successful ancestors.

Suggested Revision

When attempting to show interdependence and interaction for events in the apparently distant past, a problem arises because there cannot be laboratory demonstration and verification. Inevitably, science must resort to extrapolation of data (i.e., speculation) in order to reach many of its conclusions. This is particularly true concerning the origin of matter, life, species, and man himself since no one was there to observe and record these events. Whenever scientists are forced to reach conclusions without experimental data, far less credence can be placed in their conclusions. It cannot be overemphasized that speculation is a very poor alternative for actual observation.

All scientific evidence to date concerning the origin of life implies at least a dualism or the necessity to use several theories to fully explain relationships between established data points. This dualism is not unique to this field of study but is also appropriate in other scientific disciplines such as the physics of light.

Various theories involving evolution have been proposed to explain predictable changes in certain kinds of objects over long periods of time. Evolution could be called long-term adaptation. It is thought to result from mutations and genetic recombinations in organisms which, through natural selection, have produced a more efficient relationship with the changing environment than less successful ancestors. One theory of evolution claims that even atoms, interacting with one another, evolved over eons of time, giving rise to the present assemblage of different kinds of elements. Another evolutionary thesis describes the progress of stars all the way from young gaseous nebulae to pulsating dying stars. Still another interacting series of events may have produced the evolution of rocks from igneous to sedimentary and metamorphic.

Perhaps the most popular evolutionary theory describes interacting changes which occur in living organisms over long periods of time. From the origin of the first living particle, the evolution of living organisms is proposed to have been directed by environmental conditions and the changes occurring in them. One hypothesis suggests that a soup of amino acid-like molecules, formed in pools some 3 billion years ago, interacted with oxygen and other elemental constituents of the earth, possibly giving rise to the first organization of matter which possessed the properties of life.

Contrasted with this group of evolutionary theories which require extremely long periods of time to explain the "select-and-adapt" process are another group of creation theories.

Creation can be defined as the instantaneous formation of something out of nothing, or the occurrence of something without natural cause and effect. Instead of suggesting as one theory of evolution does that the nearly two million living species, as well as millions of extinct species, are merely descendants of an early and simple form of life via the mechanism of natural selection, mutation and adaptation, some creation theories propose, for example, that species had distinct, individual, and unique origins.

One of the creation theories is the relativistic "steady state" cosmology first proposed in 1948 by Hoyle and independently postulated by Bondi and Gold. This theory is based on the assumption that the universe as a whole is not subject to the degenerative character of the stars and galaxies. To account for the apparently wide difference in age of the universe and star clusters in our galaxy, this theory proposes that hydrogen gas is constantly in process of creation in interstellar space. Creation is also considered to be a factor in the "superdense state" theory. In this approach, all the matter in the universe suddenly appeared at one time crowded together in a relatively small space from which it was released by a tremendous explosion. The age of the universe is reckoned from the time of the explosion in this theory.

While the Bible and other philosophic treatises also mention creation, science has independently postulated the various theories of creation. Therefore, creation in scientific terms is not a religious or philosophic belief. Also note that creation and evolutionary theories are not necessarily mutual exclusives. Some of the scientific data (e. g. , the regular absence of transitional forms) may be best explained by a creation theory while other data (e. g. , transmutation of species) substantiate a process of evolution.

EXHIBIT 6: STATE BOARD OF EDUCATION DISCUSSION

Transcript of discussion by the California State Board of Education following testimony at public hearing in Los Angeles — 13 November 1969

THOMAS HARWARD: Mr. Chairman, in order to conclude this in this session of the State Board, I would make a suggestion that the *Science Framework* committee and, if there are enough of them, have the State Department of Education meet with a group of the scientists who have made their presentations today, in order to put into the *Framework* some sort of statement that would assure us that there would be both sides of the question put into the science textbooks as based upon this *Science Framework*. This is my suggestion, and if it's possible that it could be done, it could be concluded at this meeting. This is my only concern — that both sides of the question be presented in the State textbooks. I think that there's been some very eminent people speak today who have shown that there is a concern on the part of the scientific community that the opponents or critics of evolution have their say in the textbooks in the State of California. I would make this suggestion that these people could meet and form some sort of resolution or a memo in the *Science Framework* that would be satisfactory.

JOHN R. FORD: I think that since I was the one who made the original objection to the *Science Framework* at the last meeting, that I ought to say something about it. I think Dr. Grose, who was our first speaker, more than uniquely expressed the feelings I have. And that was that the full recognition of the various theories and the material that he has produced for us as a suggested change, I would like to see the *Science Framework* committee, if at all possible, to read over this carefully and see if many of these ideas or these statements, perhaps verbatim, that he has expressed here be added. I felt that the committee has done a tremendous job in certain changes in the *Framework*. Inquiry process, of course, I think is very important for our young people to have, but as far as the creation idea, I feel that he has much more ably expressed this than I could possibly have done so, and given us ideas and insight that is important. In other words, evolution certainly should be included in our public school system as a theory. The children should have an opportunity to understand all the major facets of the origin of man, the continuation of the species, the inaccuracies, the discrepancies that are apparent, and then let the children themselves decide which of the theories they wish to accept. (mild applause) As a Board, we should certainly take these considerations in mind in relationship to the adoption of our *Science Framework*.

MAX RAFFERTY: I thought it might be of interest to not only the Board Members, but also to the interested members of the audience, for me to read these paragraphs which deal with the subject now of controversy. This is the Department's recommendation as to how to resolve this difficulty and it reads as follows:

The origin of life and the evolutionary development of plants and animals have long been of interest to the layman and scientist alike. The oldest explanation is a religious one — that of special creation. Aristotle proposed a theory of spontaneous generation. In the 19th century, the concept of natural selection was proposed. This theory rests upon the idea of diversity among living organisms and the influence of the natural environment upon their survival. Fossil records indicate that hundreds of thousands of species of plants and animals have not been able to survive the conditions of a change in environment. More recently, efforts have been made to explain the origin of life in biochemical terms.

The charge which the Board gave to the Department last month was to bring in a revision of this particular segment of the *Science Framework* which would more accurately illustrate the diversity of opinions and the number of different theories which are extant in regard to the origin of life and the development of life on this planet. The Department has done this in this respect. Whether it is to the satisfaction of this Board or not, remains for the Board to conclude. But I would like to point out to you that, if the *Framework* is amended to the point where we have to have an extremely long presentation of these various points of view and theories, it's going to imbalance the *Framework* which deals only in very brief terms with this one particularly thorny question of evolution, creationism, Aristolianism, biochemistriism or what have you. And I would like to urge the Board that whatever statement be necessary, if indeed you want to change the Department's correction, that you keep this matter of time distribution and space distribution in mind.

DORMAN COMMONS. Personally, it seems to me that the revisions that have been made leave ample room not — I would think ... and I would take exception to Glen's (Harward) characterization of this as a two-sided question. It isn't really that at all. What you are really talking about is that there are many, many, many theories. At this particular point, I think the *Framework* makes it clear that you are talking in the realm of many theories. I, too, enjoyed the first presentation (Grose's) this morning. I thought it was an excellent one. But I am satisfied that the *Framework* has to survive. It seems to me that we can go on beating this over the head for a long time. I would like to see the *Framework* adopted.

THOMAS HARWARD: Mr. Chairman, I am not satisfied with that, and I don't think that it's broad enough in assuring us that there will be textbooks written that will present other theories than ... The reason I make this statement is that I think that all of us received a paper from a professor of biology at Pepperdine who had run a survey on the attitudes of students, and he came to the conclusion through the survey that 98% of them accepted evolution as a fact. And so we have got to reverse this attitude. These people are going to have both sides or all theories presented to them. We have to do them in the same amount of emphasis in order to overcome this gross discrepancy.

JOHN R. FORD: Mr. Chairman, I wonder if a little compromise here would be in order, and i think this will sell legally. If we would look at the third paragraph of the proposed revisions of Vernon Grose's materials, where he says that, "All scientific evidence to date ..." Include that paragraph plus the last paragraph of the 3-page typewritten sheets, along with the other statements listed in the first paragraph on page 125 (revised) would not take care of the situation for us?

HOWARD DAY: Why don't you read those to the audience?

JOHN R. FORD: The way it would be ... deleting the first sentence of the revised page 125 which reads, "The origin of life and the evolutionary development of plants and animals have long been of interest to the layman and the scientist ..." — substituting for that, "All scientific evidence to date concerning the origin of life implies at least a dualism or the necessity to use several theories to fully explain relationships between established data points. This dualism is not unique to this field of study but is also appropriate in other scientific disciplines such as the physics of light." To immediately follow that, "While the Bible and other philosophic treatises also mention creation, science has independently postulated the various theories of creation. Therefore, creation in scientific terms is not a religious or philosophic belief. Also note that creation and evolutionary theories are not necessarily mutual exclusives. Some of the scientific data (e.g., the regular absence of transitional forms) may be best explained by a creation theory while other data (e.g., transmutation of species) substantiate a process of evolution ... Aristotle proposed a theory of spontaneous generation ..." And then continue with the statement listed on page 125.

THOMAS HARWARD: I would accept that compromise.

JOHN R. FORD: And I would move the adoption of the *Science Framework* as amended with these notations I just made.

HOWARD DAY: Dr. Leslie?

GLENN LESLIE: Mr. Day, it seems to me that that's a significant change. At this point, I would not want to have this bear the recommendation of the Curriculum Commission, in spite of the fact that I am anxious to get this adopted. I would want to go back to the Curriculum Commission for consideration of this. And I would want that in the record.

HOWARD DAY: It is part of the record.

EUGENE N. RAGLE: If there's a motion before the house, I second it.

THOMAS HARWARD: Let's start with a little discussion. I think this is my original recommendation — that the Curriculum Commission be brought back into it. And Dr. Ford, do you feel that this is our position now?

JOHN R. FORD: I am perfectly willing to have the problem or the motion tabled until the Curriculum Commission can look at these materials that have been presented today, since we did not have them before. As I told the Curriculum Commission last week, I thought very much of the statement they made. Since then, I've talked to some of the members, telling them that there was additional information that I thought should rightly be included, but I did not have a chance to talk to them about it at the time of their meeting. I would certainly agree with you.

HOWARD DAY: Dr. Leslie, how many of your Committee are in the audience today?

GLENN LESLIE: I think there are only two beside myself.

MRS. SEYMOUR MATHIESEN: Dr. Leslie, couldn't you do that during the noon hour so that we could do something about the *Framework* this afternoon?

GLENN LESLIE: I could do it to my own satisfaction perhaps, but I think that if this is adopted, it's adopted with the recommendation of the Curriculum Commission, and I think that the change is significant enough that I would not want to speak for the Curriculum Commission at this time. I think that we should hear from the Science Committee who wrote it. You understand now that we've had three children here who have made unanimous recommendations for it. They are all your creations. You appointed — the State Board appointed the science writing committee. They did the writing. You, in turn, gave to the State Department of Education a mandate to come in with changes. You, in turn, wanted us to approve — the Curriculum Commission — to approve ... Now we're your creation also. It seems to me there may be developing a generation gap here, if we go this way.

MRS. SEYMOUR MATHIESEN: They were really created to help the Board to come to a conclusion that we do get published. We have taken your advice ... we still have ... we have taken your advice. We wouldn't mind you looking it over and telling as whether you agree or not. But I think the time has come for all of us to accept it.

GLENN LESLIE: Then I would not want to say that this is the recommendation of the Curriculum Commission. I would want the Curriculum Commission to ...

HOWARD DAY: As far as these changes are concerned, you cannot say whether the Curriculum Commission has approved ...

RUSSELL PARKS: Mr. Chairman, as the Chairman of the Curriculum Commission, I happen to agree with the statement that you just made. I think that I would agree with any of the groups. If you make a substantive change in a report — whether it's in my opinion good or evil, that's not the point — either you must, if you want to attach the names of any of your committees to it, you must take it back to them and let them decide whether or not they want to be associated with it or not. If you want to adopt it, with your revisions and without referring it back to them, then you can do so and say that the agreement that they have is only in the unamended portions. That's just being fair.

MIGUEL MONIES: I think, Mr. President, that the California State Advisory Committee on Science Education … do we have any members? I think that they should be included. This was apparently a part of their work and changes such as these certainly should go back to them. The other thing – I hope that we don't attempt to, in the *Framework*, determine what is going to be included in the books. I think that this is a danger inherent in trying to formulate a framework — to say, "This is what should be included in the textbooks in California." I think that there is a danger there. I think that, in terms of a framework, working in some general order for science is a different thing than saying, "We are going now to say what should be included or not included in textbooks through the *Framework*. " I think is a different story.

HOWARD DAY Dr. Leslie, one question. In the original presentation, you were concerned about the time element involved. Putting this off until the December meeting, how does that affect your time problem?

GLENN LESLIE: Well, I suppose I am a little bit selfish in this regard. We had hoped that this one time in the history of the Curriculum Commission that we could have a *Framework* adopted on schedule and the criteria adopted, and so on. However, I think that that point is not that important. If there are important changes to be made, then time is not that important. We can do what we did in the social sciences. We can go ahead and prepare criteria for adoption of textbooks and if we get caught in a budget again, can continue the discussion about what goes into the *Science Framework* for the next year. As it is, you know, it is going to be 7 years before the influence of this *Science Framework*, from the beginning of the time the writing group had the orders to do so until it gets into the hands of the boys and girls in the classroom. By that time, maybe we'll have another theory or two.

EUGENE N. RAGLE: Mr. President, it should be commented that the two most articulate advocates of waiting and whose comments have generated, perhaps, this discussion we have heard this morning, are both men of science and obviously men of deep religious faith. It is significant, obviously, that they are the people who raised these questions, being well versed in both these fields. It is significant also that most or all the testimony we have heard this morning has been on behalf of incorporating some rather significant changes in even the most newly-proposed revision of the *Science Framework*. Dr. Ford has proposed an amendment which I have seconded, and he has agreed, and Dr. Harward has agreed, that certainly this amendment would be suitable. However, the Curriculum Commission wants it referred to them. Are we not sure that they have no really substantive objection to waiting another month for consideration of this? Let me suggest that perhaps Dr. Ford would like to withdraw his second and refer this all back to the Curriculum Commission, with the understanding that this particular revision suggested by Dr. Ford of incorporating part of Mr. Grose's testimony, be referred to the Curriculum Commission for incorporation into a presentation next month.

HOWARD DAY: I just want to suggest that the Board has no objection to cooperation with the Curriculum Commission ...

DORMAN COMMONS: Well, Mr. Day, I do have an objection. I think that this matter has been considered long enough. I think that if the Board wants to make the amendments — and I am perfectly willing to go along with the amendments that have been suggested, we can make them and proceed. It's not unusual for this Board to take a position which in some instances is not directly in line with what the Curriculum Commission has suggested. I think we have every right to amend the guidelines as we see them right here and now. If we send this back again, it simply is going to open up and take another ... we'll have another public meeting, I suppose, because when you come up with revised guidelines, it means that you have to open it up and have more comments. This matter has been discussed, and it seems to me that the guideline as amended takes care of all of the matter we would be taking care of. I really don't believe that any good purpose is going to be served in delaying.

DONN MOOMAW: I find myself agreeing that they should have more of a say. And I also know that time is important. And I know that we have a right to do whatever we want to. Dr. Leslie, I am impressed with the buff (colored) corrections as well as the other. I would vote for it now. Maybe I'm not subtle enough to see that this addition that we have suggested from the first person who spoke in the hearing (Grose) is any more than a clarification of what the buff was trying to say. Could you help me here, just as an individual ... where this really is giving you static?

GLENN LESLIE: Personally, I don't have it in front of me.

DONN MOOMAW: Well, that's all you need to say ... I mean ... what I meant by that, I didn't mean to be rude, Dr. Leslie. I am just saying, "I'm not sure why you're opposing this." You're just doing it as a matter of principle because ...

GLENN LESLIE: ... because I am not the Curriculum Commission. I'm only one member. I came here to recommend to this Board the *Framework* that was amended, as amended. There are substantive changes, and you see, I would want us not to be on record as agreeing to it as amended until it ...

DONN MOOMAW: I was just making the point that I do not feel personally that there arethat many differences, and it was a clarification. And I think it's been helpful.

MAX RAFFERTY: I think the Board has a perfect right to recommend whatever it wants to in this matter. I too am concerned about the time. I don't really think you ought to put Dr. Leslie on the spot anymore. He is simply one member of the committee group, and he cannot speak for the committee.

DONN MOOMAW: No, I wasn't asking him to speak for the Commission, Max. I was asking him, on the basis of what he said, I thought he had saw (sic) something in what we were doing that we might be able to clear up by some change of word or two, that would satisfy the total concept. Obviously, we have some differences, but if it was just a matter of a little clarification or something that he personally could live with and change ... I didn't mean to put you on the spot.

MRS. SEYMOUR MATHIESEN: I am going to call for the question.

HOWARD DAY: May we have a reading of the motion again, please?

JOHN R. FORD: The motion was to adopt the *Science Framework*, as revised by the Curriculum Commission and *Science Framework* committee with the additional change that I read with regard to page 125 of the buff-colored sheets, striking out the first sentence — the first two sentences — substituting for them the paragraph beginning, "All scientific evidence ..." on the first page of the revision suggested by Vernon Grose, following that with the second full paragraph on page 3 beginning, "While the Bible ... " and then continuing with the buff-colored as written, "Aristotle proposed a theory ... "

HOWARD DAY: Before we have any more debate, Mike, I would like to call for the question. Ready for the question? All in favor indicate by saying "aye." All opposed by the same sign.

DONN MOOMAW: *No* ... If Mike wants to speak, I'll vote "no. "

MIGUEL MONTES: Let me explain ... (the entire Board erupted into murmuring that this was out of order) No, I want to ... let me make a particular point, Mr. Chairman. This is the privilege of being on the State Board of Education. And that point being, although we believe — and certainly myself being of the Christian conviction — we may get into difficulty, "While the Bible ..." — this although we may be Judeo-Christians and a majority — even if we accepted that a majority of us believed — that there are other philosophic beliefs ... But that particular point is what made me feel not ... we should be talking in terms of "philosophical treatises that mention creation, science has independently ..." I think that the inclusion of the Bible will include one *Framework* reference while we might want to include other references for those who do not identify themselves with the Bible ... (discussion again erupted)

HOWARD DAY: May I call again for the question? All those in favor indicate by saying "aye." Those opposed by the same sign. (it was a unanimous "aye") So ordered.

EXHIBIT 7: *LOS ANGELES TIMES* EDITORIAL-II

This editorial was published on 16 November 1969

EVOLUTION BECOMES A THEORY

Even long before Darwin, the question had been a haunting one for educators. The California State Board of Education was the latest to try for a solution. It decided that evolution will be taught as theory rather than fact.

Under a science framework adopted as a guideline for development of new science textbooks, the biblical version of God's creation of life also will be included, even though no other religious scriptures will be mentioned.

These changes in the framework - a basic outline for new textbooks slated to be adopted by the board next year - were made to meet objections of conservative board members that the theory of evolution was being presented as fact and that the theory of the world's creation by a Supreme Being was being left out entirely.

Although approval of the framework by the board had been expected to be routine, these objections last month led to a one-month postponement — and only after two paragraphs were inserted was the framework adopted.

The two paragraphs — proposed by a member of the audience and adopted unanimously by the board— contained references to both the Bible and other sources. They read:

> "All scientific evidence to date concerning the origin of life implies at least a dualism or the necessity to use several theories to fully explain relationships between established data points. This dualism is not unique to this field of study but is also appropriate in other scientific disciplines such as the physics of light.
>
> "While the Bible and other philosophic treatises also mention creation, science has independently postulated the various theories of creation. Therefore, creation in scientific terms is not a religious or philosophic belief. Also note that creation and evolutionary theories are not necessarily mutual exclusives. Some of the scientific data (e.g., the regular absence of transitional forms) may be best explained by a creation theory while other data (e.g., transmutation of species) substantiate a process of evolution. "

The two paragraphs were proposed by Vernon L. Grose, vice president of the Tustin Institute of Technology in Santa Barbara, which he described as a consulting firm that conducts management seminars for other businesses.

Grose, pleased that the board went along with his changes, said the additions to the framework would avoid a "dogmatic approach" to the theory of evolution and would "balance knowns with unknowns."

The language suggested by him replaced a single sentence that read:

"The oldest explanation is a religious one, that of special creation. "

Those few words had been the only reference to religion in the entire document.

EXHIBIT 8: LETTER TO RELIGION EDITOR, *LOS ANGELES TIMES*

26 November 1969
Dr. Dan Thrapp

Religion Editor
Los Angeles Times
Times Mirror Square
Los Angeles, California 90053

Dear Dr. Thrapp,

Our mutual friend, Dr. Ralph D. Winter at Fuller Theological Seminary, has passed on to me your request that I send you a copy of what was presented to the State Board of Education in their November meeting and that I also indicate some of the specific misunderstandings which may be found in subsequent articles in the *Times*.

I am pleased to respond to your request by enclosing both the complete transcript of my oral presentation to the Board (which was made at their request) as well as the entirety of my suggested revision for the Science Framework (of which only two paragraphs were adopted).

Since you also asked for a list of specific errors which occurred in the *Times* following the 13 November meeting, I would call your attention to:

1. On Friday, 14 November 1969, an extensive account was carried on Page 1 and contained the following errors:

a. The secondary headline of the article, "inclusion of divine creation idea in textbooks guide," is in direct contradiction with both my oral presentation (pages 7 and 8) and the portion of my suggested revision which was adopted into the Framework, "... science has *independently* postulated the various theories of creation. Therefore, creation in scientific terms is not a religious or philosophic belief. "

b. The second sentence continues the same fallacy, "The framework also will include the Biblical version of God's creation of life." This is simply not true!

c. The third paragraph definitely suggests that religious scriptures are a subject of the framework and are limited to the Bible. Please note that the only reference to the Bible and other *philosophic* treatises is in the sense of *eliminating* any such works from the framework, "While the Bible and other philosophic treatises also mention creation, science has *independently postulated* the various theories of creation."

d. Paragraph 7 refers to me as "a member of the audience" when I was an invited speaker (without solicitation) as shown by the enclosed letter of invitation from Sacramento.

2. The editorial of 16 November 1969, again repeated all the errors listed above. However, the errors were rearranged in such a manner that the false issue (i.e., teaching religion in a scientific curriculum) was now compounded, "The biblical version of God's creation of life also will be included, even though no other religious scriptures will be mentioned." This further distortion implies that, even though religion will be part of science instruction, only *one* religion will be taught!

3. An article on Sunday, 23 November 1969, on Page B is devoted to this false issue which, by now, has been raised by the *Times* to the level of an "education dispute," with top instructors vowing to repudiate the framework (which I consider to be an excellent piece of work!). The article again states the false premise; i.e., that the State Board "added the Biblical version of the origin of life as another theory to be included in classroom science instruction."

Dr. Thrapp, it takes a great amount of naivety to believe that all these errors (in one direction only) are the result of misunderstanding the English language. All of as are guilty at times of seeing only that which we want to see, but this seems to exceed inadvertent error.

When the errors are combined with peripheral comments in both the reported accounts and the editorial, a completely distorted picture is inevitable. Let me illustrate some of these comments:

1. All writings use the word "conservative" in such a way as to imply that creation is a pet thesis of conservatives (undefined as to political or religious views) while evolution is evidently "non-conservative." It is frankly most distasteful to me to mix science and politics in this way, since I am not a conservative.

2. Last Sunday, the only letter to the Editor of the *Times* on this issue which has been published since the Board's action was not only facetious but entirely based on the false premise of my concern. It states that the Board has "mandated the teaching of the Biblical interpretation of creation in elementary science classes."

3. Only in the *Times* was I personally described thus: "... Vernon L. Grose, who said that he is vice president ..." This obvious effort to reduce credibility in a position I have held since 1966 is hardly professional. The nationwide Associated Press coverage, all other California newspaper accounts, television and radio broadcasts did not find it necessary to resort to this descriptive approach.

4. In the 14 November 1969 article, one whole paragraph which followed the description (3) above was devoted to discrediting my relationship with USC. It was not only false but given in such a manner that the reader could assume that my motives were less than honorable. (To correct the false report, I might add that I have been a faculty staff member at USC

continuously since 1967, teaching both on campus and in Europe. I returned from a 5-month full-time teaching assignment for USC in Madrid and Seville, Spain in May of this year (hardly last winter!). I declined an assignment to teach for USC in Japan this fall because of other business commitments, but the enclosed memo from Dr. John LeBlanc last month proves that I am still on the roster of lecturers.)

Even though it is often difficult to be objective on issues wherein we are personally involved (and particularly maligned), my primary concern lies in a far greater problem. I am concerned that the bias of the *Times* thus far has raised a false and fallacious issue which is polarizing educators and scientists into two totally unnatural and undesirable camps. Further, this issue can only contribute to further division in our community just when we need so badly to be united.

Many church groups are undoubtedly rejoicing in error that the Bible is now a science text, and at the same time, agnostics and atheists are inflamed to irrationality. If a collision on this issue occurs; e.g., at the 4 December 1969 meeting of the State Advisory Committee on Science Education in San Francisco, I will personally hold the *Times* responsible.

I do hope that you will understand, Dr. Thrapp, that I have always respected the *Times.* It is the only newspaper to which we have subscribed during the ten years we have lived in California. I might have questioned Vice President Agnew's remarks about the press far more, however, if this issue had been fairly reported in the *Times.*

A recent *Times* editorial entitled, "Agnew vs. Press: Round Two," states:

> "We, as a medium of communication, strive constantly to evaluate our product. We seek facts. On what we sincerely judge to be the facts, we then express our editorial judgments."

Via this letter, the *Times* now has the facts. I trust that they, in fairness, will have the grace to clarify and if necessary renounce their previous position.

I would be delighted to talk to you, either in person or by telephone (213) 348-4974, if there are ambiguities in the material I am sending you. Thank you so much for taking the time and interest to become involved!

Very truly yours,
TUSTIN INSTITUTE OF TECHNOLOGY

Vernon L. Grose
Vice President

EXHIBIT 9 "TEMPO" TELEVISION INTERVIEW

A transcribed interview of Dr. Ralph W. Gerard on "TEMPO" (KHJ-TV, Los Angeles Channel 9) by Baxter Ward, "TEMPO" Moderator, 5 December 1969

Ward: Dr. Ralph Gerard, Dean of the Graduate School of the University of California at Irvine, is one of the authors of a new framework governing science instruction in our public schools. There is a controversy between the Committee's viewpoint and that of the Board of Education. We invited viewpoints from both sides, but we could find only one — Dr. Gerard's. What happened to the opposition?

Gerard: I wouldn't tell you what I think.

Ward: Well, we're glad that you are here. What is the controversy over, with regard to the new science plan for teaching?

Gerard: Well, I suppose that would depend on whom you ask. I am a member of the committee that had been appointed over four years ago by the State Board of Education to prepare guidelines for people preparing educational materials for teaching science in the 1 to 12 grades in the State of California. This Committee of about 15 — some very fine people on it, has worked hard for over four years, has produced a document of 205 pages which has been very widely praised by people who have seen it.

Ward: May I say that 205 pages hardly deserves four years of preparation. Was there a lot of argument and discussion? Why this long?

Gerard: No, no. Because a great deal of it doesn't appear. It was much longer beforehand. This is a boiled-down version. The Board of Education took out one paragraph on one page and put in another paragraph.

Ward: What paragraph did they remove?

Gerard: May I continue, and I'll answer that my own way?

Ward: Surely.

Gerard: The paragraph was actually in an Appendix and it was merely giving an example of the kind of material that one should treat in certain ways. It had to do with evolution but rather tangentially. They inserted another paragraph which, in effect, said when one thinks about the origin of living things, one must compare the theory of evolution with the theory of special creation. So this is what has become the public issue — the question of whether evolution should be pitted against natural, spontaneous, or Biblical stories or anything else. In my judgment, this is a very secondary issue. It is to me rather horrifying that a Board of Education in this great state in this year, 110 years and a few days after Darwin's "Origin of Species" was published and completely changed the way the world thinks, that at this time a group of people could have the — well, I don't know what word I want to use — stupidity, audacity ...

Ward: Or temerity ...

Gerard: Temerity — I could use a lot of words, but one has to be careful on the air.

Ward: I don't know that stupidity would be the correct word.

Gerard: I think it is!

Ward: There was a great deal of sincerity involved ...

Gerard: But one can be sincerely stupid.

Ward: Well, this would depend on the viewpoint.

Gerard: But whether we're talking about anyone in particular or not, some of the most sincere people are the most stupid because very strong positions are often possible when one knows only a small part of the story.

Ward: Do *you* know the story?

Gerard: I think I do reasonably well.

Ward: I'm not here to argue with you. I'm here to learn the basis. Did your Committee meet yesterday again?

Gerard: Our Committee met yesterday, and I would like very much to read to you and to our audience the resolution that we passed.

Ward: All right.

Gerard: You asked me to say what the real objection to the action of the Board was. I said the issue of evolution is important, but it is not the primary one in my judgment, nor for that matter, in the judgment of the Committee. "The State Advisory Committee on Science Education takes strong exception to the alteration of content on page 125 of its report on the Science Framework of California Public Schools made by the State Board of Education. It also takes exception to the manner in which these changes were introduced. The Committee, composed of over a dozen competent educators and scientists appointed more than four years ago by the Board, has worked faithfully, diligently, and carefully to forge a guide for the science education of the children of this state. Its only recompense would be in contributing to a public service. It called on additional experts for advice, utilized the efforts of national science and education groups that had earlier devoted much time and money to the problem of science education, circulated preliminary drafts of the framework to teachers, and education administration in this state ..."

Ward: So far, it's been the crying towel. Now get down to the heart. We appreciate the dedication, but what is the point?

Gerard: The point is that it is utterly unacceptable for people rather casually and without any examination to change the whole intent of a major document like that and especially, as I would assert, they are not competent to make the judgment.

Ward: But Darwin did it, as you say, a hundred and some years ago, just like that.

Gerard: Darwin took 20 years after he had his theory worked out before he published it.

Ward: But the evidence was available to all mankind for thousands of years prior to Darwin. I could argue either side, I just ...

Gerard: A lot of people had ideas about evolution before Darwin. He was the first one who really marshaled the evidence so that it was apparently incontrovertible so that the impact of it was felt but was put together so that the world had to see it and believe it. I don't quite know what we're discussing.

Ward: That's all right. I get angry too and lose my track ...

Gerard: I'm sorry if I sounded angry. I am very positive about this ...

Ward: Well, when I get positive, I find it very difficult to form sentences and continue my theory of some kind ... I was hopeful that there would be some facts rather than merely footstamping. That's what it's been so far — you're justifying the Committee.

Gerard: I am telling you what the Committee feels about the way the Board of Education acted and ...

Ward: So far, it's been how the Committee is so proud of itself, if I may say so. Now get down to the Board.

Gerard: Well, I'm sorry. The Committee is saying the document was a very substantial, carefully prepared work and it should not have been assassinated.

Ward: In the Committee's view ...

Gerard: How do I answer that? Here was a Committee of people selected because of competence in these fields, had spent a great deal of time and effort developing something, and to have one or two people, and I believe it was essentially one or two people, cavalierly, without letting the Committee have any response back, say changing your whole viewpoint.

Ward: This is the interesting part. You understand that I'm speaking because we don't have a debater here, opposite you, for another viewpoint.

Gerard: I wish the other person were here.

Ward: I'm sure he'd be far more qualified than I.

Gerard: I could react with more relevant points. But fine, go ahead ...

Ward: What is the role of the Board of Education? Don't they in instituting your Committee reserve the right to make the changes that they might choose?

Gerard: I don't know what the legal aspect of it is. There is a good deal of question as to whether throwing religion into a document on science for the teaching in our schools is legally valid or not.

Ward: How much religion are they wanting to put in now? That is, to include in your framework?

Gerard: You make me think of a story.

Ward: I'm delighted, but ... okay.

Gerard: I would think of a story that I might not be able to tell but I wouldn't mention it unless I could tell it. This goes back 30 years, a quarter of a century when the Russians were very eager to get advanced science into Russia, and an emissary from the Russian government came to talk to a distinguished chemist from the United States. He said, "We'd like you to come and work in Russia." He said, "We'd give you any inducement you'd want." "Well, that sounds interesting." "Any salary you mention, all the laboratory space, helpers. We know that people think things are tough in Russia, but you will have a ten-room house, servants, a Cadillac, a yacht by the Black Sea. Don't worry about being sealed off — three months of the year, you can go where you will. Your money will be paid in New York" and on and on and on. Listening to this, he said, "Well, that sounds wonderful. There's just one little matter ..." "What's that? We'll take care of it." "Stalin must go." It's that kind of an input.

Ward: When you took off your glasses, I thought you were going to punch me in the nose ... How badly diluted do you feel your framework will be now because of that action?

Gerard: This is the wrong question, if I may say that. It is not diluted at all. It is simply subverted, and if I may go on reading the resolution that we passed yesterday, I think it will be perfectly clear what the real conflict about it is. I was saying that it had solicited criticisms from around the state, had had meetings with groups of teachers and administrators, that it had gone to the Board once before. They'd asked for some minor changes which we made. Our revised version was considered by the Board on November 13, and the crucial changes — this paragraph — were put in (they were actually written by a non-Board member and just included practically without discussion), and the Committee members there were not allowed to make any comment at the time. "The changes, though small in extent, had the effect of entirely undercutting the thrust of the 205-page document. Our concern has been overwhelmingly with the nature of scientific evidence and conclusions and the great care needed to avoid error and dogmatism. The topic of evolution entered almost incidentally. In fact, the content of science in general is underplayed. The Board, by pitting a scientific fact and theory as to its mechanism against a particular religious belief as if they are commensurate, has thus offended the very essence of science, if not also religion." That's what the cry is about. The Committee felt strongly enough about it so the informal resolution calls (which I will not read) ... said if you will not remove this or put it under your own name or allow us to write a rebuttal to this one paragraph, we simply refuse to have our names associated with it.

Ward: Is that going to be the net result?

Gerard: Well, I don't know. It goes back to the Board now.

Ward: You stated that a non-Board member present in the Board meeting room supplied the text to the remark or the paragraph. Who is he?

Gerard: It was a man — I believe it was a Mister or Doctor Grose. I was not present. I have this by hearsay.

Ward: How much influence did that person have on the Board?

Gerard: Apparently a great deal.

Ward: Did *he* have influence, or did his *belief* have influence, do you think?

Gerard: I don't know.

Ward: I don't know either.

Gerard: I haven't met any of the Board members. I haven't met this gentleman. I have not been at one of the Board meetings.

Ward: Well, it seems to me that you are extremely angry with the Board, yet if the Board has the final responsibility, can we quarrel with that, or would you prefer they not have ... or do they have? You essentially directed me on the type of question you would like asked ... Therefore, I am just trotting out a whole variety ...

Gerard: The reason I am here today is because when this broke in the papers, a great many people, scientists and educators around the state were quite disturbed. I wrote a three-page statement which went out to the newspapers and triggered a good deal of publicity.

Ward: I think I read that, and that's why I asked you to appear ...

Gerard: Did you see the whole thing?

Ward: No, I didn't see the whole thing. I saw a portion of it.

COMMERCIAL BREAK

Ward: Dr. Gerard is a member of a Committee that is offended because of an action taken by the Board of Education. The Board felt that an approach to the teaching of science was incorrect, in that evolution should perhaps be diminished or considered in the same ...

Gerard: No, I cannot accept your formulation. We are not offended as individuals.

Ward: You are offended as a body, and that's a very touchy document you read.

Gerard: We are saying to the Board, "You had no proper right to behave in this way to us." That is personal, but not because we are angry as individuals (although we are), it is because we feel that this is a very cavalier fashion of dismissing a careful effort. This is the essential difference between the way science produces conclusions and the way other groups of the people produce conclusions.

Ward: What would you like to see done?

Gerard: I would like to see that paragraph that they inserted removed.

Ward: Would you like to have some discussion about it first? Or just arbitrarily removed?

Gerard: Well, I think it is so completely irrelevant to the whole rest of the document that the proper thing would be to remove it. Now if they feel that it is essential that this point of view be expressed, then they certainly have a right to express it. They should express it as a separate statement over their own names — not over ours.

Ward: Well, won't they do it over their names?

Gerard: I don't know. We've asked them to do that.

Ward: Just as we visit, I have a great respect for your ability and background and effort in this thing. But also it seems to me that you are more sensitive as a Committee member over their *action* than over what their action *means*.

Gerard: You are ... I'm sorry. It's because you keep emphasizing that part of it, and I read only the first third of the statement because I wanted to get the feel of it here. The Committee feels that introducing this irrelevant point of view and saying, "Look, whether you believe this production of science or whether you believe something entirely otherwise is sort of up to you and the child to make the decision." And we say this is utter nonsense. I take great exception to what Dr. Rafferty said over a TV program soon after this was done — "let the children decide."

Ward: You say the scientists should decide.

Gerard: No, I say one should have the evidence as to how one goes about reaching conclusions. One should know what you are making a judgment on, and then let's have art in one place, religion in another and science in another and politics in another. They come together, but let's not say, "Is it this or is it that?" That's improper.

Ward: I certainly can see your point as a scientist, but I can see Dr. Rafferty's point. If the matter were to be submitted to the pubic for a vote, which faction do you think would win? I realize we've kept you longer than you agreed.

Gerard: Are you telling me I'd better stop?

{Note: At this point, the tape recording was interrupted for about one minute. During this time, Dr. Gerard indicated that his mail ran about 3:1 in favor of his position. He also mentioned that those who disagreed with him were unkind.}

Gerard: Those that disagree with me are very violent about it.

(Note: The tape was temporarily intermittent at this point, but Baxter Ward emphasized that Dr. Gerard had showed plenty of emotion himself.)

Gerard: I won't allow that to pass without disagreeing with you. They call me names — I don't call them names. I'm prepared to say that I think the action of the Board was illegal, immoral, and fatheaded, but that's not altogether ...

Ward: That evens the score.

Gerard: That's the action ... I'm not saying that they are bad people. I'm not saying that they are dishonest. I'm not saying that they have done things that offend decency ...

Ward: But you also threw in "stupid" for good measure, so it comes out about right.

Gerard: I hadn't said that before.

Ward: Oh, I thought you had. All right. I think you have accomplished your purpose, which is to inform us of the view of your Committee, and in that, you were extremely successful. Thank you very much for coming in.

EXHIBIT 10: LETTER TO CALIFORNIA SUPERINTENDENT OF PUBLIC INSTRUCTION

6 December 1969
Dr. Max Rafferty
Secretary and Executive Officer
California State Board of Education
721 Capitol Mall
Sacramento, California 95814

Dear Dr. Rafferty:

Since November 13th when the State Board of Education adopted two paragraphs of a revision that I suggested for the Science Framework, there has been much hysteria about a completely fallacious issue. Perhaps it is a testimony to how much presupposition can override facts.

Nevertheless, I feel that it is imperative that you and the State Board (which as yet has not responded to the barrage of accusations) set the record straight. My intent at the November meeting, as verified by both my oral remarks and the entire suggested revision, was to specifically *exclude* any religious connotation regarding the concept of creation. The mention of the Bible in the adopted paragraph was in the sense of *denying* the Genesis account for the Framework and substituting the various theories of creation now taught in such reputable institutions as the Massachusetts Institute of Technology.

Most of the creation theories with which I am familiar are proposed by atheists. In fact, the unadopted portion of my suggested revision contains reference to two creation theories, both of which were postulated by atheists. I did this for the very purpose of offsetting the presuppositions of religionists, on one hand, who would claim, in error, a great victory for God and the Bible, and my scientific colleagues, on the other hand, who would falsely read Biblical creation (which is believed only by faith) into the Science Framework.

Yesterday, Dr. Ralph Gerard, one of the members of the State Advisory Committee on Science Education, was interviewed by Baxter Ward on KHJ-TV (Channel 9) here in Los Angeles. As I watched this telecast, I was agreeing with Dr. Gerard in his insistence that religion had no place in a science framework. Yet, I was the author of what he viewed as an anathema! To compound my frustration, Dr. Gerard's letter to the *Los Angeles Times* was published this morning, again indicating that he was laboring under a complete misconception about what the State Board had done.

To deenergize a situation which I believe that the news media deliberately perpetrated, I have just finished talking at length with Dr. Gerard and found that we are in basic accord. I apologized for any role that I might have played in the misinformation which caused him to be alarmed.

On the other hand, I feel that the State Board owes Dr. Gerard (and other educators in the state) an explanation that you did not succumb to adopting the thoughts of some "insistent ladies" at the November meeting. Rather, you adopted a statement, not only from a different source, but also on an entirely different basis. I was apparently identified in Dr. Gerard's mind (and perhaps some members of the press) as speaking to the same point as those who spoke for the Creation Research Society.

To this end of clarification then, I have drafted and enclosed a recommended statement that you might wish to read or issue to the press at the December meeting of the Board next week. Hopefully, it should allay the fears of those who, like myself, do not want to mix science and religion. If we fail to make a strong statement of this nature. I believe that the entire step of progress we have made to date stands in jeopardy.

Dr. Gerard and I have agreed today in principle that the governing criterion for any and all theories to be included in the Framework must fall within the limits which science imposes on itself, i.e., conclusions drawn strictly from objective observations and not subject to a presupposition based on faith (whether in religion or scientism). I explained that I was interested in not only including discussion of NON-RELIGIOUS CREATION but also other theories of interest to physicists, geologists, astronomers and other scientific disciplines. One of these is the theory of devolution (the inverse of evolution). Dr. Gerard commented that this is also useful in his field of biology.

Dr. Rafferty, we must avoid the perpetration of the "evolution-versus-creation" syndrome that I decried at the November meeting. I believe that the Framework as now amended avoids monolithic thinking, not just by its inclusion of scientific creation but any other concepts that may provide explanations that evolution does not. For example, the energy decay in the universe which is exhibited by the Doppler red shift in astronomy or by radioactive dating (where we measure the stability of a radioactive element by its half-life period, i.e., the time for half of its atoms to be disintegrated into some lower energy form) does not follow the evolutionary concept of complexity arising from simplicity but just the opposite! The universe is actually in a state of running down, energy-wise — not building up. This contrast with the theory of evolution must be shown to the school child just as the two opposing theories of light in physics are. The crux of this presentation must be in *comparing* the theories, not in discrediting one over another or pitting them in opposition.

As a final note, the Board should give thought to my second recommendation in the November meeting, i.e., that a person who has both understanding and interest in incorporating these additional theories be given some official status by the Board to work with the Curriculum Commission and, if necessary, Dr. Gerard's Science Education Committee. I would be pleased to serve you in this capacity.

Please forgive me for attempting to give you direction on this matter. I am only expressing my concern that we not lose the progress made by the State Advisory Committee on Science Education and the State Board thus far via a misconception.

At the suggestion of Dr. Gerard, I am sending copies of this letter to all the State Board members. Since he also requested a copy, I am pleased to honor his request.

Very truly yours,
TUSTIN INSTITUTE OF TECHNOLOGY
Vernon L. Grose
Vice President
Copies to: Ralph W. Gerard, M. D.
California State Board of Education members

EXHIBIT 11: ORIGIN OF THE UNIVERSE, MATTER, LIFE, AND MAN

Background statement for curriculum criteria prepared by Vernon L. Grose for the Curriculum Commission, advisory body to the California State Board of Education — 9 January 1970

Universal scientific laws are numerous in all areas of scientific endeavor. This does not mean that scientists do not frequently challenge such laws. However, the laws are seldom discarded under challenge but rather are most often strengthened through modification.

In contrast to widely accepted scientific laws are a group of theories concerning origins which are not universally accepted by scientists. In fact, there are several reasons why indisputable *scientific* theories for origins will probably never be realized:

1. Since the events themselves (i.e., origins) occurred in the apparently distant past when no one was present to observe and record what happened, there are no "original" data currently available from which to postulate a theory.
2. There are no incidents apparently happening today wherein matter or life is "originating," i.e., suddenly appearing without cause or predecessor. If there were only a few "originating" incidents available, then these events might provide strong clues about

primordial origins.

3. The "ground rules" that scientists have established for science are very limited. Scientists have declared "off limits" many types of thought which could possibly prove helpful in postulating a universally acceptable theory of origins. Examples of scientifically unacceptable information include:
   a. Any value system which assesses relative or absolute merit such as goodness or badness of something.
   b. All moral systerns including concepts such as justice, mercy or love.
   c. All artistic, poetic, spiritual, and philosophical thinking.
   d. Any intelligence or information that cannot be repeatedly verified by at least one of the five senses.

Therefore, the likelihood of *scientists*, while remaining scientists, postulating a single and universal theory of origins is virtually nil. If scientists wish to forego some of their self-limitations and cross over into philosophy, for example, theories come much easier.

However, it should be evident that such a merger of science and philosophy cannot possibly produce a wholly *scientific* theory. Further, no *single* theory (required for universality) will emerge from a combination of science and philosophy because philosophy is far wider in scope than science. Perhaps this realization provides a clue as to why there are many theories regarding origins, rather than one universal theory.

Scientists, if they are rigorous and disciplined, spearhead the demand for no limitation on the number of theories for origins. This is not to say that all theories postulated are equally likely to describe the "original" event. On the other hand, discarding any particular theory concerning origins, on strictly *scientific* basis, is not easy.

Without exception, all theories proposed for origins are less than scientific. They all involve a certain amount of philosophy, faith or belief that is not scientific. This is equally true, for example, for theories that propose evolution as the mechanism and those which offer creation as the explanation. Conversely, there are no scientific data that can be used to disprove predictions about either evolution or creation (as "originating" mechanisms) even though both schools of thought appear to rule out the other.

For this reason, school children must be presented with several theories for origins. This comparative presentation of theories serves at least two purposes. First, it illustrates the scientific irresolution of the issue. Secondly, it points out that scientists will *never* find the answer of origins by themselves, i.e., without the help of philosophy, faith or belief.

At the risk of oversimplification, all theories for origins could be grouped into one of two philosophic schools of thought as shown in Figure 1. The most important factor to bear in mind when comparing these two schools of thought is that they both take into consideration *exactly the same scientific data* to reach conclusions which are virtually diametric! Therefore, it must be obvious that, if such completely opposing conclusions can be reached with identical scientific data, the conclusions are primarily philosophic and only secondarily scientific.

Perhaps this deduction partially explains why such unscientific emotion and passion is evoked whenever "scientists" of one persuasion attempt to convert their colleagues of the other school. The scientist is no longer a scientist if he must persuade, sell, or convert his scientific colleague — he is a *philosopher*! Scientific activity, via the limitations discussed earlier, has been specifically designed to be overwhelmingly conclusive strictly on the basis of objective data. So if emotion overrides the data, philosophers have taken over from scientists.

Therefore, where origins are concerned, all theories appear destined to remain predominantly philosophic and only to a minor degree scientific. To avoid dogmatism, scientists need to be reminded to maintain awareness of when they leave the "objective" world of science and enter the "subjective" world of philosophy. Even the most renowned scientists occasionally err in this respect.

PHILOSOPHIC BASES FOR EXTRAPOLATION BACKWARD IN TIME (BEYOND DATA) TO ORIGINS

Parameter	Philosophy 1	Philosophy 2
Basic Construct	Uniformitarianism, modified by "quantum evolution*" for discontinuities in a simple-to-complex continuum	Discrete origins followed by uniformity
Basic Correlate	Complexity arising from simplicity	Complexity established initially at origin followed by degeneration associated with increased entropy
Origin of Universe	Spontaneous generation, without knowable cause	Discrete origins without specified cause or mechanism
Origin of Life	A miracle performed by inanimate matter via spontaneous generation	A miracle performed by an ultimate or supreme intelligence
Effect of Time	"Long-time" dependent to permit random selection to produce current complexity	Time independent
Ultimate Cosmology	Randomness or chance occurrence	Inherent design or purpose
Example of Theory	General Theory of Evolution	General Theory of Creation

"Quantum evolution" refers to Dr. G. G. Simpson's explanation for "the relatively rapid shift of a biotic population in disequilibrium, to an equilibrium distinctly unlike an ancestral condition .... It is ... believed to be the dominant and most essential process in the origin of taxonomic units of relatively high rank, such as families, orders, and classes. It is believed to include circumstances that explain the mystery that hovers over the origin of such major groups." G. G. Simpson, *Tempo and Mode in Evolution* (New York: Columbia University Press, 1944), p. 206.

EXHIBIT 12: LETTER TO THE EDITOR. *PALO ALTO (CALIFORNIA) TIMES*

This letter, by a Stanford University Professor of Materials Science, represents an early attempt to focus public attention on the actual Science Framework wording instead of news media interpretation of that wording

January 13, 1970

Editor
The Palo Alto Times
Palo Alto, California

Dear Sir:

In view of all the misconceptions that have been perpetuated in news articles and letters to the editor with respect to the current controversy concerning the guidelines for science education in California schools, it is perhaps worthwhile to attempt briefly to get the record straight. Reliable information is still the best basis for intelligent judgment, even in the area of creation and evolution.

1. The proposed changes in the guidelines do *not* suggest that the Genesis story of creation be taught as an alternative to the theory of evolution.

2. What the proposed changes do suggest is that since there are and have been various *scientific* theories involving a "creation" concept; e.g., the steady-state cosmology of Hoyle, the guidelines should indicate that we do not *know* that a continuous uniform pattern of evolution is adequate to describe all the phenomena involved in the development of the present state of man and the world.

It is only a defense of honesty and scientific integrity to present scientific theories as partial and incomplete descriptions of the world. This is true of such "well-established" theories as the quantum theory and relativity theory. It is certainly true of the theory of evolution, which by its very nature is much less susceptible to crucial tests of its validity. Yet the tendency has been to present the general theory of evolution, "amoeba-to-man", as if the books were closed and all problems were answered.

It is likewise only a defense of honesty and scientific integrity to emphasize that scientific theories by their very nature do not treat origins. Genuine origins are singularities that scientists prefer to work their way around, hence the proposal of a cyclical universe to avoid the singularity of the "big bang" origin. Education in the meaning of science should stress this characteristic of science, rather than giving the impression that because amoeba-to-man can be described, the origin of the amoeba is a matter of indifference.

Personally I think that the attempt to use the word "creation" in a scientific sense in the same context as "evolution" is semantically impossible because of the strong historical and emotional associations of these terms. When the scientist speaks about "creation" in his context, including the conservation of energy, he is using the word in a very specialized way. To expect the public to accept this use of the word and to divorce it from its common association with religious writings is in my opinion a hopeless goal.

The present discussion calls for neither a defense of the Bible nor a defense of evolution. These may be favorite topics, but they are not at issue here. The present controversy does *not* center on the question: "Should the Bible be taught as science?" Those who attempt to make it do so perform a disservice to the public.

Sincerely yours,
Richard H. Bube

EXHIBIT 13: SUGGESTED ADDITIONS TO CRITERIA FOR EVALUATING BASIC PROGRAMS IN SCIENCE: KINDERGARTEN, GRADES ONE THROUGH EIGHT

Prepared by Vernon L. Grose for the Curriculum Commission for their meeting with the Advisory Committee on Science Education on 12 February 1970 in San Francisco

SUGGESTED ADDITION #1

The Teacher Materials shall:

12. describe how science is related to other disciplines like philosophy, particularly when postulating theories like evolution and creation for origins.

SUGGESTED ADDITION #2

The Learner Materials shall:

17. provide examples of gaps in scientific knowledge as a background for stimulating and focusing inquiry processes.

SUGGESTED ADDITION #3

1.0 Attitudes of Science

The program shall provide opportunities for the learner to:

1.4 recognize the limitations of scientific modes of inquiry and the need for additional, quite different approaches to the quest for reality including answers to questions like the origin of the universe, matter, life, and man.

EXHIBIT. 14: STATEMENT ON PUBLISHER COMPLIANCE WITH STATE BOARD OF EDUCATION REQUIREMENTS

Delivered by John R. Ford, M. D., Vice President of California State Board of Education at the 8 July 1971 Board Meeting

There is a high probability that science textbooks currently being evaluated in response to the Call for Bids for science textbooks and reusable educational materials, dated March 1, 1971, were written prior to the adoption of the Science Framework of November 13, 1969. Since publishers normally write their textbooks in such a manner that they can be marketed simultaneously in several states, few publishers would consider tailoring a textbook for only one state, even for the largest state in the union.

One of the most important requirements established by the California State Board of Education in the Science Framework is that more than one theory for the origin of the universe, matter, life and man must be presented in the textbooks. To our knowledge this requirement is unique to California although several other states have subsequently indicated that they will adopt the same requirements.

Textbook publishers may have inadvertently or intentionally overlooked this unique requirement. However, the California State Board of Education was firm in its insistence that this requirement be met by more than token acknowledgement. No textbook should be considered for adoption by the State Board that has not clearly discussed at least the two major contrasting theories for origins — "chance" or the general theory of evolution, and "design" or the general theory of creation. In the event that publishers may have overlooked this important requirement, they should be advised that:

1. The requirement must be met.
2. There is no contemplation that the date for meeting this requirement will be extended.
3. Any revisions to artwork and text material that publishers wish to submit for consideration prior to the deadline date of September 15 need not be in final format and may be typewritten and keyed by page number in the textbooks already submitted.

EXHIBIT 15: ESTABLISHMENT STATUTE

California Assembly Bill Number 2800 became law on 19 October 1971 with the signature of Governor Ronald Reagan. Chapter 1188 of that enactment contains Article 3 which establishes the Curriculum Development and Supplemental Materials Commission

Article 3. Curriculum Development and Supplemental Materials Commission

583. There is in the Department of Education the Curriculum Development and Supplemental Materials Commission consisting of a Member of the Assembly appointed by the Speaker of the Assembly, a Member of the Senate appointed by the Senate Committee on Rules, one public member appointed by the Speaker of the Assembly, one public member appointed by the Governor, and 13 public members appointed by the State Board of Education upon the recommendation of the Superintendent of Public Instruction or the members of the State Board of Education.

So far as is practical and consistent with the duties assigned to the commission by the State Board of Education, at least seven of the 13 public members appointed by the State Board of Education shall be persons, who because they have taught, written, or lectured on the subject matter fields specified in Section 583. 3, in the course of public or private employment, have become recognized authorities or experienced practitioners in such fields. At least three of the 13 public members appointed by the State Board of Education shall be full-time classroom teachers assigned to teach any of grades 1 to 8, inclusive.

583. 1. The Members of the Legislature appointed to the commission pursuant to Section 583 shall have the powers and duties of a joint legislative committee on the subject of curriculum development and supplemental materials and shall meet with and participate in the work of the commission to the extent that such participation is not incompatible with their positions as Members of the Legislature.

The Members of the Legislature appointed to the commission shall serve at the pleasure of the appointing power.

583. 2. (a) Commission members shall serve for four-year terms and shall not be eligible to serve more than one full term. Prior service on the commission for a term of less than three years resulting from an initial appointment or an appointment for the remainder of an unexpired term shall not be counted as a full term.

(b) With respect to the appointment of 13 public members by the State Board of Education to the first commission, four shall be appointed for terms of two years, four shall be appointed for terms of three years, and five shall be appointed for a term of four years. 583. 3. The Superintendent of Public Instruction and the State Board of Education shall consider for membership on the commission persons representing subjects commonly taught in public schools, including:

(a) English
(b) Social sciences
(c) Foreign languages
(d) Science
(e) Mathematics
(f) Fine arts
(g) Applied arts
(h) Conservation education

583.4 The Superintendent of Public Instruction or his representative shall serve as executive secretary to the commission.

583.5 The members of the commission shall serve without compensation except that they shall receive their actual and necessary travel expenses in attending meetings of the commission and in attending meetings of any committee or sub-committee of the commission of which they are members. Expenses of the commission shall be paid out of appropriations made to the Superintendent of Public Instruction or the State Department of Education.

583.6 The commission shall select one of its members to be chairman of the commission.

583.7 Whenever an employee of any public school district, state college, or other public agency is appointed to membership on the commission, his employer shall grant him sufficient time away from his regular duties, without loss of income or other benefits to which he is entitled by reason of his employment to attend meetings of the commission and to attend to the duties imposed upon him by reason of his membership on the commission. The employer of any such member may make available such stenographic, secretarial, and staff assistance as is reasonably necessary to enable him to execute the duties imposed upon him by reason of his membership on the commission.

583.8 The commission shall study problems of courses of study in the schools of the state and shall, upon request of the State Board of Education, recommend to the State Board of Education the adoption of minimum standards for courses of study in preschool, kindergarten, elementary and secondary schools. Courses of study in the public schools shall conform to such minimum standards when adopted.

583.9 As used in this article, "commission" means the Curriculum Development and Supplemental Materials Commission.

EXHIBIT 16: SUPPORTING STUDENT INQUIRY AND FREEDOM FROM PREJUDICE IN THE PRESENTATION OF THEORIES REGARDING THE ORIGINS OF LIFE AND THE UNIVERSE

Resolution Number 197172-13 adopted by the Downey (California) Unified School District on 8 May 1972

WHEREAS, an objective of education is to provide students with the known facts associated with the subjects taught as part of our education curriculum; and

WHEREAS, students are encouraged to develop the freedom of inquiry in forming their conclusions in regard to fact and theory; and

WHEREAS, students should be helped to understand that a theory, in itself, is an idea that is based upon tangible evidence, and that theories are changed or modified as new information becomes available, and as such cannot be assimilated as scientific; and

WHEREAS, theories presently taught concerning the origin of life and the universe entail implications relating to personal beliefs under the establishment and free exercise provisions of the First Amendment, and Article 2, Section 9014, of the Education Code;

NOW, THEREFORE, BE IT RESOLVED that the Board of Education of the Downey Unified School District, hereby encourages teachers in all fields when considering or teaching the origin of life or the universe, to present the major theories, including those of creation and the evolutionary processes, to stress these as theories and not as established fact, to accord them equitable treatment, emphasis, and attitude, and to encourage research and exploration without prejudice on the part of the teacher;

BE IT FURTHER RESOLVED, that the Downey Unified School District provide adequate amounts of reference material that lends support to each theory, and that teachers should supplement presently adopted texts with approved material which attempts to provide unbiased information about the various theories of the origins of life and the universe, and that the (textbook) committees of the Downey Unified School District consider textbooks for future adoptions in light of providing information in regard to the various theories of the origin of life and the universe.

This resolution adopted on the, 8th day of May, 1972, by the Board of Education of the Downey Unified School District.

Manuel Gallegos, Ed. D., Superintendent and Secretary to the Board of Education

EXHIBIT 17: LETTER FROM APOLLO 14 ASTRONAUT EDGAR D. MITCHELL

This letter is a sequel to an extended private conversation between Dr. Mitchell and V. L. Grose on 28 August 1971 shortly after his return from the Moon

16 June 1972
Mr. Vernon L. Grose
Vice President
Tustin Institute of Technology
22556 Gilmore Street
Canoga Park, California 91304

Dear *Vern*:

Since our last conversation together on the subject of expanding the processes of scientific inquiry to include dimensions previously excluded, I have given the subject additional thought.

It seems imperative that all branches of science must objectively examine phenomena that *apparently* contradict and violate known physical laws as presently interpreted. After all, while experimentation *may* be objective, experimenters are not. Their philosophic predisposition and bias forms the basic departure point for selection and evaluation of data as well as the conclusions they reach from these data.

This predisposition is readily obvious when considering the relunctance (sic) of the scientific community to seriously study the various aspects of consciousness as they relate to current models of the origin of the universe, life and man himself. The contemporary interpretation of molecular mechanics and the electromagnetic nature of molecular fields leaves only one philosophic possibility for origins. They must be "chance" occurrences. Yet, there is a persistent and subconscious hope, if not a strong conviction, that in such an ordered universe as we observe scientifically man is more than simply a "chance" synthesis of matter. The subjective awareness of man almost demands it, in spite of scientific skepticism.

We should remember that conjecture about whether the universe happened by "chance evolution" or by "conscious evolution" does not change the reality of its unfolding pattern, and *neither* point of view has been proved convincingly with empirical techniques. That is why I strongly favor the presentation of both points of view with the added hope that such duality will ultimately lead to one of two eventualities.

First, the scientific community may modify our model of living organisms which is currently based solely on pyramiding of active and interacting atomic particles and their associated electromagnetic fields. (It is my private opinion that the current models cannot be extended far enough to encompass already proved phenomena.)

The second alternative might be that scientists will postulate a new unified field concept that will allow predictive incorporation of a distinct, energetic mechanism which interacts with the fields of matter to produce that higher order functioning of life, which we call "consciousness" — a view distinctly different from that of a "chance" origin and evolution of life. The moral, ethical and theological ramifications of this second alternative would obviously have to be treated later. However, the future ability of science to explain the Universe totally almost demands this second alternative.

I look forward to further discussion with you on this vital subject.

With regards,
Edgar D. Mitchell

EXHIBIT 18: STATEMENT REGARDING THE TEACHING OF EVOLUTION AND "CREATION THEORY" IN CALIFORNIA PUBLIC SCHOOLS - I

Presented by G. L. Stebbins, Professor of Genetics University of California at Davis to the Curriculum Development and Supplemental Materials Commission on 26 July 1972 in San Francisco

Every scientist can agree with the statement made in the booklet "Science Framework for California Public Schools" (p. 26), "that what a scientist mainly does is to build and test hypotheses." Consequently, any hypothesis that must be accepted on faith, and cannot be questioned or tested by means of observations or experiments designed to acquire new facts about it cannot be a part of scientific knowledge. It has no place in a scientific curriculum. The only faith which a scientist may have about scientific matters is that he and other scientists can obtain more facts upon which to strengthen or reject the hypotheses that he holds. The belief in the special creation of living organisms is an untestable hypothesis. Those who advocate its inclusion in the science curricula of our public schools do not permit scientists to criticize or examine it. They simply assume that if, to their satisfaction, insufficient facts can be obtained to support the theory of evolution, creation must be accepted on faith, and without examining or questioning the way in which a Supreme Being could have created life. For this reason, discussion of the faith in special creation is religion, not science, and is out of place in a scientific curriculum.

In discussing this matter, the "Science Framework" Handbook makes an error of statement that is commonly made by those who advocate so-called "creation theory. " It refers (p. 106) to "the regular absence of transitional forms" between major groups of organisms. Paleontologists now have evidence that transitional forms between amphibians and reptiles were widespread during the Carboniferous Period (Romer 1966); that animals transitional between reptiles and mammals existed for 100 million years during the Permian and Triassic Periods, when they dominated the world's fauna (Crompton and Jenkins, 1968); that Archeopteryx, in spite of statements by some "creationists" to the contrary, was in nearly every characteristic intermediate between reptiles and birds (Romer, 1966); and that some of the australopithecines were in every observable characteristic intermediate between ape-like animals and man. Furthermore, the discoveries and experiments of the past thirty years have fully justified the faith of evolutionary biologists, that man's scientific ability and progress will eventually fill in the gaps in our knowledge. For all who desire to listen and learn, new discoveries will confirm the belief that nearly all biologists now hold in the validity of evolution as an explanation for the origin of the diverse kinds of organisms on the earth.

Consequently, teachers of science in California Schools should not be hampered by a requirement to give equal time to the stories of creation while teaching evolution as the scientifically accepted explanation for the origin of kinds of organisms. At the same time, teachers should be discouraged from speaking against the religious beliefs of any pupils, whatever faith they may have. They should be encouraged to ask these pupils to discuss matters of faith and religion with their own religious leaders.

[in private correspondence with V. L. Grose, Professor Stebbins suggested that his position on this subject was expanded in "The Evolution of Design," *The American Biology Teacher*, Volume 35, Number 2, February 1973 as well as in a statement he prepared for campus debate on 24 May 1973 (copies available from him)]

EXHIBIT 19: STATEMENT REGARDING THE TEACHING OF EVOLUTION AND "CREATION THEORY" IN CALIFORNIA PUBLIC SCHOOLS - II

Presented by Claude A. Welch, President of the National Association of Biology Teachers to the Curriculum Development and Supplemental Materials Commission on 26 July 1972 in San Francisco

A theory holds a very special place in science. A theory is not just any old or new hunch that strikes the fancy. All theories, first of all, contain a series of assumptions, known as postulates, which attempt to explain known observations and predict new ones. A theory is often called an hypothesis when it is first formulated; the term theory is generally reserved for those sets of assumptions which have stood the test of time through their capacity to explain and predict.

Competition among theories in science is as old as science itself. The well known Ptolemeic-Copernican and the Spontaneous Generation-Biogenesis controversies are good examples. The assumptions of each theory were clearly stated and their explanatory power carefully argued. This process has been repeated many times in the history of science, and science is the stronger for the debates.

The biological theory of evolution is now being challenged in California, not by another scientific theory but by a religious doctrine. "Creationism" has never, to my knowledge, been expressed as a scientific theory. Never have the assumptions been specifically stated so that their consequences could be checked by observation. Scientists need a clear answer to questions like the following if, as stated in the California Science Framework, "science has independently postulated the various theories of creation":

1. What are the assumptions or postulates which make up the "theory" of creationism if, indeed, such a theory exists?
2. Should we assume that creationism is the same as Chapter 1 of the Book of Genesis in the *Old Testament*?
3. Or is creationism the same as Chapter 2 of the Book of Genesis in the *Old Testament*?
4. Approximately how old is the earth and its organisms according to creationism?

If creationism is a religious doctrine then it has no place in any science textbook used in our public schools.

If creationism is, indeed, a scientific theory then its assumptions must be stated explicitly so that scientists can examine the testable consequences of these assumptions.

The scientific community is in no way attempting to limit competition among scientific theories, because it is through competition and debate that new avenues to truth are often found. What the scientific community is *not*, repeat *not*, interested in is another round of fruitless debate between scientific theory and religious doctrine. The forced imposition of religious doctrine, disguised as science, into the science textbooks is a discredit to religion and a threat to our educational system.

EXHIBIT 20: RESPONSE TO STATEMENT BY G.L. STEBBINS PROFESSOR OF GENETICS, UNIVERSITY OF CALIFORNIA AT DAVIS *(EXHIBIT 18)*

Presented by Commissioner V. L. Grose to the Curriculum Development and Supplemental Materials Commission on 26 July 1972 in San Francisco

Dr. Stebbins' statement appears to contain three major segments: (1) the relationship of hypotheses and scientific faith, (2) discussion of the gaps in the fossil record, and (3) the equation of evolution to science and creation to religion.

Hypotheses and Scientific Faith - I not only agree with Dr. Stebbins in his quotation from the Science Framework, but I would have continued quoting from where he stopped:[1]

> Hypotheses emerge from assumptions and beliefs ... But science involves more than formulating hypotheses; each hypothesis must be tested to see how well it explains an object or event. The need for new theoretical models arises when a scientist is confronted with phenomena that he cannot adequately explain by existing theories, or as he seeks a more fundamental understanding of some aspect of nature than it is possible to attain using existing theories and models.

It is precisely because there is extensive disagreement in the scientific world with the General Theory of Evolution (the "amoeba-to-man" concept — not the idea of genetic variation — which I will refer to as the "Case for CHANCE") as a testable hypothesis for origins that the State Board of Education has called for additional theories to be presented to the science student. However, it is not just the *inadequacies* of the "Case for CHANCE" but also the *adequacies* of the "Case for DESIGN" as a useful hypothesis for the data we now have available which has caused the State Board of Education to require *both* cases to be presented.

Since I have mentioned that the "Case for DESIGN" is a viable hypothesis which is useful in a predictive manner, it should be obvious that I disagree with several of Dr. Stebbins' statements including:

1. The belief in the special creation of living organisms in an untestable hypothesis.

Undoubtedly, all scientists would agree that the origin of life (as an event), whether it occurred by *chance* (without purpose or rationale) or by *design* (with some outside intelligence involved), is an unrepeatable experiment. Evolutionists are the first to state that conditions on Earth some 3 billion years ago (when they postulate that life may have "happened" during the chance collision of inanimate substances) cannot even be *guessed* since we have no data on the Earth's surface environment at that time. Therefore, any and *all* hypotheses concerning the origin of life are untestable. Why separate special creation from the evolutionary guesses?

2. Those who advocate (special creation's) inclusion in the science curricula of our public schools do not permit scientists to criticize or examine it.

I am personally unaware of those to whom Dr. Stebbins is referring. However, I would make a two-fold comment; (a) all those scientists with whom I am acquainted who endorse the "Case for DESIGN" openly invite both criticism and examination of their position; (b) the quoted statement applies almost universally to evolutionists, as discussed in the introduction Darwin's *The Origin of Species*[2]. In fact, it is virtually impossible to find an evolutionist who can list the inadequacies of the General Theory of Evolution.

It appears that Dr. Stebbins, in his conclusions concerning faith, may have confused the issue by suggesting that faith in one hypothetical guess should be connoted as science while faith in another type of hypothetical assumption should be classified as religion when in reality they are both precisely identical; i.e., they both involve belief, trust, confidence in and reliance upon something which they cannot prove! Again, why the effort to create mutual exclusives out of identities?

Gaps in the Fossil Record - Surely Dr, Stebbins was attempting subtle humor by suggesting that "the regular absence of transitional forms in the entire fossil record" has disappeared recently. Unfortunately, his references were not precise enough to be specifically traced. But, for example, Romer (to whom Dr. Stebbins refers) does state that there are no known thecodonts which show positive indications leading toward either pterosaurs or birds.[3] Likewise, Marshall says, "The origin of birds is largely a matter of deduction. There is no fossil

---

[1] *Science Framework for California Public Schools, California State Department of Education, Sacramento (1970) 26, 27.*

[2] *Darwin, Charles, The Origin of Species, Everyman's Library, London, No. 811, xxii.*

[3] *Romer, Alfred Sherwood, Vertebrate Paleontology, The University of Chicago Press, Chicago and London, Third Edition, 140.*

of the stages through which the remarkable change from reptile to bird was achieved."[1] Goldschmidt continues the long list of experts who acknowledge systematic and large gaps in the fossil record when he says, "When a new phylum, class or order appears, there follows a quick, explosive (in terms of geological time) diversification so that practically all orders or families known appear suddenly and without any apparent transitions."[2] To belabor the consensus on this subject would be redundant. Dr. Stebbins' overwhelming faith that "man's scientific ability and progress will eventually fill in the gaps in our knowledge" appears to have swept him a bit ahead of his colleagues as well as the facts in order for him to find an "error" in the Science Framework. However, what is the significance of this "error" even if the scientific literature agreed with Dr. Stebbins? Surely, it is nothing more than a plea for continued faith, hope and patience that the gaps in the presently inadequate hypothesis will someday be filled. It hardly leads logically to the third and concluding segment.

Science and Religion - Although Dr. Stebbins leads into this segment with the word, "consequently," his argument concerning the hampering of science teachers by the State Board of Education fails to develop from the two earlier segments. Seriously, the third segment could have been just as logically written with the following transposition:

> Consequently, teachers of science in California Schools should not be hampered by a requirement to give equal time to the stories of evolution while teaching creation as the scientifically accepted explanation for the origin of kinds of organisms. At the same time, teachers should be discouraged from speaking against the scientific beliefs of any pupils, whatever faith they may have. They should be encouraged to ask these pupils to discuss matters of faith and religion with their own biologists.

Although this rewritten portion could appear to be facetious, it is meant to illustrate in a pointed manner the effect of bias on a scientist's so-called objectivity. It is well for the scientifically-uninitiated to recognize also that biology is only one subdiscipline of science, and while biologists are seeking converts to their particular faith, the findings of other branches of science such as physics, chemistry, astronomy and mathematics continue to raise serious doubts concerning the "justified faith of evolutionary biologists" to which Dr. Stebbins alludes.

EXHIBIT 21: PALEONTOLOGIST'S SUPPORT OF PROFESSOR STEBBINS

This letter was solicited by Professor G. Ledyard Stebbins to support his statement of 26 July 1972 (Exhibit 18)

THE SIMROE FOUNDATION
Tucson, Arizona
1 August 1972

To Whom It May Concern:

I have studied the "Statement regarding the teaching of evolution and 'creation theory' in California public schools" presented to the Curriculum Development and Supplemental Materials Commission by Professor G. Ledyard Stebbins of the University of California.

I have paid special attention to Professor Stebbins' discussion of transitional forms between major groups of organisms, as this is a subject that I have studied for many years and on which I have published extensively. Professor Stebbins' statement agrees with my own views and with those of virtually all paleontologists, that is, of scientists who are adequately acquainted with the facts in this field. There has been some technical discussion about details

---

[1] *Marshall, A. J. , Biology and Comparative Physiology of Birds, Academic Press, New York, 1.*

[2] *Goldschmidt, Richard B., "Evolution as Viewed by One Geneticist," American Scientist, Volume 40 (1952), 97.*

of the origins of some groups of organisms, but the question was not about whether they arose by evolution but about what particular strictly evolutionary processes were involved in their origin. All discussants agreed that origin was by evolution.

Anyone who cites me or my work in opposition to Professor Stebbins' statement is either woefully ignorant or willfully misrepresenting the facts.

Literally thousands of transitional forms are known, and more are discovered every year. All will never be found, because from the nature of deposition and preservation of remains of organisms, not all have been preserved as fossils. Nevertheless the evidence is so extensive that even on this basis alone (and there are many more lines of evidence in agreement) evolution must be considered not only as a theory but, in a correct vernacular sense, as an established fact. To teach otherwise would be to try to set history back to the Middle Ages.

I further endorse Professor Stebbins' statement as a whole. It is an admirable, concise expression of the valid and desirable relationships between science and religion in connection with public education. I would add only that in my opinion the enforced teaching in public schools of special creation, a religious doctrine peculiar to certain specific sects, violates the Constitution of the United States.

George Gaylord Simpson, Ph.D. Sc.D, D.Sc., D.h.C., LL. D., etc.
Professor of Geosciences, University of Arizona Professor Emeritus of Vertebrate
Paleontology, Harvard University.
Former Alexander Agassiz Professor, Museum of Comparative Zoology
Curator Emeritus, The American Museum of Natural History.

EXHIBIT 22: A ZOOLOGIST'S LETTER

The late Professor Alfred S. Romer wrote this letter on 4 August 1972 in support of Professor G. Ledyard Stebbins' request for scientific support for his position

MUSEUM OF COMPARATIVE ZOOLOGY
The Agassiz Museum
Harvard University
Cambridge, Massachusetts 02138

Professor G. Ledyard Stebbins
Department of Genetics
University of California Davis, California 95616

Dear Professor Stebbins:

Thank you very much for sending me information regarding the "Science Framework" booklet, particularly p. 106, of which you have sent me a copy. I was astounded to read in this the phrase "the regular absence of transitional forms". Such a statement is directly opposed to the known facts of paleontology. It is certain that we will never, of course, know all the extinct animal forms that have ever existed on the globe, and a century ago, when paleontology was in its infancy, very few truly transitional forms were known. But over the course of the past century more and more transitional forms have been discovered. If we consider, for example, the group of vertebrates, in which we are all most specifically interested, there are areas amongst the lower fishes where relationships and transitions are uncertain. But for all higher groups transitional forms are definitely known. Between fishes and amphibians, the ichthyostegids of the late Devonian of Greenland, show beautifully the transition between fish and amphibian. Work in the Carboniferous in recent decades, particularly that of Carroll of McGill University, have demonstrated the fact that there is a very fine transition between advanced amphibians and primitive reptiles; in fact the situation is such that it is very hard to know where to draw the line between the two classes. We have, of course, long known

Archeopteryx, which almost exactly splits the difference between reptiles and birds. And as regards the transition from reptiles to mammals, a very considerable number of workers (including myself) have almost completely closed the gap between these two classes. For example, it is usually considered that as an arbitrary point, the distinction between a reptile and a mammal could be put at where the old jaw joint between quadrate and articular bones was replaced by that between squamosal and dentary elements. One of my recently published discoveries from South America is that of a form which in most respects could still be considered a very advanced mammal-like reptile but in which the aquamosal socket for the lower jaw has already developed. In this case it is difficult to know whether to call the animal an advanced reptile or a primitive mammal, and we have here a perfect example of a true transitional form.

To sum up, the statement that there is a "regular absence of transitional forms" is a direct untruth which should not be stated in any document put forth by any responsible public agency.

Yours very sincerely,
Alfred S. Romer

EXHIBIT 23: LETTER FROM HOSPITAL ATTENDANT

This letter, from a "self-taught" Canadian with no formal education, was written in response to an article, "Alternative Hypothesis, " which appeared in Scientific American in August 1972.

Aug. 11, 1972
Mr. Vernon L. Grose
Vice-President of Justin (sic) Institute of Technology
Santa Barbara

Dear Mr. Grose:

In August/72 edition of *Scientific American* (P. 43) you are mentioned as "The author of the revised statement in the science framework" promoting both evolution and creation as scientific theories. If ever one were involved in a complex issue, it is you. Unfortunately, I shall now try to confuse the issue some more.

If I were a biology instructor in the public schools of California, I would be dismissed for refusing constantly to call creation theory a scientific theory. Usually fundamentalists do not consider "Creation" as a theory, but rather as a "Fact". In the philosophy of science "Theory" is never used in contra-distinction to "Fact". A contra-distinctive term would be "Hypothesis".

Obviously a scientific theory is a theory qualified by certain criteria. Not just any theory is a scientific theory. The same is true of a hypothesis. What are the basic elements of a scientific theory or a scientific hypothisis (sic)? The basic elements of both are mechanisms or processes characterized by invariant sequence. The invariant sequence give rise to the central tendencies measured in probability levels in the lab. The statement that things happen by chance alone when observing a normal distribution should be interpreted as "Things happen in the context of many or an infinite number of invariant sequences all connected by feedback".

In the scientific theory of evolution only mechanisms and processes with statistical repeatability in the lab. are resorted to in explanation of observed phenomena. Poliploidy, paedogenesis, neotony, natural selection, genetic drift, hereditary symbiosis, differential replication of molecular species and so on, are some such mechanisms and processes. This must be what Sir Charles Lyell was talking about when he said not to resort to extraordinary agents in the infancy of investigation.

It is often said by fundamentalists that natural selection is not demonstrated by the laboratory analogue, artificial selection. This is not true. If it were, astronomers could not say they have demonstrated stellar origin and evolution or energy production by the laboratory analogues gravitational-hydrodynamic waves, adiabatic compression and fusion processes. To satisfy the fundamentalist they would have to create the actual star and keep it going. Maybe in the future they shall.

In any event it will be a matter of time when you will be required to be "The author" of another "Revised statement in the science framework." Eventually you will revise the philosophy of science to include extra-ordinary agents of explanation. Who knows but that the devil theory of disease may be on its way back.

Maybe you can see why my refusal to accept creation as a scientific theory would be persistent. But I propose to allow the fundamentalists to teach their rot in the public schools. Don't make an instructor with attitudes like mine teach creation theory.

Have two separate classes and two separate instructors, one with my viewpoint and one with the fundamentalist viewpoint. Then justice will be done to both theories. Let the children decide in the final analysis. Academic freedom, and it is an important species of freedom, is the best way to learn to love freedom and guarantee the survival of democracy. If anything there must be freedom of conceptual thought.

Please forward this letter or a copy of it to the National Association of Biology Teachers. (Please send me their address). I would like an expression of your opinion and their's on my proposal.

Thank you.
Yours truly,
Lorne Albert Lenaghan.

EXHIBIT 24: LETTER FROM SPACE PIONEER WERNHER VON BRAUN

Dr. von Braun's observations regarding the origin of the universe, life and man were invited by Commissioner V. L. Grose during the adoption of science textbooks in California

August 14, 1972
Mr. Vernon L. Grose
Vice President
Tustin Institute of Technology
22556 Gilmore Street
Canoga Park, California 91304

Dear Mr. Grose:

In response to your inquiry about my personal views concerning the "Case for DESIGN" as a viable scientific theory for the origin of the universe, life and man, I am pleased to make the following observations.

For me, the idea of a creation is not conceivable without invoking the necessity of design. One cannot be exposed to the law and order of the universe without concluding that there must be design and purpose behind it all. In the world around us, we can behold the obvious manifestations of an ordered, structured plan or design. We can see the will of the species to live and propagate. And we are humbled by the powerful forces at work on a galactic scale, and the purposeful orderliness of nature that endows a tiny and ungainly seed with the ability to develop into a beautiful flower. The better we understand the intricacies of the universe and all it harbors, the more reason we have found to marvel at the inherent design upon which it is based.

While the admission of a design for the universe ultimately raises the question of a Designer (a subject outside of science), the scientific method does not allow us to exclude data which lead to the conclusion that the universe, life and man are based on design. To be forced to believe only one conclusion — that everything in the universe happened by chance — would violate the very objectivity of science itself. Certainly there are those who argue that the universe evolved out of a random process, but what random process could produce the brain of a man or the system of the human eye?

Some people say that science has been unable to prove the existence of a Designer. They admit that many of the miracles in the world around us are hard to understand, and they do not deny that the universe, as modern science sees it, is indeed a far more wondrous thing than the creation medieval man could perceive. But they still maintain that since science has provided us with so many answers, the day will soon arrive when we will be able to understand even the creation of the fundamental laws of nature without a Divine Intent. They challenge science to prove the existence of God. But, must we really light a candle to see the sun?

Many men who are intelligent and of good faith say they cannot visualize a Designer. Well, can a physicist visualize an electron? The electron is materially inconceivable and yet, it is so perfectly known through its effects that we use it to illuminate our cities, guide our airliners through the night skies and take the most accurate measurements. What strange rationale makes some physicists accept the inconceivable electron as real while refusing to accept the reality of a Designer on the ground that they cannot conceive Him? I am afraid that, although they really do not understand the electron either, they are ready to accept it because they managed to produce a rather clumsy mechanical model of it borrowed from rather limited experience in other fields, but they would not know how to begin building a model of God.

I have discussed the aspect of a Designer at some length because it might be that the primary resistance to acknowledging the "Case for DESIGN" as a viable scientific alternative to the current "Case for CHANCE" lies in the inconceivability, in some scientists' minds, of a Designer. The inconceivability of some ultimate issue (which will always lie outside scientific resolution) should not be allowed to rule out any theory that explains the interrelationship of observed data and is useful for prediction.

We in NASA were often asked what the real reason was for the amazing string of successes we had with our Apollo flights to the Moon.

I think the only honest answer we could give was that we tried to never overlook anything. It is in that same sense of scientific honesty that I endorse the presentation of alternative theories for the origin of the universe, life, and man in the science classroom. It would be an error to overlook the possibility that the universe was planned rather than happening by chance.

With kindest regards.
Sincerely yours,

Wernher von Braun

EXHIBIT 25: RESOLUTION OF THE BOARD OF EDUCATION IN ORANGE COUNTY. CALIFORNIA

Adopted at a regular meeting held on 17 August 1972 and passed by a unanimous vote of Board members present. Reprinted by permission of the Orange County Board of Education, Robert Peterson, Secretary

WHEREAS the California State Board of Education, by unanimous decision in November, 1969, adopted the "Science Framework for California Public Schools," which directs that more than one theory for the origin of the universe, life and man must be presented in the science classroom; and

WHEREAS the California State Board of Education, by unanimous decision in March, 1970, approved criteria for a program in science for Kindergarten and Grades One through Eight, which stipulated that the "case for DESIGN" (the general theory of creation), as well as the "case for CHANCE" (the general theory of evolution), must be presented when discussing the origin of the universe, life and man; and

WHEREAS the science textbooks and materials submitted by publishers for evaluation in September, 1971 were written prior to the adoption of the "Science Framework," and thereby do not address the requirement to present both the "case for DESIGN" and the "case for CHANCE" in discussion of origins; and

WHEREAS the Curriculum Development and Supplemental Materials Commission will recommend to the California State Board of Education, in their September meeting, textbooks and materials which do not discuss both the "case for DESIGN" and the "case for CHANCE"; and

WHEREAS the Orange County Board of Education concurs with the requirement, established by the California State Board of Education, to teach both the "case for DESIGN" as well as the "case for CHANCE" when discussing origins; and

WHEREAS the Orange County Board of Education intends, by its concurrence, neither to allow the introduction of religion into the science classroom, nor to discourage or offend the private religious beliefs of any student; now, therefore, be it

RESOLVED, That the Orange County Board of Education urges the California State Board of Education to withhold approval of all science textbooks and materials until such books and materials have been edited for compliance with the requirement to present both the "case for DESIGN" and the "case for CHANCE" when discussing origins.

RESOLVED further, That copies of this Resolution be sent to the State Board of Education, members of the Curriculum Development and Supplemental Materials Commission, and the Press.

EXHIBIT 26: A REBUTTAL OF A REBUTTAL

This letter rebuts a response by Commissioner V. L. Grose (See Exhibit 20) to a statement by Professor G. L. Stebbins (See Exhibit 18)

UNIVERSITY OF CALIFORNIA, DAVIS
College of Agricultural and Environmental Sciences
22 August 1972

Dr. Charles S. Terrell, Chairman
Curriculum Development and Supplemental Materials Commission

Dear Dr. Terrell:

At this time I would like to answer more fully the "rebuttal" by Dr. Grose of my statement at the July 26 hearing. Three points of his statement need to be answered. With respect to the first, the error on page 106 of the "Framework" with respect to the absence of transitional fossil forms, I can only refer you to the letters from Professes Romer and Simpson, of which you have already received copies. The second point is expressed in the sentences: "Evolutionists are the first to state that conditions on Earth some 3 billion years ago (when they postulate that life may have 'happened' during the chance collision of inanimate substances) cannot even be guessed since we have no data on the Earth's surface during that time. Therefore, any and *all* hypotheses concerning the origin of life are untestable. Why separate special creation from the evolutionary guesses."

This is a gross misrepresentation of what scientists really believe about the origin of life. I have sent it to authorities who know this field better than I, and I hope that you or a member of your Commission will hear about it from them.

The third point was not brought up in Dr. Grose's written statement, but did appear in his oral "rebuttal," particularly with reference to his quoting Dr. Werner von Braun as an authority on the origin of life. This is: What persons can be regarded as informed scientists, qualified to make an authoritative statement about a particular scientific question? This question is of particular importance for the deliberation of your Commission, for this reason. I have learned from one of your members, as well as from representatives of the Department of Education with whom I have talked in Sacramento, that you are likely to adopt one of two grade school series: that published by the American Book Company, or that of Leswing Communications, Inc. For both of these, I have been provided by Dr. Chunn's office with copies of the national editions, and of California or alternative editions, in which certain pages have been altered in line with page 106 of the "Framework."

The most common alteration has been the substitution of "Many scientists" in the altered edition for "Scientists" as in the national edition. The purpose of these alterations is presumably, to give the teachers and pupils the impression that informed scientific opinion is divided with respect to the statements made in those pages. Consequently, it is very important for your Commission,. as well as the Board, to understand what we scientists regard as informed scientific opinion.

I believe that I am speaking for the scientific community when I say that a scientist is a person who does experiments, gathers original observations, reads first hand papers in which experiments and observations are described, and goes to scientific meetings or conferences to discuss problems with his colleagues. A person who merely writes about science by quoting second hand, third hand or old, obsolete statements can by no means be regarded as a scientist. Furthermore, scientific opinion can be regarded as informed only when the scientists in question have acquired first hand information about the subject being discussed. Although I regard myself as a scientist, I would not expect anybody to regard as authoritative my opinions on the physics of ballistic missiles, or the electronics of transistors. Conversely, I do not believe opinions about the origin of life given by missile oriented physicists are authoritative; neither are statements about heredity, environment and intelligence given by the inventor of the transistor.

On this basis, I urge you and your commission to adopt formally the national edition of one of the two series in question, and to disregard both the statement of page 106 of the "Framework" and the alterations that have been proposed in compliance with it. I can assure you that, during the period between your preliminary adoption on September 15 and the final meeting on November 9-10, a number of scientists and interested people will be prepared to discuss this problem publicly. I am only one of many: I have stated my opinion; and from now on I shall collaborate with others on this matter.

Yours very sincerely,

G. Ledyard Stebbins
Professor of Genetics

EXHIBIT 27: BIRD CURATOR'S SUPPORT OF PROFESSOR STEBBINS

This letter was solicited by Professor G. Ledyard Stebbins to support his letter of 22 August 1972 (Exhibit 26)

(undated)
Dr. Charles S. Terrell, Chairman
Curriculum Development and Supplemental Materials Commission

Dear Dr. Terrell,

I am writing to urge support of the view of Professor G. Ledyard Stebbins regarding the adoption of national versions of science textbooks for grade school use, books which correctly portray the scientific facts and interpretation of the evolution of life. Modified versions which purport to describe the origin of life according to the unscientific "Creationist" philosophy represent (1) woeful ignorance of the last 200 years' painstaking accumulation of scientific data, the brilliant synthesis of diverse information from paleontology, genetics, biogeography, and biochemistry; (2) an insult to thinking citizens in the State of California, in this enlightened age; and (3) betrayal of public trust in the proper education of our children.

Thank you for your time and consideration. Sincerely,

Ned K. Johnson, PhD
Associate Professor of Zoology and Curator of Birds
University of California, Berkeley

EXHIBIT 28: PHYSIOLOGIST'S SUPPORT OF PROFESSOR STEBBINS

This letter was solicited by Professor G. Ledyard Stebbins to support his letter of 22 August 1972 (Exhibit 26)

UNIVERSITY OF CALIFORNIA, LOS ANGELES
Brain Research Institute
The Center for the Health Sciences 28 August 1972

Dr. Charles S. Terrell, Chairman
Curriculum Development and Supplemental Materials Commission

Dear Dr. Terrell:

I have become concerned about the proposed adoption of textbooks for the elementary grades which seem to place special creation as an equal partner of theories of evolution of the astronomical universe and of life. I have received from Prof. G. L. Stebbins copies of his letter to you on the background of this disagreement, though not of the original "framework". If the quotations from the framework which he gives are accurate, then the framework really does grossly distort what most scientists think, about conditions on the early earth. Of course nobody was there, but the inferences made on the basis of physics of gases, chemistry of the rocks, and other geophysical knowledge, are not just guesses. It is true that hypotheses about the origin of life are not testable in the same way that hypotheses about the densities of rocks are testable, by direct palpation — but then neither is the idea that the sun is made of hot gases; is this idea therefore no better than a guess?

Another aspect of the proposed replacement, which may not have been brought out in discussions to its fullest extent, is that if views derived from religious sources are given equal place in textbooks nominally about science, then religions other than those which believe in relatively recent creation will properly demand "equal time". I am thinking in particular of the views of the Buddhist Churches, which I understand believe in an untreated, cyclic universe. Will space need to be found in elementary science textbooks for these views, which are the religious beliefs of a substantial number of California residents?

I hope your board will not espouse creationism, and thereby create yet another schism between education and populace, over the establishment of religion in the schools.

Sincerely,

Donald O. Walter, Ph. D. Associate Professor of Physiology

EXHIBIT 29: BIOLOGIST'S SUPPORT OF PROFESSOR STEBBINS

This letter was solicited by Professor G. Ledyard Stebbins to support his letter of 22 August 1972 (Exhibit 26)

UNIVERSITY OF CALIFORNIA, LOS ANGELES
Office of the Dean
College of Letters and Science
1 September 1972

Dr. Charles S. Terrell, Chairman
Curriculum Development and Supplemental Materials Commission

Dear Dr. Terrell,

I have received from Professor G. Ledyard Stebbins a copy of his letter to you of 22 August 1972 concerning page 106 of the "Framework" and the alternate versions of the two grade school books under consideration for adoption. I am familiar with the contents of page 106 of the "Framework" and have had an opportunity to examine the national edition of the volumes published by The American Book Company and pages altered to conform with page 106 of the "Framework".

I am in complete agreement with the evaluation of these materials by Professor Stebbins and I join him in strongly urging you and your commission to adopt the national edition of these books and not the altered versions. The latter do not represent current scientific knowledge. The children of California deserve the same exposure to a scientific understanding of evolution as do children in the rest of this country.

Sincerely yours,

Harlan Lewis,
Professor Biology and Dean, Division of Life Sciences

EXHIBIT 30: LETTER FROM APOLLO 15 ASTRONAUT JAMES B. IRWIN

4 Sept. 1972
Labor Day

Dear Vernon,

I am grateful for the information that you sent. As you can guess, I too am constantly asked the same question - was it chance or design? So I know very well the type of responses that you would invoke. There are many 'hardheaded' individuals, and I hesitate to use the word but they are just that. Religious people who would close their eyes to facts if there is any disagreement with their interpretation of the Bible. There are scientists who worship man's library of knowledge as the panacea to man's problems. These would readily agree with evolution because it seems to be in vogue with man's present library of knowledge.

I had the unique opportunity to view the earth, moon, and universe from a totally new perspective. I can not imagine that the majesty of the beauty I beheld and the infinite order would ever come about from chance. No, on the other hand, I did see a Master's Design. So to those in both camps, I would say get knowledge but more importantly get wisdom to correctly interpret the knowledge so that all men may have the correct understanding.

My best wishes for your life,

Jim Irwin

EXHIBIT 31: ZOOLOGISTS SUPPORT OF PROFESSOR STEBBINS

This letter was solicited by Professor G. Ledyard Stebbins to support his letter of 22 August 1972 (Exhibit 26)

UNIVERSITY OF CALIFORNIA, BERKELEY
Museum of Vertebrate Zoology 5 September 1972

Dr. Charles S. Terrell, Chairman
Curriculum Development and Supplemental
Materials Commission

Dear Dr. Terrell:

I am writing to you in your capacity as chairman of the Curriculum Development and Supplemental Materials Commission of the State Board of Education. As a biological scientist, teacher, and parent I am very much concerned about the unfortunate intrusion, which seems about to take place, of non-scientific material into the science curriculum of the public schools in California. I am referring of course to pending textbook adoptions under the influence of the now notorious two paragraphs introduced onto page 106 of the "Science Framework" handbook in 1970.

I have read parts of two of the textbooks which I understand the Board is considering adopting for the fourth and sixth grades (the ones published by the American Book Co.), and this includes both the proposed altered versions and the national editions. Furthermore, I have seen copies of letters by Professor G. Ledyard Stebbins to you and Dr. Wilson Riles regarding this matter, as well as letters from Professors G. G. Simpson, Alfred E. Romer, and others. I am firmly in agreement with the position and arguments of Dr. Stebbins, and sincerely hope that your Commission will not be so unwise as to adopt one of these altered texts which so clearly mis-represent science and scientists. It seems ironic that this threat to educational quality should happen in California, which has for so many years provided leadership to the nation, and to the world for that matter, in educational matters. Let's not retrogress by forcing confusion among California school children regarding certain religious doctrines and science. If this is successful, I expect the Board of Education will next want us to confuse politics with science.

I would like to comment particularly on one point raised by Dr. Gross (sic) in his statement at the July 26 hearing of your Commission that was only briefly dealt with by Dr. Stebbins and others in their letters. This is Dr. Gross' (sic) contention that "Evolutionists are the first to state that conditions on Earth some three billion years ago cannot even be guessed since we have no data on the Earth's surface during that time. Therefore, any and *all* hypotheses concerning the origin of life are untestable." This is a serious error and shows a lack of understanding of scientific methodology. It is not true that scientists cannot even guess at conditions three billion years ago. They have in fact very detailed notions of what conditions were like then, and they are not based on guesses, but on evidence. Perhaps the methodology involved would be clearer if I made a comparison with historical events less far in the past. I suspect your Commission would agree that we "know" something about the events and conditions in say the 1850's, even though none of us were alive then to witness them first-hand. What we think we know about the 1850's are really hypotheses based on very substantial evidence of many, many kinds. The evidence is in fact so good that few would doubt the "reality" of some events and conditions attributed to that period. As one goes farther and farther back in time, the evidence gets scarcer and harder to acquire. Eventually, things get sufficiently difficult that experts are required to gather and interpret the evidence. Remaining unchanged, however, is the basic methodology of gathering data, constructing hypotheses which are supported by the data, searching for and acquiring more data to test the hypotheses against, and then either accepting, modifying, or abandoning those hypotheses, etc., etc. This is the methodology of both the experimental and historical sciences, as well as of historians concerned with human history. By this kind of reasoning process we now know a great deal about the origin of life and the conditions prevailing on earth three billion years ago. As new evidence becomes available, our understanding of these matters will be modified accordingly.

I hope these comments will be helpful to you and your Commission, and that you will now appreciate perhaps why scientists are exasperated by statements such as those of Dr. Gross. (sic) Other errors, such as the presumed lack of transitional forms in evolution referred to also by Dr. Gross (sic) and incorporated as well into the new version of the "Framework" (p. 106), have already been discussed by Dr. Stebbins and others.

My colleagues and I will be looking forward with interest to your Commission's forthcoming decisions on science textbooks. We trust and urge that you will adopt the national edition of one of the two series being seriously considered, and disregard completely the altered versions.

Very sincerely yours,

William Z. Lidicker, Jr. Professor of Zoology

EXHIBIT 32 : LETTER TO CALIFORNIA STATE BOARD OF EDUCATION BY NOBEL LAUREATE UREY

UNIVERSITY OF CALIFORNIA, SAN DIEGO
Department of Chemistry La Jolla, California 92037
September 8, 1972

Dr. Newton Steward
President, California State Board of Education
Sacramento, California

Dear Dr. Steward:

Professor Stebbins of the University of California at Davis has called my attention to the possibility that the science text books of the secondary and primary schools in California may be revised, suggesting that the theory of evolution is uncertain, and I have read certain possible revisions that would suggest this very strongly. I wish to say that I do not know of one single scientist of my acquaintance who does not believe that living organisms, plants and animals all grew by a (sic) evolutionary process during the long ages of geological history, and I do hope that the modifications of the text books that I have seen will not be adopted. The evidence in regard to evolution as a fact impresses me as strongly as the evidence for the theory of universal gravitation. It is just one of the firm things of science. Of course, at the same time, there are many details which we will constantly be working on. The problem of the origin of life is being investigated vigorously at the present time. However, there is no disagreement in regard to the general theory, and this should be remembered in choosing the text book. I hope you will exert your influence in this direction.

Best regards.

Very sincerely, Harold C. Urey

EXHIBIT 33: TELEGRAM TO CURRICULUM DEVELOPMENT AND SUPPLEMENTAL MATERIALS COMMISSION

Message transmitted for 10 September 1972 meeting of the Commission in Sacramento

As professional research biologists we consider the proposal to include creationism in science textbooks outrageous /stop/ Deliberate propagation of ignorant superstitions has no place in public educational systems of California /stop/ Teaching creationism will impair the mental development of our children /stop/ How about a course for 3rd graders on witchcraft or alchemy? /stop/ We enjoin the Board not to take the undignified and reprehensible step of decreeing that California children be exposed to creationism.

Dr. Erik H. Davison, Associate Professor of Biology, California Institute of Technology
Dr. Caleb E. Finch, Assistant Professor of Biology, University of Southern California

EXHIBIT 34: STATEMENT BY JOHN R. FORD, M.D., VICE-PRESIDENT OF CALIFORNIA STATE BOARD OF EDUCATION

Presented at the regular meeting of the Board in San Diego on 14 September 1972

I am frankly amazed at the wide diversity of interpretations of my public statements that I have read and heard. The press, religionists, biologists and many others have reached conclusions so widely varied that I will attempt a succinct statement which is hopefully free from ambiguity.

I am solely interested in a scientifically-valid discussion of alternatives for how the universe, life, and man began. This position might be further expanded by enumerating some negative statements:

1. I do not wish to see the Bible, God, or the Genesis account of creation mentioned in science materials used in the public classroom.
2. I am not talking about a six 24-hour day account of how every-thing began.
3. I am not proposing an origin of man in 4004 BC.
4. I am not an anti-evolutionist nor do I wish to inhibit or prohibit the teaching of any aspect of evolution currently taught as a theory.
5. I accept and endorse, as the best current explanation for variation among plants and living things, the Special Theory of Evolution (as defined by G. A. Kerkut and others).

Some positive statements are also in order:

1. I believe that all scientific data available today can be interpreted, when postulating how the universe, life, and man might have originated, in more than one way.
2. Children are taught only one idea today — that the universe, life and man are simply "accidents" that occurred by fortuitous chance without cause, purpose or reason. While I would not deny any scientist the right to believe this untestable hypothesis, I would at the same time propose that the same identical scientific data he uses to reach that conclusion will support equally well (if not more so) the hypothesis that these origins occurred by *design* with cause, purpose and reason.
3. The California Attorney General ruled in 1963 that the "chance" idea (or General Theory of Evolution) could be taught in public school classrooms *provided* there was no indoctrination of the idea. When only one idea is taught, how can we avoid indoctrination?
4. I do not desire a complete rewriting of science materials. Rather, I only seek to have any and all discussions of origins in the materials to be connoted as unresolved theories and further that two contrasting possibilities — CHANCE and DESIGN — be discussed side-by-side without bias for one over the other.

5. Recognizing that DESIGN ultimately raises the question of a Designer, I would refer the science student to other fields of study such as philosophy, religion, history or social science when he inquires about who did the design. This action is appropriate not only in this case but also in all the other areas that science classically ignores; e.g., value systems, morals, art and poetry.

EXHIBIT 35: EDITORIAL IN "SCIENCE" (OFFICIAL ORGAN OF THE AMERICAN ASSOCIATION FOR THE ADVANCEMENT OF SCIENCE)

This editorial was written by William Bevan, Publisher of Science *and appeared in Volume 177, Number 4055 (page 1155) on 29 September 1972 and was copyrighted in 1972 by the American Association for the Advancement of Science. The editorial evoked an exchange of letters between V. L. Grose and Dr. Bevan (see Exhibit 36). In 1973, Dr. Bevan was named Executive Officer of AAAS which contains 290 affiliated societies, ranging from the 120,000 member American Chemical Society to the tiny American Malacological Union. The monthly organ* Science *had a circulation of 163,000 in 1973.*

TWO COOKS FOR THE SAME KITCHEN?

There is reason to assume that the differences between religious doctrine and scientific thought in the matter of biological origins were resolved by the Huxley-Wilberforce debates of a hundred years ago, or most certainly by the Supreme Court's 1968 action in overruling a long-standing Arkansas statute against the teaching of evolution. However, the matter has not been settled, and recent events in California warrant the serious attention of every citizen—scientist, theologian, or otherwise.

In the fall of 1969 the State Advisory Committee on Science Education after several years of work and with the approval of the Curriculum Development and Special Materials Commission, presented a new *Science Framework*[1] for kindergarten through the secondary school levels to the State Board of Education. During the Board's consideration of this document, objection was raised that in dealing with the origins of life, the Committee did not so much as allude to creationism; following subsequent discussions, the *Framework*, modified to include creation theory as a complement to evolutionary theory, was adopted. The Committee vigorously protested the change, but to no avail. Subsequent statements from the Board strongly suggest that it will require that all science textbooks to be considered for adoption in California include a serious treatment of creation theory.

The implications of these actions are several and serious:

First, what is "good" for California is likely to become "good" for the rest of the nation, since California purchases 10 percent of all textbooks sold in the United States. Unless publishers are prepared to produce special California editions — and they probably are not—the standards set for California will, willy-nilly, become the standards for many other states.

Second, success in this first step will make a second, third, or fourth step toward politicizing the classroom that much easier, for if the state can dictate the content of a science, it makes little difference that its motivation is religious rather than political. The consequences will be the same. Many will recall the condition of Russian genetics during the heyday of Lysenko when Russian biologists defended an erroneous theory on the grounds that it must be true because it was Marxist.

Third, the Board's action is testimony once again that scientists have failed in their communications about science to the nonscientific public. We have taught the substance of

---

[1] *California State Department of Education, Science Framework for California Public Schools, Kindergarten-Grades One through Twelve (State Printing Office, Sacramento, 1970), xii / 1948 pp.*

science without communicating the approach, the methods, or the rationale of science. The essential requirement of scientific theory is that, in principle, it is capable of contradiction by empirical data. It is perfectible and it stands only as long as it has not been contradicted. It is in the process of becoming. Thus classical (Darwinian) evolutionary theory has been significantly transformed and enriched not only by the discoveries but also by the thinking of Mendel, and later of Weissman, and, most recently, of Wilkins and of Watson and Crick. Creationism is a theory of primordial history and, as such it responds to different rules of discourse. It is not subject to empirical test, nor does it allow of improvement. Certainly it is not a logical complement of evolution theory.

The action of the Board with regard to textbooks will be a matter of record sometime this fall. Advocates of creationism are bringing pressure for the use of creationist materials in the schools both of California and of other states, as well. Meanwhile, the National Association of Biology Teachers, stimulated by the California events has established a Fund for Freedom in Science Teaching aimed at preserving sound science education whenever it appears in jeopardy.

EXHIBIT 36: LETTER EXCHANGE

The following letters were exchanged regarding an editorial, "Two Cooks for the Same Kitchen?", which appeared in Science *on 29 September 1972 (See Exhibit 35).*

12 October 1972

Dr, William Bevan, Publisher
SCIENCE
American Association for the Advancement of Science
1515 Massachusetts Avenue NW
Washington, D. C. 20005

Dear Dr. Bevan:

With great personal interest, I have read your editorial of 29 September 1972 entitled, "Two Cooks for the Same Kitchen?" You made some extremely important and convincing points, provided your initial assumption is valid.

Unfortunately — and I use that word in the kindest sense — I believe that you have assumed a very common but ill-advised position. Since I have been quite intimately associated with the California issue from its outset, I think that I can speak with some authority about the facts of the case.

First, the Huxley-Wilberforce debates and the Supreme Court's 1968 action have absolutely no relevance, even generically, to the California issue. Second, it is not a science-versus-religion issue. The primary error that everyone makes in adopting the position expressed in your editorial is in assuming that those advocating the "Case for DESIGN" are synonymous with those who propose an origin of the earth in 4004 BC, fixity of the species, and insertion of the Book of Genesis in public schools. Because the two words, evolution and creation, appear destined to evoke eternal antipathy from advocates of either view, I have consistently attempted to use the alternative terms, chance and design, to de-focus the emotion and get to the root of the difference of opinion concerning origins.

Your editorial line of reasoning, which I accept at face value as sincere although chauvinism could be detected, suggests that anyone advocating the "Case for DESIGN" would, of necessity, have missed the true meaning of science. Surely SCIENCE or AAAS would not suggest that the currently-popular Neo-Darwinian hypothesis is by any means the only explanation for scientific data that have been observed regarding origins. If so, much of mystery, intrigue and adventure of science has already disappeared for the science student!

If you would grant the possibility that there might be another explanation for origins beside the amoeba-to-man concept of Neo-Darwinism, then I would propose that SCIENCE be kind and fair enough to publish a counter-position to its constituency so that objective dialogue, rather than hysteria, might develop. To that end, I have enclosed two statements — a brief synopsis of the "Case for DESIGN" and Textbook Review Criteria. In the former, I have incorporated part of your thoughtful editorial, indicating that I am in total agreement with your position on the meaning of science.

The "Case for DESIGN" has been advocated in California for two reasons: (1) because it is believed to present as strong an explanation, if not stronger, as "The Case for CHANCE" for scientific findings to date, and (2) because there is widespread dissatisfaction in the scientific community with certain aspects of the amoeba-to-man idea or the "Case for CHANCE" as a satisfactory explanation for scientific data concerning origins.

While I am personally a technologist (as are Dr. von Braun and astronauts Mitchell and Irwin whose endorsements are enclosed) rather than a pure scientist, I feel that those of us who work in technology have a fair grasp of the meaning and intent of science. The National Academy of Sciences, for example, has recognized and continues to recognize my capability to speak in the world of science, as indicated by my enclosed biography.

Those who would attempt to shut off debate about the "Case for CHANCE" on the basis of who proposed an alternative — whether it be the State Board of Education or individual technologists — would be wiser, it would seem, if they directed their attention to the *alternative theory* itself. After all, Hans Christian Andersen in his "Emperor's New Clothes" illustrated that truth may even come from "unqualified" sources at times. (The enclosed list of qualified scientists who are advocating an alternative theory for origins should indicate that the "Case for DESIGN" is more than a "technician's aberration of science.")

Regarding your concern about political interference in the field of science education, would you propose that scientists should be permitted total isolation from challenge outside their field? If not, what better avenue of cleansing do you propose than the traditional American democratic process of having the public express their views through their representatives; e.g., the State Board of Education? In contra-distinction to your analogy of censorship within the Soviet Union, no theory is being deleted, replaced or even threatened by dictum of the State Board of Education. They have simply permitted a second theory to be compared with an existing one!

As a person with scientific education, I become frightened when I see emotion and hysteria generated by a challenge to a pet theory. Are there *any* weaknesses in or valid criticisms of the amoeba-to-man concept in the eyes of AAAS? Has "The Case for DESIGN" been completely disproven in the eyes of AAAS? If so, what data exist for this disproof?

I sincerely thank you for calling this issue to the attention of the scientific community. As I indicated earlier, I would indeed be pleased if open dialogue could be established by SCIENCE and AAAS to ventilate this basic issue of whether everything which scientists have observed has occurred by chance or by design. To that end, I shall look forward to your reply to my letter.

Very sincerely yours,
TUSTIN INSTITUTE OF TECHNOLOGY

Vernon L. Grose
Vice President

AMERICAN ASSOCIATION FOR THE ADVANCEMENT OF SCIENCE
Washington, D. C.
October 20, 1972

Dr. Vernon L. Grose
22556 Gilmore Street
Canoga Park, California 91304

Dear Dr. Grose:

Thank you for your long and thoughtful letter of 12 October. I must say it is by all odds the kindest letter that I have received in response to my recent editorial on biological origins.

While I appreciate your taking great pains to point out that "The Case for Design" is not religious doctrine and the issue of evolution versus creation is not a science-versus-religion controversy, others who have identified themselves with the creationist position have not expressed the same point of view. They have been inclined to send along literature that can only be identified as religious in its orientation. At best, I come off as closeminded and rigid. It has even been suggested that the recently announced Fund for Freedom in Science Teaching will probably acquire its income from the National Science Foundation's budget.

Two, for me, important points of my editorial appear to have been missed. First, while you say that the action of the State Board of Education is simply arranging to permit an alternative view to be compared with an existing one, it in fact emerges in my mind as a requirement that the teacher teach both. This I regard as an infringement on the right of the teacher to determine what and how he shall teach. Requiring its inclusion furthermore constitutes dictation to scientists by laymen as to what shall constitute the corpus of their science.

My second point is that creationist theory as far as I can tell has not itself undergone the normal process of creation and evolution (no pun intended) that characterizes bona fide scientific theories. Furthermore, I have read with a sensitive attitude the materials that you sent, and other materials that have come to my hands, on creationism, and, for the life of me, I cannot find it articulated with precision sufficient to make it a useful tool in the evolution of scientific information and knowledge. Theories do not stand as "ultimate explanations." Rather, they are simply intellectual tools by which the scientist, groping to advance his science, sorts out his good from his bad guesses.

It may surprise you for me to say that I regard myself to be a religious man and, indeed, as a most conventional communicant. But I think I have sorted out for myself the ground rules that guide me as a scientist from those that guide me as a religious practitioner, and they are not the same. I have forwarded the materials you have supplied to Dr. Philip Abelson, Editor of Science. I cannot, of course, say what action he will take in this connection.

Cordially yours,

William Bevan
Executive Officer

EXHIBIT 37: EDITORIAL ADVISORY BOARD

Scientists who agreed to work with science textbook publishers on editing textbooks for objectivity regarding the origin of the universe, life and man October 1972

Name	*Educational Specialty*	*Professional Assignment*
Biedebach, Mark C.	PhD in Biophysics, University of California Los Angeles	Associate Professor of Biology, California State University Long Beach
Cameron, Roy E.	PhD in Botany, University of Arizona	Cognizant Scientist, Desert Microflora and Antarctic Programs Jet Propulsion Laboratory, California Institute of Technology
Ferm, Richard L.	PhD in Chemistry, University of Kansas	Senior Research Associate, Chevron Research Company
Fischer, Robert B.	PhD in Analytical Chemistry, University of Illinois	Dean, School of Science and Mathematics, California State University Dominguez Hills
Isensee, Robert W.	PhD in Organic Chemistry, Oregon State University	Professor of Chemistry, California State University San Diego
Lindquist, Stanley E.	PhD in Comparative Physiological Psychology, University of Chicago	Professor of Psychology, California State University Fresno
Lofgren, Norman L.	PhD in Physical Chemistry, University of California Berkeley	Professor of Chemistry, California State University Chico
Olson, Kenneth V.	PhD in Biology, University of Minnesota	Professor of Biology, Northern Colorado State College
Remmel, Ronald S.	PhD in Elementary Particle Physics, Princeton University	National Institutes of Health Post-Doctoral Fellow, University of California Berkeley
Stern, Kingsley R.	PhD in Botany, University of Minnesota	Professor of Botany, California State University Chico

EXHIBIT 38: REVIEW CRITERIA FOR SCIENCE TEXTBOOK COMPLIANCE WITH "CHANCE VERSUS DESIGN" REQUIREMENT

Background statement prepared for science textbook publishers by Vernon L. Grose to aid in screening textbooks being submitted for adoption — October 1972

Discussion of Origins - Although our prime interest concerns the myths about the origin of the universe, life and man, we are also interested in assuring that all discussion about events in the distant past where *speculation* rather than *observation* is dominant be carefully stated (almost emphatically qualified).

Prehistorical Dating - Since radioactive dating methods are frequently challenged and are far from universal in science, caution should be exercised when discussing *quantitative values* for various ages in the past. Again, *speculation* is the primary ingredient in dating. It is very difficult to avoid a tautology or circular logic when using the index-fossil method of dating:

As Professor R. S. Rastell, a geologist at Cambridge University, has pointed out, we study a series of fossil contents within a formation to determine their succession in time. Then, we proceed to determine the succession in time of the formations by means of their fossil contents!

For example, if it is assumed that *only* the simplest forms of life are contained in the oldest geological formations (a prime requisite of evolutionary thought), then the oldest formations (by definition) are those which contain simplest forms.

In actuality, it is difficult to find a formation sequence which even approximately fits this simple-to-complex theory; i.e., a sequence in which formations containing the remains of the most primitive forms of life lie directly on the crystalline base while the uppermost layers contain the highest organisms. Literally hundreds of cases have been documented which show the reversed sequence; i.e., simpler forms lying on top of complex forms.

Adaptation/Development as Sole Explanation — While adaptation has been demonstrated to have occurred within very narrow limits within species, it is not necessarily the sole explanation for the widely-diverse characteristics of plants, animals and man. The "Case for DESIGN" would stress that all of these things were designed from the beginning with those characteristics which allow us to classify and predict. When comparative analyses of man and lower animals are discussed in terms of "development," there is a biased predisposition to a sequential continuum required to support the "Case for CHANCE."

Man as an Animal — Comparative anatomy and embryology both yield the conclusion that complex forms of life contain characteristics common to simpler forms of life. At least two contrasting philosophic reasons for this similarity or commonality of characteristics can be postulated:

1. That complex forms of life evolved (randomly and sequentially, by chance, without purpose or rationale) from simpler ancestors (Case for CHANCE).
2. That all living creatures are the product of a single Designer who used a common design with variations in complexity (Case for DESIGN).

Note that (1) is time-*dependent* (limited to a *series* development) while (2) is time-*independent* (fits either *series* or *parallel* development). Only under (1) is man *required* to be an animal. The "Case for DESIGN" proposes that man's differences from all other animals demands a unique category or classification. Classification criteria are inadequate when they fail to acknowledge the quantum gap between man and animals. To illustrate this gap, we only need to note than man alone:

-can walk upright

-has tradition, customs and knowledge that is passed on

-has true intelligence (non-conditioned)

-can easily grasp and manipulate objects with his fingers

-can learn skills (writing, painting, sculpturing, music, etc.)

—can control and improve his food supply

In addition to the specific criteria discussed above, several conceptual conflicts between CHANCE and DESIGN should be acknowledged and discussed (in a manner and language appropriate to the student level wherein these conflicts arise):

1. The Concept of Randomness — In layman's language, this concept says that the incredibly orderly systems of the biological world are the result *solely* of random mutations and natural selection. In years past, biologists held that this lucky, rare, chance reaction occurred by a "hit or miss" method, and the longer the time allowed for such chance reactions to occur, the more numerous they must have become.

In using this argument, it is forgotten that the longer the time allowed for a *reversible* synthesis to occur, the more likely the *reverse* reaction, or decomposition, also becomes! Outstanding authorities like Sir Peter Medawar, V. F. Weisskopf, Professor of Physics at MIT, and M. Eden, Professor of Electrical Engineering at MIT have raised serious questions about this concept. The latter said recently, "It is our contention that if 'random' is given a serious and crucial interpretation from a probabilistic point of view, the randomness postulate is highly implausible and that an adequate scientific theory of evolution must await the discovery and elucidation of new natural laws — physical, physiochemical and biological." Science students deserve to know about this basic unresolved disagreement about the "Case for CHANCE" rather than having it hidden or ignored in order to preserve a pet theory.

2. The Entropy Conflict - The Second Law of Thermodynamics can be simply interpreted as saying that the entire Universe is in a "running down state" or another way to state it is that "entropy (the amount of disorder) is *increasing* in the Universe." Radioactive dating methods are based on the decay or *running down* concept. Yet, the very heart of the "Case for CHANCE" is the idea that complexity is *building up* from simplicity. The science student is entitled to know that these two ideas are basically *incompatible*! There are obviously cases wherein an open system can be considered to exist (and thereby the Second Law would be temporarily non-applicable), but surely this type of anomaly cannot be utilized to explain the present state of the Universe (especially since uniformitarianism — another cornerstone of the "Case for CHANCE"—would seem to rule out anomalies!).

3. Abiogenesis - This term means "the spontaneous generation of life from nonliving matter (without the mediation of previously living matter)." Even though Pasteur proved to the satisfaction of scientists that abiogenesis cannot take place today, the "Case for CHANCE" still requires this "miracle" to explain how life started. Aside from blind faith in something which can no longer occur, this concept has *no scientific* basis. The Science Framework, as originally submitted to the State Board of Education for approval, contained the following statement: "A soup of amino acid-like molecules, formed in pools some 3 billion years ago, interacted with oxygen and other elemental constituents of the earth, probably giving rise to the first organization of matter which possessed the properties of life." Interestingly, this statement (which illustrates abiogenesis) was quietly withdrawn when public attention was turned on it. G. A. Kerkut, Department of Physiology and Biochemistry at the University of Southampton, has published in *Implications of Evolution* (Pergamon Press, 1960) seven basic assumptions required for the "Case for CHANCE" and, regarding abiogenesis, he shows it is based entirely on *blind faith*. School children should be taught that science does not know and never will know how life began, because origins are philosophic or metaphysical rather than scientific.

4. Uniformitarianism - This term is given to the idea that "everything, including the geological strata, has developed slowly, uniformly and regularly through immense periods of time without catastrophes of any type." Correlative with this idea is the belief that the laws of nature have always been the same as they are today, so that the present state of nature is the explanation of its past state and of its future state too. Yet, we have as the only explanation for the origin of the universe being taught in schools today the Big Bang Theory — a *catastrophe* — not a development based on a past existence! Further, apparently it was a one-time-only type of catastrophe since nothing like it has occurred since! The science student is being deluded unless he is apprised of this type of inconsistent thinking being connoted as scientific.

EXHIBIT 39: THE CASE FOR DESIGN

A Statement prepared by Vernon L. Grose for science textbook publishers to aid in editing textbooks — October 1972

It can be scientifically hypothesized that there was a planned beginning of the universe wherein a Designer, with purpose and rationale, established a design for the universe and everything of which it would ultimately consist.

The nature and mechanism of this originative work is unknown and probably unknowable because scientists cannot duplicate it. (This unknowable aspect, however, cannot preclude the admission of design as a workable hypothesis in science because all other hypotheses for origins are likewise unknowable.)

Because of the inherent design in all of nature, scientific laws can become universal; i.e., ultimate findings of science are singular rather than diverse in character. This singularity of design permits the reproducibility and predictability upon which modern science is based. To contrast this idea of *design* with a *chance* hypothesis, there is every reason to believe that the universe should be a multi-verse, since there can be no purposive direction or rational channeling of myriads of random events if they are chance occurrences.

Proceeding then on the assumption of a designed beginning, as opposed to an unexplainable cosmic accident, the "Case for DESIGN" becomes a framework or hypothesis to not only correlate the past findings of science but also to point the way for future investigation.

In fact, it was this very concept of design which provided the foundation for modern science. It continues to this day to motivate scientists to refine and perfect concepts of the orderliness of nature.

As the *Science Framework for California Public Schools* states: "(scientists) connect one fact with another — they seek for order and relationship — and in this way they arrange the facts so that they are linked by some inner law into a coherent network." What is this "order and relationship" and the "inner law" which scientists constantly seek, if it is not the very design which was established at the beginning? Alfred North Whitehead expressed it well when he said:[1]

> There can be no living science unless there is a widespread, instinctive conviction in the existence of an *Order of Things*, and, in particular, of an *Order of Nature* (italics his).

The essential requirement of scientific theory is that, in principle, it is capable of contradiction by empirical data. It is perfectible and it stands only as long as it has not been contradicted. It is in the process of becoming. In that sense, every finding of science continues to enrich and transform the "Case for DESIGN" into a more accurate model of the inherent design of nature.

There is a point of commonality between the "Case for DESIGN" and the "Case for CHANCE." They are both seeking systematic order and relationship among the isolated facts that they observe. The primary difference in the two schools of thought (or philosophies, to be more precise) is that one group believes that the fantastic and incredibly consistent design that both groups perceive is purely an accident (Case for CHANCE). The other group believes that the design is the product of a Designer (Case for DESIGN).

Contrary to the view of those who would confuse the "Case for DESIGN" with fundamentalistic religion, the "Case for DESIGN" invites continued challenge. Rather than dodging or ignoring data that appear to threaten the viability of their theory (as many evolutionists do when faced with the laws of thermodynamics, the mathematical concept of randomness, or cataclysmal theory), those who use the "Case for DESIGN" are reasonably confident that no empirical data will contradict the concept of inherent design.

The decision to employ this hypothesis for predictive purposes should not be based on any *a priori* position, philosophic or religious, but strictly on its merits as a testable and modifiable concept. Just as any other scientific theory becomes improved upon empirical testing, the "Case for DESIGN" will undoubtedly be refined in the future. A few current predictions, concerning future scientific investigation, which have been postulated by numerous scientists when reasoning from the "Case for DESIGN" include:

1. There should not be any self-ordering of inanimate matter on an upward course from simplicity to high complexity, as though inanimate matter possessed intelligence of itself.

2. There should be a clear and unmistakable distinction, rather than a continuum, between

---

[1] *Whitehead, Alfred North, Science and the Modern World, The Macmillan Company, New York, (1926), 4.*

groups of living things since they were individually designed.

3. There should be a predictability of reproduction among living things; i.e., that cows should produce cows rather than giraffes.

4. There should be clear and unmistakable gaps in the historical record (e.g., fossils) between major groups of living things.

5. Life should only be capable of being produced from previous living matter (biogenesis), unless mediated by an outside intelligence.

6. There should be historical evidence of establishing initial complexity, rather than an emerging complexity, within a given group of living things (not to be confused with limited environmental adaptation within a given group).

7. Man should be clearly differentiated from all animal life in his ability to learn, control his environment, dominate the animal world, and think rationally.

8. There should be evidence that the general scheme of the universe is progressing from initial order to ultimate disorder.

As of this date, the "Case for DESIGN" continues to be a viable and useful scientific hypothesis. While it is not competitive with the "Case for CHANCE" (since theoretically scientists do not compete for popularity of their ideas), it does offer a contrasting view which, in the minds of many scientists, more accurately interrelates the findings of science.

EXHIBIT 40: RESOLUTION BY THE COMMISSION ON SCIENCE EDUCATION, AMERICAN ASSOCIATION FOR THE ADVANCEMENT OF SCIENCE

Statement prepared on 13 October 1972 and sent to all state boards of education and other important educational agencies influential in determining school politics

The Commission on Science Education, of the American Association for the Advancement of Science, is vigorously opposed to attempts by some boards of education, and other groups, to require that religious accounts of creation be taught in science classes.

During the past century and a half, the earth's crust and the fossils preserved in it have been intensively studied by geologists and paleontologists. Biologists have intensively studied the origin, structure, physiology, and genetics of living organisms. The conclusion of these studies is that the living species of animals and plants have evolved from different species that lived in the past. The scientists involved in these studies have built up the body of knowledge known as the biological theory of the origin and evolution of life. There is no currently acceptable alternative scientific theory to explain the phenomena.

The various accounts of creation that are part of the religious heritage of many people are not scientific statements or theories. They are statements that one may choose to believe, but if he does, this is a matter of faith, because such statements are not subject to study or verification by the procedures of science. A scientific statement must be capable of test by observation and experiment. It is acceptable only if, after repeated testing, it is found to account satisfactorily for the phenomena to which it is applied.

Thus the statements about creation that are part of many religions have no place in the domain of science and should not be regarded as reasonable alternatives to scientific explanations for the origin and evolution of life.

EXHIBIT 41: RESOLUTION BY THE NATIONAL ACADEMY OF SCIENCES

Approved on 17 October 1972 in Washington, D, C.

Whereas we understand that the California State Board of Education is considering a requirement that textbooks for use in the public schools give parallel treatment to the theory of evolution and to belief in special creation; and

Whereas the essential procedural foundations of science exclude appeal to supernatural causes as a concept not susceptible to validation by objective criteria; and

Whereas religion and science are, therefore, separate and mutually exclusive realms of human thought whose presentation in the same context leads to misunderstanding of both scientific theory and religious belief; and

Whereas, further, the proposed action would almost certainly impair the proper segregation of the teaching and understanding of science and religion nationwide, therefore

We, the members of the National Academy of Sciences, assembled at the Autumn 1972 meeting, urge that textbooks of the sciences, utilized in the public schools of the nation, be limited to the exposition of scientific matter.

EXHIBIT 42: RESOLUTION BY THE BOARD OF DIRECTORS, AMERICAN ASSOCIATION FOR THE ADVANCEMENT OF SCIENCE

Statement approved on 22 October 1972

Whereas the new *Science Framework for California Public Schools* prepared by the California State Advisory Committee on Science Education has been revised by the California State Board of Education to include the theory of creation as an alternative to evolutionary theory in discussions of the origins of life, and

Whereas the theory of creation is neither scientifically grounded nor capable of performing the roles required of scientific theories, and

Whereas the requirement that it be included in textbooks as an alternative to evolutionary theory represents a constraint upon the freedom of the science teacher in the classroom, and

Whereas its inclusion also represents dictation by a lay body of what shall be considered within the corpus of a science,

Therefore we, the members of the Board of Directors of the American Association for the Advancement of Science, present at the quarterly meeting of October 1972, strongly urge that the California State Board of Education not include reference to the theory of creation in the new *Science Framework for California Public Schools* and that it adopt the original version prepared by the California State Advisory Committee on Science Education.

EXHIBIT 43: RESOLUTION BY THE UNIVERSITY OF CALIFORNIA ACADEMIC SENATE

Statement approved by The Academic Council on 27 October 1972

It is our understanding that within the next few months the California State Board of Education will be approving many science textbooks for use in California public schools, grades K through 8. The text of the "Science Framework for California Schools" prepared in 1969, suggests that one criterion for the Board's approval of a text may be the extent to which, in the discussion of the origins of life, a "special theory of creation" is treated as a scientific theory in a manner parallel to an account of evolution. We believe that a description of special creation as a scientific theory is a gross misunderstanding of the nature of scientific inquiry.

To provide the basis of a scientific theory, an hypothesis must make testable predictions. Our ideas of biological evolution are continually being tested in the process of an enormous amount of investigation by thousands of professional biological scientists throughout the world. As in all sciences, there are many facets of the evolution picture that are not yet thoroughly understood, and researchers at the frontier of knowledge, often in disagreement with each other concerning details, continually revise their thinking. Thus, evolutionary theory itself has evolved considerably since the time of Darwin. But virtually all biological scientists are agreed on the broad features of the theory of evolution of life forms, the evidence for which is completely overwhelming.

The issue is not whether the concept of a relatively sudden special creation is true or valid, but rather than its origin lies in philosophical thought and religious beliefs, not in scientific investigation. Partly because of the wide diversity of religious opinions regarding creation, and especially because our traditional adherence to the First Amendment of the United States Constitution requires the separation of religious instruction from State-supported schools, we believe that the teaching of special creation should be avoided entirely in California public schools; certainly, it should not be presented in textbooks as a scientific theory.

We join the National Academy of Sciences, the American Association for the Advancement of Science, and other learned societies in urging the State Board of Education to reject inclusion of an account of special creation in State-approved science textbooks.

EXHIBIT 44: STATEMENT BY THE COMMITTEE ON PHYSICS IN PRE-COLLEGE EDUCATION, AMERICAN ASSOCIATION OF PHYSICS TEACHERS

Submitted on 3 November 1972 to the California State Board of Education

The Committee on Physics in Pre-College Education, a committee of the American Association of Physics Teachers, has carefully reviewed recent developments in the selection of science textbooks for the elementary grades by the State of California. For several reasons the committee feels compelled to share its concern over the matter to both the Curriculum Committee and the State Board of Education: First, California has been regarded by many as a state with an education program to be emulated by others. Second, because the number of students affected by text selections in California is extremely large, about 1016 of the comparable national population, the criteria applied to text selection in California are of significant, national concern. Third, and of extremely great importance, is that the study of any particular religious philosophy of creation as part of science programs will have a negative impact on the general public and on the progress of science itself.

It is with these considerations in mind that the Committee on Physics in Pre-College Education joins with other nationally representative scientific organizations and recommends to all individuals involved in the forthcoming decisions the following:

> The State Board of Education is urged to apply the criteria set forth in the Science Framework for California Public Schools, Kindergarten—Grades One through Twelve, 1970, in the selection of science textbooks with one major exception: The following paragraphs appearing on page 106 should *not* be used as part of these criteria.
>
> All scientific evidence to date concerning the origin of life implies at least a dualism or the necessity to use several theories to fully explain relationships between established data points. This dualism is not unique to this study but is also appropriate in other scientific disciplines, such as the physics of light.
>
> While the Bible and other philosophic treatises also mention creation, science has

independently postulated the various theories of creation. Therefore, creation in scientific terms is not a religious or philosophic belief. Also note that creation and evolutionary theories are not necessarily mutual exclusives. Some of the scientific data (e.g. the regular absence of transitional forms) may be best explained by a creation theory, while other data (e.g. transmutation of species) substantiate a process of evolution.

It is also recommended that the reports of the various county review committees as well as the positions of the Curriculum Development and Supplemental Materials Commission and the State Advisory Committee of Science Education which developed the framework be taken as *the* basis for decision by the California State Department of Education.

This committee is of the opinion that adoption of texts or supplementary materials which are contrived to meet standards which are not accepted by the scientific community will be grossly detrimental to the development of an increased public awareness and appreciation of scientific endeavor. Acceptance of texts written to conform to the paragraphs noted above will turn aside many excellent efforts on all frontiers of science education and reinforce current anti-science attitudes.

EXHIBIT 45: UNANIMOUS RECOMMENDATION OF THE CURRICULUM DEVELOPMENT AND SUPPLEMENTAL MATERIALS COMMISSION TO THE CALIFORNIA STATE BOARD OF EDUCATION

Prepared and passed on 8 November 1972 by the Commission and presented to the State Board of Education by Commissioner V. L. Grose on 9 November 1972 in Sacramento

That, on the subject of discussing origins in the science textbooks, the following editing be done prior to execution of a contract (with a publisher):

1. That dogmatism be changed to conditional statements where speculation is offered as explanation for origins.
2. That science discuss "how" and not "ultimate cause" for origins.
3. That questions yet unresolved in science be presented to the science student to stimulate interest and inquiry processes.

EXHIBIT 46: STATEMENT BY BOARD OF DIRECTORS AMERICAN CHEMICAL SOCIETY

Offered as part of the record for the California State Board of Education hearing on 9 November 1972

We have taken note of the article, "Alternative Hypothesis, " in the August 1972 issue of the *Scientific American*, p. 43, as well as of the opinion statement made by Dr. Junji Kumamoto in his letter of September 16, 1972, to Professors Stebbins and Spieth. We are in accord with the views that Dr. Kumamoto presented. Courses in science should be concerned only with scientific data and theories, not with theories from other areas of human thought. Therefore, non-scientific theories of origins, such as the "special creation" recorded in Genesis, while appropriate for presentation in classes in history or philosophy, should not be part of a science curriculum. They simply are not science. Any implication that they are within the framework of science would be misleading the students. We respectfully suggest that California science textbooks should not be distorted by inclusion of non-scientific material by legislative fiat.

Max Tishler President
President

Robert W. Cairns
Chairman, Board of Directors

EXHIBIT 47: SCIENCE AND ORIGINS

A Report delivered to the California State Board of Education by Vernon L. Grose

Curriculum Development and Supplemental Materials Commission 9 November 1972

It is three years ago, less four days, since I last addressed the State Board of Education. On that eventful day, two unrelated paragraphs were selected from approximately three pages of material that I had written regarding the need to teach more than one idea about the origin of the universe, life and man. These two paragraphs were unanimously adopted by the State Board into the *Science Framework for California Public Schools.*

In my oral presentation to the State Board on 13 November 1969, I made four major points, in no one of which had I any thought of reviving the old and outworn debate on evolution versus the Bible. I actually hoped to *eliminate* the issue! For example, in order to emphasize that I was *not* an anti-evolutionist when I suggested creation as a viable hypothesis for origins, I stated that I personally would have supported the position of Clarence Darrow (rather than that of William Jennings Bryan) in the Scopes trial of 1925.

In the intervening three years since the State Board adopted those paragraphs, there has been an untold amount of both consternation and approval, hysteria and applause, and disagreement as well as endorsement of the idea that public schools should teach more than one concept for the origin of things. I personally regret all this controversy. As it has developed, it is neither in the best interest of science itself nor of the 5 million school children in California. It may have clouded the public's understanding of the nature of science and the methods of scientific inquiry while polarizing segments of society that have no reason to be polarized.

Rather than securing a rational platform for objective discussion of origins, it appears that the two paragraphs adopted have done far more to encourage emotional, irrational thought. This type of thinking, of course, attracts the public interest, and I am sure it is one reason we see television cameras and a large representation of the press here today. Nevertheless, this publicity is not in the best interest of science, if it merely highlights an emotional conflict.

Serious controversy itself, obviously, is not new to science. It may, in fact, be the very stuff from which science is made. Science, properly conceived, is neither a consensus maker nor a field in which one should feel the least bit uncomfortable if he disagrees with a prevalent viewpoint. Had it been otherwise, Copernicus, Galileo, Einstein and many others would no doubt have abandoned the field of science. On the other hand, the kind of controversy of which science is constituted is not just ordinary argumentation. In fact, personal opinions are useless in science unless they can be validated in a consistently testable and predictable manner.

Measured by this high standard, the dialogue that has been going on in California for the past three years has at times seriously missed the mark of rational controversy. It has sometimes been based on emotional "straw men" or decoy issues. I deeply regret that I myself have at one point been in error. Until very recently, I held the private opinion that all the "straw men" were being erected in opposition to what I have connoted as the "Case for DESIGN." I have come to realize, and I now publicly confess, that I too have been inadvertently guilty of erecting a "straw man" as well. I appreciate the opportunity, therefore, of trying to state more effectively what I consider the basic question.

POPULAR POLARIZATION

I had hoped by substituting the words, *chance* and *design*, for *evolution* and *creation* that I might thereby reduce the emotional content of discussion and point to what I thought at the time was the key issue. I now confess that I have made an error. That substitution turned out to be a "straw man." The key changes in the curricula are essentially the same, but the public issue is quite different. As much as I was convinced that the issue was one that did not have to involve ultimate cause (which, of course, lies outside of science), I now believe that the issue troubling the public can only be dealt with by making reference to ultimate cause.

Although I intended no malice and was determined to be objective, I now realize that it was not fair to suggest, by placing *chance* and *design* in juxtaposition, that all those who endorse *chance* automatically reject *design* in nature. It would have been more accurate had I said that the issue is whether or not a Designer was at all involved because, as everyone knows, all scientists must believe in some universal order in nature.

Those who believe in a Designer would call that order "design," but this is only a special description of nature's order. On the other hand, those scientists who choose to acknowledge order but not to ascribe any authorship to that order should be permitted options other than having the order attributed entirely to chance.

We are faced with the prospect today of a long list of qualified people in science divided into two groups, with either group not basically disagreeing with what the other group has to say. Yet the irony is that it appears to the public as though one must choose between one group or the other. The reason for this dichotomy is that both groups are presenting partial arguments. This is the adversary process at its very best. However, scientists working in science utilize the scientific inquiry process whereby *uncertainties*, as well as *certainties*, are admitted with equal ease and candor.

On one side today, we can expect to hear a large group of expert scientists, who have been mobilized to testify, arguing with passion. All of their testimony can probably be summarized into the following five propositions:

1. The nature and methods of science are misunderstood by the lay public.
2. Religion has no place in the science classroom.
3. The essence of science is not and should not be determined by legislative action.
4. The Wilberforce-Huxley debates and the Scopes trial of 1925 clearly established the futility of debating evolution and creation.
5. The General Theory of Evolution (the "amoeba-to-man" continuum) is a widely-accepted explanation for interrelating observed scientific data.

On another side (and apparently in opposition to the first group), we can expect to hear a second group of expert scientists, likewise mobilized to testify and arguing with equal passion. Their testimony undoubtedly will emphasize one or more of five propositions:

1. All the findings of science have such incredible order that they bespeak deliberate and intelligent design.
2. Science is and must remain theologically neutral — neither anti-theistic nor theistic.
3. Much of the current teaching of science is not theologically neutral. It omits and even denies the possibility of intelligent design while advocating scientific naturalism (a form of anti-theism) as the only scientific explanation for origins.
4. The General Theory of Evolution (the "amoeba-to-man" continuum) is less than a complete explanation for all that we see in scientific observation.
5. The historical scientific data not only permit but demand more than the single hypothesis for origins now being taught; i.e., the General Theory of Evolution.

Personally, I find myself in agreement with not one but *both* of these sets of five propositions! The recent opinions expressed by such prestigious bodies as the National Academy of Sciences, the American Association for the Advancement of Science and the University of California Academic Council carry my wholehearted endorsement.

How then should the State Board reach its crucial decision? Should its vote be derived by counting the number of scientists present today on these two sides? Should it be swayed by the number of prizes, titles, degrees or political recognition held by advocates of one side or the other? Since legal action against the State Board has been threatened by advocates of both sides, even that leverage has been neutralized. Yet, a decision *is* required. Could it be that the required decision need not be in favor of one group of arguments over the other? I believe it is possible, provided a primary issue (which neither group has openly acknowledged) is exposed.

TOWARD RATIONALITY

Any attempt to resolve two poles of thought toward a position of agreement is based on the assumption that both parties are comprised of honorable people of good faith. I make that same assumption in what I propose today. It seems logical to start with principles that all scientists accept. The first such principle I would mention is that science is limited in its scope.

The Limitation of Science

Modern science, from its outset, has sought a certain type of truth by observing natural phenomena and interrelating these observations. The scope of inquiry has been limited to that which could be observed by one of the five senses. This is quite a severe limitation in view of *total* truth, since reality also includes moral values, beauty, justice, love and other characteristics which, by definition, are excluded from scientific opinion.

As Board Member Hubbard so well stated in the September 1972 Board meeting, the subject of ultimate cause is also outside the scope of science. Ultimate cause is more logically the concern of metaphysics, philosophy, and theology than science. Where scientists pass from observation to speculation and enter one of these three fields is certainly a gray area. That leads me to a second and hopefully universal premise among scientists — that science and religion perceive truth on different bases.

Science and Religion

If we accept a dictionary definition of religion as: "any specific system of beliefs, practices and ethical values involving a philosophy," then it is obvious that dialectic materialism and scientific naturalism are just as much religions as are Buddhism, Islam, Christianity or Judaism. All of these belief systems deal with ultimate cause and solicit dedication, belief, devotion and emotional response in the process of postulating answers to what are scientific unknowables. In other words, religion need not be *theistic*! As an example, Madelyn Murray O'Hair recently secured Government recognition (and thereby tax exemption) for the religion of atheism on this basis, since her atheism makes the absence of a God a matter of belief.

The California Attorney General, recognizing that scientific naturalism was a religion, ruled in 1963 that evolution could be taught in the public school only so long as it did not involve indoctrination. In other words, the anti-theism of scientific naturalism could not be taught to the exclusion of other explanations for unknowables or ultimate causation lest science no longer remain truly scientific; i.e., absolutely neutral in areas beyond its limitations.

Lest anyone assume that scientific naturalism is merely neutral rather than anti-theistic, I would quote the noted paleontologist, George Gaylord Simpson:[11]

> Adaptation (by natural selection) is real, and it is achieved by a progressive and directed process. This process is natural, and it is wholly mechanistic in its operation. This natural process achieves the aspect of purpose, *without the intervention of a purposer*, and it has produced a vast plan, *without the concurrent action of a planner*. (italics added)

In addition to being questioned by many scientists, the latter part of this statement is clearly extra-scientific. It expresses a man's personal belief which is closely akin to, if not synonymous with, a theist's belief in God. We cannot properly deny Simpson his prerogative to speak of his beliefs. However, if his personal beliefs continue to be presented to the exclusion of those other scientists whose beliefs are otherwise, we lose the necessary neutrality of science regarding ultimate cause.

Further, a serious peril for science looms in Simpson's statement. Notice the subtle change from "what he believes might be a plausible story (mythology)" to "what he declares unequivocally has happened (history)"! His beliefs, as a scientist, have no more validity

---

[1] *Simpson, George Gaylord, "The Problem of Plan and Purpose in Nature," Human Evolution, New York: Henry Holt and Company (1959), 104.*

regarding ultimate plan and purpose in the universe than a cab driver's, longshoreman's or banker's. Simpson's belief then leads me to a third area of hopeful agreement among all scientists — that science itself may be neutral concerning ultimate cause, but scientists, as persons, may not be.

Science But Not Scientists

While science is and must remain scrupulously neutral regarding the subject of religion or a belief in a Supreme Being, scientists simply are not! It goes without saying that a person, even a scientist, is not a set of compartments — rigidly defined by humanity's arbitrarily established criteria and thereafter tightly sealed against mixture. Each of us is a *totality* — simultaneously scientific, political, spiritual, philosophic, sexual, religious, and social.

Eminent scientists run the full gamut of religious belief, from those who see a Divine Being ordering each step that a person takes to the other extreme of categoric and unalterable denial of the existence of any Divine Being. Thus our textbook problem is not a question of science versus religion. It is *not* a question of whether a scientist can be a religious person or not. These questions are *non sequiturs.*

We then approach a crucial conclusion: The instinct of the public may be to create an issue by pitting "Biblical creation" against "science," but this really turns out to be an issue devoid of meaning. If we carelessly assume that all scientists believe thus and so while religious people believe something else, we have simply missed the point.

For example, if we should say that scientists surely must not have religious beliefs, how shall we categorize Michael Faraday, who preached sermons every Sunday for over 20 years and yet is acknowledged to be one of the greatest men of physics? Or how shall we classify Johannes Kepler? His written record of his derivation of those very laws of planetary motion we used to go to the Moon in Project Apollo contains numerous paragraphs of praise to Almighty God, whose designs Kepler believed that he was tracing out by his lengthy calculations. This common confusion of viewpoint between *science* and *scientists* then leads me to the discussion of the rudimentary issue facing the State Board.

THE PRIMARY ISSUE

For more than 50 years now, there has tended to be only one philosophy expressed in the science classroom regarding origins. This philosophy has implied insistently, if not stated with emphasis, that science has conclusively ruled out the possibility that a Supreme Being could have had any role in the universe. This philosophy, we now realize, is *not* neutral. It does *not* say, "We do not know." It is religiously anti-theistic, particularly on the subject of the origin of life and variations of that life. It proposes in emphatic and unqualified language that life originated by some cosmic accident and that from that point on, everything that we can see living today is the result of a continuum of development wherein inanimate matter (and later simple life forms) has possessed what Simpson alluded to in my earlier quotation as "a purposive and upwardly directive force."

This philosophy has tended to prevent us from acknowledging that there could be more than one conclusion drawn from historical scientific data. Accordingly, only one extra-scientific view has been presented. For every school child who has been taught to believe in a Supreme Being, the practical effect of this monolithic indoctrination has been automatically to negate the possibility of their belief being scientifically correct. It has forced the child to think that he has been taught to believe an error, and this on *scientific* grounds!

Herein lies the rudimentary issue at stake before the State Board today:

THE *TEACHING* OF SCIENCE HAS OVERSTEPPED THE RECOGNIZED LIMITS OF SCIENCE BY ENTERING THE AREA OF ULTIMATE CAUSE AND HAS COMPOUNDED THIS TRANSGRESSION BY BECOMING DOGMATIC AND DOCTRINAIRE IN EXCLUDING ANY ALTERNATIVE CONCEPTS WHERE THE DATA WOULD CLEARLY ALLOW ALTERNATES.

My personal concern, from the very beginning, has been to protect science from becoming scientism; i.e., I have tried to open the door to alternate ways of viewing all the scientific data that we have observed to date. In my own thinking, I believe that I too went too far in considering the "Case for DESIGN" a *scientific* theory. I realize now that I was tempted to exceed the limits of science in exactly the same way as those who have claimed the "amoeba-to-man" continuum to constitute scientific evidence against a Designer.

Undoubtedly, in both cases, the limits of science have been overstepped. Both theories are probably not scientific in the true sense, in that neither one of them is really testable. Origins are simply not occurring today under conditions where we can test their mechanisms. Therefore, these two theories serve as frameworks for interrelating the data that we see and are primarily *philosophic* rather than scientific.

There has been considerable press coverage devoted to the question of whether a body like the State Board should have any right to speak concerning an organized body of knowledge like science. This is not the first time that the ability of scientists to successfully discipline their own field of endeavor has been questioned. The foreword in a recent book on the warfare of science and scientism makes this interesting observation:[21]

> The politics of science is one of the agitating problems of the twentieth century. The issues are clear: Who determines scientific truth? Who are its high priests, and what is their warrant? How do they establish their canons? What effects do they have on the freedom of inquiry, and on public interest? In the end, some judgment must be passed upon the behavior of the scientific world and, if adverse, some remedies must be proposed.

This provocative statement leads me then to propose what types of direction I feel that the State Board is expected by the public to exercise. I refer to the realm of *teaching*, rather than the *content* of science.

PROPOSED CHANGES FOR SCIENCE TEXTBOOKS

As a charter Commissioner on the Curriculum Development and Supplemental Materials Commission as well as a member of the Science Committee of that Commission, I was officially appointed to meet individually with the 7 major publishers of science textbooks, currently under tentative adoption by the State Board, to review their materials for compliance with criteria on the subject of origins. To assure that any personal bias that I might have on the subject would be monitored for scientific accuracy, an editorial group of scientists, most of them from the California State University and Colleges system and from a wide background of scientific disciplines, was assembled. At least one member of this group met with us whenever changes were discussed with publishers. I owe this group a considerable debt in arriving at the conclusions I present today.

All the discussion concerning textbook changes could be summarized in two general ground rules:

1. We have urged a softening of dogmatism or absolutism in areas where speculation dominates observation or where inference overrides scientific evidence.
2. We have urged alternate explanations where only one explanation is offered in speculative situations such as origins.

Since this second rule for changes is the one of greatest interest to the State Board and concerned scientists, let me illustrate how it was discussed. If a textbook was explaining how plants "became" animals and proposed adaptation, or its technical equivalent "quantum evolution," (where a purposive mechanism with implied causation was suggested) as the explanation, we offered the publisher two alternatives. The first and preferable alternative, by far, was to remove *all* explanations for this supposed phenomenon, since there is absolutely no record of such an event occurring — there is simply a break in the fossil record at that point

---

[1] *de Grazia, Alfred, The Velikovsky Affair, New York: University Books (1966), 2.*

(such as discussed in Exhibit 47) and *any* explanation would be speculative rather than testable. The second alternative was to provide an optional possibility that animals may never have come from plants at all but that they may have been designed from their inception as a unique class of living things.

While the concept of design ultimately implies a Designer (as Wernher von Braun appropriately pointed out when supporting the concept of design[31]), no mention of a Designer or God or Supreme Being.

was requested or even suggested. Further, at no point (despite some reports) were the terms, *Genesis*, *creation* or *Bible*, requested or suggested. In fact, we insisted that those publishers who already had such terms in their textbooks *remove* them!

This fact should finally put to rest the false fears that have been fueled by reports which consistently made assumptions not found in the *Science Framework*, textbook criteria, or State Board Vice President Ford's statement in the September 1972 meeting.

An additional option that was discussed with the publishers was the desirability of frequent, explicit statements in each Teacher's manual concerning the inability of science ultimately to establish whether or not a Designer has played any role in the origin of the universe, life, and man himself. In other words, scientists should be both honest and wise enough to say, "We don't know," where it is applicable. Since the scientific inquiry process admits uncertainties and certainties with equal ease, why should not the textbooks list and discuss the weaknesses, difficulties, and the "many facets of the evolution picture that are not yet thoroughly understood" (mentioned in a recent University of California Academic Council policy statement[2])? This balance of unknowns would surely stimulate the inquiring minds of science students.

The nature of any proposed textbook changes will be minor. To my knowledge, not even re-pagination will be necessary for any publisher. However, for the first time in the history of teaching science in the public school in areas where speculation rather than observation has often prevailed, we have confronted speculative scientism with a balancing speculation that there may have been the participation of a Designer in various origins. Only through this symmetry can the neutrality of science be upheld.

In spite of his statement of his own belief quoted earlier, Simpson himself actually agrees that the role of a Designer cannot be denied on scientific grounds. He says:[3]

> It may be that the initiation of the (adaptation) process and the physical laws under which it functions had a Purposer and that this mechanistic way of achieving a plan is the instrument of a Planner — of this still deeper problem the scientist, as scientist, cannot speak.

It is this kind of scientific honesty which we applaud!

In summary, the proposed changes now prepared by the publishers are intended to assure the following objectives:

1. The science curriculum must remain scrupulously neutral on the subject of ultimate cause; i.e., the various philosophical or religious beliefs concerning origins.
2. By means of this strict neutrality, science students should not therefore be influenced either in favor of a Designer or against the existence of a Designer by any material that discusses origins. This neutrality very simply carries out the principle that science should not influence, in any direction, the belief systems of students on a subject which lies outside the domain of science.

---

[1] *von Braun, Wernher, personal correspondence with Vernon L. Grose, 14 August 1972.*

[2] *McCurdy, Jack, "Special Creation Theory in School Books Opposed," Los Angeles Times, Part II, November 2, 1972, 1.*

[3] *Simpson, op. cit., 104-5*

CONCLUSIONS

Regarding the remainder of the public hearing on this subject, I would offer a suggestion. It is unfortunate that many concerned scientists and others have come here today to present arguments, pro and con, on an issue which, while widely reported, simply does not exist. Since the wording in the *Science Framework* as well as specific criteria related to the subject of origins can be interpreted in a myriad of ways, it would be far more profitable if the scientific expertise here today could be utilized in evaluating the specific changes that each of the 7 publishers have prepared, rather than attempting to settle what was meant or what could be meant by the Framework or the criteria. After all, the latter were but means to an end — specific wording in textbooks. That wording is now available for analysis.

Should the State Board so desire, the Curriculum Commission would be pleased to furnish each interested party present today with a complete set of proposed changes on the subject of origins for their review and response. In this way, everyone who has demonstrated their concern by preparing statements and appearing here today could make a positive contribution to our school children.

I would hope that this three-year experience can serve as a warning to all of us that even intelligent people of goodwill can be drawn into battle over an artificial issue. However, I have great confidence that the changes which are proposed, while achieved in the midst of great controversy, represent a major step forward in science education.

To conclude, I would offer the same quotation of Charles A. Siepmann with which I closed my presentation to the State Board three years ago:[61]

> For those who go far with it, education is, on one side, a lonely business. For to be eminent in mind or in imagination is, as with all eminence, to stand alone and apart.
>
> Education is also awful, for it leads us inexorably beyond the known to the unknowable — so that, like Socrates, we end with the recognition that we know nothing. Consequently, the chief attributes of a truly educated man are humility (defined as courageous insight that refuses to seek simple answers to complicated questions) and a certain suspension of judgment (not to be confused with skepticism) in recognition, always, of "that reserve of truth beyond what the mind reaches but still knows to be behind." This quality of mind is common both to great scientists and great men of faith.

EXHIBIT 48: PALEONTOLOGIST'S DISCUSSION OF GAPS

This discussion of Figure 7 on page 105 is reprinted from George Gaylord Simpson: Tempo and Mode in Evolution, New York: Columbia University Press, 1944, (104-7, 206), by permission of the publisher

Figure 7 gives some idea of the inadequacy of the record, both systematic, with regard to the origins of the various groups, and random, within the orders once they have appeared in the fossil record.

Listing of data as to the occurrence of possible ancestry involves subjective judgment as to what constitutes a "possible ancestry," and in some cases opinions differ radically because of the magnitude of the morphological gaps between the bases of ordinal records. These data are also strongly affected by random, and in some cases also systematic, gaps in the records concerning possible ancestors. They do, however, more nearly than any other available information provide objective criteria as to the span within which the orders probably originated.

---

[1] *Drake, William E., The American School in Transition, Prentice-Hall (1955), frontispiece.*

At best, paleontological data reveal only a very small proportion of the species that have lived on the earth, and the known fossil deposits do not and never can represent anything distantly approaching adequate sampling of the diverse facies of the various zoological realms through-out geological time. Even as regards the most recent and best part of the record, for the Tertiary, and the relatively abundant larger land mammals for some of which good sequences are found, the evidence is less than 1 percent of what would be necessary to give continuous know-ledge of all the major phyla ... — and it would be pointless to emphasize once more the general incompleteness of the paleontological record, except to stress that this incompleteness is an essential datum ...

Major Systematic Discontinuities of Record

The levels to which these conclusions apply without modification are approximately those discussed as macro-evolution (under that or an equivalent term) by neozoologists and biologists. On still higher levels, those of what is here called "mega-evolution", the inferences might still apply, but caution is enjoined, because here essentially continuous transitional sequences are not merely rare, but are virtually absent.

This regular absence of transitional forms is not confined to mammals, but is an almost universal phenomenon, as has long been noted by paleontologists. It is true of almost all orders of all classes of animals, both vertebrate and invertebrate. A fortiori, it is also true of the classes, themselves, and of the major animal phyla, and it is apparently also true of analogous categories of plants. Among genera and species some apparent regularity of absence of transitional types is clearly a taxonomic artifact: artificial divisions between taxonomic units are for practical reasons established where random gaps exist. This does not adequately explain the systematic occurrence of the gaps between larger units. In the cases of the gaps that are artifacts, the effect of discovery has been to reveal their random nature and has tended to fill in now one, now another – now from the ancestral, and now from the descendent side. In most cases discoveries relating to the major breaks have produced a more or less tenuous extension backward of the descendent groups, leaving the probable contact with the ancestry a sharp boundary.

This is true of all the thirty-two orders of mammals, and in most cases the break in the record is still more striking than in the case of the perissodactyls, for which a known earlier group does at least provide a good structural ancestry. The earliest and most primitive known members of every order already have the basic ordinal characters, and in no case is an approximately continuous sequence from one order to another known. In most cases the break is so sharp and the gap so large that the origin of the order is speculative and much disputed. Of course the orders all converge backward in time, to different degrees. The earliest known members are much more alike than the latest known members, and there is little doubt, for instance, but that all the highly diverse ungulates did have a common ancestry; but the line making actual connection with such an ancestry is not known in even one instance.

Quantum Evolution

Perhaps the most important outcome of this investigation, but also the most controversial and hypothetical, is the attempted establishment of the existence and characteristics of quantum evolution ...

It is, however, believed to be the dominant and most essential process in the origin of taxonomic units of relatively high rank, such as families, orders, and classes. It is believed to include circumstances that explain the mystery that hovers over the origins of such major groups ...

For the sake of brevity, the term "quantum evolution" is here applied to the relatively rapid shift of a biotic population in disequilibrium to an equilibrium distinctly unlike an ancestral condition.

EXHIBIT 49: TYPICAL PROPOSED CHANGES TO SCIENCE TEXTBOOKS

These revisions were proposed by Commissioner V. L. Grose to the Curriculum Development and Supplemental Materials Commission to illustrate publisher compliance with State of California criteria on 17 November 1972

OVERVIEW OF PROPOSED CHANGES

1. One change for each of the following publishers is submitted to illustrate the different styles, philosophies, and formats among publishers:

American Book Company
Harcourt Brace Jovanovich, Inc. Harper & Row, Publishers
Holt, Rinehart and Winston, Inc.
Laidlaw Brothers (A division of Doubleday & Company)
Charles E. Merrill Publishing Company
Silver Burdett Company, Division of General Learning

2. All proposed changes comply with the unanimous recommendation of the Curriculum Development and Supplemental Materials Commission presented by the Science Committee to the State Board of Education on November 1972 (Exhibit 44), as well as the following specific criteria:[1]

Teacher Materials

4. (Teacher materials shall) Recognize the relationship of science to other disciplines, particularly when considering unresolved theories, such as origins, evolution, creation, and so forth.

Learner Materials

18. (Learner materials shall) Provide examples of unresolved questions in science as a means of stimulating the inquiry process.

Instructional Program

2. (The instructional program shall) Reveal science not as a set of complete, final, and unalterable truths but as an ongoing search whose inferences, hypotheses, theories, and conclusions are subject to continuous evaluation and change.

Rationale for the Criteria

1.0 Attitudes

1. 4 (The textbooks and reusable educational materials shall provide opportunities for the learner to) Recognize the limitations of scientific modes of inquiry and the need for additional, quite different approaches to the quest for reality, including the search for answers to questions like the origin of the universe, matter, life, and man.

3. All of the proposed changes were discussed with the individual publisher involved, and no objection to the changes has been raised. In several cases, the proposed version appears in the publisher's wording.

4. Every change proposed herein was approved for scientific accuracy by at least one member of the Editorial Advisory Board (see Exhibit 37).

5. All changes are presented in the same format: (a) current version, (b) proposed version, and (c) rationale for change. This enables the reader to make immediate comparison.

6. There are examples from both pupil's textbooks as well as teacher editions.

7. There are examples of deletions, additions, and rewording.

8. There are examples of discussing various origins—the earth, life, and various animals.

9. There are examples of unresolved scientific questions being raised.

---

[1] *Criteria for Evaluating Textbooks and Reusable Educational Materials in Science, approved by California State Board of Education, 13 March 1970.*

PROPOSED CHANGE NUMBER 1

Publisher: American Book Company
Textbook of Interest: SCIENCE - Exploring Ideas - Grade 4 Teacher's Annotated Edition - Pages 206-7
Current Version: The Earth is part of the solar system, made up of the sun and the planets that revolve around the sun. All scientists do not agree on when and how the Earth was formed. Some have questioned whether it had a beginning. Many scientists think the Earth was formed with the rest of the solar system billions of years ago. Others believe in a special creation of the universe not nearly so long ago. Those scientists who conclude the Earth to be billions of years old explain that many, many changes took place during the Earth's history.
Proposed Version: The Earth is part of the solar system, made up of the sun and the planets that revolve around the sun. All scientists do not agree on when and how the Earth was formed. Some have questioned whether it had a beginning. Many scientists believe that the Earth, together with the rest of the solar system, was either accidentally formed or created by design billions of years ago. Many changes have taken place during the Earth's long history.

(The following *deletion* from the teacher's notes on the same page was proposed.)

You may want to describe how *Genesis*, the first book of the Old Testament, explains the creation of the earth ... For example, *Genesis* describes the early earth as "a desolate waste, with darkness covering the abyss and a tempestuous wind raging over the surface of the waters. God then separated the light from the darkness. God divided the waters in two, dividing the waters (the oceans) below the firmament from those that were above it (the sky)." Relate this account to the separation of the atmosphere (sky) from the hydrosphere (waters).

Genesis further describes: "Let the waters below the sky be gathered into one place so that the dry land may appear. God called the dry land earth, and the gathered waters seas." Relate this to the formation of land features and water features.
Rationale for Change: This proposed change supports the second point of the Curriculum Development Commission's unanimous recommendation to the State Board regarding origins. Discussion of "ultimate cause" is inappropriate in science. This publisher had interpreted the *Science Framework* as suggesting that some religionists' views about a "young Earth" were required. Further, it would be religiously biased to use a Judeo-Christian document exclusively as illustrative material. It was never intended in the Framework or later criteria to mention the Bible in science textbooks.

PROPOSED CHANGE NUMBER 2

Publisher: Harcourt Brace Jovanovich, Inc.
Textbook of Interest: Concepts in SCIENCE - Level 3 Teacher's Edition - Page T-314
Current Version: It is known that life began in the seas. Evidence of early life is found in the Earth's rock layers, which preserve as fossils the forms of many kinds of plants and animals. The oldest rocks contain no fossils. These rocks are from periods before life began or from periods when the only forms of life were minute and soft-bodied and left no fossil remains. Rocks less old provide fossil evidence of an abundance of plant and animal life in the prehistoric seas.

Eventually, over great periods of time, plants spread to land, and land animals began to evolve. Vertebrates arrived on the scene with the earliest fish. Some fish grew lungs and left the water. Amphibians evolved, followed in turn by reptiles, birds, and finally mammals. Today animals live all over the Earth, in virtually every kind of environment.

Each kind of animal has become adapted to the environment in which it lives. Adaptation occurs as the individuals of each generation that are best fitted for life in the environment survive and reproduce their own kind. Those that are poorly adapted perish and their species becomes extinct.
Proposed Version: Many scientists believe that life may have begun in the sea. The only evidence of early life is found in the Earth's rock layers, which contain fossils of less than 1%

of the plants and animals that we believe have existed. Unfortunately, there are absolutely no fossils in older rocks. This means that we can only guess about the character of early life forms. Later rocks contain fossil evidence of an abundance of plant and animal life.

If the fossils were formed in a uniform manner (with the earliest fossils at the deepest levels and later fossils on top of them), we can assume that land animals appeared after plant life. There are hundreds of documented cases where earlier fossils lie on top of later ones, but since animals required food in the form of plants, this is a reasonable assumption. The order of appearance in the fossil record of animal life is first fish, then amphibians, reptiles, birds, and mammals. Today animals live all over the Earth, in virtually every kind of environment.

Each kind of animal has features which enable it to live in its environment. Should the environment change, individuals of each generation that are best fitted for life in that environment survive and reproduce their own kind. Those whose features are less suitable for that environment perish and their species becomes extinct.

Rationale for Change: This proposed change supports all three points of the Curriculum Development Commission's unanimous recommendation to the State Board regarding origins. First, dogmatism is replaced with conditional statements regarding the origin of life. Secondly, the student is apprised that science does not address "ultimate cause" for the origin of various animal groups. Third, the problem of uniformitarianism versus catastrophism is exposed to stimulate the student's inquiry processes.

PROPOSED CHANGE NUMBER 3

Publisher: Harper & Row, Publishers
Textbook of Interest: THE YOUNG SCIENTIST - Exploring His World Teacher's Edition - Grade 2 - Pages 104-5
Current Version:

Plants began to grow in the long-ago past. First they grew in the sea.
Then they began to grow on the land.
Plants were the first living things on the land
These ancient plants were strange and different.
They were not at all like the plants of today.
The first land plants had no leaves.
Tiny hairs held the plants to the soil.
Many, many years went by.
Then branches began to grow.
Branches grew on the plants.
Many, many years went by.
Then leaves began to grow.
Leaves grew on the plants.
Green plants soon spread across the earth.
The green plants made food for themselves.
The plants could be eaten as food.
With food to eat, animals appeared.
Animals moved from the sea to the land.
Fish began to take oxygen from the air.
They took it from the air, not the water.
The fins of the fish changed.
The fins slowly changed to feet.
Animals began to walk across the land.

Proposed Version:

Plants grew in the long-ago past.
They may have grown first in the sea.
They also grew on the land.
Plants were probably the first living things on the land.

We do not know how ancient plants looked.

These ancient plants may have been different from our plants today.

Animals probably appeared on the earth after plants.

They needed food to eat.

Plants could be eaten as food.

Rationale for Change: This proposed change supports the first two points of the Curriculum Development Commission's unanimous recommendation to the State Board regarding origins. First, the original version is a speculative myth which could easily appear dogmatic to the second-grade student. Secondly, it expresses the philosophic view or evolutionary belief that fish are the progenitors or "ultimate cause" of land animals, and there is no scientific evidence for this belief.

Harper & Row's Executive Editor for Science, Robert Hay, who flew out from Chicago to discuss changes concerning origins, acknowledged that this was an unfounded myth that needed this revision.

PROPOSED CHANGE NUMBER 4

Publisher: Holt, Rinehart and Winston, Inc.
Textbook of Interest: Modern Elementary Science - Grade 4 Basic Pupil's Edition - Pages 294-5

It is proposed that this publisher's version be modified only by deletion as shown:

Have scientists been on the right track? Some people do not think so. They say that the important changes are not gradual, but are very sudden. Some of these people say that scientists must be wrong because a book of the Scriptures says the world was made in six days. However, religious scholars, who spend much of their lives studying the Bible, point out that religious writings are not meant to be scientific. For example, they do not always refer to time in the same way that scientists do. The term "day" in Scripture may not mean our twenty-four hour day. Instead, it may simply stand for a "period" of time. Therefore, when Scripture says the world was made in six days, this could mean the world was made in six "periods of time." In this sense,) science does not seem to disagree with Scripture.

Who or what made the earth and why was it made? A Greek thinker, Plato, who was born in 427 B.C., said that the world was once a huge lump of matter without any shape. According to Plato, a "Great Artist" formed this material into the world we see and hear, taste and touch.

Scripture, on the other hand, states that God made the world. God, in this view, is not just a "Great Artist, " who uses a material which already exists – He makes the world out of nothing.

What does science say about who made the world and why? Nothing! Science tries to explain events and changes by finding causes which can be observed and tested and measured. The questions of "who" and "why," however, cannot be answered by observing, testing, and measuring. To answer them, you will have to look outside the world of science. You will have to find your own method for these questions.

Rationale for Change: This proposed change supports the second point of the Curriculum Development Commission's unanimous recommendation to the State Board regarding origins. Discussion of "ultimate cause" is inappropriate in science. While this publisher has handled the subject of compatibility between scientific and religious viewpoints on origins in an admirable manner, such discussion is ultimately inappropriate in science textbooks. Further, only one religious body's views (Judeo-Christian) are presented.

PROPOSED CHANGE NUMBER 5

Publisher: Laidlaw Brothers (A division of Doubleday & Company)
Textbooks of Interest: MODERN SCIENCE - Level Five
Teacher's Edition - Page T75

Pupil's Edition - Page 101
Current Version:
MAJOR CONCEPTS DEVELOPED
1. Scientists believe life may have begun from amino acids or viruses, neither of which is usually considered living.
2. Scientists believe life may have been transported from another planet.
SCIENCE BACKGROUND
It is possible that life began as a single molecule, which could reproduce itself. Such molecules might have evolved into single cells. Then the single cells might have evolved into soft clusters of cells. Neither of these forms of life could possibly leave fossil evidence. Therefore, it is probable that scientists will never be able to explain just how life on earth began. The best hypotheses to date are presented in the pupil' s text.
The probability is very strong that life began in shallow water along the shores of the continents. In this environment the water is well aerated, and sunlight penetrates to the bottom of the ocean. From such a beginning, little evidence of life would accumulate because shore deposits are quickly eroded.
Proposed Version:
MAJOR CONCEPTS DEVELOPED
1. Some scientists believe that life may have accidentally and spontaneously arisen from nonliving matter; e.g., amino acids or viruses, (despite Louis Pasteur's proof that spontaneous generation or abiogenesis is an impossibility today).
2. Some scientists believe that life was created by a Supreme Intelligence as a distinct act.
3. Some scientists believe that life may have been transported by a meteorite from another planet.
SCIENCE BACKGROUND
The origin of life is not subject to scientific determination. It is more properly discussed in philosophy, metaphysics or religion. There is no physical evidence such as fossils of the earliest forms of life. Therefore, it is probable that scientists will never be able to explain just how life on earth began. Three hypotheses are presented in the pupil's text.

(The following addition must also be made in the pupil's text on page 101 between the section, "Viruses," and "Organized material in meteorites. ")

Designed Creation of Life - Have you ever seen anything that was *not* designed? Could you imagine that the Apollo spacecraft just "happened" without a design or a purpose? Some scientists who have studied the chemistry of life have reached the conclusion that life could not just have been an "accident." They see such order in the many materials required for life that they believe it had to have a Designer. They do not know *how* the Designer may have done it, but their findings cause them to claim that life may have been the result of a planned design.
Rationale for Change: This proposed change supports all three points of the Curriculum Development Commission's unanimous recommendation to the State Board regarding origins. First, the speculations concerning the origin of life are expanded to include another group of scientists currently not represented in the textbook. Although the textbook already and properly points out that ultimate cause for the origin of life probably lies outside of science, this point is strengthened. Third, the science student is presented with unresolved questions in the three hypotheses to stimulate his inquiry processes.

PROPOSED CHANGE NUMBER 6

Publisher: Charles E. Merrill Publishing Company
Textbook of Interest: Principles of SCIENCE - Book One Teacher's Annotated Edition - Pages 338-339
Current Version: Scientists have developed many theories to explain the origin of life. Each theory holds that the first simple, microscopic living things came from nonliving substances. This view is not based on whim or fancy; this theory, as well as all scientific theories, is based

on the available facts. This idea is certainly a startling one! It is difficult to understand! How could non-living chemical compounds form living substances? The theory which states that the first simple, microscopic living things came from nonliving matter is explained this way. When the world was formed, nonliving inorganic elements and compounds were the only kind of material present.

It is probable that some form of energy caused simple inorganic carbon compounds to combine to form organic compounds. Section 4:6 describes an *organic compound* as one that is often associated with living things. An organic compound always contains carbon, usually contains hydrogen, and frequently contains other elements like oxygen, nitrogen, and sulfur. Intense energy, such as light, heat, radiation, or electricity may have been responsible for the chemical change.

At one time, very long ago, an organic compound may have developed the ability to duplicate itself. This means that the organic compound could make more of itself from simpler nonliving compounds. How such a thing could come about is not yet fully understood. But much of the available evidence seems to point to this theory.

Proposed Version: Scientists have proposed many ideas to explain the origin of life. When they speculate about this interesting event, scientists understand that the "first cause" or ultimate reason for the origin of life will never be proven scientifically. Since science is concerned with repeatable or testable events and the "origin" of life can never be repeated, scientists will never know just how life arose. However, scientists are curious, just like you are, about how it might have happened. They even hope to produce life in a laboratory some day. If they are successful, they will still not know how life arose the *first* time, but they will have an idea about how it *could* have happened.

Many scientists believe that the first living substance (whatever it may have been) came from nonliving substances. This idea is certainly a startling one! It is difficult to understand! How could *nonliving* chemical compounds form *living* substances? Did not Louis Pasteur prove conclusively more than one hundred years ago that "spontaneous generation" (living matter arising from nonliving matter) was an impossibility? How could nonliving matter (which has no intelligence) suddenly become alive?

Another group of scientists believe that life was created by a Supreme Intelligence. How did the Creator originate life? These scientists do not know. They are as unable to prove their idea as those who believe that life arose spontaneously from nonliving matter.

Since neither group of scientists is able to prove how or why life originated, they proceed together in their investigations to try and under-stand the chemistry of life. No one can be sure if the beginning of life ...

Rationale for Change: This proposed change supports all three points of the Curriculum Development Commission's unanimous recommendation to the State Board regarding origins. First, dogmatism is replaced with conditional statements regarding theories for the origin of life. Secondly, the student is apprised that science does not address "ultimate cause". Third, the problem of abiogenesis is exposed to stimulate the student's inquiry processes.

PROPOSED CHANGE NUMBER 7

Publisher: Silver Burdett Company, Division of General Learning
Textbook of Interest: SCIENCE - Understanding Your Environment Teacher's Edition - Grade 4 - Page 9

Current Version: The Law of the Conservation of Matter and Energy states that neither matter nor energy can be created or destroyed, although either of these can be changed into another form. The Law of Entropy, which is less well known, states basically that whenever matter or energy is used, it changes into a less usable form. Also, the objects and phenomena involved become disorganized ...

Proposed Change: (No deletion or correction of above—only an addition.) Ironically, the evolutionary idea that complex forms of life have arisen from simpler forms of life seems to violate this Law of Entropy by requiring that objects and phenomena become more *organized*—

not disorganized, and change into forms of *greater*— not lesser— usefulness! Point out this apparent conflict between a recognized and universally-accepted scientific law and a "belief" that many scientists hold for the origin of plants and animals, even man himself.

For example, are not mammals more organized than birds, birds more organized than reptiles, reptiles more organized than amphibians, amphibians more organized than fish and so forth? Yet, how could this series "evolve" under the Law of Entropy?

Rationale for Change: This proposed change supports the third point of the Curriculum Development Commission's unanimous recommendation of providing unresolved questions in science to stimulate inquiry processes in the student. It seems appropriate to expose this and other of the "many facets of the evolution picture that are not yet thoroughly understood" mentioned in the recent University of California Academic Council policy statement to the State Board (see Exhibit 41).

EXHIBIT 50: CHEMICAL EVOLUTIONIST'S EVALUATION OF PROPOSED TEXTBOOK CHANGES

These two letters were sent to the California State Department of Education by Richard M. Lemmon, Associate Director, Laboratory of Chemical Biodynarnics, Lawrence Radiation Laboratory, University of California Berkeley

UNIVERSITY OF CALIFORNIA
Laboratory of Chemical Biodynamics Lawrence Radiation Laboratory
Berkeley, California 94720

December 1, 1972
Mr. Vernon L. Grose
Tustin Institute of Technology
22556 Gilmore Street
Canoga Park, Ca. 91304

Dear Mr. Grose:

I met you briefly at the time of the Board of Education's public hearing in Sacramento on November 9th.

The enclosed copy of a letter to Dr. Hall details the reasons for my opinion that you are unqualified to act as an advisor on the content of science textbooks. It's only fair, therefore, to send you a copy.

I enclose copies of reprints on some of my recent research work.

I would be grateful if you would send me copies of yours.

Sincerely yours,

Richard M. Lemmon
Associate Director

December 1, 1972
Dr. Clarence Hall
State Department of Education
721 Capitol Mall
Sacramento, Ca. 95814

Dear Dr. Hall:

Dr. Junji Kumamoto, Chairman of the Science Committee of the State Curriculum Development and Supplemental Materials Commission, has asked for my comments on the memorandum "Typical Proposed Changes to California Science Textbooks" prepared by Mr. Vernon Grose. Dr. Kumamoto asked me to do this, I suppose, because my research work is in chemical evolution. This area of research may be defined as laboratory studies of the chemical processes taking place on the prebiological Earth that led to the appearance of the biochemical components of the living cell. I should also say that I attended the State Board of Education's public hearing on the science textbook question in Sacramento on November 9th.

My general comment on Mr. Grose's memorandum is that it shows a surprising misunderstanding of major aspects of science. Since Mr. Grose is a member of the Science Committee, such a comment requires elaboration:

(1) Mr. Grose does not understand that scientific theories of the origin of life are not built on ideas of "chance" or "accident". The accumulation of biological molecules on the pre-biotic Earth appears to be the result of the intrinsic properties of matter. The organization of those molecules into living cells is also believed to result from the same properties, examples of which are electrical-charge distributions and tendencies to form chemical bonds. These conclusions are richly supported by the results of laboratory investigations. Therefore, scientists studying these matters do not need to call upon "chance" or "accident". The fact that Mr. Grose continues to do so shows an unfamiliarity with research progress.

(2) Mr. Grose is quite wrong when he writes (proposed versions for the Laidlaw Brothers and Merrill Publishing Co. texts), (a) "despite Louis Pasteur's proof that spontaneous generation or abiogenesis is an impossibility today," and (b) "Did not Louis Pasteur prove conclusively one hundred years ago that 'spontaneous generation' was an impossibility?" The answer to the question is a resounding "NO". Let me quote from a recent article in Perspectives in Biology and Medicine (Vol. 15, p. 529 (1972): "Pasteur's carefully conducted experiments were widely believed to have refuted the ancient theory of spontaneous generation, although Pasteur himself stated that the experimental evidence taken in toto could be argued either way." In brief, "spontaneous generation" is scientifically recognized as a definite *probability* for the appearance of life on our planet, given that the "spontaneity" may have required one billion years. Mr. Grose has not followed the recent progress in scientific research on the origins of life.

(3) Related to the above is Mr. Grose's statement, in "Science Background" for the Laidlaw Bros. text, that "The origin of life is not subject to scientific determination". This is comparable to saying that, because no human was there to observe it, the origin of Yosemite Valley is not subject to scientific determination. "Determination" is a slippery word - in science, nothing is ever completely "determined". But the origin of life is a matter of active scientific investigation, whereas Mr. Grose's statement seems deliberately to imply that it is not.

(4) In the same "Science Background", Mr. Grose talks about "design" and a "Designer". It was my strong impression from the November 9th hearing in Sacramento, and from statements by Board Member John Ford, that it was agreed that "design" and "designers", like "art and poetry, value systems, etc.", are not a part of science. No one who understands science would include a reference to "design" in a science textbook.

(5) In his proposed version for the Silver Burdett Co. texts, Mr. Grose shows that he does not understand the Second Law of Thermodynamics (entropy). He says that "the evolutionary idea that complex forms of life have arisen from simpler forms of life seems to violate this Law of Entropy ..." It "seems" nothing of the sort. Does Mr. Grose really believe that the appearance of a crystal (a more-highly organized form) from a solution (less-highly organized) violates the entropy idea? In some areas of space (*e.g.*, in a crystal or a living being) entropy *decreases*, but only at the price of an increase in the total entropy of the universe. It is distressing to know that a man who doesn't understand simple thermodynamics is posing as an expert on science texts.

(6) In his "proposed version" for the Harcourt Brace Jovanovich text, Mr. Grose says that "there are hundreds of documented cases where earlier fossils lie on top of later ones", implying some discontinuity in the geological record. This seems an almost-deliberate attempt to confuse the students. These findings of earlier fossils on top of later ones result from

geological processes such as folding and thrust-faulting, and does not disturb the interpretation of the fossil record. The age of a stratum is, of course, more accurately determined by potassium-argon and rubidium-strontium dating than by relative positions of strata.

(7) In the same "proposed version", Mr. Grose says "Unfortunately, there are absolutely no fossils in older rocks." Later, in his comments on the Laidlaw Brothers text, he says "There is no physical evidence such as fossils of the earliest forms of life". He is 100% wrong. He is unaware of the work of Barghoorn and Schopf (see, for example, "Scientific American", Vol. 224, No. 5, p. 30, 1971). These researchers have found remains of algae and bacteria in rocks dated at 3. 2 billion years of age - and the oldest rocks yet found are only about 3.5 billion years old.

(8) In his "proposed version" for the Laidlaw Brothers text, Mr. Grose states that "Some scientists believe that life was created by a Supreme Intelligence as a distinct act." Some scientists, I suppose, believe that Republicans are better than Democrats. Neither belief has anything whatever to do with science.

(9) In his "rationale for change" of this Laidlaw Brothers text, Mr. Grose says that his proposals "are expanded to include another group of scientists currently not represented in the textbook." Again, his words would lead to confusion among students. This other "group of scientists" is at most a miniscule fraction of the membership of America's scholarly scientific societies. As the California Section of the American Chemical Society reported to the State Board of Education, "The overwhelming majority of our members, and of all recognized scholarly scientific societies, agree that biological (or 'Darwinian') evolution is the only scientific theory that explains the proliferation of life on our planet."

(10) As a final example of Mr. Grose's misunderstanding of science, I cite his following insertion into the "Science Framework": "All scientific evidence to date concerning the origin of life implies a dualism or the necessity to use several theories to fully explain the relationships between established data points. This dualism is not unique to this field of study, but is appropriate to other scientific disciplines such as the physics of light."

This statement implies that there are two scientific theories about the physics of light. Had Mr. Grose checked this matter with the American Physical Society he would have learned that the wave mechanical treatment developed by de Broglie, Schrodinger, and others is an entirely adequate description of the nature of light — there is no "rival" theory. Mr. Grose's confusion about two *properties* of light (wave and particulate) has led him into thinking that there are two *theories* about the nature of light.

I submit that Mr. Grose has thoroughly disqualified himself as an expert on what should, or should not be, in science textbooks. I have no doubt that this opinion would find concurrence in the representative bodies of all of America's recognized scientific societies.

In summary, I submit that Mr. Grose's advice to the Curriculum Commission reflects misunderstandings of both the meaning and content of science. I profoundly hope that these misunderstandings do not result in a weakening of our state's programs of instruction in science.

Sincerely yours,

Richard M. Lemmon
Associate Director

EXHIBIT 51: MEDICAL PHYSICIST'S SUPPORT OF DR. LEMMON

This letter was solicited by Richard M. Lemmon, Associate Director, Laboratory of Chemical Biodynamics, Lawrence Radiation Laboratory, University of California at Berkeley to support his letter of 1 December 1972 (see Exhibit 50)

UNIVERSITY OF CALIFORNIA Space Sciences Laboratory Berkeley, California
December 5, 1972
Dr. Clarence Hall
State Department of Education
721 Capitol Mall
Sacramento, California 95814

Dear Dr. Hall:

I am writing to add my comments to those expressed by Dr. Lemmon in his letter to you, December 1, regarding the textbook controversy.

I agree completely with Dr. Lemmon's criticisms and I wish to enlarge on his remarks about Louis Pasteur and spontaneous generation. Mr. Grose concludes that Pasteur's experiments are in some way related to the question of the ultimate origin of life. This is quite erroneous. Pasteur's predecessors showed that the spontaneous generation of visible forms of life did not take place. Prior to their findings, it was believed that rotting meat gave rise to blow-fly maggots by spontaneous generation.

Pasteur was confronted with the task of showing that the findings by his predecessors applied also to microscopic forms of life. He demonstrated that bacteria did not arise by spontaneous generation in meat broth. These findings had nothing to do with the question of the origin of life by slow evolution of prebiotic molecular forms.

It is ironical to note that Lepeshinskaya, a disciple of Lysenko, claimed in the 1940's that she could produce spontaneous generation of microorganisms in a broth of hay. For this quackery, she received a Stalin Prize. This is what came of the politicalization of biology, with which we are now faced in California because of the activities of the Creation Research Institute and its sponsors.

It is an intellectual affront that Mr. Grose should be attempting to enforce opinions on biological matters that would result in an undergraduate student receiving a non-passing grade. Science is hierarchical and elitist. For the same reason that automobile mechanics are not permitted to carry out abdominal surgery; we must not permit individuals who are illiterate in the biological sciences to insert their opinions and prejudices into science textbooks. I object to the presence of Mr. Grose on the State Curriculum Development and Supplemental Materials Commission for just this reason. Other examples of Mr. Grose's scientific incompetence are listed by Dr. Lemmon.

I hope that the State Board of Education will vote to delegate the responsibility for the preparation and scrutiny of science textbooks to qualified scientists.

Sincerely yours,

Thomas H. Jukes, Ph.D., D.Sc., Professor of Medical Physics, in Residence

EXHIBIT 52: OPEN LETTER TO CALIFORNIA STATE BOARD OF EDUCATION 10 December 1972

TO THE CALIFORNIA STATE BOARD OF EDUCATION:

Nineteen Nobel Laureate scientists who live in California have joined in deploring the attack on evolution and beclouding its significance in the science textbooks for our public schools.

The latest recommendation from the Curriculum Commission may seem innocuous in requiring "that dogmatism be changed to conditional statements where speculation is offered as explanation of origins." However, this is clearly tailored to make room for the "creation theory" as an alternative to evolution.

Conditional statements are appropriate when multiple theories have been proposed and none of these can be eliminated by the existing scientific evidence. However, this is not the case in the present argument. The "creation theory" of man does not stand as an alternative to the theory of evolution in this scientific sense. It is eliminated by existing data. Indeed, no alternative to the evolutionary theory of the origin of man exists today which gives an equally satisfactory explanation of the biological facts. Hence the incentive in applying this requirement selectively to the evolutionary theory, instead of, for example, the atomic theory, is clearly not scientific but religious.

None of these remarks is intended to foreclose further discussion of the theory of evolution posited as a scientific issue. Many of its details are a subject of continuing controversy. Students should by all means have the opportunity to be acquainted with the factual evidence on which scientists have based their conclusions, and, in the realm of scientific teaching, should not be biased by assertions based on arguments of faith rather than scientific evidence.

We urge you to reject modifications to the science framework which require the addition of non-scientific material to our science textbooks.

Sincerely yours,

Arthur Kornberg
Nobel Laureate 1959, Medicine
Professor of Biochemistry
Stanford University School of Medicine

EXHIBIT 53: JOINT RESOLUTION TO CALIFORNIA STATE BOARD OF EDUCATION

Statement signed on 10 December 1972 by 19 Nobel Laureates from California

To the California State Board of Education:

We are appalled that you are considering a requirement that science textbooks include creation along with evolution as an explanation of the origin of life. Creation "theory" is NOT based on science and does NOT belong in a science textbook.

We urge you to support the recommendation of your own State Advisory Committee on Science Education in rejecting this unreasonable requirement for our science textbooks.

Respectfully yours,

Luis W. Alvarez
Nobel Laureate 1968, Physics, University of California

Carl David Anderson
Nobel Laureate 1936, Physics, California Institute of Technology

Felix Bloch
Nobel Laureate 1952, Physics, Stanford University

Melvin Calvin
Nobel Laureate 1961, Chemistry, University of California, Berkeley

Donald A. Glaser
Nobel Laureate 1960, Physics, University of California, Berkeley

Robert Hofstadter
Nobel Laureate 1961, Physics, Stanford University

Robert W. Holley
Nobel Laureate 1968, Medicine, Salk Institute

Arthur Kornberg
Nobel Laureate 1959, Medicine, Stanford University

Joshua Lederberg
Nobel Laureate 1958, Medicine, Stanford University

Willard F. Libby
Nobel Laureate 1960, Chemistry, University of California at Los Angeles

Max Delbruck
Nobel Laureate 1969, Medicine, California Institute of Technology

Richard P. Feynman
Nobel Laureate 1965, Physics, California Institute of Technology

Murray Gell-Mann
Nobel Laureate 1969, Physics, California Institute of Technology

William F. Giauque
Nobel Laureate 1949, Chemistry, University of California, Berkeley

Edwin M. McMillan
Nobel Laureate 1951, Chemistry, University of California, Berkeley

Linus Pauling
Nobel Laureate 1954, Chemistry, 1962, Peace Stanford University

Glenn T. Seaborg
Nobel Laureate 1951, Chemistry, University of California, Berkeley

William Shockley
Nobel Laureate 1956, Physics, Stanford University

Harold C. Urey
Nobel Laureate 1934, Chemistry, University of California at San Diego

EXHIBIT 54: INTRODUCTION TO EDITING FOR ORIGINS OF THE UNIVERSE, LIFE, AND MAN IN SCIENCE TEXTBOOKS

Background statement prepared at the direction of the California State Board of Education by Vernon L. Grose for an editing committee appointed by the State Board — 16 December 1972

Unfortunately, there are very few clear and isolated pages in the textbooks where the subject of origins are uniquely treated and which could be easily excised, edited or replaced. Instead, the publisher's organizational profile for teaching science is predominantly based on only one philosophical predisposition or world-view — namely that, starting with one unexplained chance event (the Big Bang explosion), everything has progressed in some sort of continuum running from simple to complex over eons of time.

With this philosophic framework underlying most of the material that attempts to interrelate natural phenomena, one can only select various points in the textbooks where it is obvious that observable and testable data are *not* available and point out the belief systems that are employed to bridge across these voids in the scientific record.

There are *three* origins of interest — the universe, life, and man. It may not be immediately obvious, but these three origins actually require three approaches for editing.

The Origin of the Universe — The origin of the universe, by definition, has no precedent action or material which could provide clues as to its occurrence. Therefore, from a scientific viewpoint, all discussion about it must be speculative. Dogmatic statements are clearly inappropriate.

The Origin of Life — Even the *definition* of life is not universally resolved (e.g., whether viruses have life is highly debatable), so the *origin* of life appears in the philosophic continuum at some indefinite point. The statements in the textbooks concerning life's origin generally are categoric; e.g., "Life began in the sea." Since there is no physical evidence of that point where non-living matter became living in the historical record; i.e., fossils, the justification for converting speculation into dogmatic language lies in biochemical extrapolation wherein the chemical constituents required for life on earth today are isolated and delineated and then an encompassing explanation is offered as to how these chemicals might have been simultaneously present for the introduction of some outside source of energy (often postulated as sunlight or a lightning strike) which could transform these *non-living* chemical elements into an undefined *living* substance (often described as a single, simple molecule). By keeping this initial conversion from non-living to living status restricted to a very simple structure (even a single molecule), it is apparently thought that the credibility for this chance occurrence is enhanced. However, many serious questions are raised at this time — Did this unique event occur only once? If not, what precludes the event from being repeated many times, even today (in order to comply with the scientific requirement of being testable)? How many molecules were involved? What was the exact composition of that molecule? Beyond another chance happening, what caused this molecule to divide or reproduce? How did molecules become cells? At what point in the continuum did the basic laws of physics; e.g., the Second Law of Thermodynamics, become established (or were they operative on that eventful day when life arose)?

The Origin of Man — Man's origin is more obscure than the origin of life even though life preceded man. None of the textbooks discuss anything approaching "man's first moment of existence." The philosophic predisposition mentioned earlier demands that man be a part of the simple-to-complex continuum. At what point (how, why, or when) in this continuum man suddenly shifted from other primates and established those characteristics that today clearly set him apart from the rest of the animal kingdom is not described in the textbooks. In place of discussing this interesting but troublesome problem, the student is simply assured that the continuum of living things exists and that the "glue" that holds the entire continuum together is the phenomenon known as *adaptation.* Therefore, the origin of man is undeniably linked in the textbooks to the concept of adaptation.

A major portion of the editing regarding the origin of man should be addressed to the use of the concept of adaptation. Two discussions of this concept are included — one describing its use in the textbooks and the second describing professional biologists' treatment of it.

To illustrate one approach to editing regarding "adaptation," the following approach was favorably received by the seven major publishers during preliminary screening by the Curriculum Development Commission:

1. NEUTRAL TREATMENT - Reword typical statements like, "Fish have become adapted to living in water" to such alternatives as:

Fish are fitted for living in water.

Fish are suited to living in water.

Fish have features which permit them to live in water.

Fish have characteristics which allow them to live in water.

Reasons for this treatment:

a. These are true statements.

b. They replace a "belief system" with an observation that any school child can perform.

c. They avoid the possibility of either theism or anti-theism.

2. BALANCED TREATMENT — Reword typical statements like, "Fish have become adapted to living in water" to "Fish were designed for or have become adapted to living in water."

The Use of ADAPTATION in Science Textbooks

Some of the most perplexing yet frequent terms in evolutionary thought used by the publishers of science textbooks offered for the California adoption are the terms "adapt," "adapted," "adaptation," and "adaptations." The variety of meaning implied by the use of these terms is so broad that a collective definition seems virtually impossible.

Most of the time, the use of these terms resembles the use of "k" factors in mathematics; i.e., a means for bridging over missing, difficult, or misunderstood data in order to arrive at a preconceived answer.

Even worse, the use of these terms frequently appears to be consonant with a theory now generally rejected by biologists — Lamarckism!

There are at least 24 unique definitions (with 17 separate meanings!) offered in the textbooks reviewed:

1. *GETTING ALONG*- "To get along with things as they are" Harper & Row, Grade 3, page 239 (student edition).

2. *CHANGE* - "To change to suit the condition of the surroundings" Harper & Row, Grade 5, page 373 (student edition).

"A slight change in a plant of animal that makes it better able to live in its environment" American Book, Grade 3, page 193.

"Animals had to adapt to the changing climate, the changing terrain, and most probably to the change in food supply" Laidlaw, Grade 5, page T86 (teacher edition).

3. *ADJUSTMENT* - "To adjust to conditions" Charles Merrill, Principles of Science - Book 1, page 468 (teacher edition).

4. *FITTING* — "The fitting of oneself into an environment" Harper & Row, Grade 4, page 298 (student edition).

"Any living thing is somehow fit to live where and how it actually does live" Life Nature Library - Ecology, page 122.

"Fitted to carry on life activities in a particular environment" Harcourt Brace Jovanovich, Grade 5, page 356 (student edition).

"The fact that animals and plants are fitted to their environment" (implied) Holt Rinehart & Winston, Grade 5, page 16 (teacher edition).

5. *BEING SUITED* — "To be suited to" Harcourt Brace Jovanovich, Grade 3, page T352

(teacher edition).

6. *STRUCTURE* — "A structure or behavior that helps an organism get along in its environment" American Book, Grade 6, page 337 (Teacher edition).

7. *BEHAVIOR* — "There are generally three kinds of adaptations — structural, physiological, and behavioral" Life Nature Library - Ecology, page 123.

"A structure or behavior that enables a plant or animal to live successfully in its environment" Harcourt Brace Jovanovich, Life and Its Forms, page 539.

8. *SPECIAL PART* — "Special parts of animals or plants that usually do a special job in a special way" American Book, Grade 3, page 182 (student edition).

9. *METHOD OF ORGANIZATION* — "The way a plant or animal is organized in its structure, physiology and behavior to carry out goals such as securing food and energy it needs to survive or the ability to reproduce" Life Nature Library — Ecology, page 130.

"A living things' organ systems are organized for life activities in the environment" Harcourt Brace Jovanovich, Grade 5, page 297 (student edition).

10. *SPECIAL WAYS* — "Special ways of staying alive in an environment" Charles Merrill, Book 5, page 310 (teacher edition).

11. *TESTING PROCESS* — "A constant process of testing and rejecting — with the environment acting as the testing agent on the organisms which are evolving" Life Nature Library - Ecology, page 131.

12. *ADAPTABILITY* — "Neither the most astonishing structure nor the most rigidly patterned behavior but rather adaptability itself —to light or darkness, treetop or burrow — and the capacity to alter ways of life as the environment is altered, by nature or the drastic acts of men" Life Nature Library - Ecology, page 131.

13. *FEATURE* — "Any feature that allows the organism to exist under the conditions imposed by its habitat — to make full use of the nutrients and energy, the heat and light available to the community, to obtain protection against enemies and the variations of the climate" Life Nature Library — Ecology, page 122.

14. *TRAIT* — "A structure or trait that helps an organism survive" American Book, Grade 6, page 12 (student edition).

15. *PHYSIOLOGICAL CHARACTERISTICS* — "Physiological characteristics like body shape, skeleton, skin texture, etc." (implied) Silver Burdett, Grade 5, page 272-8 (teacher edition).

16. *PROCESS* — "The process in which the structure or function of an organism changes as a result of natural selection so that it can better survive and multiply in its environment" Harper & Row, Grade 4, page 308, (student edition).

17. *BECOMING SUITED* — "To become suitable for a condition or a place. Cactus plants adapt to the hot, dry climate of the desert by developing ways of reducing their need for water" Harper & Row, Grade 6, page 402 (student edition).

These 17 separate meanings of adaptation can probably be summarized (at some risk of oversimplification) into two general *uses* of the concept in the science textbooks:

1. The inherent *features* or *characteristics* (physiological or behavioral) of living things which permit survival in a changing environment.

2. An *intelligent-like activity* of individuals within living groups to seek out and acquire abilities to survive or achieve higher levels of capability.

The first use has widespread support in science. However, there is not universal agreement (see Simpson's descriptions of four evolutionary schools of thought on adaptation).

The second use is an objectionable one for several reasons: (a) it violates most genetic theory concerning transmissibility of acquired abilities to future generations, (b) it endorses Lamarckian thinking generally repudiated by biologists today, and (c) it serves as a crutch for leaping boundaries between groups of living things; e.g., reptiles deciding to fly or fish deciding to walk on land.

Ultimately, the second use of adaptation becomes the *mechanism* for the origin of man.

The Use of ADAPTATION in Scientific Literature

The confusion about the concept of adaptation which resulted in the wide variety of definitions appearing in the science textbooks submitted for adoption in California public schools is not necessarily removed by studying scientific literature on that subject. In contrast, to such scientific terms as mass, vaporization, gravitation, or convection, definitions are extremely scarce for "adaptation." The concept is *discussed* frequently but seldom *defined.*

One recent text proposes three definitions which, collectively, allow the reader almost enough latitude to cover the 24 definitions selected from the submitted textbooks:[1]

> Adaptation is the production and maintenance, in a population, of characteristics that fit it for a particular way of life (whether changing or not). Adaptation is brought about by natural selection of heritable variations, but this has two forms: selection directs adaptive evolution, and it prunes out maladaptive variation. The word "adaptation" is also used for the results of the process, whether structural, physiological, behavioral, or any combination of these.
>
> A third meaning of adaptation is the changes made by an individual during its life, by which it adjusts to varying conditions internally or externally.

This definition would allow that adaptation is: (1) a productive activity, (2) a maintaining activity, (3) simultaneously an individual and. a population phenomenon, (4) a set of changes, (5) structural characteristics, and (6) simultaneously cause and effect. It might almost be easier to define what adaptation is *not!*

George Gaylord Simpson summarizes four schools of thought regarding adaptation as follows:[2]

Neo-Lamarckian - Adaptation is pervasive in nature and essentially purposeful in aspect, as if the environment had forced and the organism had sought adaptation.

Neo-Darwinian - The more or less adaptive status of variations is influential in determining the parentage of a following generation.

Geneticists - New hereditary variants arise abruptly, and as far as we know and as far as adaptive status is concerned, at random.

Anti-Mechanists - Adaptation is usually an essentially directional progressive, sustained, and nonrandom process.

Note that these four positions emphasize aspects of adaptation that are in direct conflict with one another. Simpson does attempt to reconcile these differences into a "synthetic theory" but says that "it would manifestly be impossible to present an adequate statement of the synthetic theory in a simple or brief way."[3]

The more one searches for a universal definition or even a universal concept of adaptation, the more it appears to elude the searcher. Norman Macbeth devotes an entire chapter of his latest book to the ambiguity of adaptation and reaches the following conclusion:[4]

> When the most learned evolutionists can give neither the how nor the why, the marvels seem to show that adaptation is inexplicable. Yet those who cannot explain it will not admit that it is inexplicable. This is a strange situation, only partly ascribable to the rather unscientific conviction that evidence will be found in the future. It is due to a psychological quirk that Simpson ascribes with admirable self-knowledge: 'For some, adaptation was merely an inexplicable fact; these students were few, because scientists rarely are psychologically capable of accepting a phenomenon as a fact and also accepting it as inexplicable.' This observation may be correct, but it is not to the credit

---

[1] *Eaton, Theodore H ,* ***Evolution****, W. W. Norton & Company (New York) 1970, page 80.*

[2] *Simpson, George Gaylord, "The Problem of Plan and Purpose in Nature, "* ***Human Evolution****, Henry Hold and Company, Inc., (New York) 1959, p. 98*

[3] *Ibid, p. 99.*

[4] *Macbeth, Norman,* ***Darwin Retried****, Gambit Incorporated (Boston) 1971, p. 77.*

of the profession. Scientists are expected to rise above such frailties.

Some basic questions arise from reading how the term is used in the scientific literature:

1. Is adaptation an "externally-forced" phenomenon or an "internally-inherited" one or both?

2. Does not "survival of the fittest" imply that the genetic code for survival was already set at birth and therefore neither a *willful* nor *accidental* acquisition of something required to survive?

One cannot escape detecting the underlying motivation that is evident in much of the literature to use "adaptation" as the means of avoiding supernatural explanations. As Simpson says, "In science, one should never accept a metaphysical explanation if a physical explanation is possible or, indeed, *conceivable*."[1] (italics added) Conceivability, of course, has no correlation with feasibility, probability, or testability.

Very few, if any, authors are as open as Simpson in admitting the anti-theistic intent of adaptation:[2]

"Adaptation by natural selection as a creative process ... is real, and it is achieved by a progressive and directed process. This process is natural, and it is wholly mechanistic in its operation. This natural process achieves the aspect of purpose, *without the intervention of a purposer*, and it has produced a vast plan, *without the concurrent action of a planner*."

In summary, the following points concerning "adaptation" should be kept in mind while editing the science textbooks:

1. Adaptation is a naturalistic mechanism postulated to explain how the preconceived philosophy of a simple-to-complex continuum might have happened.

2. Adaptation is scientific naturalism's equivalent explanation to a theist's explanation that a Supreme Intelligence might have intervened in nature.

3. Adaptation ultimately becomes the mechanism for the origin of man in evolutionary thought.

EXHIBIT 55: KNXT EDITORIAL

The following editorial was broadcast several times on 30 and 31 December 1972 by William C. O'Donnell, Vice President and General Manager of KNXT (Channel 2), CBS Owned Television in Los Angeles

Subject: Teaching Creation in the Schools

The creation vs. evolution battle has been decided in Sacramento, and it looks like God lost again.

The State Board of Education decided not to include the theory of creation in new science textbooks. Instead, our kids will continue to get nothing in science class about the origin of life except a story about a fish climbing a tree and then beginning to walk.

We know a lot of people prefer to think that the fantastic order of the universe happened by chance. Life, they think, is a big computer.

But others would suggest that a giant computer, with its memory banks and incredible number of circuits could never fall into place by accident. And they add that even if that could happen, a computer is a kid's toy compared to the complexities of the universe.

Now it's true that the Board did hedge a bit on evolution. They decided it should be taught as a theory, not as a matter of absolute, proven fact. On that basis, they ought to be willing to give the creation theory a mention, too. There's no reason why one has to exclude the other. Until somebody can pin it all down and say precisely how the universe was constructed, there should be a role for the creation theory in a science textbook, right along with evolution.

---

[1] *Simpson, op. cit., p. 94.*

[2] *Simpson, op. cit. p. 104.*

EXHIBIT 56: ON THE SUBJECT OF ORIGINS

An introductory statement prepared for Teacher Editions of California science textbooks by Vernon L. Grose 18 January 1973

Since scientific knowledge is ultimately judged by experimental facts, a cornerstone of science is the repeatability or testability of any hypothesis, theory or law. Science is more than facts — it is the interrelationship of facts. And these *interrelationships* must be *testable.*

As the *Science Framework* states:[1]

> Science is not a mere register of facts; and indeed our minds do not (like a cash register) tabulate a series of facts in a natural sequence one after the other. Our minds connect one fact with another — they seek for *order* and *relationship* — and in this way they arrange the facts so that they are *linked by some inner law* into a coherent network ... The facts are observable, but *their organization is not.* Man invents an organization which *seems to fit the facts* he observes." (Italics added)

One of the subjects that interests scientists is that of origins. In particular, the origin of the universe, life, and man have intrigued scientists for centuries. Since actual "facts" concerning these origins (i.e., detailed phenomena noted by on-the-scene observers) are missing and since origins, by definition, can occur only once and thereby are not repeatable or testable, scientists must resort to *inference* (rather than *observation*) in order to conclude how, when, or why these origins occurred.

Since it would be virtually impossible to discuss all the various models (or interrelationships) of the few facts that may bear on these origins, this textbook presents only one such model — the general theory of evolution. This is a popular view today, but by no means is it a *final* view. In addition, it does not rule out the possibility of other models being just as valid.

To illustrate, there are no facts known to scientists today that would confirm or deny that origins involved the intervention of a Creator. Further, whether inherent "order and relationship" of nature is simply a property of matter or the product of a divine intelligence cannot be resolved by scientists.

An important attitude to retain then, when teaching science, is that the student must be urged continually to recognize the *limitations* of scientific modes of inquiry. In view of these limitations, the science student should be encouraged to augment his scientific studies with additional, quite different approaches such as philosophy or religion when searching for answers to questions like the origin of the universe, life, and man.

EXHIBIT 57: STATEMENT BY THE SOCIETY OF VERTEBRATE PALEONTOLOGY

This open letter, dated 23 February 1973, was written by Theodore Downs, President of the Society of Vertebrate Paleontology on behalf of its Executive Committee

To Whom It May Concern:

On behalf of students of the science of vertebrate paleontology (the study of ancient vertebrate life, in the fossil record), I should like to state unequivocally that there is a record, derived from the sediments of the earth's crust, that reveals a great diversity and evolution of vertebrate life. This record preserves predominantly hard parts, and it reveals changes in the form of skeletal elements of the vertebrate body through time. Living vertebrate species are

---

[1] *Science Framework for California Public Schools, California State Department of Education (1970), Sacramento: 8.*

known to transmit characters from one generation to another (including man). The same mechanism had to be operative in once living vertebrate organisms through millions of years of time.

The fossil record of vertebrate animals includes specimens transitional between each of the major classes, for example: fish to amphibian (ichthyostegids); amphibian to reptile (gephyrostegids); reptile to bird (Archaeopteryx); and reptile to mammal (Probainognathus). Within the Class Mammalia, there are relatively abundant fossil records to show evolution of groups within the families of Equidae, Rhinocerotidae, Camelidae, Felidae, Mastodontidae, to name a few. The records of many of these families can be traced back to a certain point, beyond which the earlier fossils could be assigned with equal probability to the ancestry of two or more groups and are reasonably regarded as common ancestors of both. There are thousands of fossil records that may be interpreted as transitional at either the species, generic, family, order, or class level of classification.

Because of the many processes active during the earth's history, *all* the life of the past was not preserved. Through the natural, but rare processes of burial, fossilization and later exposure to the searching paleontologist, the known fossils are a relatively meager record of life compared to the thousands of species that lived in the past but were not preserved. In spite of this, new fossil records are continuously being found and will be in the future. The available record does show (for example, refer to the comprehensive textbook record by A. S. Romer, "Vertebrate Paleontology, " 1966) that each fossil reveals some relationship, perhaps similarity and at times differences, as compared with another earlier or later relative. Stratigraphic evidence as well as radiometric dating tell us whether one form lived later (or earlier) than another. The combined evidence of comparative morphology and stratigraphy (or chronologic sequence) demonstrates the change of vertebrate life throughout the known occurrence of the fossil record, and this record does not contradict the concept of evolution.

It is therefore not a true statement to declare: "Some of the scientific data (e. g. the regular absence of transitional forms) may best be explained by a creation theory ..." as quoted in the "Science framework" guidelines, p. 106, for California Public Schools science textbooks, and as prepared by the State Board of Education.

EXHIBIT 58: QUALIFYING STATEMENT ON EVOLUTION FOR SCIENCE TEXTBOOKS

The California State Board of Education adopted the following introductory statement on 8 March 1973 and ordered that all textbooks containing extensive discussion of evolution must be edited to include the statement.

The subject of origins – how things began long ago – has always been fascinating. Certain questions about how things began science cannot answer. Where the first matter and energy came from is such a question, because it cannot be treated by the accepted methods of science. However, other questions of origins are appropriate for scientific investigation. For example, what are the physical mechanisms involved in the origin of life or the origin of specific living creatures? Considerations extending beyond a natural description of the physical universe, even as to whether any supernatural reality exists, are "non-scientific," i.e., they lie beyond the reach of science and belong to other disciplines such as philosophy and religion. That such considerations are "non-scientific" does not mean that they are untrue or unimportant, but only that they cannot be evaluated by the scientific method.

The term "evolution" may be used in a number of ways. One use of the term describes processes that can be observed at present. These processes can be described with great accuracy. Another use of the term "evolution" refers to the theory that includes the hypotheses that (1) all life forms now living have come from a much smaller number of life forms in the past; this may have been just one or a few original sources of life; and (2) the great variety now in existence has developed by slow changes over long periods of time in response to hereditary and environmental factors. This theory, commonly called the "theory of organic evolution," aims to tie together all living creatures and to explain similarities between living creatures in terms of slow change from one form to another form better suited to survive in the local environment. The accuracy of this theory, like that of all scientific theories, depends largely upon the validity of the observations and assumptions on which it is based.

Most scientists agree that the theory of organic evolution is the best scientific description we have to account for the complex forms of life in the past and present. The historical reconstructions of life in the past described in this book are presented in terms of this theory of organic evolution.

EXHIBIT 59: APPROVED DEFINITIONS OF "ADAPT"

The following definitions of the various forms of the word "adapt" were approved by the California State Board of Education on 8 March 1973 and referred to all science textbook publishers for rewriting in language appropriate for the grade levels of the books in which they appear. (Compare these definitions with the existing ones listed in Exhibit 54)

adapt, verb as in "to adapt"	To change structure, physiology or behavior in such a way as to increase the probability of survival and reproduction in a specific environment
adapted, adjective as in "to be adapted"	To have a structure, physiology or behavior which increases the probability of survival and reproduction in a changing environment
adaptation, noun 1. As in "adaptation occurred"	a. That process by which physiological changes occur in a given population in response to the conditions of a changing environment
	b. That process by which an evolving sequence of living things undergoes evolutionary (genetic) change in such a way as to increase the probability of survival and reproduction in a changing environment
2. As in "to have an adaptation"	That structure, physiology or behavior which enables an organism to survive in a changing environment

EXHIBIT 60: CHEMIST'S OBJECTION TO THE RESOLUTION BY THE AAAS COMMITTEE ON SCIENCE EDUCATION (EXHIBIT 40)

This letter by a Professor of Chemistry at Valparaiso University was printed in the June 1973 issue of AAAS Bulletin

The Editor
AAAS Bulletin

Dear Sir;

The enclosed clipping (Resolution on creation by AAAS Commission on Education – Ed.) from the *AAAS Bulletin* (Feb. 1973) is hardly a scientific statement. How about requesting the Commission on Science Education to reexamine its position on the origin of life and come up with a statement regarding the so-called controversy between the evolution and creation account that would be worthy of a group that labels itself scientific. (I am presuming that objectivity is a fundamental characteristic of scientists.)

Thank you.
T. C. Schwan

EXHIBIT 61: NUTRITIONIST'S OBJECTION TO THE RESOLUTION BY THE AAAS COMMITTEE ON SCIENCE EDUCATION (EXHIBIT 40)

This letter by the Chairman of the Department of Animal and Poultry Science at the University of Guelph (Ontario) was printed in the June 1973 issue of AAAS Bulletin

The Editor
AAAS Bulletin

Dear Sir:

I object to the resolution passed by the Commission on Science Education as published in the *AAAS Bulletin* (Feb. 1973). Those involved in passing this resolution are hoisted in their own petard when they state that a scientific statement must be capable of test by observation and experiment and then proceed to indicate such procedures are possible for the theory of evolution.

There is indeed a theory which deals with the origin and evolution of life just as there is a theory which deals with the creation of life.

Neither of these can be proved by observation or experiment and ultimately both are accepted by faith. One has as much right in a textbook as the other. If theories on origins are to be discussed there must be room for all.

Sincerely yours,

W. D. Morrison

EXHIBIT 62: *CONFRONTATION ON CREATION* by Rob Kanigel

This article is reprinted with permission of "California Living," the magazine of the San Francisco Sunday Examiner & Chronicle, Copyright 1973, San Francisco Examiner. It appeared in the 16 September 1973 issue on page 28.

The day they debated how California school children should learn about the origins of life was a topsy-turvy day, filled with contradictions. It was a day which showed that the Scientific Establishment, big and powerful though it was, could still be checked — a relief for anyone fearful of Big Science, or at least the part of it that seemed possessed of Truth.

The debate — and its aftermath, the decision to treat evolution as theory rather than fact in science textbooks used in state schools — was a setback for scientists and friends of Science, not accustomed in recent years to having their faith questioned.

The debate, held in Sacramento last November, was part of a State Board of Education hearing on the science textbooks a million California schoolchildren will be using this fall, Rooted in a controversy going back ten years, the debate had grown out of the State's adoption in 1969 of a science framework (an outline for textbook publishers to follow in preparing science texts for State approval) which implied a change in the presentation of theories of the origins of life. The Board of Education, going over the head of its State Advisory Committee on Science Education, had added to the framework two key paragraphs:

> "All scientific evidence to date concerning the origin of life implies at least a dualism or the necessity to use several theories to fully explain relationships between established data points ...
>
> "While the Bible and other philosophic treatises also mention creation, science has independently postulated the various theories of creation. Therefore, creation in scientific terms is not a religious or philosophic belief."

In short, elementary science textbooks could not present evolution as positive, proven fact in accounting for the origins of life. Moreover, argued creationists who had urged the change, the "theory of special creation" (its status as a bona fide scientific theory would itself be contested during the hearing) was not, after all, so wildly out of bounds in the schools and might actually warrant equal stature with evolution.

Of the science texts previously recommended by the State's Curriculum Commission, none had included the "creationist" viewpoint while many had included evolution — some, as accepted scientific fact.

So today, the Board — which could overrule the Curriculum Commission — was hearing from the public before making its final textbook selections. God and Man and Science fighting it out together in a public auditorium before an audience of five hundred, plus a dozen TV cameras and thirty reporters.

Fifty speakers in all were slated to appear in the afternoon. In the morning the opposing arguments were summarized in open statements by two members of the Curriculum Commission. First was Dr. Junji Kumamoto of the Department of Plant Science at the University of California at Riverside.

"The issue of creation by Divine design lies in the domain of theology or philosophy and is inappropriate for inclusion in science," Dr. Kumamoto, a small, intense-looking man, began. "Science is limited by universal principles," he went on. "There can be no exception to (Science's 'laws' of conservation) and any notion to the contrary cannot be part of science."

Employing such language as "universal principles," "no exceptions," "cannot be part of science," "scientific impossibility," and "no such alternative theory," his statement seemed to admit no doubt. Fundamentalist ministers like those the public linked to the creationists were known for the absolute certainty with which they took the Gospel as literal truth; now, it seemed, Dr. Kumamoto was equally sure of Science's truth.

Most of Kumamoto's statement was devoted not to his argument per se but to the marshalling of "competent scientific authority" against the creationists. Statements had been issued, said Kumamoto, by the President of the National Association of Biology Teachers, American Association of Physics Teachers, the Committee on Physics in Pre-College Education, the Council of the Academic Senate of the University of California. There'd been a resolution by the Board of Directors of the American Association for the Advancement of Science and by the American Chemical Society. Even, Kumamoto reported, one by the National Academy of Sciences — breaking a hundred-year tradition of staying out of textbook controversies.

It was an impressive display of power and prestige the evolutionists had assembled; could the creationists possibly stand up to it? Earlier in the century, Before Science, in a social climate friendlier to Faith, perhaps. But this was 1972 and Man had sent men to the Moon. What place, one wondered, in an age of lasers and transistors, Astro Turf and artificial life, for the Creator and His work?

The site of the hearing itself spoke for Science, its methods, its tools, its accomplishments. It was in one of those bright, new office structures — the State Resources Building — where much of the technological work of the State was done. Eight floors of fluorescent lighting, high-speed elevators and computer-efficient layout all spelling out the twentieth century. Inside the hearing room was the Press, relying on the hardware of science to report on the Science State: Television crews with their cameras and bright lights, wires dangling from control boxes, microphones. Even some of the print journalists — traditionally hold-outs against Progress — sported Sony tape recorders.

In the lobby was further testament to Science and its hand-maiden, Technology. The Department of Water Resources had filled a row of display cases with big, full-color diagrams of a salt water conversion plant. One wall sported a bright, handsome canvas entitled *Penstock Intake Structure, Oroville Power Plant.* Off in a corner was a display of geological samples with one display card reading: "ANDESITE: Fresh, massive. U.C.S.=28,100 psi, E=5,050,000 psi. u=0.17. Abbey Bridge Dam Site, Hole 12R, depth = 31.1 feet." Hole 12R, depth = 31.1 feet ... what could any priest of God say that could stand up to the perfect precision of these priests of Science?

After Dr. Kumamoto had finished, Vernon L. Grose made his way to the podium. Hailing from Tustin Institute of Technology in Santa Barbara and wearing his blond hair in a crewcut, Grose was starting out with two strikes against him. But as he began, in tones of quiet confidence, to address the Board, it soon became clear that this was no wild-eyed religious fanatic. No, far from launching into an uncompromising, Bible-quoting harangue, Grose, his tall frame draped in a smartly tailored navy blue suit, led off his statement with a series of admissions and concessions.

He, too, he confessed, had once "been inadvertently guilty of erecting a 'straw man'" and thus muddling an already difficult issue. Yes, he agreed, he could subscribe to the statement of the National Academy of Sciences and the other prestigious scientific societies that religion — any form of religion, he would later point out — had no place in science. And yes, he stood by his statement of three years before that at the Scopes Trial he would have supported the legal position of atheist Clarence Darrow against fundamentalist William Jennings Bryan.

But now was the time, he went on, to stop polarizing the issue and to begin anew with "principles that all scientists accept." The first of these was that "science is limited in its scope" to what can be observed through the five senses. "Since reality also includes moral values, beauty, justice, love," he noted, this was "quite a severe limitation." Also properly outside the realm of science, he went on, was ultimate cause, "more logically the concern of metaphysics, philosophy and theology than science."

But, he reminded the Board, exactly "where scientists pass from observation to speculation and enter one of these three fields is certainly a grey area."

Then Grose pointed out that "religion need not be theistic." Scientific naturalism, for example — the system of beliefs that rejects the supernatural in Nature and thus paves the way for Darwin's Natural Selection — was also a religion. Thus, Grose concluded, "the antitheism of scientific naturalism (can) not be taught to the exclusion of other explanations" for ultimate cause lest science, by intruding on territory not its own, "no longer remain truly scientific."

Grose was at once contesting Science's sometime claim to a monopoly on truth and insisting it remain true to itself by staying neutral in non-scientific areas. Later he would observe that "a person, even a scientist, is not a set of compartments ... Each of us is a totality — simultaneously scientific, political, spiritual, philosophic, sexual, religious and social" — a position more in line with Alan Watts, Eastern mysticism and humanistic psychology than any form of backwoods fundamentalism.

After Grose's statement came a lunch break, but before Grose could make it out of the auditorium and up to the eighth floor cafeteria, he was waylaid by Dr. Richard Lemmon, a chemist from UC Berkeley's Lawrence Radiation Laboratory. Lemmon, a tall, thin man with furrowed brow and black, bushy eyebrows, intimated that scientifically Grose didn't know what he was talking about, that his scientific credentials were no doubt second-rate, and that perhaps he was pushing so hard for the creation model because he had his own "pipeline to heaven".

But Grose was used to having his credentials impugned, it seemed. He took the sarcasm good naturedly. He quietly suggested that Lemmon hadn't quite understood the gist of his argument, that he would be glad to go over it again and ...

But Lemmon — who would later warn the Board that it risked loss of its "national, even world" leadership were it to entertain the creationist view — had had his say and headed off for lunch.

Grose's casual encounter with Lemmon as well as his more formal one earlier with Dr. Kumamoto, while presumably only preludes to the afternoon events, reflected well the tone of the day as a whole, suggested what the creationists were up against. What lay between them and the minds of California schoolchildren, of course, was nothing less than the nation's Scientific Establishment — graduates of the most prestigious universities, holders of the best credentials, the kind of people author David Halberstam has called (in a wholly different context) "the best and the brightest." And so supremely confident of the validity of their views did these men and women seem to be, that the kind of emotional appeals they were loathe to forgive in others became today, in a number of instances, acceptable.

For instance, one Sacramento surgeon, Dr. John E. Summers, suggested that the creationists were "demanding that ancient Bible stories be included in our science textbooks." Not three hours earlier, Dr. Grose had stated explicitly that in his talks with publishers, "no mention of a Designer or God or Supreme Being was requested or even suggested ... (and) at no point were the terms Genesis, creation or Bible requested or suggested."

Dr. David S. Hogness, a biochemist at Stanford University Medical Center, went so far as to ask the Board: "Who among you would advocate today that we teach the flat earth theory in our schools today? To insist that the biblical viewpoint of the creation of man be taught in science classes introduces this same order of fantasy."

With the weight of scientific opinion in the other camp, the creationists seemed consumed with appearing at least as thorough and rigidly scientific as their opponents.

Mrs. Nell Segraves of the Creation Science Research Center, who started it all in 1963 by questioning the constitutionality of teaching the absence of God in the schools, had prepared a systematic analysis, complete with legal citations and excerpts from relevant laws.

Dr. Ronald S. Remmel of the Physiology and Anatomy Department at Berkeley used his experiments with disintegrating k-meson particles to argue, not that God necessarily existed, but that the idea of a creator was compatible with the results of his and other experiments.

Dr. Richard L. Ferm, a Chevron Research Company chemist, based his "case for design" on the Second Law of Thermodynamics' insistence that nature is "running down" — not spontaneously ordering itself.

Like others with unpopular views, the creationists appealed frequently to the civil libertarian principle of academic freedom — a plea more associated with left wing radicals or war protestors, say, than God-fearing Christians. "Good science is an open-ended search for truth," noted Dr. Ariel Roth, professor of biology at Loma Linda University. "To exclude the creation model because it is associated with religion is to say that science arbitrarily excludes certain hypotheses and thus is not an open-ended search for truth."

Said Dr. Leonard R. Brand, also of Loma Linda University: "We are not asking for evolution to be left out of the textbooks ... and hope that the rest of the scientific community will demonstrate equal open-mindedness."

The hearings had been billed as a debate on the origins of life. But to anyone schooled in the mainstream of Western scientific thinking, it must have seemed like an assault on Progress itself. Was it possible that fifty years after the Scopes Trial had established evolution as sure and certain scientific fact, the State's schoolchildren were still threatened by fundamentalism's ravings? And more, could the State possibly consent, in the seventh decade of the twentieth century, to placing on an equal footing with Darwin the notion that the world had come into being through Divine Will?

Well, the State didn't do that. But, as it turned out, the Curriculum Commission did finally settle on a compromise (approved by the Board a month later) which, while not mandating the teaching of creationism, at least demanded that evolution not be taught as fact. And that alone shocked many and elicited a flood of angry letters to the editor, newspaper cartoons and editorials lambasting the Board's decision over the following weeks.

I went to Sacramento with at least a tinge of bias against the Southern California fundamentalists and their Old Time Religion. And yet I had come, too, with little faith in Science. I hated its certainty, abhorred its arrogant monopoly of Truth, rejected the notion that its predictive powers, its say-so over Nature, were as supreme as some of its practitioners claimed. And so I had found a part of me rooting for the fundamentalists and their God.

As an agnostic, I certainly held no brief for religion, distrusted its certainty, its sureness. But just as I hated to see Man made weak and little for the glory of God, so, too, I feared even more to see him diminished by Science and its iron laws, its strait jacket constraints, its denial of myth and magic. And so I left Sacramento well pleased — pleased, this once, to see Science humbled.

EXHIBIT 63: BOOK REVIEW - "SYMPOSIUM ON THE ORIGIN OF THE SOLAR SYSTEM"

This book review appeared in Science, *Volume 183, Number 4124, page 504, 8 February 1974. Copyright 1974 by the American Association for the Advancement of Science*

L'Origine du Systeme Solaire. Symposium on the Origin of the Solar System. Nice, France, April 1972. Edition du Centre National de la Recherche Scientifique, Paris, 1972. x, 384 pp. , illus $8.

It is impossible to contemplate the spectacle of the starry universe without wondering how it was formed; perhaps we ought to wait to look for a solution until we have patiently assembled the elements, and until we have thereby acquired some hope of finding a solution; but if we were so reasonable, if we were curious without impatience, it is probable that we would never have created Science and that we would always have been content with a trivial existence. — Poincare (1913)

The scientific investigation of the origin of the solar system dates back to the early history of civilization. In modern times, numerous articles, reviews, and even very substantial books have been devoted to this subject. Thus the addition of yet another volume to this extensive literature makes us wonder whether it is presenting as with just another discussion of the same problem along the same old lines – or does it bring as closer to a real solution?

L'Origine du Systeme Solaire, in some sense, does both. In Section A, which constitutes the first quarter of the book, distinguished scholars review the well-known models for the origin and evolution of the solar system. Section B, comprising the bulk of this large-format (8 by 11-1/2 inches) book, treats in very considerable detail a substantial number of the important recent results in astrophysics, solar physics, meteoritics, and planetology that have direct bearing on the formulation of cosmogonical hypotheses.

This leads naturally to section C, the most interesting part of the book. Entitled Conclusions and Anticonclusions, it presents some valid criticisms of several of the leading models for the evolution of the solar nebula. Thus the present, still weakly developed state of the Genesis art is highlighted in a constructive fashion that offers guidance for future approaches to the problem.

This appears most succinctly in H. Reeve's concluding article, aptly called "Some unwritten chapters of our book." Yet to be discussed adequately is the detailed fragmentation of the massive cloud, in which protostars are born. Also in question are the hydrodynamics and the stability considerations of the protosun nebula. Most important, there remain to be specified (and carried out!) the crucial experimental tests that can distinguish between the available viable theories. It is particularly disappointing that we have almost no useful information on the specific solid state processes at work in the accretion phase.

In another contribution to the volume, editor Reeves poses seven fundamental questions:

Do the sun and planets originate in the same interstellar cloud?

If so, how was the planetary matter separated from the solar gas? How massive was the nebula?

How did the collapsing cloud cross the thermal, magnetic and angular momentum barriers?

What were the physical conditions in the nebula ?

What was the mechanism of condensation and accretion?

How did the planets, with their present properties and solar distances, form?

The contrast among the models is especially evident in the postulated accretion processes. A fundamental problem here is to keep the relative velocities of the colliding bodies sufficiently low. In Safronov's theory the motions of the small solids are determined by the gas flow. Alfven and Arrhenius suggest a mechanism that produces "jet streams" of orbiting solids with small relative motions. Cameron invokes turbulence in the nebular collapse phase, indicating an earlier origin for the accreting masses. Cold welding is crucial to several hypotheses.

New observations and experiments should help. The detection of the "interstellar" molecules methyl cyanide and hydrogen cyanide in Comet Kohoutek, which has just been reported, and the further analyses that will be based on the vast amount of data being collected on this presumed sample of primordial matter, should provide real insight into chemical and solid state investigations of the protosolar material. Pioneer 10 and 11 investigations of the massive planets (already we have found helium on Jupiter) will lead to better estimates of the composition of the photoplanetary gas in a somewhat later stage of evolution. Perhaps Mariner 10 can find on Mercury's scarred face some clues to the radial gradient (taking into account what we already know of the lunar and Martian surfaces) of large collisions at a still later stage, after the planets reached roughly their present sizes. The next book on the origin of the solar system will, it is to be hoped, integrate much of this new material. But the present volume is still an extremely useful reference and certainly is the best starting point for the reader who is intrigued to join us in this fascinating, still very speculative pursuit. At $8 it is a Best Buy.

Ichtiaque Rasool
National Aeronautics and Space Administration, Washington, D. C.

EXHIBIT 64: DISCUSSION OF ORIGINS IN CALIFORNIA TEXTBOOKS

A statement prepared for the California State Board of Education on 25 February 1974 by Vernon L. Grose

For several years, the California State Board of Education and its advisory bodies including the Curriculum Development & Supplemental Materials Commission have wrestled with the question of proper discussion of the origin of the universe, life, and man in school textbooks. The arguments have ranged back and forth regarding how, when, where, and why such discussion should take place. At times, it seems more controversial than sex education!

Two factors that should put all minds to rest on the *necessity* of discussing origins are:

1. All the science textbooks which were submitted for the recent (1973) adoption process (and undoubtedly, all *future* science textbooks) already discuss origins! So, it is not even germane to try and decide *if* they should be discussed. Many of the books even have *four-color pictures* of the origin of the universe and earth; e.g., American Book Company's *Science: Exploring Ideas* - Grade 4, pp. 206-7 or Harcourt Brace Jovanovich's *Concepts in Science* - Grade 5, pp. 26-7!

2. Anyone who has had children in the home or classroom recognizes the basic inquisitiveness that early produces the question, "Where did it all begin?" So there is no need to hide from the fact that origins do need to be discussed.

There has been thought, since the State Board of Education chose (not from scientific accuracy but from expedience of publishing science textbooks on an arbitrary schedule) in 1973 to avoid addressing origins in science textbooks and to defer discussion of origins to social science books, that the issue is one never discussed by scientists. This is a fallacy of great proportion.

To provide some measure of such a fallacy, a book review which is reprinted from a recent issue of *Science*, the official organ of the American Association for the Advancement of Science (the largest professional scientific society in the world) is attached. Please note the following points from the book review (Exhibit 63):

a. It speaks of the Symposium on the Origin of the Solar System which was held in Nice, France in April 1972. This was a *scientific* meeting, not a religious or metaphysical one. It continues a long-standing tradition of such scientific symposia.

b. Note the first two sentences: "The scientific investigation of the origin of the solar system dates back to the early history of civilization. In modern times, numerous articles, reviews, and even very substantial books have been devoted to this subject."

c. Note that there are 7 fundamental questions about the origin of the solar system which are both troublesome and unanswered at this time. (One cannot help but wonder a bit on how an artist can show a fourth grader how it happened — in four-color press yet!)

d. Note the phrase, "... valid criticism of several of the leading models for the evolution of the solar nebula." (Of course, this type of questioning is a basic part of science understood by most scientists but often misunderstood by laymen who are overawed by some who would be dogmatic and absolutist.)

e. The crowning point in the review may be the quote from the renowned author on the philosophy of science and one of the greatest mathematicians of modern time, Jules Henri Poincare, who says, "It is impossible to contemplate the spectacle of the starry universe without wondering how it was formed."

3. As the Curriculum Development Commission and the State Board of Education consider how they will direct discussion of origins in the social science context (as well as in the upcoming adoption of a revised *Science Framework*), it might be well to remember several vital points;

a. The appropriateness of discussing origins has not been, is not, and will not be decided by either the Commission or the Board. It appears to have already been determined that such discussion is appropriate in both science and social science textbooks.

b. An exact or agreed-upon version of origins has not yet been formulated in either science, philosophy or religion. Therefore, humility rather than arrogance should be the governing context in which these interesting but unresolved topics are approached. For example, artist's renditions of a loose hypothesis such as the Big Bang theory are clearly inappropriate unless the tentative nature of such pictures is heavily emphasized.

c. Avoiding discussion of a topic such as origins simply because there are unresolved theories, systems or accounts seems to be both unrealistic and short-sighted in light of the increasing candor that children expect from adults in this age.

d. The cause of science, in particular, is enhanced by pointing out the areas of scientific investigation which are yet unresolved, thereby providing young minds with mystery and frontiers of exploration available and waiting for their contribution.

EXHIBIT 65: 1974 REVISION TO SCIENCE FRAMEWORK

The following revision to paragraphs 2, 3 and 4 on page 106 of Science Framework for California Public Schools (California State Department of Education, Sacramento, 1970) was approved by the State Board of Education on 14 March 1974

Interactions between organisms and their environment produce changes in both. Changes in the environment are readily demonstrable on a short-term basis; i.e., over the period of recorded history (circa 5,000 years). These changes have been inferred from geologic evidence over a greatly entended (sic) period of time (billions of years), although the further back we go, the less certain we can be. Prehistoric processes were not observed, and replication is difficult. During the past century and a half, the earth's crust and the fossils preserved in it have been studied intensively by scientists. Fossil evidence shows that organisms populating the earth have not always been structurally the same. The differences are consistent with the theory that anatomical changes have taken place through time. The process of change through time is termed evolution. The Darwinian theory of organic evolution postulates a genetic basis for the biological development of complex forms of life in the past and present and the changes noted through time.

The concepts that are the basic foundation for this theory are (1) that inheritable variations exist among members of a population of like organisms; and (2) that differential successful reproduction (i.e., survival) is occasioned by the composite of environmental factors impinging generation after generation upon the population. The theory is used to explain the many similarities and differences that exist between diverse kinds of organisms.

The theory of organic evolution, its limitations notwithstanding, provides a structural framework upon which many seemingly unrelated observations can be brought into more meaningful relationships. Biologists also have developed, from experiments and observations, hypotheses concerning the origination of life from nonliving matter (e.g., the heterotroph hypothesis).

Philosophic and religious considerations pertaining to the origin, meaning, and values of life are not within the realm of science, because they cannot be analyzed or measured by the present methods of science.

EXHIBIT 66: THE ORIGIN OF LIFE FOR SCHOOL CHILDREN

This is what a sixth-grade child is taught in California about the beginning of life on earth.

Reprinted from *The Story of Mankind* by Henrick Willem Van Loon by permission of Liveright Publishing Corporation, New York, Copyright 1967, 1951 by Liveright Publishing Corporation

The first living cell floated upon the waters of the sea.

For millions of years it drifted aimlessly with the currents. But during all that time it was developing certain habits that it might survive more easily upon the inhospitable earth. Some of these cells were happiest in the dark depths of the lakes and the pools. They took root in the slimy sediments which had been carried down from the tops of the hills and they became plants. Others preferred to move about and they grew strange jointed legs, like scorpions, and began to crawl along the bottom of the sea amidst the plants and the pale green things that looked like jellyfishes. Still others (covered with scales) depended upon a swimming motion to go from place to place in their search for food, and gradually they populated the ocean with myriads of fishes.

Meanwhile, the plants had increased in number and they had to search for new dwelling places. There was no more room for them at the bottom of the sea. Reluctantly they left the water and made a new home in the marshes and on the mudbanks that lay at the foot of the mountains. Twice a day the tides of the ocean covered them with their brine. For the rest of the time, the plants made the best of their uncomfortable situation and tried to survive in the thin air which surrounded the surface of the planet. After centuries of training, they learned how to live as comfortably in the air as they had done in the water. They increased in size and became shrubs and trees and at last they learned how to grow lovely flowers, which attracted the attention of the busy big bumble-bees and the birds who carried the seeds far and wide until the whole earth had become covered with green pastures, or lay dark under the shadow of the big trees.

But some of the fishes too had begun to leave the sea, and they had learned how to breathe with lungs as well as with gills. We call such creatures amphibious, which means that they are able to live with equal ease on the land and in the water. The first frog who crosses your path can tell you all about the pleasures of the double existence of the amphibian.

Once outside of the water, these animals gradually adapted themselves more and more to life on land. Some became reptiles (creatures who crawl like lizards) and they shared the silence of the forests with the insects. That they might move faster through the soft soil, they improved upon their legs, and their size increased until the world was populated with gigantic forms (which the handbooks of biology list under the names of Ichthyosaurus and Megalosaurus and Brontosaurus) who grew to be thirty to forty feet long and who could have played with elephants as a full grown cat plays with her kittens.

Some of the members of this reptilian family began to live in the tops of the trees, which were then often more than a hundred feet high. They no longer needed their legs for the purpose of walking, but it was necessary for them to move quickly from branch to branch. And so they changed a part of their skin into a sort of parachute, which stretched between the sides of their bodies and the small toes of their forefeet, and gradually they covered this skinny parachute with feathers and made their tails into a steering gear and flew from tree to tree and developed into true birds.

Then a strange thing happened. All the gigantic reptiles died within a short time. We do not know the reason. Perhaps it was due to a sudden change in climate. Perhaps they had grown so large that they could neither swim nor walk nor crawl, and they starved to death within sight but not within reach of the big ferns and trees. Whatever the cause, the million-year-old world empire of the big reptiles was over.

The world now began to be occupied by very different creatures. They were the descendants of the reptiles, but they were quite unlike these because they fed their young from the "mammae" or the breasts of the mother. Wherefore modern science calls these animals "mammals." They had shed the scales of the fish. They did not adopt the feathers of the bird, but they covered their bodies with hair. The mammals, however, developed other habits which gave their race a great advantage over the other animals. The female of the species carried the eggs of the young inside her body until they were hatched and while all other living beings, up to that time, had left their children exposed to the dangers of cold and heat, and the attacks of wild beasts, the mammals kept their young with them while they were still too weak to fight their enemies. In this way the young mammals were given a much better chance to survive, because they learned many things from their mothers, as you will know if you have ever watched a cat teaching her kittens to take care of themselves and how to wash their faces and how to catch mice.

Index

C

D

E

F

G

H

I

J

K

L

M

N

O

P

Q

R

S

T

U

Lightning Source Inc.
LaVergne, TN USA
14 August 2009

154791LV00001B/97/A